Spirit calls Nature

A Guide to Science and Spirituality,
Consciousness and Evolution
for a Post-Material Synthesis of Knowledge

Marco Masi Ph.D.

Third Full Edition 2025

Contents

Dear reader, first and foremost, a piece of reading advice.

Today, terms like 'multidisciplinary,' 'interdisciplinary,' and 'transdisciplinary' have become buzzwords, frequently invoked in academic discourse. They are often used to signal innovation, relevance, or a progressive approach to complex problems. However, despite their widespread use, their actual, sustained application in practice remains limited and often superficial, without fostering genuine integration of methods, theories, or epistemologies, with each discipline operating in its own silo and contributing fragments rather than co-constructing knowledge in a collaborative synthesis.

Within the pages of this book, you will embark on an extraordinary journey that integrates scientific, philosophical, and spiritual perspectives. It challenges you to step outside your comfort zone and learn to reconcile often diverse, yet not incompatible, worldviews. You will have to develop the ability to be not only inter-, multi-, and transdisciplinary through engagement with concepts, methods, and perspectives drawn from such diverse fields and cultures that you might not easily reconcile but will also need to be willing to go even a step beyond, embracing true integrality. The text invites readers to navigate across boundaries—intellectual, methodological, and especially ideological. Readers will be asked not just to recognize different disciplinary voices but to actively synthesize them—integrating insights from science, philosophy, spirituality, psychology, sociology, and much more into a coherent understanding. To fully grasp the significance of this work's message, you will have to adopt the perspectives of the objective and reductionist physicist, the holistic and subjective psychologist, the abstract philosopher, the trans-rational mystic, the Western analytic mind, and the Eastern spiritual intuition at once. Thus, this will not be your easiest reading.

However, the reading resists offering definitive conclusions; instead, it suggests a form of inquiry that values complexity and emphasizes that being integral is not merely about the accumulation of knowledge from multiple domains, but also from multiple perspectives and understandings.

This doesn't imply that you will have to digest every piece of this treatise. On the contrary, though the content has been ordered according to a logical structure and later sections sometimes refer to previous ones, you don't need to read every chapter and study every section systematically in sequence. The chapters are largely self-contained, and you may also read selectively by identifying the most relevant parts that attract your attention, prioritizing essential information to optimize your time.

Introduction

We live in a time when the lack of more comprehensive and integral approaches, visions, and paradigms that could make sense of the world is felt with an increasing necessity. Physicists are searching for a theory of quantum gravity, popularly (but inappropriately) also known as the 'theory of everything.' Psychologists are looking for an 'integral theory' that can capture the human's mental, emotional, and inner dimensions in their entirety. Others speak of integral ecology, integral life practices, holistic approaches to medicine, sciences, or philosophy. We speak of global vs. local issues that need a broader perspective and new 'global solutions.' The ethnic, cultural, and geographical local diversities inside a globalized world enter into conflict with an emergent unified vision of the human race and Nature. Current social and economic models based on individualistic competition and strife are felt to be increasingly inadequate and are contrasted with more integral visions of unity based on principles of collaboration and cooperation aiming at general wellbeing.

These are just some examples that emphasize the tension between fragmenting and polarizing conceptions which place, at the center, the individual or the part and, at best, see the whole as a mere sum of these parts, and an enlarged futuristic vision attempting a grand synthesis expressing the whole without negating the function and role of its parts.

'Integral theorists' are looking for the big picture containing the totality as an expression of an emergent process. However, a new approach that was able to make a real synthesis of knowledge between science and consciousness, reason, intuition, and spirituality, East and West, is still lacking. We contend that this couldn't be otherwise because these attempts were founded on false premises or an insufficient awareness. The reason standing behind this failure has three main causes.

First of all, because we take (more or less implicitly) for granted that this unification must be achieved only through reason, the intellect, what we refer to as the 'analytic mind,' and without realizing that, by doing so, we posit a priori a world that can be described only by the mental, which, in its origin and nature, is already a separating and polarizing form of cognition, thereby preventing an authentic unification from the outset. Backed by the success of physical sciences and mathematics, logical positivism sought to achieve in the first half of the twentieth century the elimination of metaphysics resorting to purely logical, and empiric means. But it failed to leave behind metaphysics. Unfortunately, we still struggle to learn the lesson from this failure: Pure rationalism is an insufficient mean to understand the World.

Secondly, we have an innate tendency to believe that our ideas, inspirations, insights, or 'illuminations' are our own. Most of the great ideas, intuitions, and visions that we believe are original and recent discoveries of modernity frequently turn out to be botched attempts to reframe centuries-old (if not even millennia-old) inspirations and realizations. Isaac Newton was aware of this when he said: *"If I have seen further, it is by standing on the shoulders of Giants."* We won't be able to see further if we don't rediscover some ancient wisdom and don't realize how we are permanently reinventing the wheel.

Thirdly, we will have to adopt (and rediscover) a new (or, more precisely, a forgotten) way of seeing the world, the cosmos, Nature, and especially ourselves. An integral theory or synthesis can't be limited to an intellectual exercise that naively tries to make sense of the surface waves of an ocean and, at the same time, denies the existence of the ocean itself. The transition from a purely analytic to an intuitive knowledge is a requirement for an integral vision that doesn't want to remain yet another abstract mental construct but also becomes a living and lived practice. Without a change of consciousness, there can't be any 'synthesis,' 'integrality,' 'holism,' or whatever kind of grand vision of things, life, and ourselves.

Therefore, the overall aim of this treatise is to bring us a step closer to this new state of consciousness by establishing a 'self-inquiry guide' to science, spirituality, consciousness, evolution, mind, philosophy, yoga, and reality. This treatise is a series of essays that overview the limits of science and reason, the mystery of consciousness, the nature of reality, man's search for meaning and purpose, and the future spiritual evolution of mankind. An invitation to look beyond the straitjacket of reason, science, and materialism. A series of reflections on the limitation of our *'sense-mind'*–that is, the mind that sees only objects, external actions, draws its ideas from the data given by external things, infers from them only and knows no other truth, until it is enlightened from above.

The acknowledgment that our evolution can't be only technological but will also and especially be spiritual. A project for a better humanity whose destiny will be neither that of a hyper-technological materialistic society nor a naïve spiritual grand-commune but an adventure of consciousness that will be guided by an inner spiritual evolutionary force that unites Spirit, Nature, and matter.

You will be guided through a critical analysis of *'physicalism,'* the current dominant materialistic paradigm in which everything is reduced to physical entities, processes, and laws of physics in a supposedly purely mechanical universe that is posited from the start as being devoid of meaning and purpose, with consciousness nothing more than a 'side effect', a bunch of particles ruled by a set of differential equations.

This collection is also a synthesis of knowledge between East and West in the frame of a cosmic evolutionary philosophy where life, consciousness, and matter no longer represent irreconcilable mysteries but, rather, are self-evident spontaneous outcomes of an evolutionary Nature that subconsciously perceives the call of the Spirit. An overview in which science, materialism, spirituality, or idealism reveal themselves to be a limited view of reality that can and must be transcended in order to merge into a more complete paradigm – one that will be neither scientific, materialistic, rational, or philosophical, nor spiritual or mystic, but will become another form of knowledge containing all of them and yet surpassing them to become something which will be none of them.

Only then will what appeared to be so mysterious in consciousness, life, matter, and mind, or so weird in quantum physics, in the material universe or the workings of Nature and evolution, appear in a new 'trans-rational' light in which all the paradoxes or seemingly irreconcilable contraries will acquire a new and almost obvious and self-evident meaning, purpose, and aim in a universe that otherwise appears meaningless, purposeless, and aimless.

We will pave the way, first to a deeper understanding of what science, reason, and materialism can and especially cannot do. Secondly, open the doors to integration between matter, mind, and spirit. A 'spiritual materialism' that recognizes the significance of humankind's social and psychological evolution by reconnecting a more human science with Nature, the soul, and the cosmos. Something we will call throughout this treatise a *post-material* worldview. Something that, in our common conceptions, we feel being as opposed to '*naturalism*'–that is, the doctrine according to which everything can be studied and explained inside a scientific paradigm and in terms of natural phenomena, without any recourse to non-material entities. However, since we will expand the notion of 'science' and 'Nature', this dichotomy dissipates and, to avoid confusion, we will avoid the use of this term, or just call it 'materialistic naturalism.'

Because, despite all of the attention paid to this topic, and contrary to popular belief, there is no such thing as a 'spiritual science' or a widely accepted idea that unites science and spirituality. We are still at the very beginning and nobody can claim to know what it will look like. The future of science and spirituality won't be realized by simply adding the latter to the former. Neither will it be realized by explaining one in terms of the other. It will have to be something that transcends both and that creates a third form of inquiry and knowledge, an 'integral science', that will be neither scientific nor spiritual in the ordinary sense. The next step of the evolution of consciousness will be that of the end of the curve of reason and materialism, which will be followed by the coming of a spiritual age that coalesces science and spirituality into something that will surpass both, towards a third state of awareness and cognition.

It will be distinguished by the realization that evolution is not a blind, mechanical, and meaningless process with no aim and purpose, as our anthropomorphic, limited, superficial, materialistic, and analytic mind is compelled to believe, but rather a beautiful goal-driven material as well as a spiritual process that conceals a powerful secret we intimately and intuitively already know and have always known.

Analytic materialism has reached an evolutionary apex beyond which it will have to make room for more advanced and deeper understandings of reality or succumb to its inflexibility and refusal to open itself to higher visions and insights. This is not about abolishing science and reason, but rather about placing them in their proper context, within a larger paradigm where they will retain their value and potential for change, but will also be recognized for what they are: a piece of limited knowledge and a useful tool for transformation that, however, should not be extended beyond certain domains they can neither understand nor control. It is about becoming aware that the mind alone is incapable of comprehending itself and about deconstructing an all-pervasive and dominant scientism. It is time to transcend the illusion that matter and rationality are the ultimate arbiters of truth because physical reality is also trans-physical and trans-rational.

But to rise beyond and above our present simplistic paradigm, we will have to recognize how we are limiting our thoughts and models to a mono-dimensional or, at best, two-dimensional, theoretical framework of the world which, however, is inherently multi-dimensional. What the current rational materialistic and reductionistic science, as well as most parts of the philosophy of mind and, not least, most forms of spirituality, are attempting to achieve is the reduction of all of reality to a monistic 'flatland' conceptual or experiential framework that, however, is neither flat nor monistic.

To understand the deeper motivations that should lead us to a new understanding that rises above the dominating paradigms, we must first become fully aware, not only of its limitations but even more of its unaware premises and underlying fallacies with which they work. It is necessary to become aware of why and in what sense the nature of the universe goes beyond something that human reason can grasp. There is a gap between what we think, conceptualize, and articulate and what is.

This will be the purpose and the overall conceptual framework underlying this treatise. Now let us see what the single parts of this journey will focus on.

This first part is organized as follows:

Chapter I sets the stage by introducing the reader to an evolutionary perspective on reason and science. Both are powerful tools of knowledge but not the last word of Nature. We will take a closer look at materialism and reason as transitional aspects of the evolution of consciousness. A special emphasis on '*Occam's razor*,' a methodological principle of parsimony that scientists and philosophers favor, will be addressed, critically highlighting how it prevents further progress and blocks us into a lower-dimensional worldview that leads to inconsistencies and contradictions.

A significant distinction between mind, consciousness, and emotions follows. The philosophy of mind, in particular, conflates the first two, which, when combined with a thoughtless application of the previously mentioned razor, results in a complex of thought patterns that inevitably leads to fallacies and superficial models, and ultimately leads in the wrong direction. Once this prolegomenon has been internalized, it becomes easier to realize what the real trouble is with consciousness and why science didn't make any tangible progress when it comes to questions related to it.

Chapter II will discuss how there are things whose existence no one denies but which remain elusive to scientific explanation, such as our subjective phenomenal experience, the so-called *'hard problem of consciousness,'* and the *'binding problem'* – that is, our minds' ability to bind qualitative properties and construct integrated meanings in a semantic hierarchy. The question of whether the mind is computational and if and how computers may become conscious, or at least be able to mimic human intelligence, is discussed. Modern approaches to the *'mind-body problem'* fall short of providing a satisfactory account of how our minds work, let alone answering the question of what consciousness is.

Chapter III will try to make sense of the different levels of consciousness. Clarifying and distinguishing is essential if we do not want to confuse ourselves. This implies a conceptual clarification of the so-called *'subconscious,'* which is far too frequently misunderstood as also encompassing the 'subliminal' and 'unconscious.' This will allow us to tackle the longstanding question of free will by reviewing modern findings in the spirit of Benjamin Libet's pioneering experiments. Do we have free will, or is it an illusion?

With this understanding, chapter IV will deliver us a different basis for dealing with the findings of modern neuroscience. The mind-body problem will be analyzed from a perspective that does not take for granted the purely reductionist and materialistic approach but will maintain an analytic and scientific standpoint. Once we become aware of our *'correlation-causation fallacies,'* we will not need to resort to metaphysical theories to show how the 'neurocentric' third-person descriptions of our first-person inner subjective experiences are incomplete, if not flawed. Special attention will be paid to recent scientific findings that correlate our conscious, subliminal, subconscious, or unconscious experience with brain activity. The search for the center generating our conscious awareness in the brain will be illustrated by several findings in neuroscience that ultimately didn't furnish the expected answer and why the seat of consciousness remains elusive. A special section is devoted to the empirical data that suggests the existence of a *'plant intelligence'* and how also unicellular organisms must have a primitive form of cognition, a *'basal cognition'*, questioning the dogma that presupposes neural activity as the source of any form of cognition and sentience.

In the second part, we will introduce the reader to *'philosophical idealism,'* first through some explanatory examples, and then by determining the true nature and distinction between *'primary and secondary qualities,'* revealing the unaware assumptions with which we understand reality. Then we will deal specifically with '*quantum physics,*' which, we will argue, is the most solid empirical evidence for a worldview based on the recognition that everything we see, perceive, and think about the world is an illusion - a figment that does not represent an objective reality in and of itself but is nonetheless 'real.' Questions such as "What are matter, space, and time?" and more generally "What is reality?" will be addressed.

We will then quickly review some aspects of Western philosophy that showed clear signs of attempts to transcend the intellect, trying to see the world from a higher perspective than the limited rational and materialistic paradigm, and which, however, have been forgotten or, at best, deemed obfuscating mysticism by intellectual mindsets that couldn't free themselves from their own limited and narrow analytic horizon. From Plato to Spinoza to the philosophers of modernity such as Husserl and Deleuze, an emphasis will be placed on phenomenology from an evolutionary perspective of Western thinkers. Particular emphasis will be placed on the most recent metaphysical, post-materialistic understanding of consciousness, life, and the universe. Our picks are the universalistic outgrowths of '*panpsychism,*' namely, '*cosmopsychism,*' '*analytic idealism,*' and '*panspiritism,*' which embrace a fundamental and irreducible universal consciousness to close the explanatory gap between consciousness and matter. This will be an overview of the Western spiritual and intuitive philosophy that also has the purpose of showing how all the apparent divergencies, contradictions, or inconsistencies resolve once a broader perspective is embraced.

However, before moving towards the spiritual perspective, we take up an old but potentially very actual approach to the investigations of Nature; this was the first attempt to create a non-reductionist science, namely, Goethe's way of seeing. We argue that the famous German poet and writer developed a sort of higher-mind natural philosophy that considers the multiplicity in the unity of Nature and that could become a starting base, or at least a source of inspiration, for a discipline connecting science and intuition.

This will conclude the second part, which was devoted mainly to Western culture's science, philosophy, and metaphysics.

The third part will guide us towards a more spiritual approach. It will point out that reality can be seized only by something that transcends our capacity to rationalize by moving towards new states of consciousness. It focuses first on the question of the meaning and purpose of our lives, the universe, and existence as a whole, which is reviewed from a spiritual and evolutionary angle. The notion of 'progress' itself will be seen from a quite different perspective, going beyond that of a mere scientific or technological viewpoint. Humans are only transitional beings in the history of evolution, bridging the infra-rational with the supra-rational. This is the visible, almost imminent evolutionary transition that is already taking place and in which several previously inexplicable facts become practically self-explanatory.

At this point, we will step out of the mono- or, at best, two-dimensional understanding of reality and begin to explore the *'planes and parts of being'* in light of the cosmology of the *'integral yoga'* of the Indian mystic and poet Sri Aurobindo. A description of higher states of consciousness, from mind to what he called *'Supermind,'* where all existence reveals itself as a Oneness, a multiplicity in unity, and only apparently manifesting itself in polarities. This paves the way for an understanding where neither the monistic physicalist idea that 'all is matter' nor the equally monistic idealist ontology that 'all is mental' is satisfactory. Rather, it is a comprehensive view of reality in which matter and mind are only two terms of a much vaster and richer reality. That will close many explanatory gaps with which modern science and philosophy are struggling. Some speculations will follow on the coming of a spiritual age and what the next step in evolution upon Earth might look like by the emergence of a new species, the *'gnostic being'*.

We then will set forth the discussion on the same line, delineating an *'integral cosmology'* which connects matter and spirit with the layers that exist between them. What are matter, space, time, and forces when apprehended from a higher state of consciousness that most humans still do not even imagine might exist? Is there a direction in evolution, or is it just a play of random mutations, natural selection, and genetic processes? Why is the universe built by an infinite variety of things while it retains in itself a secret unity in diversity? What is life? These and several other questions will be answered in the context of a supramental vision.

The disconnect with the preceding parts is only apparent. Philosophical idealism indeed helps us to step out of the illusion of the purely material objectification process that the conventional scientific mind attaches to sense-data. However, idealism is only a step in between that will help us go beyond a rationalistic and materialistic naturalism; it is itself a transitionary conception of reality that must be expanded. It does not even consider the eventuality that the Nature we observe with our limited sense-mind awareness is only a surface appearance of a much vaster Nature and, thereby, something that a broader notion of naturalism and idealism must embrace. This aspect will be identified in a *'central principle of evolution,'* which, by a widened conception of an *'integral teleology,'* will allow us to describe the emergence of life in a more extended but also more coherent manner than what idealism, let alone physicalism, can do.

Some remarks follow on the current world situation, its science, and how a materialist mindset conditions not only our thinking but also our doing. We will argue that the philosophical, intuitional, and spiritual account we have illustrated is not an abstract rumination but, on the contrary, should help us towards a necessary change of mind and soul. We will reevaluate modern science trends from this viewpoint, highlighting how, despite all the appearances of technological and scientific progress, deep down, modern science is stagnating. It will become clear how the seemingly promising research approaches, such as AI, new medicine, and transhumanism, or other futuristic hyper-technological materialistic and rationalistic approaches, will fail to deliver on what they are supposed to do. And yet we will guard the reader: The mind is not dead.

A final section will be devoted to the visible coming of a '*spiritual age*' that will surpass that of the Age of Enlightenment or, as Sri Aurobindo called it by borrowing a term coined by Karl Lamprecht, a '*subjective age*.' The future of reason is augmented by an inner call in the collective of the human race, leading to a new and broader, more harmonious life in the frame of an '*ideal of human unity*.' This will end the third part and complete the present treatise.

However, the overall integral vision and spiritual cosmology emerging from it, which bridges consciousness, science, evolution, and spirituality by a synthesis of knowledge, has no pretension to be exhaustive, let alone the final word. Nonetheless, we believe that it is, to date, the most comprehensive framework which includes all the directions of human knowledge that have been outlined so far. It is more than a critical assessment of materialism or a summary of human scientific, philosophical, and spiritual knowledge. It is an outline that aims to make these so-diverse wisdom meet – not as an effort to make them simply interact with each other but, instead, as a first step to losing their identity in order to coalesce in a new unity in diversity.

Part I
The End of the Curve of Reason

I. Prolegomenon to Reason's Self-deception

1. Mind and Materialism: Powerful but Transitional

"Man is far too clever to be able to survive without wisdom."
E. F. Schumacher

Probably nothing has changed human society as much as Galilean-Newtonian science did. Modern materialistic science, which is based on empirical data, experiments, and logical, rational, and mathematical analysis to objectify and quantify through testable predictions, has not only radically changed our material existence but, perhaps more importantly, our culture and the way we think about the world and ourselves.

The beginning of modern science can be, loosely speaking, identified by Galileo Galilei's experiments with inclined planes or his observations with his primitive telescope, which disclosed to him an entirely unknown new universe. Isaac Newton and many others followed with revolutionary discoveries. However, more than anything else, it was the disruptive power of the human mind—its rational thinking and ability for intellectual analysis applied to matter—that changed everything. There is nothing within reach of our instruments, from the tiniest bits of matter to the cosmic domains, that has not been measured, quantified, dissected, reduced, and geometrized according to a bottom-up approach. Most notably, the sciences of life, such as biology and medicine, did—and continue to do—their utmost to reduce everything into inanimate material elementary components.

By doing so, reason could unleash its full power. After over four centuries, the human material condition has been radically transformed by scientific discoveries and technological breakthroughs. Knowledge about the material realm, from the subatomic particle to the cosmic infinity of the universe, expanded our awareness and understanding of our place in Nature.

Modern materialistic science surfaced as an instinctive response to the philosophy and doctrines adopted and followed by the Christian Church during the Middle Ages. The Renaissance Period sparked the restoration of the spirit of free philosophical inquiry unconditioned by theological dogma. This contributed to a greater tolerance for the development of secular ideology. It was a reaction against the monopoly of the Church over knowledge of the world. By the end of the 17th-century, elementary education had become more widespread, imposing itself in the 20th-century as the norm.

Humankind realized that religions, dogmas, or (worse) superstitions were certainly not the ways to a better life. The medieval atrocities of the Inquisition resonate more or less subconsciously in our minds to the present day. The failure of religion to furnish us with an intellectual understanding of life, the universe, and our existence as a whole became increasingly more than evident.

We are not at the center of the universe, and, yes, evolution is true. Religion may have prevented the collapse of human civilization by imposing some collective moral and ethical rules of conduct. However, blind faith, obedience to religious scriptures and authority, and theological discourse did not project humankind to a higher material wellbeing or more comprehensive understanding of reality. Worse, it did not even boost its spiritual status—something which it promised and was supposed to do.

Science surpassed religious faith and elevated the discriminating skills of the mind to new heights as the supreme means of knowledge. Understanding the nature of reality and man's capacity to uncover it with reason has made it clear that science and reason are better tools in terms of capturing the workings of the material world than any scripture or authoritative clerical dogma.

The mind is an intellectual mechanism that coordinates, organizes, observes, distinguishes, separates, and categorizes in accordance with a rational order. It naturally looks for order in the world around it. And because the physical universe displays an order in forces and forms, naturally, the mind didn't fail to find it. The analytic description of this order became synonymously known as 'knowledge.' The old Pythagorean dream of describing the reality of experience perceived through the senses and organized by the mind in the form of a mathematical language became an established and universally accepted form of communication and mutual understanding. The Renaissance period elevated mathematics to one of the most cherished and highest forms of knowledge. Accordingly, the restoration of interest in mathematics and empiric approaches led to a widespread emphasis on the study of the quantitative aspects of existence. Measurements and numbers became the ultimate tools and expressions for 'objectivity.'

The new widespread scientific approach led to revolutionary inventions such as the telescope and the microscope—pieces of equipment that had never been seen or heard of before. This demonstrated the extent of the support that this new conception of knowledge received. Ground-breaking observations made with these very instruments questioned and eventually discredited the ancient beliefs and initiated a new spirit of skepticism. This prompted people to avoid accepting ideas based on faith but encouraging the idea that, instead, everything should be systematically observed, measured, or examined anew, regardless of existing dogma or philosophy.

Intellectual analysis of natural occurrences in the outside world, where the human senses were replaced by the 'senses' of mechanical measurement devices, became the new paradigm for searching for knowledge. An inherent aspect of this quest was the theory assuming that knowledge consists of the conclusions obtained from observation and examination via the instrumentation of the physical senses onto physical things. Anything that cannot be reduced to physical entities, material objects, or particles or, at least, some abstract mathematical concept, is 'non-real,' 'immaterial,' or 'unphysical and, as such, must be branded as an illusion or an emotional figment without value. Innovative scientific apparatuses enabled the senses to reach out as far as space and down to minuscule proportions. Celestial mechanics, which allowed for the precise prediction of planetary orbits and even the prediction of the existence of a then-unknown planet, Neptune, seemed only to confirm the idea and principle of a mathematically flawless, mechanical, and deterministic universe.

The systematic examination of matter by physical senses and its mental organization into a mathematical theory led to the overwhelming success of what is nowadays called *'Newtonian or Galilean physics,'* thus widening the schism induced by the division of science and theology. Secular humanism was born. The world was recognized as the exclusive territory for study by science, and knowledge was defined as that which could be proven only by procedures that were abstracted as far as possible from human experience. An analytic and mathematical conception of reality was elevated to a central position and inspired a mechanistic philosophy of the universe entirely justifiable in terms of logical and numerical principles. The universe became a gigantic soulless clockwork with no meaning or aim.

The Darwinian revolution represented yet another backlash to any non-materialistic and teleological hypothesis[1] – that is, ideas involving a purpose or aim behind the evolutionary processes. The first humans did not appear on Earth because someone was playing around with dust; instead, they were the result of evolution, as were other species. Later, we learned that we had a common ancestor with the chimpanzees and gorillas. What a shock! Not only are we not at the center of the universe, but we are only one animal among many others.

What seemed to be the final nail in the coffin came with the progress of the 20th-century biological sciences, which apparently became able to describe all life in terms of purely biochemical reactions. The discovery of the DNA code, the structure of cells and their metabolic functions, the significant advances of biology and medicine, and the recent breakthroughs of neurosciences all pointed in one direction: Living organisms are biological machines. There is no need to invoke a ghostly 'life principle' other than those dictated by matter

---

[1]Teleology (from the Greek word τέλος, 'télos' meaning 'end', 'aim', or 'goal,' and λόγος, 'logos' meaning 'explanation' or 'reason'), not to be confused with theology, is a theory or an understanding of processes implying the existence of an aim, purpose, goal, or finality in it. It seeks to explain phenomena in terms of design and purpose and sees the causes by which they arise as the mean and function for that purpose rather than originating, more or less accidentally, from purely aimless and mechanical processes. In particular 'natural teleology' considers the possibility that natural processes are the extrinsic manifestation guided by final causes. The typical example of a naturalistic teleology is the belief that evolution is a goal-driven process, something which the present neo-Darwinian evolutionary theories deny.

and the laws of physics. The reduction of everything to matter seemed to be unstoppable.

René Descartes, the well-known French philosopher and mathematician of the 17th-century and founding father of modern science, conceived of a universe in terms of purely mechanical and physical laws. With the exception of the human soul and God, which are separate and have nothing to do with Nature, everything that exists, including our bodies and brains, must be described using mechanical processes. All the world is a machine, and Nature is only about material particles interacting with each other according to the laws of physics. Nevertheless, Descartes still conceived of the existence of some non-physical stuff: Inside the human body is a soul (in contrast to animals which, he believed, have no souls and are mere automatons), and outside the universe, there is a God. His mind-body dualism envisaged a *'res extensa'* and a *'res cogitans,'* an extended and a thinking substance, the latter being distinguished from our bodies and brains, an immaterial mind.

Francis Bacon, the English philosopher and contemporary of Descartes, also conceived of what we call in modern parlance 'experimentation' or 'empiricism,' as the *"torture of Nature."* Nature must be put on the rack to reveal its secrets.[2] Descartes' and Bacon's almost purely mechanistic and material conception of the world became the historical source and origin of modern 'physicalism,'[3] which goes even further by getting rid of the last non-physical remnants, reducing all mental process to material ones and rejecting any non-physical conceptions like a soul or God as an 'unnecessary hypothesis.'

And yet, almost no modern scientist, including those who consider themselves dualists, admits to being inspired by the nearly soulless Cartesian worldview or Bacon's executionary thoughts. In reality, this self-distancing relies only on a minor technical detail: Descartes envisaged the pineal gland as the 'portal' of the soul or mind to the material world. This hypothesis no longer withstands the scrutiny of modern findings. However, apart from this or other minor aspects, modern scientists, especially physicalists, fully embrace Descartes' and Bacon's worldview. That their mindset, and understanding of the human's place in Nature, is alive and well is also testified to by the fact that never in all of Earth's natural history has the destruction and 'torture' of the

---

[2] This creepy analogy was no coincidence: Bacon was also a statesman who actively advocated for the use of torture on humans in legal proceedings as a means *"for discovery, and not for evidence"*. [366] [367]

[3] Physicalism is a form of *'ontological monism'*–that is, a view of reality that reduces everything to matter, the laws of physics, space, and time. It is a more precise term than 'materialism', which seems to suggest only matter as the ultimate foundation of everything. In the literature, however, the two are frequently used synonymously, as we will also sometimes do here.

natural environment and its species perpetrated by another species taken on such dimensions.

The success of the materialistic scientific inquiry into the mathematical order of a mechanical universe led to the belief that the mind and consciousness are just an epiphenomenon of the brain. There seemed to be no reason whatsoever to conjecture otherwise. Science has been so tremendously successful in explaining the workings of Nature that there was no reason to invoke non-material substances to explain the mind and consciousness.

Strangely enough, not only is this still a stumbling block for materialism, but science did not make an inch of progress in four centuries after Descartes' speculations. Consciousness, the mind, and our subjective experiences appear to have something irreducible.

However, it is only a matter of time, so goes the belief that science will explain that as well. We shouldn't unnecessarily multiply entities to explain the observed universe every time there is a mystery that science still can't solve. There is no reason to conjecture that the mind and consciousness are nothing but mechanical processes dominated by mathematical laws. This way of thinking eliminates the dualism between mind and matter and wants to reduce consciousness and all of life to biochemical reactions. Our experiences of pain and pleasure, hate and love, joy and grief are only an emergent epiphenomenal manifestation of the biochemistry of that gray matter in our skull. We would like to believe otherwise but, let's face it, we are just 'biorobots.' Physicalism became the new paradigm.

This world and life conception were further strengthened in the collective consciousness by the practical application of science. From the industrial revolution (with its steam engines and telegraphs) to the modern digital revolution (with its computers, the internet, and smartphones), the power and efficacy of materialistic science stand in its full glory and splendor in front of us all. Technology—the utilitarian outcome of the theoretical aspect of science—dominates our lives not only materially but also financially. There is almost no enterprise or job that does not depend on the ups and downs of the market of technologically based products. Science and technology have become, first and foremost, a financial power that dominates the world. Moreover, since the times when Galileo had to struggle against the obscurantism of a Catholic Church and its Inquisition, Galilean science has become the central paradigm of every academic institution. The dramatic modernization of science to achieve a distinguished position of authority has finally sealed, once and forever, the clear demarcation line between science and religion. Science became truth, per definition.

This made us all smarter, right? The Flynn effect attests to the fact that education contributed to an increase in population intelligence. This is the long-term increase in IQ test scores that have been measured in the general population worldwide over the 20th-century, indicating that, on average, humanity has become smarter. It is known that throughout the last century, every generation of newborn children has performed, by some significant number of IQ points, better than its parents.

However, it is also known that around the turn of the millennium, this trend slowed down and that it might even invert its direction. Children are performing only moderately better than their parents; humanity's IQ is stabilizing. What is the cause of this evolutionary stagnation? Environmental or economic factors? Our lifestyle? The new technologies that have detrimental effects on children's development? Maybe. Nobody knows for sure.

On top of that, we are experiencing a post-truth era. Scientific facts have become an opinion, not an accepted truth. Critical thought seems to have become unnecessary. An increasing number of people consider it a waste of time to conduct a sober and objective assessment of facts before forming an opinion. Climate change denialism is rampant, flat-earthers have become trendy, and the quantum woo has even infected academic levels.

What is happening? Are we 'involving' back to the dark ages? Or is it a reaction to something? If so, to what?

It is clear that if this regression establishes itself, the result will be a global disaster. If we do not recognize its genuine character as an expression of a thirst for more, and not less, it will lead us to an involution. It is necessary and urgent to recognize this apparent crisis of reason and analytic thinking as something that demands, neither the abolition nor the amplification of the mind, but something new that asks to manifest itself but is not allowed to do so.

Describing the undeniable success of materialism and reason furnishes us with a still-too-superficial understanding of what was and still is going on. We must expand our vision and encompass the history of science and the application of thought and reason as a dynamic whole, recognizing its less obvious but equally important function. We should look retrospectively and see what the role and function of science and reason were in the evolutionary history of humankind.

Had we not disengaged from the obscure and primitive forms of beliefs and religious dogma, separating the objects of study from the ideas and beliefs of the subject, reason wouldn't have had the opportunity to develop as it did. The intellectual faculties of the species were greatly favored by the concentration of its cognitive skills into a rational enterprise that science demands. Without the *'Age of Enlightenment'* (also called the *'Age of Reason'*), almost all of humanity would have remained stuck in a state of development that didn't go much further than blind religious beliefs and Medieval superstition or, at best, would have been forever satisfied with some glimmerings of the ancient Greek philosophers.

Fig. 1 Painting of the salon of Madame Geoffrin which represented an imaginary meeting in 1755 of the most prominent figures of the Enlightenment such as Rousseau, Voltaire, Montesquieu and several others.

This is the greatest service that science has provided to humanity. It has allowed us to exercise and further enhance our mental cognitive function. Science, with its theories and discoveries, not only delivered to us an incredible amount of knowledge about the world and ourselves and not only allowed for an enormous increase in material wealth but also, first and foremost, established the mind in the collective. Rational, analytic, and logical thinking might not be something that is equally distributed throughout the population. Still, as compared to the Middle Ages, on average, it has become a more widespread common good. The mind has rooted itself in the human race much more deeply than what it was when Copernicus dared to conjecture that we are not at the center of the universe.

Therefore, firstly, we contend that science and reason are a means by which Nature advances an evolutionary purpose. Secondly, we should not forget how every stage of the progression is never the final one but is always the preparation for the next evolutionary leap. Mind, reason, the physicalist's temptation to reduce everything to matter, and physical laws were all preparatory exercises to foster the skills of an infant humanity. However, once they hit their limits or are even exhausted, they must be recognized for what they are. The human intellect must inevitably be only a transitional cognitive function, not the last rung of the ladder. Physicalism has its truth and value and was a necessary starting point that provided a firm basis upon which a still-not-fully-developed mind had to start. Although, it is only a transitional philosophy and methodology that served a purpose. It is not the final frontier of knowledge.

And that's where we encounter the resistance to change. As is so typical of the mind itself, once it finds the solution to a problem or makes a new discovery, it has an innate tendency to universalize it to everything it knows, rendering it as a fixed and eternal law once and forever. Because the mind was able to develop the scientific method, empiricism, data analysis, and mathematical descriptions to explain appearances (though only to a limited extent) and dominate the material world, abstracting it from any religious myth and leading to such extraordinary material progress for humankind, the very same mind now feels justified to enthrone itself as the ultimate and absolute ruler and tool of knowledge. If reducing everything to inert chunks of matter has delivered such a powerful description, like the standard model of particle physics, it is only a matter of time before physicalism will explain away everything that exists. If the application of reason and rational thought worked so well in giving us so many answers and efficient tools to modify and control the environment, we instinctively extend reason's power beyond any boundary as something that must necessarily be the final answer to everything and good for all.

However, the answers to the fundamental questions about the source and essence of things like matter, life, and mind remain hidden. Quantum physics defies any classical concept of 'reality' bound to mental concepts of local realism and determinism. The complexity of life—be it physical or biological, let alone social and psychological—increasingly looks beyond our rational and intellectual capacity for scientific systematization. The appearance of that immaterial thing like consciousness in an apparently inert material universe remains a mystery more than ever. The faith that, someday, new ideas concerning discoveries in molecular biology, the origin of the universe, or a unified field theory where matter is the sole and ultimate *'ontological primitive'*[4] may solve these mysteries, is becoming less convincing. We must seriously consider whether the failure to find adequate answers to these fundamental questions is not only a matter of time and scientific progress but also an unavoidable result of the intrinsic limitations in the conception and methodology of the scientific quest for knowledge and the unwillingness to go beyond a strict physicalist approach. The mind has done a pretty good job of charting the territory but has not told us much about the territory itself.

Moreover, as history had shown much too often, when those who once were oppressed dethrone the old ruling power and become the new dominant class, they tend to behave as a new oppressive force that hampers advancement and evolutionary progress. Especially from the 20th-century onward, modern science has allowed for the emergence of attitudes that are typical of the religious authority against which it developed in response to. There is a

---

[4] An 'ontological primitive' is any entity that is considered to be fundamental and not reducible to something else.

reverence for established truths, a ruling of new thought founded on the position or academic status of its author rather than on the rationality of its conception. It expresses itself in an aggressive refusal of new viewpoints that challenge the fundamental belief system on which modern science has been founded.

After science's rapid expansion, it has become increasingly clear that the old belief in science and reason as the solution to everything is an illusion. Science alone can't explain everything about material reality. Additionally, reason and technological progress alone aren't the magic wand that will fix every material problem humanity is facing. The original hope of the fathers of the Age of Enlightenment that reason is the ultimate tool that will explain everything and elevate humanity's condition to a higher status of existence has been quite convincing for a long time but is now crumbling and coming to an end.

It is true that, at least from the material point of view, humanity has undoubtedly made notable progress compared to the Middle Ages (though one might argue, this is true only for some nations and social classes, while others are still struggling with poverty and material underdevelopment). This view dominated the 18th and 19th centuries and was somewhat weakened by the destruction of two world wars in the 20th-century but still dominates our thinking and beliefs today. This trickled down into our educational system, fixated on materialistic achievements and a physicalist worldview. Education is good for finding a job. STEM education has become the mantra of all educational programs, and subjects like music, the arts, and philosophy are seen almost as a waste of time. Obviously, if you believe that your children are only biological robots, as you believe yourself to be, you are forced and must also believe (consciously or subconsciously) that their brains are only computers that must be fed with information and exercised like machines.

Then, one has only to observe how much we invest in research and technologies that adopt a strictly physicalist and reductionist paradigm and that is supposed to lead us towards a bright future. We are told that it is only a matter of time before genetic medicine can eradicate genetic diseases, before neurobiology heals Parkinson's or Alzheimer's, before stem cells regrow organs, before genomics wins the war against cancer, and before AI frees us from stressful work, allowing us to nap in self-driving cars that will drive us autonomously and safely to our destinations. And, in the event that this human race would be so dumb as to destroy itself using the very same science and reason itself (for example, with weapons of mass destruction in a third world war) or annihilate its own self-sustainment, destroying the ecological environment, science will be the magic solution to this again: Let us send humans to Mars and terraform the red planet. In the case of extinction on one planet, the other will be the 'spare tire' and save humanity. Only science, technology, and the application of the intellect are the solutions. What else could there possibly be?

But even if all that will become a reality (and, as we will argue, there are good reasons to think that these techno-scientific marvels will become a reality only very partially), the question remains: Will this fill the sense of void, emptiness, and lack of meaning that, despite all these material advances, is growing, especially in young generations? We gradually realize that the current framework of materialistic and strictly analytic science cannot resolve both practical and social issues and several conceptual problems of a more philosophical nature. This scientism did not lead to a happier existence for the human race and did not make us less savage and wiser. At the end of the day, it is even making us dumber.

We have replaced the repressive nature of superstition with the tyranny of science absolutism in a dead, post-modern mechanistic age that reduces the human race to a bunch of walking talking algorithms. If science remains a purely materialistic search for knowledge, it will hamper the evolutionary process of humankind. If physicalism does not take its place as a complement to science, rather than all of science, Nature may throw us back to the days of alchemy or magic. If reason does not open itself to something that transcends it, the mind might regress back to irrationality and superstitions. We will have to recognize the sign of the times; otherwise, we will have to go through new dark ages and learn the lesson all over again.

It is time to look forward and go beyond our intellectual childhood—of course, not by returning to some nostalgic past but by recognizing the limits of the present paradigm, which posits its foundations on matter and logic alone. The age of pure reason, in which the intellect is the sole ruler and arbiter, is coming to an end, with technology serving as its 'longa manus.' This is not a call to throw the baby out with the bathwater. Reason, rational thinking, empiricism, the scientific method, and technology will continue to play a central role and offer new solutions to old problems. The mind and its material transforming power will continue to be a central dynamic force. But, it will have to be kept inside the dominion of expertise and competence of its own, without invading territories that it can't understand, not even in principle. What we need now is a more profound and integrated synthesis of knowledge by recognizing the spiritual component of life, mind, consciousness, and Nature.

Science, reason, and the rational physicalism that followed from the Renaissance and the Age of Enlightenment served an evolutionary purpose. However, precisely for this reason, they must be considered transitional. The desire to fix them, once and forever, as the ultimate tools to achieve knowledge and truth in every domain is an anthropocentric ideology that stands in stark contrast to the very same evolutionary process that science itself always cherishes so much. Science is a species-specific mode of cognition. The mind is incredibly efficient in deceiving itself. We will see how the idea that it can impartially and objectively assess the inherent Nature of the world around us, independent of its own mental projections, is its greatest self-deception.

And yet, despite its shortcomings, there is a wonderful aspect of human reason that makes it extremely interesting and tremendously powerful: If it is honest and humble and does not tell itself fairytales, it has the exquisite ability to become aware of its own limitations and to draw a clear demarcation line between what it can know and do and what it cannot. And, even better, once it can clearly see this boundary, the mind can be the decisive tool with which it can transcend itself by exponentially accelerating an evolutionary process that otherwise would take ages. That is the path we will follow here.

2. When Occam's Razor Cuts Too Deep

'Occam's razor' (or *'Ockham's razor'*) is a principle also known as the *'law of parsimony'* according to which *"pluralities [entities] should not be posited [multiplied] without necessity"* (*"pluralitas non est ponenda sine necessitate"*). This principle was first introduced by the English Franciscan friar William of Ockham (1287–1347), an academic philosopher and theologian. It states that when confronted with two or more competing theories that are supposed to explain the phenomena, one should favor the simplest one. Equivalently, it states that the simplest solution to a problem should be considered the most likely one.

This is frequently applied in science. If different hypotheses make the same predictions or describe the same reality and facts, one should take that which is endowed with the fewest assumptions. In the scientific method, Occam's razor cuts out all the seemingly unnecessary assumptions, postulates, ad hoc hypotheses, and, eventually, empirically untestable statements that are considered to be unnecessary to explain what we observe. It is a methodological minimalism that looks after the most parsimonious ontology that requires the smallest number of pluralities and entities whilst maintaining sufficient explanatory power to account for all the known facts.

This rule has dominated the research methodology of scientists and philosophers for a long time. For example, Isaac Newton was asked what gravity is, in itself? Is the gravitational field something immaterial? How can it influence, by a contactless action at a distance throughout empty space, other material bodies? His answer appealed to principles of parsimony and simplicity: *"We are to admit no more causes of natural things than such as are both true and sufficient to explain their appearances."* [1] We should not try to explain what things are in themselves, but rather develop a theory that describes the observation, especially with mathematical means, and without adding further 'entities' or 'substances' supposed to explain what gravity is.

In brief: Occam prescribes that simpler theories and conjectures should be favored over more complicated ones.

This all sounds very reasonable. And it is, in some context, to some extent, and in particular conditions.

The problem is that, especially when we are dealing with scientific theories that involve philosophical questions, what has to be considered 'parsimonious' or 'simple' is a matter of subjective preference. Be they materialists, idealists, dualists, or from whatever intellectual, technical, scientific, philosophical, or ideological background they come—they claim that it is their theory, not the rival one, that is the most parsimonious. For example, on the one hand, contemporary skeptics and atheist militant movements invoke Occam's principle to support their physicalist ideas–an application that a friar could hardly have had in mind. On the other hand, some theists tend to argue for a 'God out of the gaps'–that is, God as being the most parsimonious hypothesis whenever there is a phenomenon that our theories can't explain–a line of argument that can hardly be considered scientific.

Moreover, the history of science frequently showed also that deeper truths turned out to be less parsimonious and much more complex than what we would like them to be, and that an intellectual and philosophical rigor assumed, and, thereby, led to false conclusions or prevented further progress.

Thus, this principle of methodological minimalism has too often been misinterpreted and twisted into a modern form of philosophical minimalism, which cuts out not only entities but also everything that does not conform to our belief system.

Let us unpack this.

Indeed, the success of Occam's razor dates back to the inception of science itself. The Copernican revolution, which switched our worldview from a geocentric model to a heliocentric model, did not come about because the natural philosophers had any proof that the Earth is orbiting the Sun. The final proof that the heliocentric system accurately represents reality came much later (due to the observation at the beginning of the 19th-century of the stellar parallax).

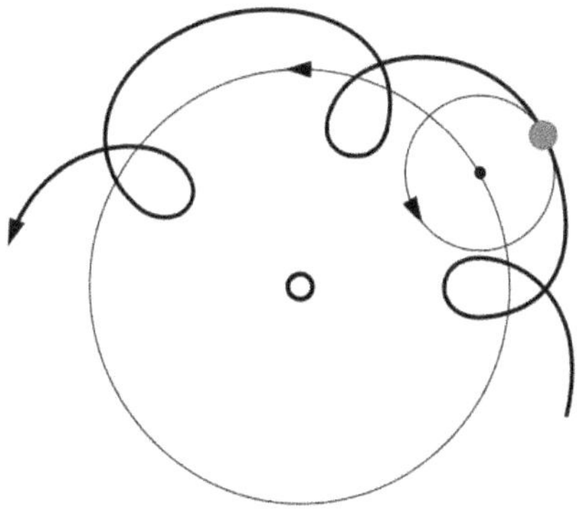

Fig. 2 Retrograde motion explained with epicycles.

Nevertheless, people began to accept Copernicus' suggestion because it is the simplest model that does not lead to 'pluralities' and 'multiplication of entities.' By contrast, the geocentric model of Ptolemy, which insisted on maintaining the Earth at the center of the universe, was much more complicated because it had to resort to a plethora of '*epicycles*' — that is, a number of circles along which planets were supposed to move and whose center moves around a larger circle, which again moves along another circle, and so on. Contrary to common belief, the epicycle theory could indeed trace, with a high degree of precision, all the orbital paths on the celestial sphere of all the known planets without having to posit the Sun at the center of the universe. Suppose one adds a sufficient number of epicycles, one on top of the other, each with an appropriate size and moving with the right angular velocity. Then one can approximate whatever kind of path of objects moving in the sky.[5]

But this was all very complicated. One had to multiply by a considerable number the entities—namely, the epicycles—to have the theory work in accordance with the observations. It is here where the principle of parsimony prevailed: Even though at the time it was not at all clear whether it was the true representation of the Solar System, it made much more sense to adopt the heliocentric model because it was much simpler and it 'saved appearances' with only one circle (or ellipse): The planet's orbit around the Sun. This was, and remains, the most paradigmatic and successful example of the application of Occam's razor favoring a conjecture over the other that was still in need of final proof.

However, an aspect that is frequently overlooked is that, historically, this was not the only reason — and probably not even the main reason — why people opted for the heliocentric model. Heliocentrism was not just a different interpretation of reality. Especially with the advent and application of Newton's theory of universal gravitation, heliocentrism became a theory that also had enormous explanatory and predictive power. By strictly mathematical proof, Newton could explain where Kepler's famous three laws of orbital motion come from. They arise as a natural consequence of the gravitational interaction between a massive central body (the Sun) and another smaller one (the Earth). Gravity, in a heliocentric system, tells us when we have to expect the next passage of a comet once we have measured its orbital parameters. Moreover, as already mentioned, the application of Newton's law of gravity to the observed anomaly of the motion of the planet Uranus allowed, in 1846, the French astronomer and mathematician Urbain Le Verrier to predict the existence of another planet, Neptune, simply by calculating with pencil and paper, without even looking through a telescope. The existence of Neptune was confirmed shortly after by the observations based on Le Verrier's suggestion regarding where to look. Such an explanatory and predicting power of the theory of gravitation in the frame of the heliocentric system was something that adherents to the Ptolemaic geocentric model could only dream of. The epicycle theory could still correctly describe the observed astronomical motion of all the celestial bodies, but it didn't predict or explain so much. It is this latter aspect that made heliocentrism the much more appealing option with its simplicity only an extra bonus.

---

[5] For a more in-depth analysis of epicycles and how their spirit survives even in modern interpretations of quantum mechanics, see [36] ch. II.8.

This doesn't mean that principles of parsimony did not inspire the scientists of the time; rather, the historical efficacy of Occam's razor has been overemphasized. In fact, one could find many other historical examples in which the razor cut too deep and led science in the wrong direction or stopped it from progressing. Ironically, already before Occam, Ptolemy stated that "*we consider it a good principle to explain the phenomena by the simplest hypothesis possible*." [2] Considering the Sun, the central celestial body was an unnecessary hypothesis and was therefore ignored for another 14 centuries. In geology, continental drift was long considered an unnecessary and too contrived conjecture to explain the dispersal of species (for an in-depth analysis of Occam's razor misapplications in biogeography, see [3].) It was thought for a long time that atoms do not exist because they were considered a superfluous metaphysical assumption. For years, Max Planck refrained from taking seriously his own idea about the discreteness of the energy quanta (which led to the inception of quantum theory) because he considered it a weird and unnecessary assumption that should be regarded only as a provisional working hypothesis (an assumption that the great Ludwig Boltzmann did not dare to embrace). If you didn't know anything about quantum physics and relativity, these would appear to be superfluous theories, and Occam's razor would opt for classical physics as the preferred theory. In fact, classical physics once seemed able to describe the entire universe by positing only particles and the classical laws of mechanics and electromagnetism. Nowadays, we know how the theory of relativity—and, especially, quantum mechanics—turned this worldview upside down. The existence of nuclear forces is an entirely unnecessary hypothesis from the perspective of classical physics as well. Proteins, rather than DNA, were once thought to be the carriers of genetic information in biology because they have a more complicated structure that can carry more information than the DNA molecule and, therefore, appeared to be a more plausible genetic information carrier at the time. Paradoxically, a hypothesis appeared to be simpler because it opted for a more complicated object.

In 1813, the German mathematician Carl Friedrich Gauss extended Euclidian geometry to non-Euclidian geometry by relaxing Euclid's postulate of parallel lines. To put it bluntly: In Euclidean geometry, two parallel lines remain at a constant distance from each other, while in non-Euclidian geometry, they can 'curve towards' and cross at some point or 'curve away,' increasing their distance to infinity. This could have been seen as a completely unnecessary and absurd extension–that is, a non-parsimonious hypothesis– without any practical applications whatsoever. After all, we perceive space following the Euclidian axioms, and there was no evidence suggesting otherwise. Fortunately, Gauss and many others who worked on this new geometry were not intimidated by methodological prescriptions a la Occam. They delivered one of the most beautiful theoretical frameworks of modern geometry, which, almost a century later, was applied by Einstein to formalize his theory of general relativity, itself one of the most impressive intellectual realizations of the 20th-century. It might also be noteworthy that the development of non-Euclidian geometry was not coincidentally embedded in a cultural context in which German idealism was in full swing and which did not uncritically embrace the materialistic naturalism (something we will discuss further in Pt.II-II.2; for a more in-depth account, see also [4]).

These were only a few examples of a long list that should make it clear how Occam's razor is no more than a heuristic principle, a rule of thumb, and should not be elevated to a scientific criterion, let alone a proof for or against a conjecture.

Unfortunately, this is what too many scientists do (more or less subconsciously). Whatever authority praises Occam, we must be aware of how any conclusion reached through the principle of parsimony can be only a suggestion or a pointer that can lead us in the right direction, though it might not. What appears to us as 'simpler' or 'unnecessary' is often a matter of personal opinion that strongly depends on what we know and, especially, what we don't know. It depends on whether we have a complete understanding of the complexity of the phenomenon we want to describe or whether we still don't. If one misses that complexity and does not know all the underlying laws, components, and variables that determine a phenomenon, invoking Occam's razor will almost certainly cut them out leading to an oversimplification, wrong conclusions, and therefore, a lack of progress. A quotation credited to Einstein says: "*Everything should be made as simple as possible, but not simpler.*" A dogmatic application of the principle of parsimony may result in something too simple — that is, something simply incorrect.

The point is, that we are frequently unaware of our ignorance. It is a quite normal state of affairs to believe that one knows all the laws and basic entities that make up the observed reality; however, we realize later—sometimes much later—that we have in mind a much too superficial model of reality. For example, the more we study the structure and function of living organisms, even of a single cell, the more we remain surprised by the until-then unimagined level of complexity. It often turns out that we repeatedly underestimate our ability to grasp how tremendously complex life, mind, and the brain are.

Therefore, Occam's razor may sometimes be useful for narrowing the search, for example, when you have to perform experiments furnishing raw data that don't need particular interpretations. But in many cases, especially when we are looking for deeper truths that involve philosophical speculations, the solution may lie way beyond those narrow limits. Ruling things out of consideration only because they do not conform to the principle of parsimony can backfire.

We felt it necessary to make this preamble on Occam's razor because, as we will see, a too rigorous application of it has not only frequently led the philosophy of mind to wrong conclusions, it has also made it blind and prevented it from achieving further progress. The author could not find a single historical case in which it settled a debate. In science, the explanatory and predictive power of a theory confirmed by experimental evidence has always been the ruler in the court, whereas in philosophy, cutting razors were never the criterion that allowed for the prediction of the fates of rival theories.

Principles of parsimony and simplicity can be personal guiding principles but are no less effective than principles of 'beauty' or 'elegance' which, indeed, can also lead to discovery. But all these can't be put on par with principles of logical inference, or mathematical consistency, let alone replace experimentation. Like it or not, at the end of the story, we choose a worldview because of our personal preferences, our belief systems, and our unaware (necessary or unnecessary) assumptions. And, sometimes, the reality isn't as simple as we would like it to be. Posting simplicity a priori as a working hypothesis can blind us from seeing reality.

Therefore, here we will not obey Occam's dictatorship. Of course, we will not multiply entities or resort to unnecessarily complicated assumptions without justification either. Instead, we will point out the existence of some entities that are there, that can be ascertained and tested almost empirically, by a first-person approach and, therefore, that can't be dismissed only on the grounds of an abstract and limited principle. That is what we are now going to do with mind, consciousness, and emotions.

3. Mind, Consciousness and Emotions: An Experiential Introspective First-person Investigation

"There are three things extremely hard:
steel, a diamond, and to know one's self."
Benjamin Franklin, Poor Richard's Almanac, 1750.

In the current philosophical debate, you will find people using words like 'consciousness,' 'mind,' 'awareness,' 'feelings,' 'emotions,' 'existence,' 'reality,' 'unconscious,' 'subconscious' and all sorts of immaterial, indefinable and ineffable 'objects' all over the place, taking for granted that we all know what they mean and that, by the same word, everyone means the same thing. The truth, however, is that the opposite is almost always the case. Even professional philosophers and scientists use these words, sometimes interchangeably and synonymously throughout their papers, books, or talks, and only rarely feel the need to explain what each term means *to them*. It turns out that, in most of the cases, when we are dealing with complex metaphysical issues involving concepts that, besides an analytical characterization, also require an introspective ability to determine their nature, people tend to conflate words and meanings, throwing everything and the contrary of everything into the same pot. This sometimes leads to a tremendous conceptual confusion in which everyone is using the same terminology but often means something completely different from what others understand.

It occurs much too often that, despite every effort to clarify an idea, after long and painstaking intellectual exchanges between two contending philosophical positions, we realize how our counterpart misinterprets and misrepresents it or remains confused at our attempt to convey a message. This state of affairs is not only common among the ordinary uneducated man but is also a regular and almost everywhere-accepted slippery lack of clarity among high-ranking academics who play around with a vocabulary that, if not clarified, can have two or even three different (sometimes opposite) meanings and get it through in peer-reviewed journals or other scientific publications. Therefore, it is necessary, first and foremost, to clarify what the words 'mind,' 'consciousness,' and 'emotions' themselves are supposed to mean, at least in the context of the present treatise.

On the other hand, the history of philosophy has indeed taught us another important lesson. At the beginning of the 20th-century, a movement of logical positivism and the philosophy of language tried to pin down knowledge into a rigorously defined set of logical rules and linguistic definitions. It failed. This is because, when it comes to evanescent and non-objectifiable entities like 'consciousness' and 'mind,' there is no way, not even in principle, to converge onto a universally acceptable definition. Definitions themselves rely on words that indicate qualitative perceptions of phenomenal and mental content in the first place and can't tell us anything about it, if not through our own direct experience. One can't define an experience without resorting to terms that presuppose the knowledge of the lived experience itself. In the end, attempting to define consciousness, mind, or other subjective experience using words is akin to defining water vapor from ice, the molecular structure of H_2O in terms of wetness, or milk from cheese. Strict and rigorous mental definitions of first-person experiences that others cannot have direct access to — other than by their own experience — can't work, not even in principle.

This became clear when Ludwig Wittgenstein, the Austrian-British philosopher of the 20th-century, pointed out how we can analyze, dissect, and reformulate — with words — the structure, grammar, and meaning of a word like 'toothache'; however, that will mean nothing to someone who has never had a toothache. In his *Tractatus Logico-Philosophicus,* Wittgenstein could show how the positivist's dream to describe a relationship between the essence of the world and a realist theory of language leads to self-referential paradoxes. Such a theory would have to be so general that it must also encompass itself.

Therefore, the best approach to these matters is the good old middle-way: We must make an effort to clarify, as much as possible, the meaning of words while, at the same time, avoiding caging them into unique and inflexible definitions. We will focus our attention on consciousness, mind, and emotions. Other expressions and concepts require clarification, and we will eventually do that. However, more than anything else, these require being saved from falling into the bottomless pit of an endless confusion that affects the so-called "philosophy of mind" (and that is already an incorrect expression in itself, as it assumes that mind and consciousness are synonymous).

So, what do we mean by 'consciousness,' 'mind,' and 'emotions'? Are 'consciousness' and 'mind' synonymous or distinct? Are emotions a product of the mind? What about the distinction between the waking consciousness and the sleeping consciousness? Do awareness and consciousness indicate the same thing? Can someone be conscious without being conscious *of* something? Different people with different cultural and professional backgrounds would answer these questions very differently.

For example, most of us, including many professionals, consider what we experience in our waking state as the only meaningful concept of consciousness. Suppose you don't react to an external stimulus from the physical environment, you are said to be 'unconscious.' If you sleep, you are considered 'unconscious.' The simple observation that we can be conscious of our dreams and that there is still a subjective and experiential state of existence in a sleeping state, at least in the dream state, does not bother most physicians. And how do we know that we are truly unconscious in the seemingly unconscious state of deep sleep or general anesthesia? When we awake from a deep sleep, it is as if we have experienced nothing, though that's due to memory. It's like when you get black-out drunk and don't remember all the embarrassing stuff you did the night before. One reckons that there are still some experiential images and appearances while you are 'knocked out,' but they simply cannot be recollected. Similarly, we don't remember anything about our lives as newborn babies. Then, let us be parsimonious: to believe that we, as babies, were conscious is an unnecessary hypothesis. Mothers would probably disagree, but, you know, mothers aren't usually busy with philosophical questions.

For physicians, these are unnecessary hypotheses and distinctions that they cut out with Occam's razor because they do not serve their purposes. And this is also fine; their duty is to anesthetize people for surgical intervention or to wake them up from a coma, not to explain the philosophical subtleties of what consciousness is. Surgical general anesthesia induces analgesia–that is, it blocks the pain pathways. It avoids voluntary muscle movements, causes amnesia, and leads to pharmacological paralysis. But we have no idea if and to what extent conscious awareness exists during anesthesia. At least from the philosophical standpoint that searches for deeper truths, defining this amnestic state as 'unconscious' is an unwarranted extrapolation. These definitions make sense as long as such an understanding of consciousness is kept in its proper operational domains without having the pretension of being a universal definition that could be extended beyond those boundaries.

Regrettably, however, this is what many do. People tend to extend such simplistic notions of waking consciousness as a universal concept without further thought and extrapolate and build their theories and intellectual constructs upon this superficial assumption. Fortunately, scientists are slowly becoming aware of this misconception. Modern findings begin to cover up the fallacy: Unresponsiveness and the fading of conscious awareness of the surrounding world do not denote unconsciousness. Interestingly in this regard is the so-called *twilight anesthesia,*' an anesthetic technique that sedates patients only mildly and induces amnesia but no loss of waking consciousness [5]. During this 'twilight state,' they are responsive and can be asked to perform some tasks that they would not be able to recollect after the surgery. This fact alone speaks volumes: The inability to recall events during sedation is no proof of unconsciousness.

Others may have a completely different understanding of consciousness. Many equate the concept of consciousness with 'self-awareness' — that is, the ability of humans (together with some other animals like chimpanzees, orangutans, and possibly others) to recognize themselves in a mirror. This is the famous *'mirror test'* or *'self-recognition test,'* which supposedly demonstrates self-awareness. Interestingly enough, after half a century, scientists still disagree on the meaning of the term 'self-awareness' and what this test truly proves (for an interesting review, see [6]). The American evolutionary psychologist, Gordon Gallup who invented the test in 1970, defines it as "the ability to become the object of one's own attention." However, much too often, you will find specialists in the field, more or less explicitly, equating it simplistically with 'consciousness' or 'self-consciousness.' Humans are then said to be 'conscious' because they pass the mirror test. In contrast, cats, dogs, and birds are not self-aware – that is, 'unconscious' – because they do not possess the ability to recognize themselves in a mirror.[6] If one points out that this shows only a bodily self-recognition and an ability to link the internal awareness of muscle movements with an image, and that, according to this definition, cats, dogs, or birds must be considered 'unconscious,' without sentience, feelings and inner life, like Descartes did, who held that animals were only machines, those who adopt such a simplified understanding of consciousness will reply that they don't care about these subtleties. They are dealing only with humans, and the notion of self-awareness serves well as a description of the psychological states they are interested in; operationally, it

---

[6] To add further confusion, it turns out that at least some fish pass the mirror test. [392] Note how in the article's title, the word 'consciousness' appears but is never mentioned again in the text, as if it is implicitly assumed to be synonymous with 'self-awareness'.

works. There is no reason to put things upside-down, invoking other conceptions — that is, Occam's razor is invoked.

This would, again, be fine and well if we did not tacitly posit precisely such a sort of conception and apply it in contexts where it becomes meaningless or highly misleading, to say the least. What about children not recognizing themselves in a mirror before the 18th month? Are they unconscious zombies? Or simply infants unable to identify their body? And that's where conceptual problems grow massively. These sorts of simplified concepts of consciousness are then taken straightforwardly into a metaphysical domain, where the concept of consciousness assumes an entirely different and much deeper significance. This can obviously lead only to the confusion and misunderstandings we mentioned above.

The inappropriate usage of words like 'consciousness' and 'mind' arise not only because of an intellectually slippery way of working with semantics. It relies on the fact that humans do not know themselves. It might be time to recover the good old Greek maxim, "know thyself" ('γνῶθι σεαυτόν,' gnothi seauton). In school and college, we have been drilled to employ only the intellect and look at everything through the third-person glasses of reason. The human race has lost its innate ability to look inward by a self-inquiring first-person approach. An inward subjective investigation is even discouraged by the mentalized academic environment, which is regarded as the ultimate arbiter for what should be considered right or wrong, true or false. Children are permanently hammered out of their contemplative state because, so goes the theory, "They should not dream and should learn to cope with the real world." We project our inner consciousness and mind, externalizing it into the material world, and have forgotten to rest calmly in ourselves and know ourselves by looking inward. We have even forgotten that we can do that and, when others invite us to do so, we wonder what they are talking about. It is one thing to talk about something material outside of us — say, a mountain, a flower, or a starry sky. It is a completely different thing to talk about the inner 'objects' of awareness inside of us. Some types of emotions, usually the most violent and distressing ones, are still in our conventional reach of understanding. We can, to some degree, relate to them and capture their dynamics and play of forces when they are intense and common enough. However, most of our emotional state is hidden to us. That's why people speak of the 'subconscious.'

And, yet, a little bit of exercise — and an 'inner movement' that goes inward — would reveal to us that they are not subconscious at all. We don't notice it; we are not accustomed to paying attention to it. This is not because these things are out of our reach but, rather, because our cultural and educational environment doesn't care about and, in fact, repressed some of the natural faculties we all have. Then, obviously, once we try to dig deep into ourselves, trying to find out what consciousness and mind are, we can no longer find anything. We forgot how to do it and can relate only to material things outside of ourselves. We can't even conceive what it means to 'look inside'; it sounds only like a play on words to us. The purely analytic empiricist approach has blinded us so much that we have become 'spiritual analphabets.' Schools, society, and jobs have drilled us to develop our minds, but they pay insufficient attention to the development of our inner intuitive and spiritual faculties. Everything has been built on the principles of physical and mental hygiene, but there hasn't been much progress in terms of teaching and learning about ourselves. That's why so many feel the urge to resort to yoga, meditation, or practices of mindfulness. We feel that something vital to our wellbeing has gone missing — something that our society does not value. In particular, highly developed minds — those who exercised their rational skills to the utmost, got a Ph.D., and became university professors — are frequently the most inept in going inward by taking a first-person perspective and knowing themselves. Paradoxically those are precisely the authorities we ask for answers to metaphysical questions.

Therefore, we will not adopt a superficial understanding of the word 'consciousness.' We will refer to phenomenal consciousness (see later for the precise meaning) and will take an experiential self-inquiry first-person approach towards investigating it, rather than relying on convoluted philosophical or complicated scientific explanations. This is the only serious approach and means to avoid confusion: knowing directly via a first-person investigation what one means by consciousness is independent of authority and our ideological backgrounds.

At any rate, though this does not want to be a meditation guide for inner exploration (there are so many, and we don't feel the need to provide another one), we will occasionally invite the reader to look inside and take a more experiential, rather than intellectual, approach. Looking inside means searching for answers not only by a rational and logical inquiry — which, of course, maintains its value — but also by observing one's own thoughts, feelings, and sensations. Asking for the nature of consciousness is, first and foremost, a matter of experience, not of logical deduction. This is why it appears so mysterious at times: We try to understand the origin of its essence and workings through logical reasoning and external references, while it can be captured only by an introspective first-person approach.

Now, having said that, to set the stage, let us first take up the somewhat easier object of our inner knowledge — namely that which goes by the word 'mind.'

Etymologically, it originates from the word *'mynd,'* from the Old English *'gemynd,'* which was taken from the proto-Germanic branch of the Indo-European languages and which means "to think, remember." It is no coincidence that in Sanskrit it was called *'manas'* (मनस्), which meant the inner organ of thought and memory. Interestingly, the modern German word for mind is *'Geist,'* which does not retain this original etymology and adopted it from the old Greek 'ψυχή,' psyche, or the Latin *'spiritus'* or *'mens,'* having different possible meanings: that of thought or intellect, that of Spirit or Soul, and even meaning an immaterial ghost. Each of the various interpretations depends on the context. However, this ambiguity had frequently added another source of confusion when texts had to be translated, especially from the German philosophers.

Though the old wisdom that set the mind on par with something immaterial was perhaps not so misplaced, as the modern physicalist science would like us to believe, we will, of course, keep the identification of mind with a cognitive thought process. There is no need to ascribe to the word 'mind' a different meaning than is conventional to do — that is, the cognitive faculty of thought, rational and logical thinking, analytical reasoning, and intellectual discrimination that are proper, especially to the human species.

Several authors conflate the mind with consciousness or emotions or both. We feel that this is neither useful nor in line with the facts revealed to us by an inner subject investigation. Just consider your personal experience, revealing to you the qualitative difference between thinking about a mathematical problem and being in love. The former is a thought process of the mind, while the latter is an emotional state. You don't need neuroscientific knowledge or logical reasoning to understand that. One 'knows' this experientially.

As a further example, which needs a bit more introspection, observe how you perceive the thought and the one who perceives that thought as distinct. This is something we have always intimately known but rarely noticed. We somehow experience in ourselves a difference between a thought that appears in our minds for a fleeting moment and the subject that witnesses that thought and that persists after that thought has gone from our 'screen of apprehension.' Again, the former is just a thought, while the latter is an entity, a subject, a state of existence we call "I."[7] All the thoughts come and go against a base that is ever stable. That base is a silent and very concrete presence. This experiential knowledge reveals us that the mind is an object of experience, not that which experiences.

We must learn to clearly distinguish between intellectual knowledge and experiential knowledge. In our society (which, after the Age of Enlightenment, became increasingly fixed on exclusively intellectual knowledge), experiential knowledge has been downgraded to something irrational and of no value. And yet, we should realize (we have always known that but we forgot) how it is one thing to have a mental representation of a rainbow in terms of physical science, such as describing how a prism separates light into different bands of wavelengths of electromagnetic radiation, while it is another thing to

---

[7] According to several contemplative traditions, this "I", the ego, also reveals itself as a thought of a 'universal witness,' so to speak, but let us not go too far here. We will take this up again in Pt.III I-4.

experience the beauty of the rainbow's color spectrum without knowing anything about physics. This is not just the difference between science and poetry; it is also the difference between mind and experience. The former is a mental conceptual description of a physical phenomenon, whereas the latter is a subjective qualitative experience in our consciousness that manifests itself following sensory observation. Conflating the two things — the rational cognition of the mind with the experience of a phenomenon — makes no sense. The two things are ontologically different and cannot be taken as synonymous — no more and no less than we can call an apple a banana. Invoking principles of parsimony that should allow us to throw everything into the same pot only underlines our lack of subjective internal discernment. These things appear to be mixed together only to the surface awareness, though by going inward, it becomes almost apparent and self-evident how they must be separated. That all consciousness can be spoken as 'mind' has become accepted and common malpractice, leading to endless misinterpretations and confusion.

We might summarize a characterization of the mind as follows:

__The (limited finite human) mind__ is an instrumental entity of consciousness (which, however, is not consciousness) with the cognitive function of thinking and perceiving. It is the rational intellect that organizes and structures sensory perceptions and translates them into representations, ideas, concepts, and meanings based on symbols, forms, shapes, and numbers. It is the thinking reason that interprets sense perceptions, analyzes data, judges, reflects on concepts, decides, discerns, selects, discriminates, makes logical inferences, and stores its acquired knowledge into memory. It is the intellectual instrument that allows the homo sapiens to think rationally and logically by an analytic, deductive and inductive thought process.

The above does not pretend to be a rigorous definition. In fact, it is not a definition at all. It wants only to be considered a description whose sole intent is to clarify the meaning of the words we are using. It is supposed to serve as a conceptual clarification that should allow us to distinguish the mind from consciousness, emotions, or other 'objects' of our inner field of experience.

Note that we didn't use the word 'intelligence,' as it is an ambiguous concept. Is a high IQ score a measure of intelligence?

Fig. 3 Keywords for mind: Reason, thought, intellect, rationality, logic, representations, ideas, concepts, meanings, symbols, forms, shapes, numbers, interpretations, modeling, data analysis, deduction, induction, judgement, reflections, decision making, discernment,

Maybe it is in some contexts, while not in others. The IQ measures analytic and computational skills but not so much other forms of intelligence, such as musical and artistic intelligence. People also speak of the 'intelligence of the heart' or 'emotional intelligence.' Is this an intelligence without value? We don't think so, but we will refrain from using 'intelligence' and 'mind' as synonymous. Intelligence is defined as the ability to process data without, necessarily, the ability to comprehend meaning. Computers can be, in some sense, very 'intelligent' but do they understand meaning? Is there an entity and a mind that knows and understands? These are questions that we will address further on. Meanwhile, because of these uncertainties and this vagueness, we prefer not to resort to the word 'intelligence' as a label for the mind.

Before we proceed to the characterization of the concept of consciousness, let us briefly digress on the evolutionary aspect of the mind. Frequently, the question that arises in relation to mind, reason, and intellectual faculties is: Should we ascribe the same cognitive ability to other, non-human animals? We can say for sure that animals don't have the same mental and rational abilities as humans because, for example, none of them has been shown to be able to invent the wheel, maintain fire, craft complex tools, develop artistic sensibility, painting, music, drawing, engraving, and sculpture, let alone build houses, machines, rockets, spaceships, or computers. However, it is also true that animals have mental abilities, though at a lower level and with less practical consequences. Also, animals can learn, make decisions, and, especially the more evolved species like dolphins or chimpanzees, have remarkable abilities in terms of logical and inferential reasoning. Much less certain is whether we might ascribe a mind to a worm, a plant, or a cell. In these cases, we tend to see a purely reactive organism that has no mind. While we will later contend that there are good reasons to conclude how this cannot be entirely correct, what we will point out here is that a cognitive function like mind and reason cannot be seen as just a block that exists or does not exist, that is on or off. Also, mental faculties can be experienced at different levels of efficiency and quality.

In the evolutionary process, the mind has evolved in steps and appears with different gradations. The human mind is the last in appearance, but it would contradict the logic of evolution to take human reason as the ultimate tool for understanding reality and the final step in the evolution of cognitive functions, extending it to all reality. It would be a terribly anthropocentric idea to put mind — that is, the human mind — at the center of the universe and eventually conceive of a Deity that 'thinks' as the human does.

The central point we are trying to make here is that the mind, in the sense of the cognitive ability to think rationally and intellectually, is a relatively recent appearance in the evolutionary process of the physical world. It is the most potent cognitive function that has, so far, appeared on Earth, but it should not be universalized and overemphasized as the rationalist does, mostly being unaware that it contradicts evolutionary principles itself. Reason is simply a transitory event in evolution. After all, a still simple form of cognition that we should not enthrone to some ultimate form of cognition or universal ontological primitive that is supposed to explain everything. The mind is transitional, not final.

So, then, what is consciousness?

In its earliest uses in the 1500s, the English word derives from the Latin *'conscius,'* which means "knowing together" or "knowing with" ('con' means "together" or "with" while 'scio' means "to know"). It could mean "knowing with others"—that is, shared knowledge or "knowing with oneself," in the latter case meaning "knowing that one knows." Later, John Locke, the famous 17th-century British philosopher of the Enlightenment, defined consciousness as "*the perception of what passes in a man's own mind.*" [7] It also took on a moral connotation in the sense of moral conscience. René Descartes used the word 'consciousness' as the "internal testimony" ("conscientia, *vel interno testimonio*"). Descartes was wrong about many things, but we regard this idea of consciousness as the 'pure witness' remarkably true [8]. In the modern philosophy of mind, consciousness is usually regarded as the ability to have any experience or feeling–what the American philosopher Thomas Nagel called "*something it is like to be*" in his famous 1974 paper "*What is it like to be a bat?*" [9] If there is something, someone, or anything it is like to have an experience in and of itself, then one is said to be conscious. We won't reject this characterization of consciousness but will differentiate and try to avoid dangerous conceptual conflations that might follow from it.

In fact, we already hinted at the nature of consciousness by contrasting it with the mind — that is, thought processes. If we step back and position ourselves internally as pure witnesses who observe their thoughts coming and going without any attempt or desire to intervene, we can become aware that there are images, representations — a 'film,' so to speak — playing on an internal background, a screen of consciousness. This 'witnessing screen' is not the thought it witnesses, just as a cinema screen is not the projector's light that shines a sequence of images constituting the movie. Consciousness becomes aware of thoughts and the innumerable mental representations that the mind constructs, but it is not these intellectual figments. Thoughts change, come, and go continuously, though, in the background, there is an immutable observer that does not change. Only this simple internal experiential, subjective acknowledgment, which does not need philosophies, complicated logical reasoning, or neuroscientific data, should make it clear how the mind and consciousness must obviously be two distinct entities. Consciousness does not need to think, nor even perceive, either in the mental or sensory way. While the mind is a function, consciousness does not 'function'; it simply *is*.

We may also point out how it is a well-known fact that, through meditation techniques, one can diminish the flux of thoughts and images in one's own mind. Advanced practitioners also claim to be able to completely stop the mind — that is, the thought processes — and report states of consciousness of pure awareness and self-existence without any mental content. Meditative conscious states of free awareness — that is, free of any phenomenal content excluding external sensory stimuli and internal mentation from conscious experience — are described in several contemplative traditions and, in rare instances, have even been investigated by modern science through EEG and brain scan technologies [10]. The descriptions of such deep meditative states, of a bare state of self-less and object-less awareness, may vary from culture to culture, as well as among times and personalities. However, they all agree that, contrary to what the mind would like us to believe, conscious awareness is heightened—and not diminished—by the lack of mental content. The less the mind intervenes, the more something else shines through. One directly and clearly self-experiences something distinct from the mind that transcends it. That is consciousness.

These facts and realities can become clear and almost self-evident only through an introspective first-person investigation. The physicalist dismisses the first-person view because the third-person approach so successfully described our external reality and, thus, assumes that it will be equally successful in describing our internal experiential reality. However, that can't work, not even in principle, because the conceptual entities through which we described the external reality are already internal symbols in our' screen of consciousness.' It is an attempt to reconstruct our inner reality with these very same projections and figments. The abstract third-person perspective is unaware that it is already trying to reduce consciousness to consciousness's own abstractions.

We will take up this latter aspect in much more detail later. What is essential at this point is to avoid creating a potpourri of mind and consciousness, justifying this conflation with a supposedly scientific attitude that refrains from the 'multiplication of entities' but then fails to recognize the existence of two very concrete and distinct entities. A closer introspection reveals to us that these are ontologically different realities as a matter of experiential fact.

The failure to recognize this does not come about because of a logical fallacy but instead happens because we are misguided by our own superficial awareness of things. This typical tendency to reach for the melting pot arises due to the fact that humans do not know themselves. We are mentalized beings and think in terms of the formations of the mind. We do not even realize that it is possible to 'see' beyond them. However, when we step back from the superficial appearances of our mentalized perceptions, which misinterpret the things as they appear to us at first glance for the objective and ultimate truth, we can avoid this fundamental category mistake.

At the end of the story, we intimately know it: There is a subject in us that does not change despite all the thoughts and external happenings. Consciousness is a separate power or force from the mind because the mind is almost always restless, while consciousness is silent. Through our inner introspection, we can become aware of something that is a witness. It is in itself being, existence, and self-aware. There is something prior to mental events. Consciousness can be aware of mental content but is not that 'mental excitation.' It may identify with the mind but is not that with which it identifies. In us, it is a prior unmovable witness to the mutable 'mental ripples' but is not these ripples, just as the ocean is not its waves. It is a blank page of 'beingness,' the testimony of the mental and other qualitative phenomena that it can apprehend.

Therefore, consciousness cannot be defined by words, like we can define a unit of length that measures external physical objects or a number like $\pi = 3.1415$ ... which can be derived from the relation between the circumference and the diameter of a circle. Consciousness is an internal self-existent, and self-explanatory entity. If here, we are questioning consciousness, this is only because of our lost ability to introspect. Physicalists are afraid to introspect because they believe it means losing their minds. It is like recognizing ourselves in a mirror image. Humans can do that, together with a few other animals. However, the latter are able to recognize themselves in a mirror only if they don't panic; the others will have to create a fiction. It is the same with introspection. If we do not want to recognize ourselves, our inner essence, we must create the physicalist fiction.

Therefore, we can loosely try to characterize consciousness with words in order to converge on the understanding that avoids confusion when we use the word 'consciousness.' The primary aspect of consciousness, as its 'beingness,' 'existence,' or quality of 'self-awareness' might be the foundation.

Consciousness is an inherent witnessing sense of existence, presence, and sentience, the knower who knows oneself, a self-existent reality, pure 'self-aware beingness' aware of phenomenal existence which, however, does not necessitate an object of which to be aware. It exists before any mental, emotional, or physical perceptions. It is that in which all experience appears and is known as pure and undifferentiated irreducible awareness and sense of existence which witnesses phenomena but remains unaffected by it. It is a changeless, silent, space-less, featureless, and immobile reality that stands as a pure and undisturbed background behind all our activities.

If you look at them with the mind only, these sorts of qualitative characterizations of something in us result in confusing, recursive, and eventually also self-contradictory descriptions. However, they become obvious, self-explanatory, and self-evident if you can step back and simply experience, be, witness, observe, and introspect without resorting to conceptualizations.

Here the nature of the witness, the knower, or the subjective identity we call "I" and "me" is left open. As a side note, it may be said that the non-dual teachings of the Indian Advaita Vedanta or of some Buddhist traditions would, tell us that also that individuality is an illusion, a non-existent identity, a thought. The real 'knower' is beyond that "me", and is infinite awareness, the Brahman, or ultimate Absolute. In fact, they contend that there can be even thought without identity. At this juncture, we don't dig deeper into this aspect. Non-duality will be discussed later in more detail.

We resorted to the analogy of the cinema screen and the projected light on the very same screen. As the blank screen does not need the movie to exist, so does pure consciousness not need the contents of experience to exist. However, the fact is that this pure consciousness — this pure self-awareness and 'self-existent beingness' in us — is not just there experiencing nothing. Consciousness is also conscious *of* something, the film of life, the movie of thoughts, the impression of emotions, etc. Our consciousness is sentience and has perceptions of phenomena and events of which it becomes aware. This is, however, not a quantitative experience but, rather, an exquisitely qualitative experience. Consciousness allows for a subjective qualitative experience of phenomenal events, such as the feelings of pain and pleasure, the perceptions of qualities like the redness of a rose, the sweetness of sugar, the lived perception of emotions like love and anger.

When Thomas Nagel asked, "*What is it like to be a bat?*" he argued that the distinctive character of phenomenal consciousness is that there is something incommunicable about what it is like for a subject to have that particular experience, namely, being a bat [9]. The 'what-it's-likeness' of experience of being, feeling, knowing, understanding, intending, and so on are distinct experienced qualities proper to any conscious creature. There is something about what it is like for humans to be humans or for bats to be bats, for me to be me and for you to be you. This is unlike stones, mountains, chairs, or an air molecule, for which we suppose that nothing is experiencing something or being something.

Deep down, pure consciousness is the ultimate primitive, the 'screen' on which all impermanent and changing experiences are displayed. However, the sentience of the smell of flowers or the feeling of pain when one puts one's hand in a fire are all qualitative experiences that impress upon consciousness, the so-called '*qualia*.' Qualia are phenomenal properties, sensations arising due to a direct experience that goes beyond mere data processing. In our brains, we don't perceive the message, "This electromagnetic wave has a wavelength of 700 nanometers"; as living and sentient beings, we simply perceive a color — the qualia we call 'red.' If you put your hand into the fire, you won't receive the information, "The temperature of this exothermic reaction is 223 degrees Celsius"; you simply feel pain. Sentience always has something qualitative to it that can't be reduced to a purely quantitative statement. One must live the experience to know what it is like. You can never ever, not even in principle, know what it feels like to experience qualia only by a quantitative technical description. You must have the experience from a first-person perspective.

Qualia are intrinsically first-person observable properties but not third-person inherent properties of the world.

This is what is referred to as '*phenomenal consciousness*.' It is the aspect of consciousness capable of experiencing phenomenal properties or qualia. It also goes beyond the five senses. Emotions, like anger or happiness, joy or grief, etc., are internal qualitative first-person experiential phenomena that a quantitative intellectual description can't convey to someone else who never had these experiences and wants to know it only by a third-person perspective.

However, that does not imply that intellectual constructs aren't phenomenal experience as well. Also, experiencing a thought is a phenomenon in our minds. The conceptual organization of our experience of the world is also based on phenomenal properties that we perceive and conceive of in terms of space, time, forms and numbers, or other more or less abstract properties.

Broadly speaking, we can characterize it as follows:

__Phenomenal consciousness__ is consciousness having subjective experiences and perceptions of experiential properties, qualia (the experience of seeing, hearing, smelling, tasting, touching, and having pains or pleasures). It is the experiencing of consciousness that knows 'what it is like' to have that experience. These subjective qualitative perceptions of phenomena can be of a sensory, physical nature — that is, the experiences arising from the five senses, or that of an inner emotional state, or the apprehension of mental states as concepts, ideas, images, imaginations, etc.

The coming sections and chapters will elucidate the nature of the whole issue of phenomenal consciousness further. For the time being, what we would like you to become aware of is the distinction between mind, consciousness (or just 'pure consciousness'), and phenomenal consciousness. The distinction between consciousness and phenomenal consciousness is more of a dynamical character. We might also say that phenomenal consciousness is consciousness that has the experience of something rather than nothing. It is always the same conscious being but in an intentional state that focuses on and attends to a phenomenon by experiencing it. If you like analogies; Consciousness stands to phenomenal consciousness like an ocean with and without waves, but it is always the same ocean. Meanwhile, the distinction between mind and consciousness is of a more fundamental ontological character.[8] If these concepts are not distinguished and clearly differentiated, there will be significant confusion and misunderstanding.

Now, what about emotions? Are they a state of mind? A state of consciousness? Is an emotion a thought?

---

[8] This doesn't mean that mind might also be something derivative from consciousness. But, for the time being, let us leave this for later.

The word "emotion" has its origin in the 16th-century French word *'émouvoir,'* taken from the Latin *'emovere,'* which means "to stir up," "to move out," or "to agitate." In fact, emotions have a strong dynamical component, internally as well as externally. Emotions can move us to action and, eventually, to psychological and physical violence. And one can't deny that there is an intimate relationship between our emotions, mental states, and conscious experience. After all, a thought that recalls a fearful event can trigger in us a sensation of fear. A person's remembrance can induce in us emotions of love or anger, sympathy or antipathy. There is probably never a thought in us that is not more or less subconsciously related to an emotion, or at least to some feeling of attraction or repulsion. Thoughts and emotions are intrinsically intertwined and always go hand in hand.

However, this does not imply that they are the same thing. There is a fundamental, qualitative, experiential difference between an analytic and rational thought process that, for instance, aims to solve a logical, mathematical problem and the sensation of an emotion like love, hate, anger, fear, joy, or grief. One can think of an emotion and, eventually, even arouse its sensation, but the mind's thought *of* that emotion is still not the emotion itself. An idea and a mental construct, a conceptual representation or a rational discourse, is one thing; an emotion is another. Thoughts are imagined, apprehended, and comprehended, while emotions are felt. It is no coincidence that we speak of "reasons clouded by emotions" or equate irrational thinking with emotional thinking. Because, intimately, we know that the nature of reason and the intellect is not that of an emotional process, though we might have a hard time keeping the two separate. It is no coincidence that we highly rate a rational and analytical theory that tries to be detached from emotional impulses, as we intuitively know that the former works best when the latter are distinguished and set aside.

Therefore, a thought is not an emotion, and an emotion is not a thought. They are two distinct and qualitatively very different powers of manifestation in our field of experience. To classify an emotion as a 'mental state' (unfortunately a common practice) is no less absurd like categorizing a thought as an 'emotional state'. Nevertheless, like a thought, an emotion can come and go in front of the witnessing background that does not change — the immutable consciousness. Again, without having the pretention to systemize by definitions, just to clarify meanings that should help us avoid falling into fallacies or misunderstandings, we might loosely describe emotions as follows:

Emotions *are sensations of desire, impulses of passions, sentiments, affections, and instincts of attraction or repulsion. They are an interplay of more or less intense polar perceptions of joy or anger, happiness or sadness, courage or fear, love or hate, etc.*

Notice where thoughts and emotions are felt in our bodily sensations. We associate the center of our thinking mind in our head, in a particular location between our eyebrows in the center of our skull. However, emotions express themselves physically elsewhere, in our belly region. When we say, "You should not think with your belly," we mean precisely that: Misinterpreting our irrational emotions for our rational thoughts leads us astray. The fact that thoughts and emotions manifest physically in completely different regions of our bodies is also a sign of their very different nature.

Last but not least, where does the body stand in all that? Of course, our bodies are material, made of stuff we call 'matter.' They do not at all have that immaterial and fleeting character of a thought or an emotion, let alone consciousness. Life would become quite complicated if our bodily existence were as impermanent as our thoughts and emotions. But are the properties of permanence and impermanence a conclusive criterion for reality?

In any case, we know our bodies not only by an external observation (say, our eyes looking at our hands) but also — and even more than that — by an inner perception and feeling of its presence and existence as something that appears in the same field of our conscious awareness, as thoughts and emotions do. If we close our eyes, we have the experience of the body as a sensation that appears in ourselves within our field of consciousness. All our bodily perceptions, such as taste, sound, sight, touch, and smell, are physical signals that we experience *in* us. We will go into greater detail in this latter aspect in the following sections. It suffices to say that the body is not just a clump of matter but, rather, serves as the interface, a junction between our inner and outer worlds.

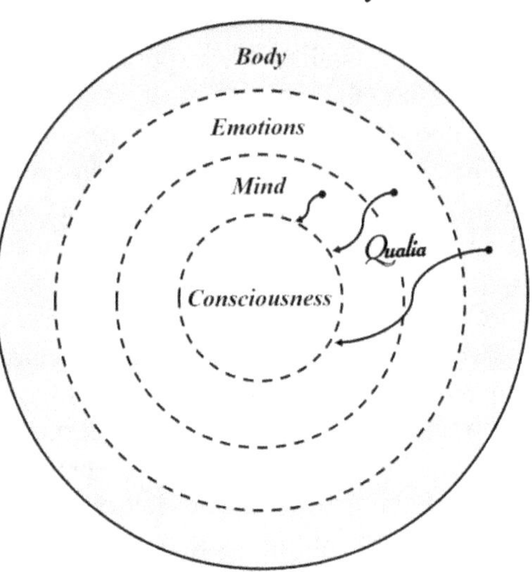

Fig. 4 The layers of our being.

We have finally reached the point that is the horror of every philosopher of mind. The most dreadful result, which scientists and philosophers try to avoid like death. The outcome that they reject with skepticism and disgust. Yes, we multiplied entities, and, no, we didn't follow a principle of parsimony by applying Occam's razor. While the physicalist wants to reduce everything to matter, namely only to our body/brain, the dualist admits two entities: body and mind. But we are so irresponsible as to distinguish (at least!) four entities: body, mind, consciousness, and emotions. What a waste of intellectual energies! And this is only the first step. We will extend the structure of Fig. 4

much further. Until then, it will serve us as a first sketch and guiding principle which will enable us to look beyond the conventional horizon.

We didn't reach this conclusion simply by some abstract reasoning, or because we wanted to save appearances, or due to preferences aimed at preserving some ideological belief system. We invited the reader to go inward and check the facts experientially, by a method of self-inquiry and an experiential investigation that looks at what things are and how they present themselves in ourselves beyond mental constructs. We identified mind, consciousness, and emotions as interrelated but ontologically different entities by an internally oriented empiric observation. We could confirm the separate function and existence of the mind, consciousness, and emotions as a matter of fact, just as one can distinguish between two colors or tastes or between the sounds of two musical instruments. If I hear the music of a violin on my right and a piano on my left, I will not say that the violin and the piano are the same instruments because I want to be parsimonious with terminology. If rainbows display a spectrum of different colors, I won't go around telling people that rainbows are all white because I don't want to 'multiply entities.' If I see apples, oranges, tomatoes, and bananas on a table, I won't feel any urge to resort to Occam's razor and call them all 'bananas.'

It is in this spirit and logic that we will proceed. If reality requires the multiplication of entities, so be it. What is essential is that we position ourselves by the truth, not by some easy, short-cut intellectual construct that, in the end, wants only to please one or the other belief system. In particular, Western philosophy did not distinguish or discriminate between very different ontological entities such as the mind and consciousness. It is time to recognize this as a category mistake.

II. The Trouble with Consciousness

*"Science says the first word on everything, and
the last word on nothing."*
Victor Hugo

1. The No-progress Quests

Despite the overwhelming success of the application of reason through the adoption of a purely materialistic approach of the sciences to explain, control, and increase our dominion over matter, there remains a strong sense of dissatisfaction with the human condition. Particle physics, chemistry, biology, medicine, the neurosciences, and many other reductionist intellectual approaches of science to Nature allowed us to gain invaluable insights into the world and, to a certain extent, also a bit into ourselves. These sciences and the human mind will continue to do so in the future. Nevertheless, when it comes to the fundamental issues of our existence, the search for meaning, the search for an answer, if it is meaningful to ask for a meaning in the first place, or the question about the intrinsic nature of things, then reason keeps silent. There are easy problems and hard problems that science and reason must still tackle. Let us begin with the former.

There are questions for which science and reason have been shown to deliver, if not a final answer, at least a progressive clarification. These questions may still not have found a final systematization inside an intellectual framework, but progress has been made, and it is reasonable to assume that they will find a resolution inside a strictly materialistic and mental paradigm. We have already made progress in some areas of knowledge about the world, and we will continue to better understand these things in the future through the cumulative advancement of science and technology. Science is a cumulative process as Thomas Kuhn, the renowned 20th-century American philosopher of science, characterized in his magnum opus, '*The Structure of Scientific Revolutions*.' New observations are made, methods are improved, new and more sophisticated theories are developed, and new technologies enable us to discover and understand an increasing number of facts progressively. Inside this paradigm, we expect to find the answers to some difficult questions. It is only a matter of time and further progress before we will get there. Let us call these the '*progressive-quests*' — that is, those problems we will find stepwise and progressively clarifying answers.

For example, we don't know whether extraterrestrial life exists. However, it is plausible that with much more powerful optical telescopes, radio-telescopes, or technologically more advanced astronomical observatories on satellites, or with space probes visiting other planets or moons in the solar system, we will find — in one form or another — traces of life beyond Earth.

After all, progress has already been made: Modern telescopes can look far beyond what was once considered the limit of the visible universe, and astronomers have discovered that our galaxy is full of stars that have planets — some of which have many similarities to Earth and orbit in a habitable zone, thereby potentially harboring other forms of life.

Another example could be that of the search for a theory of quantum gravity. Physicists have been searching since the time of Einstein (which is about three-quarters of a century now) for a theory that unifies quantum mechanics with general relativity, and that is supposed to describe all the fundamental interactions–that is, gravity, electromagnetism, and the two nuclear forces–as a single 'super-force.' Despite all the efforts and intense research conducted by thousands of physicists and the generations of engineers who have built increasingly more powerful particle accelerators, a theory of quantum gravity remains more elusive than ever. But, again, considerable progress has been made since Einstein's days: The standard model of particle physics is at least able to unify quantum mechanics with special relativity and brings under the same umbrella three of the four fundamental forces, leaving out only the force of gravity.

It is expected, but far from certain, that a new quantum gravity theory might also solve the so-called *'fine-tuning problem.'* It looks like our universe has been 'fine-tuned' to allow for the emergence of life. If the universal physical constants, such as the gravitational constant, the electric charge of the electron, the strength of the electromagnetic and nuclear forces, and many others, had been slightly different from what they are, then, most probably, stars could not have formed and planetary systems capable of harboring life would not have been possible. Is it just a coincidence, or does this have a deeper meaning? One day, could all this be explained away by a theory of quantum gravity that explains why the physical constants had to be exactly as they are? Or should we interpret this fine-tuning in a teleological sense which suggests an intelligent design? The fine-tuning problem was, and remains, an extremely controversial academic quest, and not much progress has been made. The author has no strong opinion on this and, despite his sympathy for the idea that final causes in Nature exist, it seems to me, nevertheless, plausible that one day the value of these physical magnitudes, which actually seem to appear by pure chance, might be determined by a yet-unknown mechanism that we still don't understand. Therefore, let's be conservative and consider that the fine-tuning problem is a progressive-quest.

From the technological point of view, there have been lines of research that we thought could successfully be accomplished in a few years or a decade, but that, instead, became a never-ending story with no light at the end of the tunnel. A case of this sort that is worth a mention is controlled nuclear fusion. After the explosion of the first hydrogen bombs in 1956, demonstrating that the creation of uncontrolled nuclear fusion reactions was possible, it was believed that controlled nuclear fusion power plants for energy production for civil and commercial purposes would soon become a reality. After all, this was what happened with nuclear fission; only nine years elapsed between the bombs in Hiroshima and Nagasaki and the world's first nuclear power station to generate electricity (which came into service in Russia in 1954). Not so for nuclear fusion. The technology necessary to isolate a plasma at a temperature of millions of degrees to reach a controlled nuclear fusion process turned out to be much more complicated than expected. Since the detonation of the first hydrogen bomb, there has been no technology available to efficiently replicate the same controlled process for energy production purposes. This, however, doesn't mean that we will never get there. It is a matter of technological advancement, further progress, and more research. However, for some reason I can't explain, I have a bad gut feeling about it. Nonetheless, it is plausible to believe that the creation of the first nuclear fusion power plant is a progressive-quest. Despite all the setbacks, we can maintain an optimistic attitude and hope that we will eventually get there and be able to reproduce, technically, the energy of the stars. The question, however, is whether it will become commercially profitable.

The final example we like to make comes from the field of biology. We have no idea how life began about 3.8 billion years ago. We know that, for some unfathomable reason, chemicals arranged themselves into the first replicating cells, which started the adventure of life on this planet. And yet, again, much progress has been made: Biology, chemistry, and evolutionary sciences have made tremendous advances in understanding the molecular basis of evolution. It is not the last page of this process that looks so mysterious but, rather, the first one. I agree with the materialists that this is still not a sufficiently convincing reason which substantiates a theological or teleological hypothesis. Further progress could reveal how a slew of chemicals came together to form the first cells, DNA, and all the other components required for the emergence of life.

These examples illustrate the case for '*progressive-quests*'. They are 'progressive' in the sense that, while we still do not know the complete answer to these questions, we could nevertheless observe how science advanced stepwise (or, in some cases, with sudden and dramatic leaps) in the direction of a resolution. We have learned many things since the time of the beginning of science, about four centuries ago, and while it may still take decades (perhaps even centuries), we can be confident that in the future, with further progress, an increasing amount of facts and knowledge science will finally settle these questions.

However, after four centuries – that is, since Galileo established the basis of modern empirical sciences – we must also admit that there are questions that not only remain unanswered but that any attempt to answer them did not result in any progress or advancement at all. These quests not only formulate a lack of knowledge but are also affected by a complete lack of progress despite centuries of scientific research or philosophical debate. We will call these *'no-progress quests'*.

Most of the no-progress quests have a more philosophical or even trans-physical character. The primary and most obvious one, which conditioned human history and social behavior, is the quest for God, a Creator, higher Being, and Intelligence (or whatever one might call it) and which is supposed to be almighty, omnipotent, and omniscient. Science has made progress in describing how the universe works; it tells us that everything originated about 13.8 billion years ago from the Big Bang and can explain the formation of galaxies and stellar systems. We know something about how evolution led to the formation of life. However, when it comes to the question of whether there is an intelligent design[9], a plan, an aim, and meaning behind these cosmic processes, science and reason remain completely mute. No progress at all has been made despite centuries of scientific advances and philosophical thought. The belief or disbelief in God remains a matter of personal choice and preference. There is nothing that science can tell us, and it has never even tried to express itself on these matters because it knows that this is a no-progress quest. It makes no sense — or, at least, it appears incredibly naive — to claim that it is only a matter of time before science will tell us whether or not God exists. Nobody really believes that because, deep down, we know that this is a question that can't be categorized as a progressive-quest. Otherwise, we would have observed some advance. What science eventually told us is that past anthropomorphic theories of creation, such as the Genesis in the Bible, were completely wrong or, to put it more diplomatically, must not be interpreted literally. Science delivered a more realistic theory of creation but remains utterly unable to rule out a Creator. If the question of the existence of God had been a progressive-quest, we should have noticed how science and reason would have rendered this hypothesis progressively unreasonable. Indeed, in the progress of science, many see precisely this — namely, that the hypothesis and assumption of God have become something increasingly improbable. But guess what? Which principle do they resort to in their philosophical arguments? Obviously, Occam's razor. As we have elucidated sufficiently, resorting to Occam's razor isn't a scientific criterion. It serves only the purpose of supporting one's own ideological and metaphysical belief system from the outset and has never been shown to be a valid criterion for the advancement of science and the discovery of truth.

At any rate, the fact is that religion, spirituality, and the belief in the existence of a Deity haven't been affected by science, even not in many well-

---

[9] The author has no affiliation and is not a supporter of the philosophical movement also known as 'Intelligent Design'. If not stated otherwise we will use the term 'intelligent design' in its more general connotation.

educated intellectuals. Quite the contrary: In recent times, we have witnessed a spiritual revival that is probably, among other reasons, caused by an understanding of how science and reason have made no progress whatsoever and failed to furnish the expected answers to many questions of a spiritual and intuitive nature and that something in us longs for.

Other no-progress quests, which are closely related to the above, are the questions of the existence of life after death, whether there is a soul that survives the death of the body, and several other similar, purely metaphysical issues.

However, we do not need to delve so deeply into such an immaterial and, so to speak, 'esoteric' domain. It is the search for meaning that remains a no-progress quest.

We know there is a universe that, to some, appears to be run like clockwork, without meaning. But what is reason's and science's test, the proof or logical inference that can tell us something about the meaning, purpose, or aim of anything? In science, the question of meaning itself is meaningless from the outset. Because science is inherently unable to find a method that tells us about the meaning of things, it posits a universe without meaning as an a priori axiom, not as fundamental knowledge. It does so because it already knows that the quest for meaning is a no-progress quest. However, this does not imply that it is a nonsensical quest.

And, why is there anything rather than nothing? If there is no purpose, why is there something at all? Shouldn't there be nothing instead? If there is no meaning, why is it so hard to find out? The question of whether there are final causes and an intelligent design at the root of all existence remains unanswered, and there has been no tangible progress in the domains of science or reason's philosophical investigations. The inherent agnosticism of science, which seems unable to progress in providing an answer and even refuses to answer these questions, is an enigma in itself. If there is no God, no purpose, and no meaning, shouldn't science and reason come closer to revealing this truth to us? What is it that limits our mind and its tools to progress on the question of what this universe really is? If the mind, reason, our rational faculties, and a purely materialistic science are the ultimate tools for achieving knowledge of things that are supposed to lead us to a progressive revelation of truth, why is it that the Eastern and Western mystics of the Middle Ages can tell us much more about these matters than modern science does?

But let us go down to Earth, literally. A very physical no-progress quest is the question of the intrinsic essence of things. For example, what is matter? What is it beyond a quantitative description, intrinsically, in its essence, in and of itself? Stating that matter is made of molecules, atoms, and elementary particles, like electrons, protons, neutrons, and quarks, does not tell us much from this perspective. This only shifts the problem from a macroscopic scale to a microscopic one. We just continue to imagine chunks of microscopic matter/particles building up macroscopic material objects. This doesn't tell us anything more about the ultimate nature of matter itself. Later, from the beginning of the 20th-century onward, quantum physics revealed that these particles are not solid point-like objects with precise position and momentum but rather ephemeral entities with non-deterministic and slippery behavior. Of course, one can exhaustively characterize particles with properties like mass, charge, and spin; however, this, again, only shifts the problem and does not solve it because then one wonders what mass, charge, and spin inherently are, in their nature and essence beyond conceptual and mathematical abstractions. Physical properties are purely relational and dispositional concepts that don't tell us what physical things are in themselves. These properties can only be defined in relation to other things or properties of things (for example, in physics, mass is a measure of inertia, inertia is a measure of a body's resistance to a change in motion, motion is a change in space coordinates – that is, a body's position in space, and time, while space is ...?)

In modern quantum-field theories, matter is conceived of as a force field. What we call 'material' — that is, the property of objects being made of hard and impenetrable stuff—is deep down, at the microscopic scale, the manifestation of force exchanges. Therefore, one can reduce the concept of what we call 'matter' to the action of electromagnetic and nuclear force fields. But what, then, are these forces? Beyond the purely abstract mathematical description, we have no clue. Furthermore, quantum field theory tells us that one can ascribe a quantum field to each elementary particle. For example, an electron is the excitation of an electron field, a sort of ripple in a universal space-time quantum field. And, again, this leads to the question: What are the essence and inherent reality of such a universal quantum field beyond a mathematical abstraction? Modern quantum theories do not answer these questions (they can't and shouldn't), so we end up with a complicated mathematical representation of matter that is reminiscent of a Platonic metaphysical *'hyperuranion'* where only abstract mathematical entities like scalars, vectors, tensors, and functionals in a space-time coordinate system have a reality. At the end of the story, we are left with even less than what classical mechanics told us, and that at least satisfied our simplistic human intuition of a universe that could be described as a myriad of tiny marbles kicking each other around. The nature and ultimate essence of matter remain as mysterious as they were at the times of the Greek philosophers. In this regard, no significant progress can be seen; it is yet another no-progress quest.

At this point, it makes no sense to insist on invoking the progress of science, which will eventually become advanced enough to tell us about these things. This only fails to recognize the difference between a progressive and no-progress quest. The latter will never find a resolution, not even in principle, at least not in the present scientific rationalistic paradigm. It has nothing to do with missing data or an insufficiently powerful understanding and control of complexity. Science is intrinsically lacking something that allows it to proceed further than that, and the mind is intrinsically limited to sensory abstractions that are inherently unable to capture the essence of things. There has been no progress in the last 20 centuries on this, and there can't be even in the next 20 centuries if our cognition remains limited inside the boundaries of a superficial sense-mind awareness of the present low-level rational materialism.

Nevertheless, to justify this lack of progress, people frequently resort to what we shall call the *'complexity argument.'* This argument goes as follows: Many questions have not yet found answers because things have become much more complicated than expected. Give us time, allow science to progress and technology to advance, and we will finally be able to grasp the entire intricacy of all the phenomena. We will also be able to manage to handle increasing degrees of complexity. Modern supercomputers, gigantic data storage devices, complex AI software, and other kinds of futuristic tools will finally allow us to compute and determine the structure and function of everything and reveal to us their most intimate nature.

This thinking has its roots in the conception of a purely deterministic and, at least in principle, completely predictable universe. One can backtrack this belief system to *'Laplace's demon'* thought experiment, proposed in 1814 by French mathematician and philosopher Pierre-Simon Laplace. It became and still is the (undeclared and implicit) assumption of modern materialism and all the sciences as an articulation of strict determinism and reductionism in conceptual form. Laplace wrote:

"We may regard the present state of the universe as the effect of its past and the cause of its future. An intellect which at a certain moment would know all forces that set Nature in motion, and all positions of all items of which Nature is composed, if this intellect were also vast enough to submit these data to analysis, it would embrace in a single formula the movements of the greatest bodies of the universe and those of the tiniest atom; for such an intellect nothing would be uncertain and the future just like the past would be present before its eyes." [11]

Fig. 5 Pierre Simon Laplace (1749-1827).

In the 20th century, quantum mechanics would shatter this conception into pieces; however, it remains the (more or less unaware) assumption that drives all the reductionist scientific thinking and is at the base of the complexity argument.

A striking example of this was the Human Genome Project, the international research project that successfully sequenced the entire human DNA — that is, it determined all of the genes of the human genome. Once the goal was achieved in 2003, it was initially believed that this would open the way to the development of new therapies and a 'personalized medicine' to heal genetic diseases. However, unfortunately, it did not meet those expectations. Things turned out to be much more complicated than expected. Genes are not just single and independent blocks of information as previously thought but express their function in a far more complex manner. *"The more biologists look, the more complexity there seems to be,"* was the disillusioned subtitle of an article in the renewed journal Nature which recognized the obvious: *'Life is complicated.'* [12]

We stubbornly continue to forget an old bit of wisdom: Mapping the territory is far from sufficient for understanding it. This is a lesson that reductionist science has still not learned. Several other so-called 'Big Science' projects did not meet the expectations for the same reason but continue to be advertised as incumbent revolutions that will change everything. An example is the anticipated bioengineering revolution of the 1990s, which never really took place, stem cell research, and the slow progress of the so-called 'war on cancer' [13]. I maintain that, despite the present excitement, we will experience similar disappointments with regard to the bright future that AI and self-driving cars promise us. Many research projects have become only slow additions or minor adjustments to the already known, which do not represent real progress which justifies the expectations.

We often deceive ourselves into believing that some facts, discoveries, data, or measurements tell us something about reality when they actually don't. In the neurosciences, this becomes particularly clear: The mass media are full of nice images of brain scans that show us the so-called *'neural correlates'* – that is, the relationship between brain activity and our conscious experiences, cognitive and emotional states, or motoric actions or even where love occurs—hence leading us to believe that we are quickly advancing in the knowledge of what thoughts, emotions, and even love are. The general public has been seduced into believing that all of these studies that display the brain images and precisely explain the neural pathways of our conscious lives have an explanatory power that, upon closer inspection, they do not. There is a general obsession, a sort of reductionist *'neuromania,'* with words like 'neurons,' 'neural networks,' 'prefrontal cortex,' 'fMRI brain imaging,' 'genetic,' 'molecular,' etc. that are supposed to explain everything. We tend to believe that if we can describe a phenomenon by mapping its microscopic interactions among fundamental units, this is the ultimate and final explanation of what it is and what it does. If we know the molecular workings of a disease, we feel more confident: We tend to believe that we have a better understanding of what it is, where it comes from, how it works, and how it can be medicated. Despite this being only rarely the case (but, who knows, this belief alone may help us heal as a placebo effect) and having led to many disappointments, including a lack of expected progress and failed research lines, we insist on this deep physicalist and reductionist approach and worldview without questioning them regarding all their significant cultural effects.

But those who defend this method of doing research resort to the complexity argument, which tells us that things are complicated, tremendously complicated and that it is only a matter of time before we finally get things under control. "Let us even further dissect all things, life, and the cells into its elementary components, down to the molecular or atomic scale, and we will have a crisp, clear understanding of its workings. Give us even more powerful supercomputers that will simulate, analyze, map, and represent genes, cells, neurons, and all matter with its functions and interactions, and we will be able to understand how Nature works. Give us another 10 or 20 years and multi-billions in funding, and we'll get there — you'll see." That's the sort of rationale permanently lurking in the background. Instead of admitting that we are dealing with levels of complexity that the analytic mind can no longer understand and control, one again hears that things were more complicated than expected but that another 10 or 20 years of research will get us there.

Another no-progress quest that only rarely gets attention is the attempt to build software, some type of deep-learning neural network, or whatever kind of machine that would be able to understand meaning — that is, a process with a semantic awareness of the symbols and data it is analyzing. It is one thing to have an image, a description, or representation of an object — say, an apple. It is another thing to grasp its significance, meaning, and the semantic content of the word 'apple,' being able to understand what it is and in what context it exists.

The point is that syntax refers to the form of a sentence (grammar), while semantics refers to whether a sentence has meaning. AI operates on the syntax but not on semantics. And this makes an enormous difference.

Of course, significant progress has been made in recognition of natural language, speech, image, and pattern recognition, in pattern retrieval by data mining or other impressive AI computation skills. Google's celebrated Alpha Go computer beat the world champion of the Go board game. It ignited a worldwide emotional wave, making us believe that the intelligent sci-fi cyborg is coming. Yet what does a computer based on whatever smart and sophisticated AI software or hardware understand about what it is doing? Does it understand the meaning of the chess or Go play in which it is engaging? Beyond a mechanical iterated best-guess procedure, is there something that has a conceptual and semantic notion of what it is doing? The problem is easily recognized when, for example, we use automatic translations or speech-to-text software. These tools have become much more efficient in the last few decades. However, they are still far from attaining human efficiency because one clearly recognizes how they misinterpret words or fall into translation or linguistic misunderstandings primarily because they lack an understanding of meaning, that they are incapable of recognizing the semantic context to which the words refer, that they cannot understand the intent of the characters in the text and that they lack empathy. The same can be said of image and pattern recognition. The machine classifies patterns but does not understand their meanings. This makes it clear that there is nothing 'in there,' in the machine, that 'understands' and has a perception of meaning, let alone a conscious subject that grasps some reality.

The question then is where, how, and, especially *if* semantics is comprehended in the brain?

We will take up these points in greater detail later. Here, we point out that in this sense, there has been no progress. There has been no progress in making machines aware of meaning. We don't even exactly know what meaning is and how meaning emerges in our own minds. I contend that there has been zero progress in realizing a real semantic and conceptual awareness in AI. The awareness of meaning in AI is a no-progress quest.

Progressive quests	No-progress quests
Search for ET life	The existence of God and the afterlife
Quantum gravity	The meaning and purpose of the universe
Controlled nuclear fusion	The essence of matter
The molecular origin of life	AI 'awareness of meaning'
The fine-tuning problem	The hard problem of consciousness
⋮	⋮

However, having said that, what we are implying with this is not that, in general, research projects in the fields of genetics, AI, or cancer treatment are necessarily no-progress quests. They are, at least in part, progressive-quests, as a great deal of progress has been made in these fields, and they have furnished us with an invaluable amount of information about life or allowed for technological breakthroughs. These approaches, however, hinder the progress of science itself when they are almost obsessively fixated on a purely reductionist paradigm that reduces everything to genes, cells, neurons, and molecules and that disallows other approaches and research methodologies based on less-reductionist conceptions. It is with this molecular Laplacian worldview in mind that, much too often, people resort to the complexity argument to hide the failure of a method of research or–and this is the point we want to make here–to make us believe that a no-progress quest is a progressive-quest. If, for centuries, science and reason made no progress at all in some specific topics, it makes no sense to believe that they will do so in the next decades, invoking complexity arguments.

2. The Hard Problem of Consciousness

The most paradigmatic no-progress quest that science has been bumping into since its beginnings is the question of the origin and nature of consciousness. Some aspects of our conscious experience seem to be irreducible and do not allow for a description and explanation inside a materialist paradigm. There are questions regarding consciousness on which science has not made an inch of progress since Descartes' thoughts on the mind-body problem.

This might sound like an outrageous and even silly claim, as one might argue that, especially in the last decades, there has been enormous progress in the neurosciences, especially due to groundbreaking technological advancements such as the invention of powerful scanning imaging tools and measuring devices like functional magnetic resonance imaging (fMRI), positron emission tomography (PET), transcranial magnetic stimulation (TMS), magnetoencephalography (MEG), and new generation electroencephalography (EEG), not to mention all the advances in microscopy that have allowed us to observe the structure and function of nerve cells down to almost the molecular level. In the last thirty years, neuroscience has progressed like no other science did. The structure of our brains has been mapped with the utmost precision, and many secrets of its functions have been unveiled and mapped in great detail. We know much more about Parkinson's disease, Alzheimer's disease, and epilepsy than we did forty years ago. We can correlate our emotions, our thoughts, and the activities of our bodies to the activation of specific brain regions. We know, in great detail, how neurons work and can simulate their workings by implementing neural networks in computer programs. Thousands of scientific peer-reviewed papers on the subject are published every year.

Therefore, how can someone dare to decree that no progress has been made in the study of the nature of consciousness in the last four hundred years?

As usual, words have meaning, and everything depends on what we mean by the word 'consciousness.' We have pointed out how a naïve interpretation of the word 'consciousness' might lead to misunderstandings. The neuroscientist whose primary task is to cure mental diseases or brain impairments has no interest in philosophical musings; this neuroscientist usually takes consciousness as synonymous with the 'waking consciousness,' conflating it with mind and cuts out all the rest.

We will, instead, refer to phenomenal consciousness, as described in Pt.I-I.3, because it is this aspect of our conscious existence and inner life that is the most enigmatic one. It is, after all, the very essence of what we feel to be.

A more in-depth analysis of this aspect will be the main theme filling a long and detailed account in the next chapters of this book. However, to begin establishing a basic understanding of the trouble with consciousness, let us address the so-called *'hard problem of consciousness.'* This problem isn't new; it was more or less explicitly *Fig. 6 David Chalmers.* present in many philosophical writings of the thinkers of the past (such as in *'Leibniz's Mill argument'*, see Pt.II-II.2k .) However, in an article in 1995 [14], the Australian philosopher of mind David Chalmers pinpointed it as relating to the modern neuroscientific knowledge by distinguishing between the *'easy problems'* and the *'hard problems'* of consciousness.

The 'easy problems' of consciousness aren't easy at all but, rather, are those kinds of problems that we have categorized as progressive-quests — that is, questions regarding our brain functions that we may not yet have answers to but that we can be reasonably confident (due to further advancements in brain imaging techniques, biology, biochemistry, biophysics, and other technological advancements that allow for further insights into our brains) will finally be resolved. The easy problems of consciousness deal with phenomena in the brain that are functionally definable, such as how the brain integrates information, categorizes and discriminates environmental stimuli, focuses attention, etc. If we find a mechanism that explains how a brain region and its related nerve cells — that is, its neural networks — process visual information and categorize it into objects, forms, motion, and properties, then one can say that this 'easy problem' would be solved. It is about finding the neural correlate and describing its function due to an external stimulus or thought process that corresponds to a particular experience. For example, the neural correlates of visual perception — say, a flash of light — can be found in the activation of neurons in the visual cortex, the part of the *'cerebral cortex'* (the outer layer of the brain) located in the *'occipital lobe'* (one of the four lobes of the cerebral cortex and located in the rearmost portion of the skull).

Note that this corresponds to an exclusively third-person investigation: Project a flash of light onto someone's retina and then use a brain scan to determine which region of his/her brain will light up.

Discovering the mechanisms in the brain leading to an experience or allowing for the execution of a task isn't really an easy problem because it might still take decades of research to map all the functions and discover all the neural correlates of our conscious experience. However, it is plausible that these 'easy problems' will sooner or later be explained away by the progress of science (which is why we preferred to call them 'progressive-quests' instead of easy problems). For the easy problems of consciousness, once the mechanisms are well understood, little or no explanatory work is left to do.

Chalmers then contrasted these easy problems with at least one difficult problem of consciousness. It becomes a problem when we conceive consciousness and mind only as an epiphenomenon of the brain.

The hard problem of consciousness is not so much about the functions, the mechanism, or the neural correlate of an experience; it is about the question of why a physical process, such as neurons firing in a neural network, is supposed to lead to a subjective qualitative experience at all. There is an *'explanatory gap'* between the description, however detailed and complicated it might be, of a physical process with its material operation in a system and the emergence of a qualitative experience such as the sentience of the sweetness of sugar, the redness of a tomato, the smell of a flower, or the feeling of pain when one puts a hand in a burning fire. It is about the qualia, the 'objects of sensation' perceivable from inside of us and arising due to a direct experience that goes beyond mere data processing.

Therefore, the question is: How does a biochemical reaction in our brains translate into a subjective qualitative perception? Why and how does insentient matter generate sentience–that is, the qualities of experience–when arranged in a certain complex way, as it does in our brains?

The hard problem is that we have not even the faintest understanding of why some neural events are accompanied by a conscious experience of a subject that perceives phenomenal qualitative contents. We still don't even have the beginnings of an explanation of how complex electrochemical processes somehow give rise to our inner subjective world of colors, sounds, smells, and tastes. What we know about ourselves from the inside subjectively contradicts what science tells us about our brains from the outside and which is supposed to describe 'objective truths.'

As a curious side note it might be interesting to point out how in 1998, neuroscientist Christof Koch bet Chalmers that the mechanism by which the brain's neurons produce consciousness would be discovered in 25 years from then. In 2023 Koch admitted defeat but doubled down [15]: *"Twenty-five years from now is realistic, because the techniques are getting better and, you know, I can't wait much longer than 25 years, given my age."* I wish him good luck, but am pretty sure he will lose this bet as well.

The problem can also be illustrated by the *'philosophical zombie'* or simply *'p-zombie'* argument, according to which it is conceivable that there exist brains that perform all the same neural and sensory activities with all the decision-making functions and, seen from the outside, behave exactly as conscious beings but have no conscious experience. That is, there could be brains that react to stimuli that cause us, humans, pain or pleasure, sadness or happiness, etc., but that, to these 'zombie brains,' don't cause any pain or pleasure, sadness or happiness. They simply react to sensory data as we do and externally mimic sentient life but, internally, feel nothing. These would be brains that can measure the wavelengths of light but don't see any color, brains with bodies that eat certain foods but don't taste them. They don't know what it is like to be and what it feels like to be a conscious being. They simply imitate our behavior, relying on the same sensory data as that of a sentient lifeform.

The p-zombie is frequently illustrated in sci-fi movies in which androids behave like humans but, in the story, the character plays the role of a non-sentient humanoid. An example that may have gone unnoticed to most, but that exemplified the difference, was a scene from Terminator 2, in which John Connor asked the terminator (Arnold Schwarzenegger): *"Does it hurt when you get shot?"* The terminator replied: *"I sense injuries. The data could be* called 'pain.'" The terminator records and analyses data with which it does not associate any feeling and perception of pain; it only labels it as 'pain.' It is just data, not pain. In the same scene, the difference between a living and sentient lifeform and a machine is also reflected in the hesitant language of John when he asks: *"How long do you live? I mean, last, whatever...?"*

Fig. 7 From the Terminator 2 scene: Sarah Connor extracts the bullets from the terminator. Does it 'feel' pain?

One wonders why evolution imbued life with consciousness at all. Does this have an evolutionary advantage? If every task and behavior could be performed without conscious experience, perhaps even better, just by creating biological robots with no inner life and subjective experience, then what is the meaning and evolutionary function of consciousness in the first place? Why is there nevertheless something that allows us to feel what it's like to be conscious?

At any rate, the real question that the hard problem of consciousness raises is why a complex aggregate of matter in the form of neurons and other complicated stuff in our skull gives rise to this subjective qualitative inner life we experience. One can establish the neural correlates of qualia, say, of the sour taste of a lemon and certain brain states, and neuroscientists could also arouse such experiences by directly stimulating some areas of the brain, but why this creates, out of the blue, almost like magic, a subjective conscious experience in an apparently unconscious physical world, remains a profound mystery. How can consciousness emerge from the brain if the two have nothing in common?

To clarify further, let us take up Thomas Nagel's characterization of consciousness. Suppose you know everything about bats. Do you then understand what it is like to be a bat? Of course not. And why not? Because if you do not have the subjective experience of a bat, you will never know. You can know intellectually everything about bats, about every neuron, molecule, the atom of their brain, and yet this is far from giving you a real understanding of how a bat perceives itself and the world. This demonstrates how mere intellectual and mental knowledge, no matter how rigorous, analytic, and scientific, cannot convey the subjective experiences of bats.

Another way to illustrate this state of affairs is through a thought experiment proposed by Australian philosopher Franck Jackson, known as *'Mary's room'* or also the *'knowledge argument.'* [16]

Imagine Mary, a neurophysiologist who knows everything about color perception but has been confined from birth in a room without color, a perfectly black and white environment. Despite her long and detailed study of color perception – that is, despite knowing everything about the biochemistry, the neural pathways, and the brain activity correlating with color perception –

Mary will never know what it is like to perceive colors. She can spend her entire life reading books on physics, biology, chemistry, or neurophysiology, but she will learn what it is like to experience seeing colors only if she is released from the room and allowed to see the blue sky, green trees, or redness of tomatoes. One could also argue the other way around: Even if Mary knows nothing about the physiology of color perception, once released into the colored world, she would immediately 'learn' something new that an intellectual analysis could not convey – namely, she will 'know' what it is like to perceive the color qualia.

With this thought experiment, Jackson argued against physicalism. It shows how science based on the physical knowledge of things, however advanced and complex, can't furnish a complete explanation because it can't lead to the phenomenal experience of qualia. A first-person experience can't be explained by physical facts. Science, as it is, is in this sense incomplete.

Fig. 8 Mary's room thought experiment.

One of the most common objections to Jackson's thought experiment is that Mary doesn't gain any 'knowledge' in the scientific sense but gains something else. She doesn't arrive at a new propositional knowledge–that is, a knowledge based on statements that are either true or false. Therefore, arguing that Mary has gained new 'knowledge' is wrong. Another objection is that this is not even a real thought experiment because it is unlikely that Mary would never perceive colors in a totally black and white room because we can experience colors in dreams. This thought experiment shows only that even if language can't capture the whole spectrum of our subjective experiences, this doesn't mean that materialism is false. We simply have to resort to more moderate forms of physicalism that do not pretend from language what it can't convey.

Others, like American philosopher Daniel Dennett, go so far as to refute the knowledge argument altogether by maintaining that if Mary knows everything about the functional neurophysiological description of the physical states of how color perception occurs in a brain, then she is also able to distinguish the different percepts that would allow her to distinguish between one color and another. Therefore, even if her own brain is not in that physical state, she can nevertheless 'imagine' what it is like to be in it.

I find these objections weak. The former objections simply point out minor linguistic inaccuracies in Jackson's original argument, which do not invalidate the point in question. Restricting the concept of 'knowledge' by definition to only bookish propositional statements while excluding our whole spectrum of experiences is highly questionable and, at any rate, a matter of convenience, as it does not invalidate the point raised by this thought experiment. Frankly, in my opinion, Dennett's argument is the weakest: It pretends that an intellectual 'imagination' can lead to the subjective experience of color. My experience is not a form of 'imagination,' no more and no less than one 'imagines' pleasure or pain, smells or tastes. Experience is not a classification or a mere matter of choice of words and vocabulary.

Another common argument that tries to downplay the hard problem of consciousness comes from '*emergentism*'. We know that the properties of the world are emergent from the aggregation and interaction of particles, atoms, molecules, cells, etc., into higher-order macroscopic complex systems. The common example is that of the wetness of water. People say that wetness is an emergent property from the combination of oxygen and hydrogen atoms. Neither the oxygen nor the hydrogen atoms are 'wet', it is their combination and interaction in huge numbers that makes wetness emerge on a macroscopic scale as a fluid we call 'water'. There is nothing mysterious in the emergence of wetness from non-wet H_2O molecules. Armed with this example, the emergentist contends that, analogously, there is nothing mysterious in the emergence of a sentient conscious subject from the combination and interaction of a huge number of non-sentient and non-conscious neurons making up our brains.

This is a nice example of a logical and philosophical fallacy that shows all too well how we, as humans, do not know ourselves. Because wetness is not a physical property, it is already a quality that is linked to a conscious experience. It is a qualia, that we experience *in us*, like a color or a taste, or a sensation of pleasure or pain. In physics, there is no such thing as 'wetness.' The physical property is viscosity – that is, the measure of the friction a liquid opposes flowing on a surface. Without being aware of one's mental reifications, one equates the quality of an internal subjective experience (wetness) with an external physical property (viscosity). The hard problem of consciousness is that we have no clue how the former could be explained from the latter. The emergentist shows to be unaware of one's mind trick that has already subconsciously conflated two completely different ontologies. One unwittingly posits something from the outset coming into existence like magic. For this reason, emergentism is not a useful concept in the philosophy of mind because it does not add anything that could have an explanatory power.

In this sense, science did not make any progress for centuries and, despite all the fantastic developments of the neurosciences, did not proceed by an inch. Chalmers' hard problem of consciousness is clearly a no-progress quest. Plus, it makes no sense to resort to a complexity argument. We might be able to establish, at a molecular level, every biochemical reaction in our brains and every process and interaction of the brain's neural networks, as well as map every brain function and become able to understand all its complexity up to the smallest detail. However, that wouldn't tell us anything about why this complexity is supposed to give rise to a subjective experience. The hard problem of consciousness has nothing to do with our ignorance about the functional description of the brain.

Note, again, how one can understand the hard problem of consciousness only by taking a first-person approach. From the third-person perspective, phenomenal consciousness could not exist, as it would be contrary to any scientific evidence. From the strictly scientific third-perspective approach, there is no evidence for conscious experience at all. In principle, one should resort to Occam's razor and dismiss it all as irrational mumbo jumbo. Nonetheless, if almost all scientists agree that consciousness exists[10] in the sense that we experience subjective phenomenal events, it is only because they know they are sentient beings from their first-person experience and are willing to admit that others are as well.

This was only an appetizer to introduce the issues that modern science and the philosophy of mind have in handling the concept of consciousness. The section that follows will seek to exemplify the point in question by forwarding several arguments that make it clear how there is no hope, not even in principle, that within an exclusively physicalist third-person perspective, science will ever be able to progress on this subject. The aim of this section was first and foremost to elucidate how rational third-person materialism is doomed to deal with progressive-quests only and will never be able to handle no-progress quests (or 'hard problems'). The no-progress quests aren't intractable problems because they are too complicated; they are inherently out of reach of a purely materialistic science because, as we are going to argue, they are, in their intrinsic nature, neither material nor comprehensible by reason alone. Insisting on the progress of science as the magic wand that will lead us to answer the no-progress quests is like failing to recognize the simple arithmetic rule that kids learn in elementary school: Zero times a gazillion is still zero.

---

[10] An exception are the *'eliminative materialists'* that go so far as to deny their own consciousness, branding it as an 'illusion' or a 'false belief'–we will shortly readdress them in Pt.I- IV.5.

3. The Binding Problem and the Emergence of Meaning

Let us ponder an interesting feature of our conscious life: Binding.

Binding remains a deep, unexplained mystery that surrounds our conscious perception and could not find a definite resolution, at least not until our present day. What philosophers of mind call the *'binding problem'* relates to the fact that the brain is made of billions of cells and different working units which analyze signals and biochemical events in different regions and sometimes even at different times but, nevertheless, presents all this to our conscious awareness as a unified and single experience which does not represent itself as being composed of innumerable signals and events.

For example, although our brains process visual features such as colors, motion, and form in different specialized cortical areas using different neural pathways, we nevertheless experience these features united together into a single meaningful whole. The brain does not compute them in sequential order but distributes information in parallel to different specialized areas where the characteristics of a single object are elaborated. When we see a colored object, with a specific form, moving through space, its color, form, texture, and direction of movement are 'disassembled,' in the sense that different brain cells are activated for these different properties, independently from the other properties and the neighboring cells or brain areas. Not only that, but the result of this separate analysis of the object's properties seems not to be transmitted to some central receiving station that might unify them again in a single unique representation but, instead, circulates throughout the brain. And yet, it is a quite common experience among everyone that a unified vision lastly occurs somehow. We don't see the redness of a tomato as a property separate from its form.

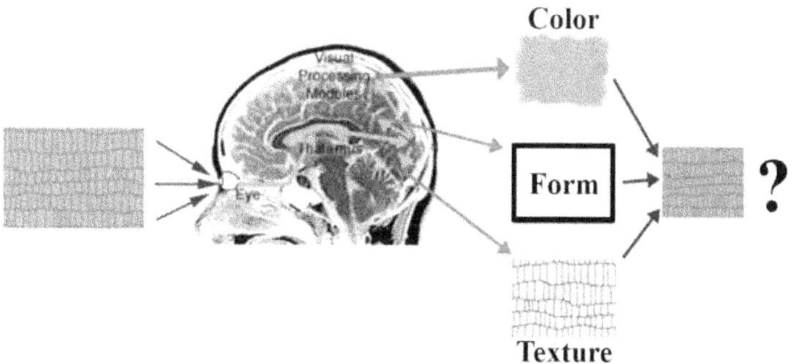

Fig. 9 An illustration of the binding problem. Credit: [17]

A visual binding problem emerges: If visual properties of things, such as shape, texture, and color, are distributed and analyzed separately throughout the brain regions, where, how, and when does the perception of a shape, with the perception of its color and texture become unified into a single experience of a colored object, come in? The binding problem emerges because of an unexplained feature integration where 'it all comes together'. While neurophysiological and neuropsychological findings show the existence of multiple brain areas that respond to primary features of objects and their spatial locations, scattering colors, movements, forms, and textures throughout the cortex, our phenomenal experience nevertheless coalesces into a perceptual unity of bounded feature representation.

These findings of modern neuroscience make it clear how the Cartesian brain model, which René Descartes advanced to explain a supposed mind-body dualism, no longer holds. In this model, he advanced the hypotheses that all the stimuli from the outer world are somehow converted into impulses and codified in the brain conveying them all, as a unique stream of information, towards the pineal gland, which he believed to be the center of perception and understanding, as well as the gateway to an immaterial mind or soul. We can nowadays certainly dismiss such a simplistic theory. The pineal gland is an endocrine gland that produces the melatonin hormone modulating sleep patterns and has nothing to do with sensory or cognitive functions.

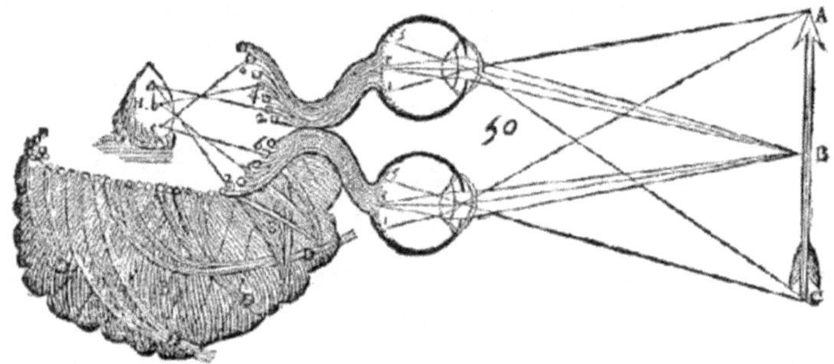

Fig. 10 Descartes' pineal gland brain model.

However, the binding of visual features is only an example of a much vaster and complex ability of our conscious experience. As a further example, binding can also arise in a semantic context in our language. For example, read this sentence:

"A sweet ice-cream."

The meaning of this sentence is immediate to everyone, and there seems to be no problem at all. But is the meaning of this sentence the resultant act of summing up the meaning of its components? How does the word "sweet" relate to "ice-cream"? The words "sweet" and "ice-cream," separated from each other, give us a completely abstract (but at the same time very concrete) perception which is very different from the sentence above. When these two semantic objects, "sweet" and "ice-cream," come together, a new qualitative subjective 'perception of meaning' comes into being.

Furthermore, things like 'sweetness' and the properties of an ice-cream are intimately correlated to a subjective experience, and it is impossible to isolate them as single logical constructs devoid of any experiential perception and information. What this suggests is that to understand the meaning of a sentence, which appears from a conscious binding of the words in a single and unique semantic whole, the subjective experience must somehow enter the scene. A theory of a logical and relational construct of the words alone with which we build sentences is insufficient and not the real 'engine' that produces such binding and the process of meaning.

Another interesting phrase to analyze, and suggested by William Seager, is this [18]:

"Why did Tom betray Mary so suddenly?"

Is the meaning of this sentence clear and unique? It isn't. Because, depending on where we place the emphasis, we can get three different meanings:

*1) Why did **Tom** betray Mary so suddenly?*
*2) Why did Tom betray **Mary** so suddenly?*
*3) Why did Tom betray Mary **so suddenly**?*

Here, we see that the meaning of a sentence isn't simply in the words. We realize how the emphasis places the sentence in a specific context that can vary from time to time. We have exactly the same words and the same sentences; however, completely different semantic contents emerge. It should, therefore, come as no surprise that developing a theory of semantics that allows an AI software to understand meaning has been shown to be an extraordinarily difficult task. Such a theory could not take single words and combine them; it must also consider the whole context, which means the whole world in the experiential dimension of a subject.

The sense of the whole sentence is also determined by the position and choice of a single word or a couple of words and may lead to completely different, even opposite, meanings. This semantic effect is called the *'Winograd schema,'* in which a pair of almost identical sentences, differing in only one or two words, have the opposite meaning. One can infer the precise meaning only by knowing how the world is built and the overall context in which the objects and concepts to which the sentences refer are known. An example of a Winograd schema was developed by the American economist Gary Smith [19]. Consider these three sentences.

*"I can't cut that tree down with that axe. It is too **small**."*

*"I can't cut that tree down with that axe. It is too **thick**."*

*"I can't cut that tree down with that axe. It is too **late**."*

In the first sentence, the pronoun "it" refers to the axe, while in the second sentence, it refers to the tree. In the last sentence, it refers to neither but instead indicates a temporal delay. But how do we humans immediately know that? Because we know the meaning of the single words, the meaning of the objects of the world they indicate, their relation to each other, and, most importantly, all these objects and the world arise in us as a subjective meaningful experience. It is only when this happens that we can bind the words into an overall context with semantic content – that is, become aware of the meaning of the whole sentence. Machines fail miserably in this task, even with these simple and short sentences. There is nothing that understands, binds, apprehends, or experiences and, thus, can construct any model of the world, let alone a significant semantic content.

Many have been tempted to believe that sooner or later, we would have discovered a logical relation between words that would have revealed to us how their coalescence emerges in a clear and unique 'element of meaning.' But, curiously, after decades of research in artificial intelligence, the philosophy of language, and models of the brain, it seems that behind the construction of a unique and single meaning from such a simple and apparently trivial set of words stands a tremendously complex process of cognition. Any attempt to gain the meaning of a sentence considering it as the resultant act of bottom-up summation or aggregation of some elementary cognitive or logical 'atoms' did not succeed. Much more sophisticated deep learning neural networks in computer science have been developed. However, we are still far from building machines that understand the meaning that words and sentences convey. As already mentioned, nowadays, translation software programs are common tools in every PC, but these are by no means able to replace human interpreters. It is easy to see where automatic translation tools or voice recognition systems fail most frequently: they don't recognize the meaning and try to guess.

To show that we are not just playing with words and that the phenomenon under discussion – that is, the relation between binding and meaning – is indeed a very serious one and a general feature of our mind and our conscious perception, we can create further examples going over from words to figures. Typical are the famous figures of Gestalt psychologists. Here, we have a perceptual object – in this case, not a sentence but a figure, whose structure comprises a unified whole incapable of being expressed simply in terms of its parts.

The Rubin figure is the most famous example because of its immediate and easy recognition: We can see two faces on a white background or a vase (or a goblet) on a black background. By focusing our attention on the figure, we can discover how our conscious perception of meaning switches from moment to moment into one or another: two faces or the vase. What happens here is that our brains organize the whole set of perceptual elements into a spatial configuration that forms a conceptual whole. All the pixels are bound in a final, simple, and – for us – precise meaning which is unique in the present moment but can change a second later.

How does the brain bind all the elements into a final meaning, and on the basis of what does it decide to switch to one or the other cognitive content? What is, after all, this subjective experience we call 'meaning'? The semantic process can't simply arise by the sum of the parts. Something much more complex and elaborated must occur in our conscious perception.

Fig. 11 The Rubin Gestalt face-vase figure.

Indeed, it is not an evident fact at all that when we look at, say, a chair, we don't see simply brute facts and data of visual content, but we perceive only after such visual experience a reconstruction *in us* in the form of an 'element of meaning' we call 'chair.' Visual stimuli alone do not convey the meaning of what is seen in the world.

Not only is this well outlined by the Gestalt figures, but it can also be experienced in several other instances. For example, the fact that there is more to seeing than meets the eye can be appreciated in the little artwork of illusion of Henri Bortoft, a physicist and philosopher of science, in Fig. 12. [20] At first, one might only see a random patchwork of black and white areas, but if you take your time and look further, you will soon recognize something emerging from apparent meaningless chaos. Try to do that until you see something before you continue to read.

Fig. 12 A random patchwork of black and white areas? Credit: [20]

The visual experience is almost instantaneous, as if meaningful content is suddenly 'switched' on like a light bulb. You will recognize the head and upper neck of a giraffe. This shows that meaning is not in the pure sensory stimulus, the raw data, as the visual information on our retina. Things have no meaning in itself 'out there.' Meaning is not in the patterns on the page you are reading;

they acquire a meaning *in you*. Our common assumption, which is that of science too, that we can acknowledge brute facts of the world independently from our mental constructs, is an anthropomorphic understanding of the world.

This conclusion is reinforced by experience with the well-known ambiguous figures used by the gestalt psychologists, such as the reversing *'Necker cube'* or the *'duck/rabbit illusion'*. The Necker cube can be interpreted as having either the lower-left or the upper-right square as its front side, while on the right of Fig. 13, the duck or the rabbit appear alternatively. In these cases, two different objects can be seen alternately, and yet the sensory experience is the same in both cases.

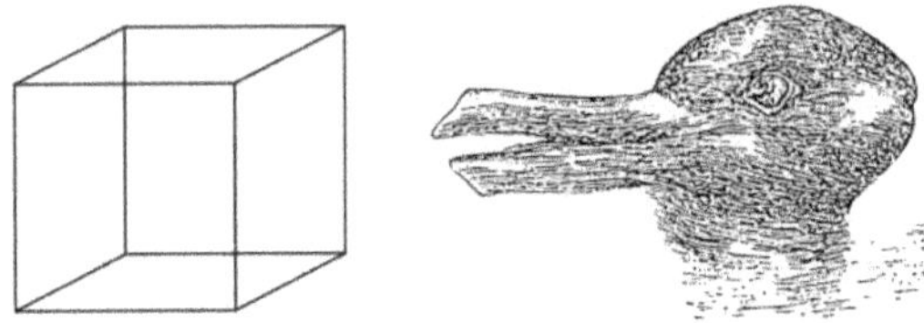

Fig. 13 The Necker cube and duck/rabbit illusion.

These examples show that visual experience alone, raw data, or empiric collection of measurements, as science does, is meaningless in itself. If these were not followed by a subjective and conscious analysis that goes beyond a mere collection of stimuli, we would be left with nothing but an unintelligible configuration of shapes, a random patchwork of black and white area. Some form of inner conscious organization and meaningful association is necessary. To use Bortoft's words: *"there is a non-sensory factor in perception,"* which translates into a *"non-sensory wholeness or unity."* This unification factor is the *"perception of meaning."* Despite being triggered by an external stimulus, the perception of meaning arises internally in our field of conscious awareness. A purely sensory experience alone, without some form of juxtaposed interpretation from the perceiving subject, is meaningless until an instance of non-sensory perception of meaning sets in. The meaning of the words you are reading now is not in the words printed on the paper of this book. Instead, it emerges due to a cognitive act that binds the sensory data into a unique and undivided 'semantic whole.' Even things – material objects, such as chairs, houses, trees, and mountains – do not possess meaning in themselves. When we say that we are 'seeing' things, we are not aware of the fact that we are *not* referring to the mere sensory experience of an external world. Rather, we are pointing at our internal perceptions of meaning. A purely sensory experience is meaningless but, nevertheless, is an experience that corresponds to a subjective state of awareness. Awareness without meaning tells us, again, how misleading the conceptual conflation of mind with consciousness can be: consciousness and perception are something separate from mentation.

The question at this point is also: Do the things in the world have a meaning in themselves, or is meaning something that comes into being *in us*? In what sense does the effect of the Gestalt figures differ from Seager's phrase or Winograd schema? The difference is, of course, the perceptual element of cognition: In one case, we have the words which form a sentence, while in the other case, pixels, and points composing a figure. However, the cognitive binding process that stands behind the construction of meaning is substantially very similar, if not the same.

This implies that we see the world depending on how we organize our perceptions. In other words, we do not discover reality as it is, but we build it through a process of integration and binding.

Moreover, American psychologist Dean Buonomano pointed out how comprehension in speech perception is also discontinuous and can be delayed [21]. Consider as an example the sentences "The mouse was broken" and "The mouse was dead." Until we read or hear the last word of the sentence, we can't determine its meaning—that is, if one is speaking about a computer mouse or an animal.

This happens not only with words or figures; we also bind and 'fuse' features by a delay in time. One can set up experiments that clearly show how we perceive and reconstruct phenomena from our perceptions by *'delayed feature fusion.'* In fact, one can show that when some experimental conditions are met, we can observe colors that aren't really there at all.

Fig. 14 Binding colors $(T_2 - T_1 < 200ms)$.

For example, if a short flash of red light, at time T_1, falls on your retina, you will obviously perceive a red color. If, after a time interval longer than 200 milliseconds, the red light is turned off and a green flash is projected, you will see the green color. So, nothing special occurred.

But what happens at time T_2 of the second flash if the interval between the two flashes is shorter than 200 milliseconds? You might slowly have a feeling for this and guess the answer: Indeed, our cerebral cortex binds and integrates the red and green colors and what we get is the perception of a yellow flash.

We see a color that has never really been shown. Indeed, it is known that the process of integration of colors already begins in the neurons of the retina.

So, we acknowledge that we see things that don't exist, at least in some special conditions. What we see is not what we get, or at least not always.

The most familiar visual binding experience is the stroboscopic movement. This is the very well-known effect we observe every time we watch a movie. All of us know that what is displayed on the screen is certainly not a continuous set of pictures but, rather, a discrete and finite number of still images. What is happening in your conscious perception of the vision of a film is a binding process that constructs an inexistent movement. You are building something which is essentially false in your brain, which is not displayed on the screen: continuous motion.

It is possible to make the stroboscopic movements more visible with appropriate experiments that show how we construct the illusion of continuous movement as translation and rotation. It is worth taking a look at some classical experiments of visual psychology that date back to the beginning of the last century. For example, Max Wertheimer, an Austro-Hungarian psychologist considered to be the founder of Gestalt psychology, realized, already in 1912, how we construct 'stories' of visual experience – something he called the *'phi phenomenon.'*

In the case of the experimental situation in Fig. 15 a, if the switching between two luminous structures in a dark environment representing a vertical and a horizontal rectangle is fast enough, we will perceive a continuous movement from the initial to the final position – that is, a rotation. This means that we perceive a fictitious rectangle also in places and positions that it never acquired, say, in the 45-degrees inclination state. We are the ones who construct a reality that doesn't exist. Similarly, you will perceive the flipping of the perpendicular structure of Fig. 15 b, though this time no longer rotating in a bidimensional space but, rather, making a rotation in depth.

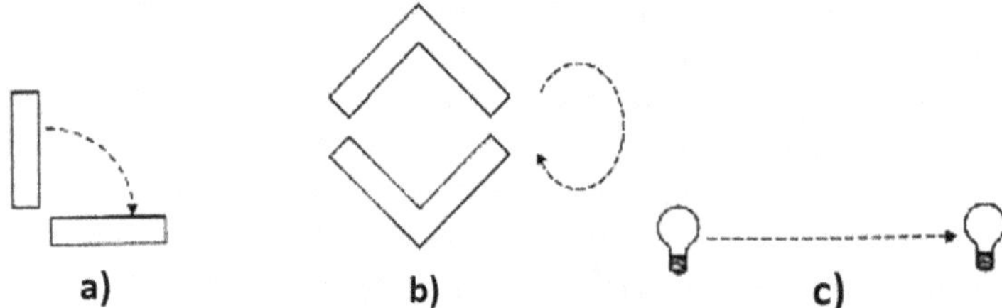

Fig. 15 Max Wertheimer's experiments of visual psychology.

Similarly, consider the two luminous sources of Fig. 15 c which are quickly turned on/off and are separated by an angular distance of 10 degrees. If the source on the right side is turned on after the source on the left side has been turned off in an interval longer than 100-150 milliseconds, we will see what was expected: The left lamp turns on and off, and only after this does the right lamp turn on. However, if the time interval between the two flashings is shortened to about 100 milliseconds, we perceive a slightly different version of 'reality': The two lamps will appear as if they are both turned on and off at the same time. We can also construct another reality if we shorten the time interval to 50 milliseconds or less, inducing an apparent movement in our conscious experience from immobility: You will not see two still luminous points but a single light source traveling from the left to the right side.

An extended version of the latter experiment is the *'color phi phenomenon'* discovered by Canadian Psychologists P.A. Kolers and M. von Grünau in 1976 [22]. In this case, the two light sources are colored. First a blue light source on the left (see Fig. 16) is shown for a short time, just a flash of blue light.

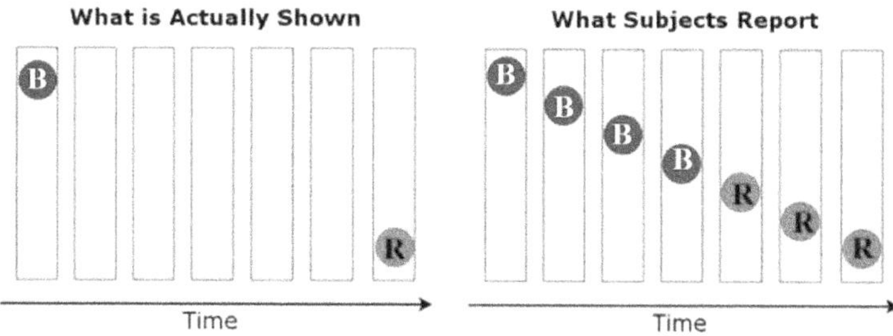

Fig. 16 The 'color phi phenomenon'.

Then, in a time-lapse shorter than 200 milliseconds, a second red light source lights up on the right. Subjects report that they see the blue spot traveling from the left to the right and changing color in between.

Again, our brains 'fill in the gaps' and build yet another 'reality' with a different meaning compared to the previous ones. These effects are not so different from the effects we can observe with the Gestalt figures. The meaning our consciousness associates with the things happening in the world 'out there' depends on how they present themselves to our senses. The color phi phenomenon is good evidence against *'Cartesian materialism.'* It is the idea that the brain internally represents truthfully everything that we consciously experience and that any other non-conscious observation is actualized outside of it. These experiments show that lots of processes are going on in our brains that we are not directly aware of.

Binding and the consequent construction of a particular meaning is a cognitive phenomenon that goes beyond text or visual appearances. It is active in other perceptions, such as touch. Also, a *'stroboscopic binding effect'* occurs at the touch sense level. This is a tactile illusion called *'the cutaneous rabbit illusion'* (or *'cutaneous saltation'*), evoked by tapping two or more separate regions of the skin in rapid succession and first discovered by F. Geldard and C. Sherrick of Princeton University in 1972 [23] [24].

The experiments use two piezoelectric skin contactors on different points of the skin. On one side, a contactor touches, at time T_{P1}, the skin once (P_1); then, after 0.8 seconds, at time T_{P2}, it hits the skin again at the same place (P_2) (see Fig. 17). Another contactor on the other side (P3) touches the skin at time T_{P3}, after a time interval $T_{P2}-T_{P3}$, ranging from 200 - 25 milliseconds.

The position at which P_2 was perceived corresponded to its real position, that of P_1, only in the case of a time interval $T_{P2}-T_{P3}$ larger than 200 milliseconds. At about 110 milliseconds, P2 was perceived in the middle of the position between P_1 and P_3. For $T_{P2}-T_{P3}$, less than 25 milliseconds resulted in a complete shift towards P_3.

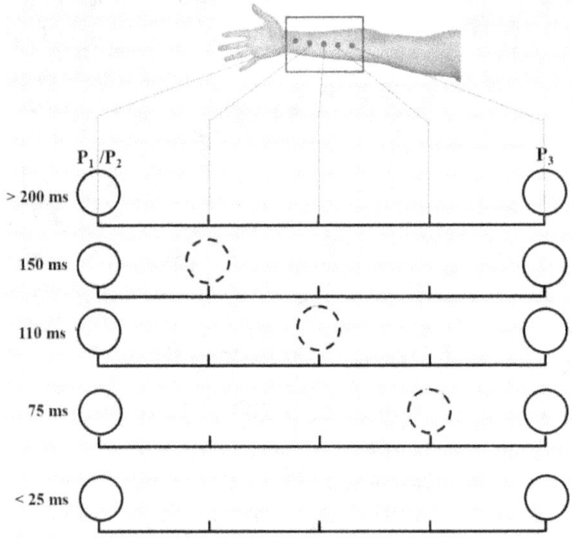

Fig. 17 The cutaneous rabbit illusion.

Again, we are in front of another illusion resulting from the binding of a set of perceptual events occurring in time and space.

Similar phi-phenomena have been shown to arise in auditory experience as well (such as 'hearing' a sound source 'travel' from one position to another, while the real sound sources are immobile) or even with illusory reversal of temporal order (such as reporting two light sources with different intensities as flashing in the inverted temporal order) [25].

This tells us that the binding power of our brains (or mind or consciousness?) is not limited to some specific functions or senses but is instead a very general feature of our conscious perceptual processes. We can find it everywhere at work.

Now, if we carefully reflect on what has been said so far about the binding of events, we would see that another interesting fact makes things appear even more unusual and bizarre. The experiments with the two lamps and that of the contractors make it clear how our conscious perception of the events occurring in the reality that surrounds us isn't direct. For example, in the case of Wertheimer's experiment, how can my brain know that in less than 200 milliseconds, the right lamp will flash and, therefore, it must construct a perception of simultaneity (the two lamps flash together). Obviously, it can't foresee this. This means that the conscious experience of reality is halted and 'waits' for input for some time before becoming a conscious state: Our perception is *'postdictive'* – that is, it first gathers data and then 'replays' it into our conscious awareness. These experiments suggest that we are always experiencing reality with a little time delay of about 0.2 seconds. This wouldn't be so surprising after all. A delay of a couple of tenths of a second seems to still be acceptable for our daily experience.

However, things are not as easy as that. One can construct other experiments which appear to suggest the opposite: In some cases, our perceptions might also be *'predictive.'* An example of this is the *'flash lag illusion,'* which displays two objects that are in the same location but that are perceived as displaced from one another. Fig. 18 top shows a red square that moves along the horizontal axis with timesteps (T_1, T_2, T_3, ...) separated by 200 milliseconds. At a specific time, say T_5, a green square appears and disappears at the next timestep. Then, the red square continues undisturbed. The sequence of events is integrated by our visual perception like a continuous movement, but, contrary to reality, the green flashing square is not perceived as being aligned with its real position; rather, it is displaced, lagging behind the moving object, as shown in Fig. 18 below.

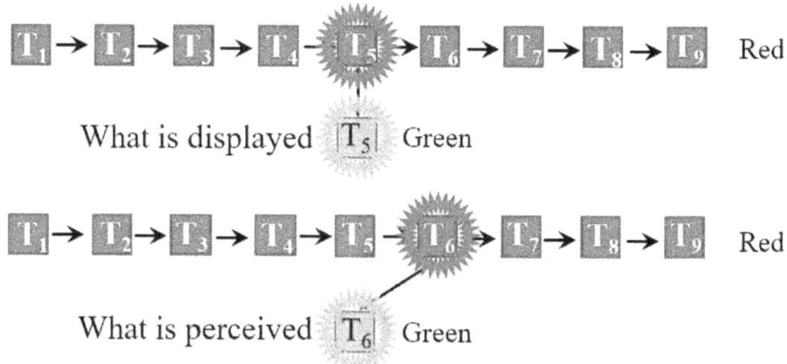

Fig. 18 The 'flash lag illusion'.

This optical illusion can be interpreted in two ways. One interpretation is that, once the green square flashes, the position of the red square is extrapolated, and the brain's visual cortex predicts the motion forward in time– that is, the perception is predictive. The other interpretation is that once the green square flashes its stimulus is recorded first subliminally, and only after a time interval, the green square enters our visual awareness–that is, the perception is postdictive.

There has been some debate over whether the first or second interpretation is the correct one, but psychologists are converging on the latter one or a combination of both ([26], [27]). A two-stage discrete model, in which periods of continuous unconscious processing precedes discrete conscious percepts, seems the most plausible [28]. After all, it is hard to entirely reject the postdictive theory because this is in line with the previous phi phenomena. Nevertheless, it can be shown that in special circumstances, predictions drive neural representations of visual events ahead of incoming sensory information. For example, when a motion sequence – say, a dot running on a circle – is suddenly stopped, our brains predict the next position of a stimulus ahead even though it was never presented [29]. This is a phenomenon we will have to take into account in the next section.

It might be of some interest to see how some specific brain damages, injuries, or neurological malfunctions can lead to a loss of binding and integration capacities. This sheds some light on how this process works.

An example is the *'Balint's syndrome'* that results from damage to both parietal lobes, the parts of the brain involved in receiving and processing sensory information. Causes can be disorders such as tumors, HIV encephalitis, trauma, neurodegenerative diseases such as Alzheimer's, etc. It is a neurological condition in which one becomes unable to intentionally move one's eyes towards an object (*'oculomotor apraxia'*), accurately reach for something at which one is looking (*'optic ataxia'*), and see the whole picture (*'visual simultagnosia'*).

The latter impairment allows one to see only parts of the whole. It manifests as the incapacity to see no more than one object at a time: Patients can accurately perceive individual details of a complex scene but can't synthesize and 'bind' the overall meaning of it.

For example, when shown a picture of a house, people affected by visual simultagnosia see only a window, a door, a wall, and so on, but would not be able to recognize that it is a building. Someone with this syndrome can recognize simple objects resembling a barbell–that is, two colored circles attached by a line–but when the line is removed, they see only one of the circles. It is a reduced ability to bind features. One can perceive no more than one object at a time in a scene in which many objects are related by properties of closure, connectedness, commonality, etc.

G. W. Humphreys and M. J. Riddoch, researchers from Canadian and American institutions, showed that not only did their patients lose this natural binding skill, but also an abnormal feature binding occurs which mismatches color, size, and motion with the perceived shapes [30]. For instance, when presented with a red X and a blue O, some of their patients reported seeing the X as blue and the O as red.

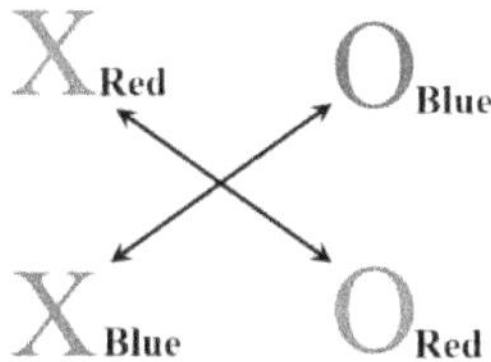

Fig. 19 Balint's syndrome: a red X and a blue O are perceived as a blue X and a red O.

Balint's syndrome also highlights the intimate relationship between binding and the experience of meaning in visual agnosics. *'Visual agnosia'* is the inability to recognize objects even though no optical impairment is present. For example, someone affected by visual agnosia may be able to verbally describe every detail and property that makes up a chair but cannot tell what the object is. Features are perceived, but the ability to bind them together into a meaningful object is lost.

Moreover, the perceptions of visual agnosics are highly conditioned by local continuity. For example, in Fig. 20, agnosics see only the number 7415 but are unable to read the word 'THIS' [31].

Fig. 20 'THIS' or 7415? Credit: *[31]*

This is reminiscent of the Gestalt figures but with the subject stuck into only one of the possible interpretations.[11]

We see again how the meaning of an object is something that comes into being in our consciousness in us and is not an intrinsic property of the object we perceive itself. This shows that a local analysis is not sufficient to build a global construct. The common assumption that global knowledge is the outcome of a local examination misses a step in between. Binding is more than the combination of local features. It requires the ability to produce meaning.

The world we see is a construct that arises from a complex binding process causing a subjective perception of meaning. What we see is not what we get. The idea of a world existent as it is with definite properties and with some semantic content external to our conscious representation and independent from an observer's subjective experience is a naive form of realism that needs a more careful and closer examination by all those who take scientific realism as the ultimate source of the discovery of truth.

That we don't 'see' the meaning *of* (or *in*) the world, let alone the world as it is, but that the perception of meaning is an acquired skill that we learn from

---

[11] For several other cases of this sort, see also Oliver Sacks' bestseller "The Man Who Mistook His Wife for a Hat" [364].

birth is well known from the fact that it takes several weeks until newborns start to recognize their parents' faces. They don't move their eyes between two images, and they see only objects that are 20-30 cm away, holding their gaze for only a few seconds. For babies, the world is only a kaleidoscope of fuzzy images without meaning.

Interestingly, we can gain a glimpse into what it is like to be in such a state of 'meaningless awareness' from those grownups affected by congenital cataracts, and that could be treated later. *'Congenital cataract'* is an organic anomaly present at birth that clouds the eye's natural lens and can result in *'amblyopia'* – a disorder of sight in which the brain fails to process visual stimuli. In the 1930s, Marius von Senden described, for the first time, the perception of space and shape in the congenitally blind before and after the operation [32]. When the sight of previously blind patients was restored and their bandages were taken off, the patients did not see the world as assumed. Instead, what they experienced was only a blotch of chaotic colored patches that meant nothing to them. They needed time and exercise to make any sense out of it.

The issue is not new. It was already debated in the 17th century and is known as the *'Molyneux problem.'* In 1689, the French philosopher William Molyneux conceived of the following thought experiment.

Suppose a man is born blind and has learned to recognize objects only by touch. For example, this person can distinguish, by the sense of touch, a sphere from a cube but has never seen them by the sense of sight. Now suppose that he suddenly can see. Molyneux questioned whether, if the sphere and cube were placed on a table, the man would be able to say which was the sphere and which was the cube *without* touching them.

The question was debated among bright minds, like the 17-18th-century English philosopher John Locke and Irish bishop and philosopher George Berkeley, who essentially agreed that when our mind can't build a relation between the tactile and sight worlds, the answer to Molyneux's question must be negative. The connection between the two worlds must be established by experience.

In 2011, neuroscience could furnish an answer. An Indian research project, *'Project Prakash,'* that treats blind children, and with their help, tries to find answers to scientific questions about how the brain develops and learns to see, showed that Locke and Berkeley were right ([33], [34]). Indeed, it turns out that in the treatment of congenitally blind children (8-17 years old), once they gained sight, they failed to visually match objects that were previously known to them only in the form of tactile information. Therefore, transferring tactile knowledge to visual knowledge is not an innate ability. However, this ability developed quite rapidly. Their skills in relating the vision-touch information had improved a few days after sight onset and were almost restored in the range of months.

These were only some of the many anecdotes which make it clear how data is not understanding. Meaning and sensory input are ontologically two distinct categories. If we don't realize this point, then a plethora of paradoxes, apparent inconsistencies, and strange consequences appear.

The question is: In what sense is science immune to these perceptual illusions? Isn't science a construct of meanings too? In what sense are the basic elements of meaning of the most exact sciences, say, for example, the concept of number, quantity, probability waves, mathematical notions, etc., different from the objects a visual agnosics perceives as "7415" instead of "THIS"? Are things like matter, particles, strings, neurons, etc., objectively real in the world, or are they on the same ontological footing as any subjective experience of meaning, such as the vase and the two faces of the Gestalt figure?

The only thing we can be sure of is that we have perceptions. We construct our world representations, not by knowing something which is out of us but always and inevitably by some processing that takes a subjective experience of perception at its foundation. It all comes down to perception. The world around us is constructed in our minds from sensory information of perception. One thing is sensory perception as such, and another is the emergence of a mentally bounded semantic world which arises from these phenomenal events.

Notice also how binding is something which already presents itself in our inmost and intimate essence: subjectivity. For example, we perceive the color, shape, and taste of an apple as a unique unified experience of a one and single subject. How does this feeling of being a single unified subject that perceives a taste which is instantly attributed to something having a shape and a color come into existence when all these properties are scattered around inside our brains? Not only do all these attributes become bound together in an object but, on top of that, they are also experienced by an undivided subject–an individual, an ego, a person who feels and says "I" and is not "you."

This is also an aspect of the binding problem with science having no clear idea what the answer could be. Some neuroscientists advance the hypothesis that binding comes into being because of coordinating mechanisms of the synchronization of neuronal activity, which phase-lock self-generated network oscillations. Though some evidence points in that direction [35], it would explain nothing. Why should whatever intricate network with 'self-generated' and however contrived 'phase-locked oscillations' lead to a personal identity? These speculations do not even begin to address the explanatory gap.

As a sidenote it might also be worth pointing out a strange analogy between our brain's ability to separate qualities and analyse it in different brain areas, with a strange quantum effect called the *quantum Cheshire cat paradox* where one can 'disembody' quantum properties, such as a photon's position and polarization in different locations. If this is only a coincidental metaphor or hides a deeper truth, I don't know. For more information, see Vol. II of my book on quantum physics [36].

4. Is Mind Computational?

To substantiate the claim that meaning is more than something contained in symbols, raw data, and symbolic information processing, the American philosopher John Searle, in 1980, proposed the so-called *'Chinese room argument'* in a now-famous paper [37]. Searle intended to put forward an argument by a thought experiment that shows that a computer program (running with a classical algorithm, *not* other forms of computation such as deep learning neural networks or quantum computation, etc.) cannot understand meaning, not even in principle.

The thought experiment imagines Searle himself in a room where he receives questions in Chinese characters through an input slot. Without understanding Chinese but following the instructions of a digital computer program that knows everything about the syntactic rules of the Chinese language by manipulating symbols, Searle sends out the answer in the form of another string of Chinese characters through an output slot.

He simply obeys the instructions of the computer without understanding neither the questions nor the answers. If the computer can pass the so-called *'Turing test'* – that is, a test which posits that a machine is intelligent like a human if its answers are indistinguishable from those of a human – an external observer, say, a native Chinese speaker, would mistakenly believe that in the room is a Chinese speaker, while in reality, it is the computer that is answering in Chinese.

Fig. 21   *Searle's Chinese room argument*   *John Searle*

Searle points out, however, that in principle, he could himself follow the rules of the algorithm step-by-step (say, by consulting the printout of the program library or database which tells him exactly all the possible rules with which he has to manage the Chinese ideograms) and that would allow him to pass the Turing test even without the computer. Due to the fact that from this point of view, there is no difference between the syntactic information processing of the computer or Searle mimicking it without understanding anything of the meaning the string of symbols represents, one must conclude that the computer can't understand meaning either. It only appears that a symbol-crunching machine understands meaning, but it does not, and the Turing test is not an appropriate tool for determining whether a machine understands meaning and semantics as humans do.

Later, Searle's thought experiment was further enforced by the so-called *"symbol grounding problem,"* discussed by Stevan Harnad in 1990 [38]. Essentially it points out that it remains a highly problematic issue wherefrom symbols (words, numbers, streams of bits, signals, etc.) get their meaning. There is a hiatus between, on one side, symbols, signals, number crunching, and, on the other side, semantic awareness, intuition, understanding, and knowing, both remaining outside a formal description of any computational model. One can translate the same general problem in the more specific context of neural coding [39].

Another implication of Searle's argument is that this also clarifies how a code, say, a binary code such ASCII or the old Morse code has no meaning or function in itself and is not something physical and existent in itself. A code always needs a 'semantic agent'–that is, a mind that understands the connection between this humanly pre-defined code with letters and symbols and its concatenation into words and sentences and is capable of working with it according to grammatical rules, a dictionary of words, and an association with semantic content. For example, a binary code sets into relation a string of ones and zeros to other symbols. Say the binary string 1010 corresponds to the number 10 in the decimal numeral system. However, it could also signify the 10th letter of the alphabet or whatever other kind of symbol. A binary string doesn't *have* or *possess* any meaning until a mind comes along and makes it meaningful through a standardized and semantic code system. One needs a being giving meaning to symbols. Bits, bytes, letters, or symbols are nothing meaningful without a conscious observer. A string of bits in a computer is nothing other than a series of electric potentials in a microchip. The string 1010 could also be created by aligning two white stones separated by two grey stones in a dusty desert with no meaning whatsoever. As long as a semantic agent does not read out a code, it is just a 'displacement' or 'puncture' in matter.

In fact, what does the word 'information' mean? The etymology of words is often wiser than our understanding of them. The verb 'in-form' tells us that something has been 'formed' by molding, carving, shaping, or puncturing an object into some pattern or modifying its physical internal, or external state. It is about forming something, making it a medium for symbols that convey a message expressing our thoughts to someone else who can understand those thoughts. However, forming patterns or modifying internal states in objects has no meaning whatsoever if there are no minds to receive and transfer the thoughts expressed. Saying that the 'fundamental nature of the universe is information,' as considered by some modern conceptions of the foundations of physics, isn't very informative either, as it becomes logically circular because it implicitly demands that something be formed which incorporates that information and must, therefore, be prior and more fundamental than the information itself.

Based on these arguments and other implications, Searle also introduced the distinction between '*strong AI*' (also called '*Artificial general intelligence*' (AGI)) and '*weak AI*' [40]. The former is a hypothetical sci-fi AI in which machines truly understand meaning as humans do. The latter indicates machines that only simulate humans' intelligent behavior through smart inferences or probabilistic guesses but have no understanding of what they are doing.

It is doubtful that we will ever have real thinking and conscious machines if they remain utterly unable to have a semantic comprehension of symbols, data, information, images, perceptions, etc. For example, it is questionable as to whether there will ever be a fully autonomous self-driving car if its deep learning neural networks (or whatever kind of AI brain it has) remain unable to grasp the significance of the environment it registers in its memory chips.

In fact, the correctness of the Chinese room argument becomes more apparent as AI progresses. For example, automatic translation has made some progress in the last decades. Computer translation from one language to another is nowadays much more efficient than it was in the 1980s. But, as everyone could sooner or later realize, when a computer (be it algorithmic or based on deep learning software) gets it wrong, this is, in most cases, due to its inherent inability to understand meaning.

At the time of writing, natural language processing is based on a set of rules that assign semantic values to words, so-called '*word embeddings,*' which is essentially an assignment of numbered values that represent some information about a word's meaning. For example, it indicates, through numerical representation, the relationship between an animal and its aggressiveness, such as (tiger, 0.99) and (bird, 0.05), meaning that a tiger is more dangerous than a bird. It also works by comparison, such as assigning a similarity index or 'angle' to words, like (tiger, lion, 5°) and (tiger, bird, 90°), meaning that there is a less divergent semantic similarity between a tiger and a lion, as both are felines than between a tiger and a bird, which have no semantic relation. Translation algorithms then work by word contextuality–that is, by counting the frequency with which, in the natural language, a word is associated with a neighboring term–and which could tell what a word means. For example, the sentence "I went to the bank, and I read the newspaper" has different meanings if, to the current word 'bank' (having two possible meanings), one associates the context word 'to' or 'the' or 'and' or 'I,' respectively. Automated translation machines maximize the likelihood that a given current word composes part of a specific semantic context.

In 2018, Google introduced a method nicknamed *BERT (Bidirectional Encoder Representations from Transformers)* that pre-trains and fine-tunes neural networks by using a sort of Chinese room reference book, figures out which features of a sentence look more relevant, and reads texts bidirectionally–that is, from left to right and also from right to left–to infer the contextuality better. This allowed for a better quality in machine semantic recognition and translation but confirmed that nothing in the machine goes beyond Searle's approach. That, in the end, is all about guessing, not understanding.

Moreover, notice also how because of a slight verbal ambiguity, a touch of humor, a cynical remark, or irony in a sentence, the translation algorithm will easily fall into an interpretational error, not rarely with hilarious reactions from the human side. The meaning of a message we receive verbally from others also depends on knowing their intentions. It is about empathy on the side of the listener and, eventually, even the ability to interpret body language. Meanwhile, computers only calculate the maximum statistical likelihood of the meaning and context of words; they know nothing about it and the world which creates that context.

This is, of course, an oversimplification of what goes on internally in a chatbot, such as ChatGPT of Open AI (we will take up this later, in Pt.II-III.4) or automated translation software, like that of or Google Translate (which works with representations of up to hundred-dimensional vectors and is trained using billions of sentences on the web to predict meaning and context for any given word). However, it makes amply clear why we have the impression that the most sophisticated translation tools soon reveal their utter semantic inability. Especially if you are reading a translated text into your native language. A computer, endowed with whatever complex and up-to-date AI technology, never understands anything. It only 'sees' strings, numbers, and vectors, which have no meaning in themselves unless a human reads them. There is a fundamental difference between translating and interpreting spoken language. There are good reasons why, despite the impressive skills of ChatGPT, Google Translate, or DeepL, and despite all the predictions to the contrary, professional human translators didn't lose their job.

We won't delve at length into describing the same issue modern AI has with image recognition. Deep learning neural networks and advanced AI software can indeed correctly recognize objects, faces, or data structures with a high degree of accuracy – and sometimes even better than humans can, especially in finding patterns in big data sets. This is quite an impressive achievement that has and will have interesting applications. However, deep-learning AI is easy to fool (for a review, see [41]), and when it fails, it suddenly mistakes a chair for an elephant, a glass of water for a machine gun, or a mouse for an ocean liner. There is nothing or no 'ghost' in the machine that understands anything. Engaging in sophisticated guessing based on a maximal likelihood match is one thing; understanding and associating meaning is an entirely different task. The problem of meaning in AI is still with us after all these years [42].

These AI applications are very good at mimicking human understanding. But, despite the impressive advancements in the last years, I maintain that a semantic awareness in a computer was and still is a no-progress quest.

Searle's Chinese argument received several critical replies and remains controversial. It shows, however, that algorithmic processing of symbols alone is not sufficient to explain a human's mind. There is an explanatory gap between syntactic rules and semantics as humans perceive it. The perception of meaning cannot be reduced to information processing alone. Symbols in themselves, like the letters in a book, have the ability to cause meaning to emerge, but only if there is a conscious being capable of understanding meaning in the first place. Meaning is a subjective experiential phenomenon *in us*, not an intrinsic property of symbols or material objects out of us and can't emerge by computation alone.

5. The Dysfunctional Functionalism

Searle's argument is also that the human mind can't rely only on a classical information processing algorithmic computation – that is, that form of functionalism supporting a computational theory of the mind.

Functionalism posits that mental states could be explained solely by their function, independent of the physical and structural substratum with which the function is performed. A frequent analogy is the function of a valve. It doesn't matter which material it is made of or what its internal structure is. Its function of allowing or blocking the flow of a liquid is what makes the device a valve. Similarly, mental states are constituted by their functional or causal role; it doesn't matter if your brain works with biological neurons or is a silicon-based alien brain. That is, what elicits mental states are programming rules and their effects, not the biochemical structure on which they are implemented and its related physical phenomenon. In this sense, the mind is just a 'platform-independent' functional system. Mental states are functional states which can but must not necessarily be realized with a biological neural network.

However, Searle's Chinese room thought experiment undermines functionalism because it shows that it is, in principle, possible to conceive of an experiment in which a system can functionally mimic a mental state with semantic content when there is none.

Another interesting objection against the functional approach to the problem of mental causation and phenomenal consciousness is the *'China brain'* thought experiment (not to be confused with Searle's Chinese room), which we present here in a slightly modified but equivalent form.

In its original form, it is formulated as follows. Suppose every person in China agrees to participate in an experiment that tries to simulate a brain made of the same numbers of neurons as the Chinese population–that is, about 1.4 billion neurons – which is about the number of neurons in the brain of a parrot. Each Chinese person simulates the function of a neuron and, according to the input signals he/she receives, outputs another signal to his/her neighbors. The neurons and their connections – that is, China's population and its communication lines–are arranged as the neuronal network in a brain.

Fig. 22 The 'China brain thought experiment': If every Chinese person simulates a neuron, could all China simulate the brain of a parrot?

Say each one communicates through a telephone line, which replaces the function of the axons and dendrites that connect the neurons in a real brain. The overall connectional structure might be rather complex and cover the entire territory of China but, at least in principle, can be considered the functional reconstruction of the entire brain, as, after all, a brain is nothing other than a huge number of nodes that work as relay stations (the neurons) linked together by receiving and transmission lines (the dendrites and axons, respectively) according to a specific network structure.

Notice that as long as the functionality of each neuron and all its connections is preserved, one can think of whatever kind of replacements for the neurons, dendrites, and axons. Thus, let us go a step further and replace every neuron (or Chinese inhabitant) with a more or less complicated system of water pipes and pumps which are functionally equivalent to a neuron and the dendrites and axons (or the telephone line) with an intricate system of water channels, say, covering all the territory of China.[12] The whole complex of water pipes, pumps, and channel systems could functionally simulate, in every detail, the brain of a parrot or, with the addition of an even larger number of nodes and connections, an entire human brain.

The natural question, then, is: If this large-scale Chinese *'connectome'* – that is, an exact mapping of all the neural connections of a brain–functionally reproduces the brain, would it, indeed, become a conscious sentient being having the same mental states as those of a parrot or human?

The functionalist – that is, several intellectuals at the top of the high-ranking philosophy departments – would answer affirmatively. Whatever material you use to build a brain, be it water pipes, chocolate, or toilet paper, if it is functionally equivalent to the brain of a conscious being, it will become

---

[12] We were inspired to create this modification by Dutch philosopher Bernardo Kastrup, who uses this analogy in his online interviews. [371]

conscious, with all of the same mental states, subjective experiences, and qualia.

Of course, we might never know if that is the case, as it is quite difficult, if not impossible, to realize in practice such a thought experiment. However, in all honesty, the author thinks this is less than unconvincing. In what sense is such a hypothesis more credible than miraculous thinking that posits the existence of Santa Claus or that storks bring babies? In my humble opinion, the China brain thought experiment only exposes the absurdity of the functionalist approach to phenomenal consciousness and the nature of mentation.

At any rate, a further issue with functionalism emerges with other thought experiments. For example, suppose a super-technology exists that can make a perfect copy of your brain. Say a scanner can map, neuron by neuron, even molecule by molecule, and reconstruct the entire map of your brain with utmost precision, then send this information to an incredibly precise sci-fi 3D printer that prints out a new brain which is an exact copy of yours. The question arises: Does this exact copy of your brain 'produce' your consciousness, your mind, and also 'you' as the same subject? The physicalist may still believe that because these are physically two different brains, they will elicit two different minds, consciousness and separate feeling of 'I-ness.' However, the functionalist who sticks to the idea that only the collective functions of all the subparts of the whole system determine the emergence of mind, consciousness, and subjectivity, independent of the physical substrate in which they are embedded, cannot resort to the physicalist's escapade. Applying this strict form of functionalism leads to a paradox: The copied brain must produce the very same subject as the original brain, contradicting the very same notion and feeling of subjectivity itself.

A more foundational question is: What is a 'function' in the first place? If we suppose that the function of the parts of a whole determines, independently from its substrate, the properties of the whole, this suggests a subtle form of dualism.[13]

In fact, note how a single object can have several potential functions at the same time. A screwdriver can be used to screw in screws and wedge a door closed, serve as a radio antenna, scratch your back, or kill someone. The existence of a function can also be context-dependent. Try to use the same screwdriver on a space station in the absence of gravity, and you will discover that it has lost its functionality, as the torque will cause your whole body to rotate around the screw. And yet, it is still the same object.

Things can also change their functions over time. This can be observed in evolutionary biology with the so-called *'Darwinian preadaptation'* (or *'exaptation'*)–that is, the adaptation of an organ that serves a different purpose

---

[13] We follow a line of reasoning inspired by system biologist Stuart Kauffman [383].

from the one which it evolved. For example, the lungs of some fishes evolved into a swim bladder which now serves to control their buoyancy. Nature invents new functions in evolution all the time.

This should make it clear how the word 'function' is just an anthropomorphic abstract descriptive designation that tells us what things do in a specific context and at a specific time. It is not a material thing inherent in the part or the thing itself. The function of a screwdriver isn't a property of the screwdriver, such as its size or mass. A function is neither a primary nor a secondary quality (we will discuss these in more detail later in Pt.II-I.2); it is an anthropomorphic conceptual construct with which we describe what things do and defines *for us* only the causal role a part plays in sustaining the whole. Notice also that a function is determined not only by a bottom-up causation but also by a top-down one: Only when the whole exists is the function of the part determined. For example, the heart's function is to pump blood, but the heart can't exist if it is not embedded in a living organism. Only the latter defines the function of the former.

All this shows that the naïve exteriorized mental conception of the world we perceive can't be the whole story. We have mistakenly projected an abstract concept into the world out of us instead of realizing how it is an emergent semantic property in us. The functionalist tends to project the abstract notion of a 'function' of an object into the object without being aware that it is a mental construct in the first place. That is because the very concept of a 'function' is a meaning. And, as we have seen, meaning emerges in us by a cognitive binding process that requires a conscious observer a priori. Because there can't be anything such as a function without consciousness, the attempt to explain consciousness by functions makes the whole thing circular and even self-contradictory. Functionalism becomes dysfunctional.

Nevertheless, despite his powerful argument, Searle is a biological naturalist and continues to believe that it is the biology of the brain which produces consciousness and the capacity to perceive meaning from symbols. One may ask, then, if this doesn't amount to falling into the same fallacy and false assumptions which the thought experiment of the Chinese room tried to unveil? In what sense is matter other than a symbolic reconstruction from our perceptions? As we conceive it, matter is already an abstract symbol, even if we are seldom aware of this. Insisting on matter as the point of departure for explaining mind and consciousness means, again, to believe that symbols in themselves have a power of causation.

We will unpack this in more detail later, but let us first review the hard facts of neuroscience as regards consciousness and mind – that is, the famous 'mind-body problem' from a strictly empiric third-person perspective.

III. States of Consciousness and Free Will

1. About Blind People Who See

The bottom line of the preceding chapter is that there is mounting evidence which suggests that our perception of the world must be largely subliminal or subconscious[14]. It seems very likely that we must accept the fact that the world we perceive is an interpretation and that the things we do and think are based on the 'stories' that our brains or minds make up.

The fact that many, if not the majority (or all?), of our thoughts, actions, and desires are driven by processes that seem to escape our conscious awareness is no news. The notion of the 'subconscious' or the 'unconscious' was introduced in the 19th century by the German philosopher Friedrich Schelling. However, it made its way to the popular audience thanks to the well-known Austrian psychoanalyst Sigmund Freud. Since then, it has become accepted wisdom, including in modern science, that we are largely conditioned by internal underlying mental, perceptual, and emotional mechanisms which we are not consciously aware of.

However, we do not always react instinctually or without reasoning but can make very rational choices without being aware of it. A striking example of this state of affairs comes from a neurological impairment called *'blindsight,'* which is the ability to see things without being aware of them at a conscious level. There have been several reported instances of people with damaged primary visual cortex (area V1), which impairs part of the visual input from the eyes to the brain. These people can see only one side or region of the visual field. It could be shown that some patients are nevertheless able to detect visual information within this blind region, though they are not directly aware of it.

For example, if the image of a straight line is presented inside the lost visual region, the patient reports seeing nothing. However, if asked to guess the orientation of the line, say, if it is presented with a horizontal or vertical orientation, some patients can guess correctly in 90% of cases. Others can 'guess' colors, move their eyes to follow the movement of an object inside the blind region, grab for something they cannot consciously see, mimic the emotions of a face they report not seeing, and even safely find their way along a path with hurdles they had to bypass. Because this can't be just a coincidence, one must conclude that some form of visual information must still be present and that it is perceived and processed subliminally. The impairment of the brain's visual cortex area V1 causes a lack of awareness but must still, at least

---

[14] We will later distinguish more rigorously between the concept of the subliminal and the subconscious.

partially, retain some form of vision that can't be reported. Patients affected by blindsight show that one can be attentive to something without knowing what one attends to (for a technical summary on the subject, see [43] or for a more popular short introduction, see also [44]).

A similar phenomenon to blindsight is the *'no-report paradigm'* in which healthy subjects, without any neurological impairment or disorder, nevertheless seem to 'see' something without knowing it and, therefore, without reporting it. Some visual content can be decoded accurately, with the subject being unable to communicate it verbally. For example, it can be shown via fMRI analysis that the perceptual recognition of a *'Kanizsa figure,'* such as the 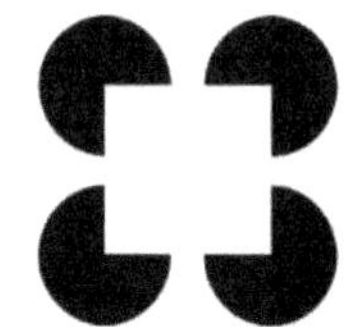

Fig. 23 The Kanizsa square illusion.

alternating visual inference of Fig. 23 of the white square vs. the four dark fragmented disks (the four 'Pac-Men'), depends on the activation of functionally different visual brain areas. That is, it is possible to determine during a brain scan which of the two figures one sees. But, if someone is distracted while looking at a Kanizsa figure with attentionally demanding tasks (such as being asked to identify letters flashing in quick sequence while looking at the figure), one is no longer able to consciously identify the two alternatives. Nevertheless, the neural signature of the distracted subjects did not differ from that of a control group not subjected to the distraction task. For example, it looks like the subjects who see only the fragmented dark disks but are unable to report the existence of the white square, nevertheless subliminally 'see' the white square. They 'see without knowing to see' [45].

One might, however, question the extent to which, in these no-report cases, such neural processing reflects any conscious experience at all as opposed to completely unconscious processing [46].

At any rate, this empiric evidence suggests that, for anyone without a neurological condition, the area V1 of the visual cortex is responsible for bringing the visual information into our conscious awareness, while a good deal of our visual perceptions and processing must occur first elsewhere and only later raised into our awareness.

2. Subliminal, Subconscious, or Unconscious?

This was only one of many examples one could provide and that we don't enumerate here, but which raise the question: Contrary to our belief and subjective perceptions, are we driven by a non-conscious mind, with us being no more than puppets guided by unconscious processes, perceptions, and thoughts and living in an illusion of free will?

First of all, the question of free will is intimately connected, not only to our conscious existence but also to the entire realm of our subconscious, subliminal, and unconscious existence. One can't deny the obvious fact that most of our actions, thoughts, and internal cognitive processes must be more or less subconscious or even completely unconscious. It is not necessary to prove that with fMRI images of people looking at Kanizsa figures. For example, otherwise, we couldn't even walk. Consciously walking from spot A to spot B implies a whole set of complex neurobiological, physiological, and cognitive processes that don't require direct attention. Fortunately, you don't have to think about every muscle contraction mechanism you have to perform while walking. After all, that's what toddlers learn: how to transform an action that first requires strong will and conscious attention into an automatic motoric skill.

But what do 'conscious,' 'subliminally conscious,' 'subconscious,' and 'unconscious' mean? People usually have only a superficial understanding of the meaning of these words and like to believe (subconsciously?) that their ill-formulated concepts are the only right ones. They don't even feel the need to check whether the meaning they have in mind is that which others imply. We already saw how this is the case with the conceptual conflation of mind and consciousness. Sometimes the terms 'subliminal consciousness', 'subconscious,' and 'unconscious' are also used interchangeably, leading to even more confusion. Interestingly, the academic definitions and characterizations also vary from source to source. There is no generally accepted characterization. We already clarified the meaning of the concept of consciousness and phenomenal consciousness in Pt.I-I.3. Philosophers also speak of '*access consciousness*' or '*meta-cognition*', which refers to all of the conscious perceptions, thoughts, and emotions that we can access because we are directly aware of them and do not escape our attention. Loosely speaking, we will adopt the following meanings.

The ***subliminal*** does *not* stand for the sub- or unconscious but, to the contrary, is a veiled mental and more intuitive inner activity which eventually can be even smarter, more creative, and knowledgeable than our access consciousness – that is, the directly accessible surface conscious mind of which we are aware. It is a larger and more efficient mind behind the surface mind and opens up to higher degrees of intuition and insight – the sort of 'inner voice' that so often has more wisdom, and that tells us what is right or wrong, true or false, contrary to what we believe with our conscious mind, and that we sometimes regret not having followed or having paid more attention to. The subliminal possesses all (and perhaps more than the) high-level functions such as thought, problem-solving, decision-making, creativity, etc. It is an unaware and yet a higher-cognition function.

The **subconscious** stands on the opposite pole and is that part that is active in the waking as in the sleep state but that, contrary to the intelligence and foresight of the subliminal, expresses itself with incoherent thoughts or images and unorganized reactions. It receives the impressions of all things and stores them up. From it, all sorts of impulses and habitual movements emerge, sometimes repeated mechanically and uncontrollably in actions, emotions, or thoughts, which can arise in the waking state but express themselves, especially in the dreaming sleep state. The subconscious action, thought, emotion, and will are often opposed to their conscious counterparts. It is full of impressions, associations, fixed notions, automatic reactions, and desires formed by our past, which govern much of our personality. Desires, motivations, and fears can often be subconscious (more or less habitual) impulses that drive us into action without us knowing the true reason for it, or even against our will. In a dream state, these become particularly vivid, though even in the waking state, we can have incoherent thoughts without knowing why. For example, we become jealous for reasons other than those expressed verbally or become captured by irrational fears we can't explain. Several negative feelings or even traumas experienced in childhood and forgotten by the waking mind are stored subconsciously and resurface in adulthood, compelling us to engage in apparently unexplainable actions or causing us irrational emotions without necessarily being sustained by an external cause. The subconscious is not unconscious inasmuch as it can take the form of a sub-mental cognition that can still control our thoughts and eventually, also unexpectedly, reemerge in the sphere of our accessible consciousness.

Therefore, here we use the subliminal and subconscious as polar aspects, though they have in common the fact that both escape our direct awareness. Our surface consciousness constantly receives something of these inner touches, communications, or influences from both of these sources but doesn't realize, for the most part, from whence they come. The subliminal and the subconscious are both introspectively inaccessible domains to the ordinary mind but express themselves in the form of high-level and low-level cognition, respectively.

The **unconscious** is neither subconsciously nor subliminally conscious of anything. A stone, a chunk of matter, or even a machine can be labeled as unconscious because there is no thought, emotion, or awareness of something phenomenal at all.[15] No one or nothing experiences what it is like to be. Contrary to what you might find elsewhere, here we will not confuse the word 'unconscious' with 'subconscious,' let alone 'subliminal.'

Therefore, here we maintain that we are *not* unconscious but subconscious when we are sleeping with all kinds of incoherent and confusing dreams. Whereas, most parts of our brains are presumably completely unconscious, in the sense that most of their neural activity does not give rise to any phenomenal experience at all and that they are preoccupied with regulating and preserving the material and mechanical aspects of our organic functions, say, our digestion or heartbeat.

---

[15] Or, at least, that is what we suppose to plausibly be the case. Panpsychists may not agree. For the time being, we defer the discussion of this point to later considerations in Pt.II.

By the way, are our bodies unconscious? After all, it is a material object, a biochemical machine which, when subjected to illness, we 'repair' with drugs, physical therapies, or even surgery, isn't it? Later, we will put forward several arguments and facts that suggest we should reconsider this. For the time being, you can also take this mechanistic point of view, if you wish.

As you can see, we do not adhere to the interpretation accepted by most people, including in the academic environment in psychology or in the philosophy of mind, which shuffle these terms interchangeably or even formulate it in a black and white representation whereby only our waking state is 'conscious,' and all the rest is 'subconscious' or 'unconscious.' If we want to avoid the confusion and misunderstandings deriving from such conflations, we must work with more precise and distinct significances, as we would like to show that this is not an unnecessary multiplication of 'pluralities' or just a matter of semantics or a wordplay. It is a necessary 'conceptual hygiene' that, if not practiced, leads to tremendous confusion and unwarranted conclusions.

These distinctions refer to very concrete first-person experiences. To clarify, consider the following.

It is a common experience that sometimes, when we wake up from sleep, we don't recall our dreams. Sometimes we don't even recall any dream at all. In this case, we usually tell ourselves that we were 'unconscious' because we have that feeling of having gone through a void, a state of unconsciousness without any experiences, feelings, thoughts, and nothing we can relate to with our mind in the waking states. There seemed to be no subject, no individuality, no "I" experiencing anything. We could even say that we didn't 'exist,' that there was no one 'experiencing existence.' We had the sensation of having fallen out from existence into a totally unconscious state and later returned into that 'existence' by waking up. When we sleep, especially when we don't have any dreams, we say we were 'unconscious.' This is the commonly accepted wisdom.

However, sooner or later, many also had the experience that after a while (this can vary from a few seconds to minutes after waking up, or in extreme cases also hours), a fragment, a part of a dream, or eventually a whole bunch of dreams suddenly literally 'bubble up' into our access consciousness. Suddenly, we recall that, indeed, we were not unconscious but went through all sorts of dreams and experiences of an inner life during that sleep state that previously seemed to be so void. Our sense of 'non-existence' and 'unconsciousness' is abruptly replaced by a recalled experience that, while not truly conscious as in the waking state, reveals itself at least as a subconscious experience or, in the case of dreams with greater clarity and lucidity, as a subliminal experience. What we previously remembered as an unfathomable and totally unconscious void in which we had lost our identity and subjective feeling, reveals itself as a lived experience of an "I" that was well present.

This shows one thing: We should never leave out of the equation the crucial role of our 'access memory'. The forgetfulness of an experience isn't evidence of the absence of experience. It makes no sense to declare someone as being 'unconscious' only because there is no report of a clear recollection of any experience. We don't say that babies are unconscious beings without feelings and experiences only because, once they grow up, as adults, they won't remember their early life. We don't consider those suffering from amnesia as having been unconscious all their life simply because they are no longer able to remember it. The reason we consider the waking consciousness as the only 'consciousness' while labeling someone as 'unconscious' during sleep is that, from a third-person perspective, we can't check on whether or not someone in that state is having a conscious experience, contrary to the case of newborns or people suffering from amnesia. Such an oversimplification is a non sequitur and, frankly, an unintelligent assumption. Let us unpack this further.

Having 'no recollection of an experience' doesn't really mean there is no memory of it; rather, our consciousness fails to access introspectively and retrospectively past experiences, which, therefore, the subject is unable to report. Inaccessible subconscious or subliminal mental processes may well be 'conscious' in the sense that, while taking place, an individual phenomenally experiences them and is also fully aware of them but, once the experience ceases, the mnemonic retention fades away almost instantly.

For example, what about lucid dreams? In lucid dreaming, one is aware of being in a sleep state and recognizes that dreams are just dreams. Many have had such an experience at least once in their lifetimes. It is a much more common occurrence than usually assumed. Interestingly, nobody considers lucid dreaming to be a state of unconsciousness. However, this is only because, if a subject does not misinterpret his/her own dreams as the 'real world' and implicitly continues to consider the physical world as the only 'true reality,' then we are also more willing to concede this state of awareness to be more 'real' – that is, we do not declassify it as an 'unconscious' state.

What can be said about sleepwalking, also called somnambulism or, technically, *'parasomnia'*? This is something many parents begin to deal with when they discover their child sitting up in bed with eyes open, or even screaming, talking, walking, or consuming food while being asleep. In adults, this can go so far as to cook, drive a motor vehicle, or even have sex or commit homicide without recalling anything later.

Are sleepwalkers unconscious? We tend to answer affirmatively only because, since there is no recollection of an experience, there must have been no experience and experiencer. However, further scrutiny shows that more than two-thirds of sleepwalkers have reported at least one dreamlike mentation associated with their sleepwalk [47]. One may call somnambulism something subconscious or a 'lower' state of consciousness, but there is no reason to believe that there isn't anybody who has an experience during the sleepwalking event. The fact that this experience can be remembered only as an incoherent dreamlike state when one returns to the ordinary waking state is no evidence of the contrary.

Another example of a strange form of awareness is the hypnotic state. A hypnotic suggestion is a trance-like state with heightened focus and concentration. In a trance state, one is said to be 'not self-aware' and unresponsive to most external stimuli but responsive to the directions of someone. In this case, psychologists refrain from classifying a trance as an unconscious state and speak of an 'altered state of consciousness' or 'reduced consciousness' because they hear their patients speaking, and eventually see them also walking. The third-person identification of conscious awareness with an active physical body betrays the materialist prejudice again.

It might be interesting to point out how hypnotic subjects can recall or forget events that took place during the hypnotic session by command. During hypnosis, one can give individuals a suggestion to forget something when ordered to do so. This is an induced *'post-hypnotic amnesia.'* However, their memories can be retrieved once they are presented with a cue, a counterorder, or a prearranged stimulus. This disrupting and restoring memory retrieval process on command is further evidence of the decisive difference between a conscious memory of the experience and the conscious experience as such.

However, one might assume that for someone in a state of sleep without dreams or under general anesthesia, or in the deepest level of a coma, there is no consciousness, no experience, and that's what can be finally said to be really 'unconscious.' Right?

From a scientific and intellectually coherent point of view, we should not jump to unwarranted conclusions. Instead, we should honestly admit that we simply don't know. The fact that one wakes up from a deep sleep or coma and doesn't remember a thing means nothing. Contrary to what most believe, there are no certainties about sleep without dreams. Usually, the dreaming state is identified with the *'rapid eye movement'* phase, the famous REM sleeping phase, which is always accompanied by dreams. Nowadays, we have gone so far that some elementary real-time dialogue between experimenters and lucid dreamers during REM sleep has become possible [48]. It was therefore assumed that the EEG trace of a non-REM phase is the signature for a state of consciousness of sleep without dreams. The exclusively third-person science is slowly becoming aware of its oversimplified assumptions, which took for granted that consciousness 'disappears' in deep or dreamless sleep [49]. It turns out that subjects awaken from non-REM phases, and without having gone through a previous REM phase, report remembering their dreams [50]. This means that there is, to date, no known method of establishing whether one is in an 'experienceless' state of consciousness. There might be no method at all, not even in principle because these findings are ultimately based on subjective reports of dreams. Looking at EEG correlates of dreaming is useful to establish the existence of an inner experiential life, but the converse isn't true. Contrary to what is almost always tacitly assumed, the lack of a signal coupled with the absence of a personal report isn't evidence of the lack of an experiential state of consciousness.

It is, therefore, questionable to suppose that during deep sleep, anesthesia, or coma, there is a complete absence of subjective experience. Anesthetic agents do not shut down brain functions globally. They modulate the activity-specific brain areas depending on their dose and substance (which is obvious, as otherwise even the basic autonomic functions, such as breathing and cardiac functions, would fail and lead to death).

The brains of subjects in a vegetative state aren't inactive either. They even undergo waking and sleep cycles. A *'vegetative state'* is characterized by the absence (or minimal presence) of responsiveness to external stimuli and a lack of awareness of the outer environment due to brain dysfunction or impairment but with preserved sleep-wake cycles.

The results of research that was published in 2006 by British neuroscientist Adrian Owen, and that was in the media spotlight for some time, showed substantial evidence of states of awareness in patients in a vegetative state [51]. His group presented a study of a 23-year-old woman in a vegetative state and presented compelling evidence that she retained some cognitive abilities previously thought to be impossible to have in such a clinical condition. The neural responses measured with fMRI imaging during the presentation of spoken sentences matched those of healthy volunteers listening to the same stimuli. Sentences that made no sense or that had ambiguous semantic content did not activate the same brain areas, just as in the healthy control group, indicating that her brain was not just reacting to acoustic stimuli but was actively understanding the meaning of the sentences. Furthermore, when she was asked to imagine playing tennis or walking in the rooms of her house, her brain's motor cortex fMRI image was indistinguishable from the images of the control group asked to perform the same mental task. All indicated that she was consciously aware of herself and retained much more than just a residual cognitive ability to understand spoken language and to respond through her brain activity by cooperating intentionally.

Further studies confirmed that this is the case in at least 20% of people diagnosed (or misdiagnosed?) as being in a vegetative state. They were shown to be capable of cognitive functions that we believed to be possible only in the waking state and were even able to answer questions, such as responding with a 'yes' or 'no' neural signal about who and where they were ([51], [52]). Moreover, their neural networks correlating with painful experiences were activated when they were subjected to painful stimuli (for a review of these and other recent findings, see [53], references therein, and further more recent findings [54]).

Similar considerations can be made about the state of consciousness under anesthesia. Findings suggest that the state of consciousness induced by anesthetics can be similar to natural sleep with dreams. Anesthetic-induced unresponsiveness does not induce unconsciousness or necessarily even disconnectedness – that is, a residual subjective experience of the external world is still present ([55], [56]). While, at least in 15% of unresponsive patients with acute brain injury, brain activation to external stimuli is nevertheless present [57].

The bottom line is that taking superficial outer bodily signs such as being awake and aware of the physical world as the only 'true consciousness' is, from the philosophical perspective, a simplistic and unreflective attitude. Relying on brain scanning images or EEG traces coupled with personal subjective reports can be useful to say something about specific states of consciousness, but it is not indicative of the absence of conscious experience. In these regards, science should be much more careful, or at least take an agnostic stance instead of branding Occam's razor. But, so goes the theory, since in these patients there isn't any external sign of awareness, one should not multiply 'pluralities' by assuming that there could be an internal awareness. Many scientists, when pressed, will begin to tell you via an automatic reflex that 'there is no reason to believe,' 'there is no need to conjecture,' 'let us be parsimonious,' and to declare altogether as 'unconscious' everybody unresponsive to external physical stimulations because it is not in line with our waking consciousness prejudices.

This is, after all, yet another (subconscious?) attitude of blind logical positivism which anthropomorphizes reality that not only is philosophically questionable but, much more importantly, has huge moral and ethical implications, first and foremost in the case of the informed decisions on withdrawal of life support to coma patients who deserve more dignity. Especially in the case of those trapped in a vegetative or coma state, to consider these unfortunate people as 'unconscious' is an insult to them and their loved ones!

Let us return to more mundane examples to which everyone can relate. To show how a subliminal or subconscious mental, emotional, and physical activity isn't 'unconscious,' but something lived by a conscious subject experiencing phenomenal events on different levels of awareness, consider the following anecdotal examples of common first-person experiences.

Only ten minutes ago, you entered your home completely absorbed in your thoughts and with your attention focused on something entirely different–something unrelated to your home environment and present actions. You got into your apartment and left your keys somewhere. Now, ten minutes later, you are looking for your keys but have forgotten where exactly you left them. When you find them again, you might then remember the instant when you took the action of leaving the keys precisely in the place you found them. If your actions have not been completely erased from your short-term memory (which can happen in extreme cases), you suddenly have a vague remembrance that you left your keys in that location by a semi-automatic action. You can recall and reconstruct, at least partially, the scene in which you can see yourself leaving the keys there ten minutes ago. It was such a fleeting action that was performed almost automatically and that quickly faded from your memory of past events. That's why we tend to describe such actions as 'unconscious.' It could be described as an altered state of consciousness in which your attention was elsewhere, but you were not 'unconscious.' You were performing a higher cognitive function and, instead of unconscious, we will describe these automatic tasks as subliminal or almost subliminal. There was full awareness that, however, faded almost instantly from your access consciousness and short-term memory.

By the way, the other interesting alternative, besides complete or partial obnubilation of the remembered experience, is the opposite: You swear that you left the keys on your entrance console table but find them no longer there. Then, to your surprise, you find them on the desktop in your living room. You can't believe that to be true and may even suspect that someone else placed the keys there. But you are alone, and your rationality suggests thinking about it twice. Not only did you perform a subliminal action, but your mind invented a story and presented you with a sort of illusionary fiction, replacing the real fact with a false narrative. Those who know themselves a bit and take notice know that these sorts of illusions occur much more frequently than commonly assumed (another is that of replacing in our mind the colors of objects, sort of what Balint syndrome patients experience).

Let us consider another example. As is well known, especially if you are an experienced driver, one can perform quite complex tasks such as driving a car in such a state of awareness (eventually complemented by false images). If one asks you to describe the surroundings you were driving through a minute or even only a few seconds ago, you won't be able to recollect many details unless, for some reason, something caused you to focus your attention on it. Only a vague and defocused image remains, if any. This is because we pick up only a few details from the environment that are necessary to perform the desired task. Only a dull image with many details missing is left. Were you driving 'unconsciously'? Nobody would describe this situation in such a manner. You may have been unconscious of your surroundings, but you were not unconscious. The subliminal selected out the unnecessary features, which, in most cases, were properties of the environment unnecessary to perform the task.

Another similar example, but this time on the subconscious level, could be that of your perceptions in the sleeping state. When you sleep, you must be somehow subconsciously aware of the environment, as otherwise nobody could be awakened by an alarm clock or external stimulus. Moreover, we must still have some unaware perception of the bed's boundaries, as otherwise, we wouldn't be able to turn over without falling from the bed.

To better clarify what being subconscious means in this context, consider that you are sleeping and your room is getting colder. Your physical body registers a temperature decrease, and your brain (more precisely, your hypothalamus, whose job is to maintain homeostasis) sends you a signal while asleep. You begin to feel how cold it is becoming. However, you still don't rationalize this. Maybe you are still dreaming, and this freezing sensation may be reflected even in your dreams (we will see that dreams are the mind's interpretation of the perceptions it receives from the plane it dwells in). Feeling an uneasy or annoying physical sensation of something coming from the external world while asleep is a quite common experience. Try to recall how you felt. You may recognize how, in these conditions, we feel ourselves in a somewhat dumb and reactive state of consciousness. We simply feel how it is getting cold and for some time, before waking up, we suffer internally and somehow battle against this unpleasant sensation. However, we don't rationalize it, and we don't think, 'The room temperature is getting colder. I have to reach for a blanket'. It is not painful but, rather, just a pure feeling of cold unease in the present moment, and you are not awake. What is in charge is not our higher-level cognition but, rather, the limbic system, which is responsible for identifying threat conditions and is active when we are in a lower emotional state, compelling us to react instinctively, irrationally – that is, subconsciously. That's why we move around our bodies, react nervously, or even battle internally with incoherent dreams to get rid of this uneasy sensation of coldness. Then, suddenly, we wake up, the higher-level functions take back control, and we realize what's happening. We get a blanket or put on a pair of warmer pajamas and then go back to sleep. Problem solved.

What we would like to point out with these examples is that we already know, from our own experiences, that there is a qualitative difference between the subliminal and the subconscious function mode. The former works in a higher-cognitive function mode, such as that in which one leaves the keys in a place, forgetting immediately about their whereabouts and eventually ruminating about complex philosophical issues or driving a car, while the latter is a lower-cognitive function mode, such as that in which one nervously reacts against freeze sensations in a dream state.

Nevertheless, both states of awareness are not 'unconscious' because there is always the same subject having very concrete phenomenal experiences. We were not unconscious as a stone when these experiences were taking place. What makes them seem so 'unconscious' in the popular understanding is that they both became inaccessible (for a while or forever) to our surface awareness. We are no longer able to report them accurately. Furthermore, the state of awareness was qualitatively different, as a subliminal thought or action is guided by a fully functional mind, whereas the subconscious feeling in a sleepy state is usually perceived in a state of confusion and an incoherent mental dream state (unless, of course, one is having a lucid dream).

These were only some examples of the different states of consciousness one can experience. The following list summarizes them in three groups;

Waking state

Lucid dreaming
Hypnotic state
Deep mediation and trance
Psychoactive induced altered state
Psychotic hallucination
Dissociated personality disorder
Dementia state of consciousness

Half-sleep
Sleep with dreams
Deep sleep
Sleepwalking
Blindsight subliminal consciousness
Anesthesia induced (loss of) consciousness
Vegetative state
Coma

Notice that there is a tendency, also among professionals, to consider only the waking states as 'conscious,' label the second group as 'altered states of consciousness,' and brand the third group of cognitive states as 'unconscious.' It is especially this third distinction that is problematic. It would be much more accurate to call them *'no-report'* or *'non-metacognitive'* states of consciousness.

The reason why we dwelled a bit on these subtleties is to exercise our thinking by an inward turn with an introspective investigation that reveals our exteriorized thinking fallacies. The coarse-grained physicalist thinking that subdivides all our conscious existence into black and white blocks, such as being 'conscious' or 'unconscious,' without further distinction and critical assessment, is an unreflective instinct we must get rid of. Moreover, this furnishes us with a firmer basis upon which we can look with a more critical eye at the experiments on free will.

3. Free will? What's That?

Before proceeding to the science of free will, let us first put this into a historical and sociological perspective.

According to the philosophical view of *'incompatibilism'*, free will and determinism are logically mutually exclusive.[16] One can't believe in both without incurring a logical fallacy. In fact, those who adhere to a strict physicalist doctrine can't believe in free will. From the purely materialistic perspective, only physical processes are 'true,' the rest being only derivative processes from it. If only physical processes determine reality, the old-styled physicalist (who, more or less without being aware of it, still clings to the Newtonian worldview of classical physics) likes to believe that, ultimately, every event and material object in the universe is ruled by a deterministic causal chain of causes and effects. Nowadays, no serious physicists would claim that the Newtonian classical deterministic laws furnish a complete description of reality. However, when it comes to questions of a more philosophical character, most are not willing to give up the deterministic belief system and switch back (subconsciously?) to a strict Laplacian causality which still firmly remains the underlying working premise. That quantum mechanics strongly suggests abandoning a local determinism being only a human and almost anthropocentric concept in favor of a non-local non-deterministic theoretical framework remains an idea and worldview that the reductionist mind resists. This is understandable. Taking the verdict of quantum physics seriously would also entail accepting a worldview that is at odds with the Laplacian form of physicalism and that so many still maintain as the ultimate underlying reality of all things. That is also the main reason why so many try hard to reintroduce determinism through the backdoor with contrived and extravagant interpretations of quantum mechanics that have to save an 18th to a 19th-century conception of reality and which is supposed to save the human-centered deterministic appearances[17].

This (more or less unaware) premise is what conditions the ideas people have on matters like free will. Maintaining the view of a classical deterministic, reductionist physicalism, everything could, in principle, be derived by mechanical unconscious material particles, forces, and fields acting in a given space-time background which, like in a gigantic clockwork, determine everything in the universe without any space for free will, let alone non-physical phenomena. In this view, what we call 'free will' is only the outer appearance resulting from a tremendously complicated chain of events inside the neural network of our brain and conditioned by external stimuli that we cannot trace back to a concatenation of causes and effects and, therefore, seems

---

[16] To distinguish from the so-called 'libertarianism' which considers free will compatible with determinism.

[17] Such as the many world interpretation or super-determinism, just to mention some. For a more in-depth discussion, see also [36] ch. II. We will discuss in more detail quantum physics and the underlying assumptions on the role of randomness in Pt.II and Pt.III respectively.

'free' only because unpredictable. We don't really make choices; also, every choice we make is an illusion that was predetermined by neuronal activity, perhaps already before we felt the desire to choose. Also, neurons can be reduced to a complex interplay of interactions of chemical compounds and these, in turn, to molecules, atoms, and subatomic particles. At the bottom, everything is a complicated foam of particles in which past, present, and future were already predetermined from the beginning of the universe, as a long drama printed in a book whose every page is a step in time. According to this purely deterministic perspective, we are, then, nothing but complicated biochemical machines in which everything can be reduced to mutual interactions between these particles and fields in space and time.

Obviously, then, this mechanical determinism leaves no room for free will. Free will is only an illusion emerging in our minds, and a belief determined by these chemical reactions in our brains as well. Free will is nothing more than our wishful thinking that has itself been predetermined by a chain of mechanical causes and effects. Also, your reading of these lines is nothing but a process that was predetermined long ago by the initial physical condition of a bunch of elementary particles and then, ruled by the law of physics, led to exactly that physical condition of your brain you are experiencing at this precise moment. Like the acceleration and speed at a given time of a stone falling from a given height are predetermined by a set of differential equations, so, too, is your life, at every instant in the past, present, and future. Our personalities are nothing other than the result of complicated physical processes in our brains, which can also be reduced to nothing beyond an aggregate of interacting neurons, molecules, atoms, and particles subjected to the same physics and chemistry. And because we are nothing but our brains, and because the brain's activity qualitatively maintains the same mechanics as those of falling rock, only following a much more complicated chain of predetermined causes and effects, the logical conclusion is that, at the bottom, we are nothing other than soulless bio-robots that believe to possess free will, but do not. Every thought or emotion we have and every choice we make was already determined from the outset a long time ago, no more and no less than any chemical reaction.

The complexity of all these interactions is, however, far beyond our control. It is even worse than most assume. Besides the processes occurring in our brains and bodies, one should take into account all the interactions with the environment. Laplace's demon could not predict my future behavior beyond a certain time interval, probably not beyond a few seconds. By knowing only the physical state of every atom, molecule, and neuron of my body and brain, he must also take into account the environment, which determines the next state of all these parts. And the fact that the environment also depends on a hierarchy of bottom-up and top-down causal layers forces us to conclude that one must know the state of the whole universe to predict the actions of a terrestrial being subjected to purely deterministic laws (an aspect we will shortly readdress in Pt.III-II.5g).

This deterministic view between matter and mind is more or less the implicitly acknowledged belief of most scientists and has its roots in the Laplacian demon paradigm. Even if this mechanistic worldview got in trouble due to the findings in quantum physics, it has nevertheless been quite successfully applied and has therefore been expanded to all sciences, including the human sciences. It is a mindset that has ideological and pragmatic reasons but no scientific evidence supporting it.

On the other side, the religious or spiritual thinker tends to believe that we have free will because God has created us as souls free to choose between virtue or sin, good or evil, and light or darkness. The biblical narrative of Adam and Eve, who chose to eat the forbidden fruit, is the most eminent symbolic representation in Western culture of the belief in humans' free will. It is a millennia-old, deep-rooted religious belief that clashed with the materialistic determinism that the very same Western culture developed more recently. This created a sort of 'philosophical short-circuit' that scientists, philosophers, and theologians are still struggling to come to terms with.

We will see later that both positions – that which claims we are subjected to the laws of Nature and therefore have no free will, and that which claims that there is something in us free from it – are both correct (or both false, if you prefer). As usual, it is the mind's one-dimensional understanding of reality that sees irreconcilable contradictions where there are none.

In fact, a little bit of further thinking could be beneficial. First of all, the question is: Who is supposed to be free from what? Are we talking of our consciousness, our minds, or our bodies being free from the physical laws or from something else which is unphysical? Can we tell if our consciousness and/or mind are free from whatever if we don't really know what consciousness and mind are in the first place? And, what is 'will'? Isn't there also a subconscious and subliminal will that leads us into action? Is the subconscious and/or subliminal free? When we talk about our freedom, are we pointing only at our conscious subjectivity, that little ego, or is there more than this little "I" that could enter into the equation? Who or what is this "I" which is supposed to be free? An ineffable transcendent soul or the emotional personality in us with all its desires, cravings, appetites and instincts? If there is no free will why does the illusion of it emerge? Moreover, is there only one possible answer to the question if we have free will, namely, 'yes' or 'no'? Or could reality turn out to be somewhat more creative and original than human's binary black or white thinking?

If we do not first clarify what we really are talking about, all these endless debates on free will lead us nowhere. We will offer a more articulated first-person perspective that might help lead us a step further. To do so, we must first reconsider and think through some premises and unaware assumptions with which we work so unreflectively and which, if not brought to our awareness, can only mislead us. First, let us take a look at what modern neuroscience tells us about the subject of free will.

4. Libet's Pioneering Experiments on Free Will

The question of whether we have free will is as old as humanity. From Plato and Aristotle through Kant and Descartes until the speculations of Bertrand Russel into modernity, this has always been a topic fervently debated among philosophers. Interestingly, however, scarce were the attempts to investigate this question with modern neuroscientific means (the reason for that, we can only speculate, though we will try to furnish an answer).

Fig. 24 Benjamin Libet (1916 – 2007)

The pioneer in this field was the American neuropsychologist Benjamin Libet, who performed a famous experiment conducted between 1979 and 1985. The first experiment showed how a simple touch on a finger becomes a conscious sensorial experience only after a relatively long time duration of about a couple tenths of a second [58]. While the most notorious experiment of Libet was set up as follows ([59], [60]).

Subjects were asked to freely decide when to flex a finger (or press a button or flex their wrist) and report the moment at which they were consciously aware of the decision. The brain activity was recorded by an EEG that measured the signal and the time at which the mounting motor cortex activity occurred, the so-called *'readiness potential'* (from the German *'Bereitschaftspotential'*). At the same time, an *'electromyograph'* (EMG) measured the electric potential generated by muscle cells. The subjects could use, as a timing device, an oscilloscope with a dot quickly circulating like the hands on a clock. The dot traveled 360° in 2.56 seconds by 6° steps, which implies that it stepped forward every 43 milliseconds (and we know that this leads to a continuous visual experience). At the moment where they were aware of the wish or urge to act, the participants reported the position of the circulating dot.

The time steps describing Libet's experiment in Fig. 26 can be illustrated as follows:

Illustration (0): Before the readiness potential.

Illustration (1): Onset of the readiness potential – that is, the time at which the subject's brain signaled the intention to act – as recorded by the EEG (at $T = -550$msec in Fig. 25).

Illustration (2): Time of awareness the proband reports as the moment at which one feels the 'urge to move' by identifying the position of the circulating dot on the oscilloscope (at $T = -200$msec in Fig. 25).

Illustration (3): Movement onset: The moment at which the EMG registers the electric potential of the muscles flexing the finger (at T = 0msec in Fig. 25).

The time lag between the moment of the first brain activity observed at the EEG and the time of the decision to act indicated by the subjects observing the dot's position on the oscilloscope amounts to a considerable 350msec. If we also take into account the postdictive interpretation of the phi phenomenon or flash-lag effect induced by the moving dots on the oscilloscope, one should even add another 200 msec on the top of that (recall how our visual system always delays its 'judgment' by 0.2s). The dots flashing up on the oscilloscope's clock in Libet's experiment (the dots on the oscilloscopes in Fig. 26) display essentially the same optical stimulus of the two flashing light sources in Wertheimer's experiment (the two light sources in Fig. 15c) or the flash-lag illusion of Fig. 18. It should therefore come as no surprise that in Libet's experiment, the subject's time readout may be delayed backward by 0.2s. But then, the overall real delay would be even about half a second!

This seems to indicate that, contrary to what we tend to believe from everyday experience, the awareness of our volition does not coincide with the brain's activity — that is, by the readiness potential's initial motor cortex buildup. On average, it takes longer than half a second between the initiation of the readiness potential and the finger flexing.

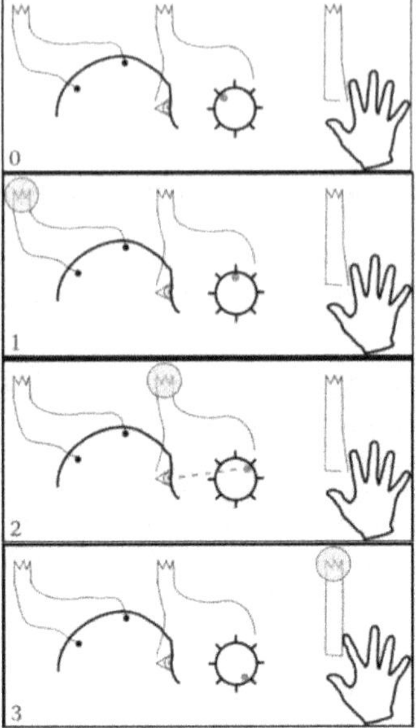

Fig. 26 Libet's experiment.

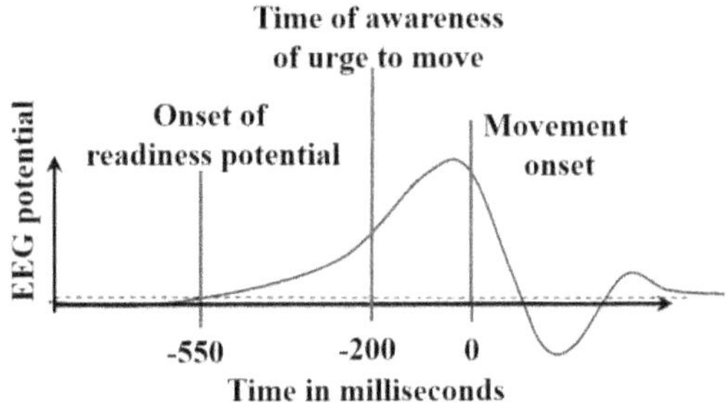

Fig. 25 EEG readiness potential in Libet's experiment.

The interpretation of Libet's experiment was and remains very controversial. In it, many see the proof that it is an unconscious brain activity that determines our volitional acts first and that becomes conscious only later. According to this interpretation, we first make a decision without being aware of it; only later does it enter into our awareness, and do we misinterpret it as our free choice. In other words: Free will is an illusion.

However, it turns out that things aren't that easy. Libet repeated his experiment in 1985, with subjects instructed to consciously veto their own urge by a *"free won't"* act at about 100-200msec between the movement onset and the urge to move. This showed that a readiness potential does not necessarily lead to action but can still be interrupted up to about 50milliseconds before the muscle activation. He could thereby support the thesis that there is a time window of about 100–150milliseconds within which the conscious will can still prevent an action that has already been initiated. In this sense, *"the role of conscious will would be not to initiate a specific voluntary act but rather to select and control a volitional outcome"* [61].

Libet's findings were confirmed in an experiment in 1999 in a modified form by British and German psychologists P. Haggard and M. Eimer. The readiness potential of the left and right hemispheres was measured by asking subjects to make voluntary movements with the left or right index finger during each trial [62]. In this experiment too, the activation of the motor cortex was, on average, before the reported time of the conscious decision to act. However, Haggard and Eimer pointed out that the reported time of awareness varied considerably. In 25% of the cases, the subjects indicated that their time of awareness preceded the rise of the readiness potential. This means that Libet's styled experiments are subjected to gross statistical errors that render its truthfulness questionable. The time at which we perceive and report our willingness to act is conditioned by a highly subjective perception that depends on our concentration and attention, and that is difficult to pin down on a ticking clock.

Moreover, one can turn things upside down again and question whether the free won't vetoing action itself might be determined by an unconscious decision that escapes our conscious volition as well? In 2009, experiments on the awareness of voluntary decisions, performed by the German and Belgian psychologists S. Kühn and M. Brass [63], indicated that the veto decisions are also made unconsciously and are only subsequently perceived as free decisions, arguing against Libet's free won't interpretation. However, the experiment of Kühn and Brass referred to a different type of vetoing than Libet's one. In Libet's experiment, the vetoing act was endogenous — that is, it originated from within the subjects — while in Kühn's and Brass experiment, it was dictated by external stimuli (upon seeing an initial go-signal, the participant had to press the "go" button but cancel their immediate intention if they saw a stop signal). This calls into question whether the two experiments effectively measured the same volitional phenomenon.

In 2016, other experiments ([64], [65]) investigated this ambiguity and confirmed that vetoing after the start of the readiness potential for movement is possible until the stop signals occurred earlier than 200 milliseconds before the start of the movement. Researchers put a red light under the electronic control of the participants' readiness potential. Thereby, subjects could literally 'see' their own readiness potential mounting and were asked to press a foot pedal as quickly as possible whenever they saw a green light but to refrain from doing so whenever the green light turned into a red light (which signaled the unconscious intention to press the pedal). If the readiness potential would truly signal an unconscious action beyond conscious control, the participants should have been incapable of responding to these sudden red lights. But many participants were able to ignore the nonconscious preparatory brain activity and stop their foot movement before it even began.

After nearly four decades of Libet's experiments being regarded as scientific proof against free will, 2019 marked a turning point where many began to recognize that the prevailing interpretations are questionable and do not seriously challenge our intuition regarding free will. The brain activity observed prior to conscious decisions reflects the conditional intentions that participants establish at the beginning of the experiment, rather than the decisions themselves [66].

This seems to suggest that the readiness potential cannot be identified with the commitment to act. The question whether the readiness potential truly signals an unconscious action beyond conscious control did, so far, not receive an unambiguous answer. Moreover, the neurophysiological interpretation of the readiness potential itself has been questioned [67], [68]. It is not even clear if it reflects a real neural signal in the first place or, rather is determined by ongoing stochastic fluctuations in neural activity. And, if anything, it indicates the brain's preparedness to act, waiting for the conscious will approval or rejection. To complicate things further it turns out that there is no readiness potential at all when, instead of making unreasoned, purposeless, and arbitrary decisions (such as arbitrarily move either the right or left hand), we make reasoned, purposeful, and deliberate decisions (such as which clothes to wear, what route to take to work, career choices, and so on.) There are different neural mechanisms underlying deliberate and arbitrary decisions, challenging the generalizability of studies that argue for no causal role for consciousness in decision-making to real-life decisions [69].

To sum up, the freedom of human-will decisions is much less restricted than assumed, and Libet's experiment can't be taken as final proof of the absence of free will. We can consciously intervene and veto processes that were previously considered automatic and beyond willful control. Maybe we have no free will, but everything indicates that we still have at least a free won't.

In hindsight, this is not a very surprising result. After all, it fits with our life experience. If we would really react in our daily lives to events with delays much longer than 0.2–0.3 seconds, many things we normally do would become impossible tasks or end in disaster. Activities which require quick reflexes, such as driving a car, would not be very safe, and car traffic accidents would be much more frequent than they already are. Playing tennis or table tennis would turn out to be an impossible amusement, and in boxing, the knockout would be awarded at the first strike.

Moreover, it might be worth taking a first-person perspective pondering what the short time interval felt like in which we had to react as fast as possible to avoid an accident. Say, the almost zeroed time lapse between someone stepping into a road and you slamming on the brakes. There is pure awareness, will, and intention to avoid an accident — a pure beingness and pure conscious awareness undiscernible by pure will. Recalling these experiences unfolding in the blink of an eye is a good exercise to realize what our intimate nature is. Though we would support a theory that denies us free will, depicting us as mere automatons, the first-person experience confirms, nevertheless, that a witness-consciousness is still there. An inexplicable phenomenal consciousness observing the film of life is present. Contrary to common belief, even if we think that there is no such thing as free will, this would not imply that we have to embrace physicalism. On the contrary, it would only deepen the dichotomy between the pretense of consciousness being an epiphenomenon of the brain and the hard problem of consciousness. Because what is that passive witness useful then? If consciousness plays no role in determining our actions, its usefulness and existence become even more mysterious from the Darwinian materialist perspective. There would have been no evolutionary advantage or any reason whatsoever in a purely material universe to bring it into existence. From the perspective which denies any role of free will, not only is the emergence of consciousness an inexplicable epiphenomenon of matter, but also its function and reason to exist become utterly mysterious.

There is another aspect that confirms good old wisdom. We, as humans, are animals subjected to instincts and more or less uncontrollable urges. The instinct of survival, craving for food, sex, and sleep, selfish impulses, emotional states of anger, wrath, and violence, etc., are something humans can control somewhat better than animals, though we are far from having a complete mastery over these primal inclinations. In what sense does the vetoing of a readiness potential differ from the rejection to give in to the instinctive desire to enjoy the pleasure of food when you are on a diet? How many times must someone try to quit smoking by vetoing the desire to reach for a cigarette until the urge begins to fade? Anyone out there who can honestly say having never been dominated by some impulsive and unreflective behavior that led to some inconsiderate action that one had to regret later?

It is clear that we are partially dominated by our more or less subconscious instincts and drives, to which we often fall prey and that we do not always find the *will* and psychological force to resist. Nobody can be truly said to be completely free until they have perfect and complete control over their primitive impulses. In this sense, we might say that the question of whether we have free will cannot be settled with a simple 'yes' or 'no' answer. There must be different degrees and levels of free will. An insect has less free will, if any, because it is a life form subjected to a complete, or almost complete, mechanical dependence from its natural and primal instincts, while humans might have a higher degree of free will because, if they desire to veto their own impulses, they can. However, we are not really good at doing this, either, and therefore we can't really state that we are completely free, as much too often, 'the spirit is willing, but the flesh is weak.' It is in this sense that we might conjecture that our will is not completely free but, nevertheless, may be rendered free. Perhaps a future, more evolved species that transcends homo sapiens might possess this complete free will. In Pt.III we will see how the reports of mystics validate this further.

However it is, after Libet's first experiment in 1983, these contrasting experiments and interpretations make it clear that there is still no general agreement on what they really suggest about free will and volition. Given the several methodological issues and measurement uncertainties (such as relying on the subject reporting the precise position of a quickly circulating luminous dot), and the ambiguous interpretations of what the measured signals really represent or indicate (such as identifying the readiness potential as the cause of the will to act), one cannot infer from Libet's experiments anything final and conclusive.

Nevertheless, Libet's experiment in its original form and interpretation was immediately adopted by almost all physicalists as the ultimate proof of the illusion of conscious will. All the later tests, which highlighted its methodological and interpretational weaknesses and made it clear that this experimental approach was not strong enough to show what it wanted to prove, have simply been ignored by the deniers of free will for decades. Libet's findings have routinely been misrepresented as final proof of the absence of free will, despite the opinion of Libet himself, who endorsed the opposite view. Although the question remains one of the most debated philosophical issues, the physicalist establishment considered the case closed without further appeal. The few 'Libet-styled' incomplete and defective experiments were accepted uncritically for decades, without much further thought, because, in its misinterpreted version, and by ignoring the part of the free won't choice, it seems to support the materialist ideology. This might also explain why, after almost four decades, these sorts of experiments received only scarce attention. Is it because once an experiment *seems* to confirm your ideological background, you tend to dismiss any new contrary evidence and refrain from checking things further?

But the sociologically interesting question is also: Why do so many enthusiastically embrace such a flawed experiment, animated by an almost religious desire to stick to a strictly materialistic worldview in the first place? Why are so many scientists and philosophers so eager to prove to themselves that they have no free will? To put it in the words of the British mathematician and philosopher Alfred North Whitehead: *"Scientists animated by the purpose of proving that they are purposeless constitute an interesting subject for study."* [70]

IV. The Mind-Body Problem and Neurocentrism

1. Correlation, Causation, and Confirmation Bias

The mind-body problem was, and remains, one of the central debates in the philosophy of mind. While it was already discussed among the natural philosophers of ancient Greece and was implicitly addressed in most Asian traditions, in Western philosophy, it dates back to the thoughts of René Descartes in the 17th century. Descartes posited what is nowadays called the *'Cartesian dualism'* between mind and matter – that is, between mind and the brain. In its somewhat more modern formulation, one includes, along with mind, also consciousness, with both being supposedly non-material entities independent from the laws of physics and both distinct from that gray material object which is our physical brain.

Dualism is opposed by *'material monism,'* which, instead, considers mind and consciousness as nothing other than emergent epiphenomena arising from the collective interaction of the neuronal activity in the brain. According to this mind-body monism, which is the bedrock of physicalism, there is only matter and no immaterial mind or consciousness. The narrative is that, sooner or later, these too will be explained away as the result of complex material and physical processes in our brains, and that there is no reason to resort to immaterial explanations. The belief is that dualism, in whatever form, will fade away, as was the case with many biological functions of life that could be explained in terms of purely biochemical processes without resorting to theories that pose immaterial and non-physical entities.

For example, *'vitalism'* or the theory of an immaterial *'Élan vital,'* a *'vital impetus'* or *'life force,'* which was thought to be responsible for the emergence of complex life forms and their organic functions, is nowadays considered, at best, a curious historic anecdote. In fact, modern biology can explain a whole variety of facts by a purely materialistic reductionist approach, from genetics to cell biology, without resorting to any notion of immaterial life force supposedly pervading living organisms. We will show, however, how also the disbelief in vitalism is based much more on an ideological than a real scientific basis.

Moreover, assuming dualism to be true, the *'mind-body interaction problem'* of mental causation must then be addressed. If dualism holds, one must explain how mind and consciousness, which are presumed to be immaterial, can have material effects, namely, determine a brain state. If something is immaterial – that is, supposedly unphysical – there is no known physical mechanism by which the former can affect the latter, and vice versa.

In Pt.II-II.3j we will see how this problem can find an elegant solution once we give up some unaware working assumptions.

There are different schools of thought among both dualists and monists about which we will not go into detail here. It may only be mentioned that we will embrace a sort of *'dual-aspect monism'* which finds a meeting point between the two, maintaining that mind and matter are the two aspects of the same 'substance.' Despite the variety of opinions, it is fair to say that most philosophers of mind and neuroscientists declare themselves, with little or no doubt, to be material monists.

After all, it is undeniable that there is a direct relation between the physical state of our brains and our subjective experiences. Everyone knows that there is a correlation between the chemistry of our brains and our perceptions, mental functions, and states of consciousness. It is quite a common experience how those tiny alcohol molecules, once they get into the blood flow of our brains, can alter our state of awareness. We know that psychedelic drugs can lead to intense subjective effects. Ever heard of dopamine? It is a neurotransmitter – that is, a molecule that enables biochemical transmission among neurons – which is responsible for the effects of a drug like cocaine. It is a well-known fact that brain damage can lead to severe cognitive impairments. If the *'Broca's area,'* a left cerebral hemisphere area, is disrupted, one loses the ability to speak (interestingly, however, not the ability to comprehend language). Someone being anesthetized using anesthetic drugs (seemingly) 'loses' consciousness. And nowadays, we have a whole bunch of sophisticated brain scan technologies that make it clear, beyond any reasonable doubt, how for every conscious experience, there exists a neural correlate in our brains. With these, one can show how any emotional state is mediated by the *'limbic system.'* For example, the amygdala, a tiny collection of nuclei about 1 cm across and deep inside the *'temporal lobe,'* lights up whenever we experience negative emotions like fear, anxiety, or stress. So does the *'hypothalamus,'* another little portion of the brain, responsible mainly for regulation (such as body temperature, circadian rhythms, etc.) but also containing our emotional reward, fear, and defensive behavior circuits. And what about that wonderful emotion we call 'love'? Neuroimaging techniques show that, when we are in love, dopamine is released in the brain reward system. Therefore, the narrative goes that love is just a bunch of hormones wreaking havoc with your brain activity – something similar to what cocaine does. Just a neurobiological chemical cocktail. Nothing else.

These are only a few of the many possible examples. Modern neuroscience has written tomes on the correlation between particular physical, emotional, or cognitive functions with specific brain areas. Therefore, apparently, no place is left for any form of dualism. Mind emerges from matter; there is no distinction. Our personalities, identities, moods, and states of consciousness seem to depend on the material biochemical state of our brains.

According to the material monists, this proves how mind must emerge from a complex organization of matter, and there is no reason to believe otherwise. The implicit narrative is that if you have a problem with that, this is due only to your belief system, still anchored in a pre-scientific world that seems to be disconfirmed by these hard facts. These are your backward religious beliefs in the existence of an immaterial soul and because of your fear of death, therefrom resulting in an irrational hope in a life after death. If you were a rational and scientific-minded person, you would not lose yourself in a ridiculous mystic woo and would recognize that matter explains away everything. Accept the empiric facts: You are nothing more than your brain. Case closed.

You might not hear physicalists expressing this so directly, but that's essentially what they think. And I'm not claiming that every scientist thinks so radically. In private discussions, one finds that there are several who are beginning to question their own assumptions and look more carefully at the premises they took for granted until then. Indeed, times are changing, and the physicalist mindset is facing more pressure from its own findings. However, let's face it: At the time of this writing, the materialist's paradigm is still the prevailing one and is rarely challenged openly in an official academic context.

At any rate, those honestly willing to look further, making an effort to not delude themselves, must conclude that the case is far from closed. In this section, we will show, as was the case with Libet's experiments, how this argument is based on selective attention to the very same scientific evidence that seemingly confirms one's ideological worldview and allows one to quickly jump to these apparently logical conclusions while ignoring the mounting evidence to the contrary. Physicalism stops there in the reasoning because it wants to stop there due to its confirmation bias. Any evidence to the contrary and deeper thought that scrutinizes this seemingly stringent logic of cause and effect is ignored, if not even repressed, because otherwise the whole strongly anthropocentric conception of the analytic-mind as being the ultimate inquirer for truth, which is the foundation of materialistic science and the mindset emerging from the Age of Enlightenment, would have to be questioned all over again.

Let us first question some basic assumptions. Does the physical change of a brain state leading to a change of states of mind, consciousness, and phenomenal experience provide logical proof that mind and consciousness are an epiphenomenon of the brain?

The physicalist's answer is affirmative because, as usual, Occam's razor is invoked: Don't let us multiply entities; there is only matter, and there is no need to look and think further. Resorting so instinctively to Occam's razor without discrimination and distinction is, unfortunately, a very common malpractice, including among high-ranking scientists. And yet, this has always led to a well-known logical fallacy into which the human mind has fallen over and over throughout the history of science: the belief that correlation implies causation. We tend unconsciously to mix up a correlation of two observed phenomena as proof of one being the cause and the other being the effect.

A first down-to-earth example that clarifies this fallacy is the correlation between ice cream sales and sunglasses sold. One doesn't need to be a rocket scientist to understand how this correlation doesn't authorize us to conclude that there is any causal

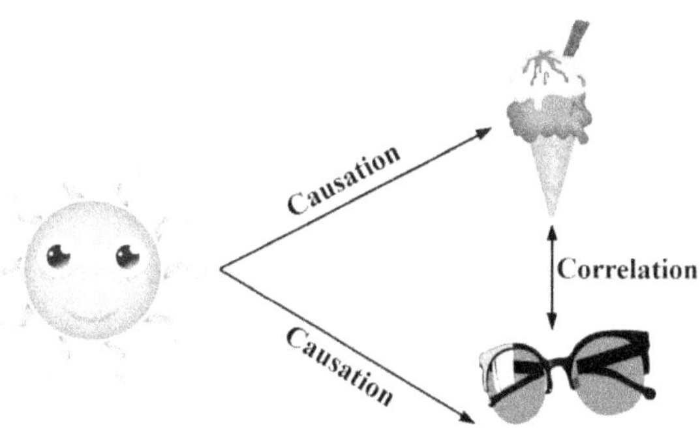

Fig. 27 No causal relationship between sunglasses and ice cream sales!

relationship. Wearing sunglasses doesn't make people more willing to buy ice cream.

Consequently, the fact that two events are always temporally coincidental or always happen shortly, one after the other, doesn't imply that the first event caused the second event to happen. These sorts of logical fallacies are known as *'post-hoc fallacies'* (from the Latin *"post hoc ergo propter hoc"–"after this, therefore because of this"*).

This might sound like a ridiculous example, but when one deals with complex systems, things can become much more subtle, even luring very smart people into these sorts of logical fallacies, especially if it confirms their expectations.

For example, in medical research, one must always be on guard, especially in randomized, double-blind trials in which subjects affected by some disease are separated into a control group that doesn't get any treatment and another group that receives a specific medication and in which neither the participants nor the experimenters know who is receiving the treatment. This is the so-called *'double-blind testing,'* referred to as the *'gold standard of testing,'* and is considered a standard practice to avoid observers' effects and eliminate participants' expectations or other confirmation biases that might lead to experimental biases. With such a rigorous testing procedure, it may sound obvious that, if on average, the group which receives the medication recovers from the illness better than that which didn't, we might feel assured of the effectiveness of the treatment.

The truth, however, is that this can be the case, but not necessarily. The separation into control and treatment groups is a random process, and there is a non-negligible likelihood that, by chance, the treatment group included a larger number of less severely ill patients or, for whatever reason completely independent from the treatment, patients who are more likely to recover by themselves. This can lead to the misleading conclusion that the trial has shown the effectiveness of the treatment, which despite all appearances, actually isn't effective at all.

It is therefore unsurprising that several "proven" drugs show no or only weak effects once released on the market. It is as if, over time, the treatment's effectiveness declines. This pattern has been observed so many times that medical research has given it a name: the *decline effect.'*

That's why serious scientists, to avoid 'false positives,' always work with the *'statistical hypothesis test,'* which is a method that, instead of delivering certainties, evaluates the probability that a hypothesis (say, the treatment is effective) compared to an alternative hypothesis (say, the treatment is not effective) is statistically significant. Only those treatments that pass the test with a high likelihood should be officially considered effective.

Stepping up the level of complexity let us make another typical example of a wrong correlation-causation identification in which a non-scientific mind may easily fall. Say that, in a specific geographical area, people are shown to become ill more frequently than elsewhere, and in the same area, a particular substance has higher concentrations than it does elsewhere. One might be tempted to believe that the substance must be toxic and the cause of the illness. Again, this might be true, but it might also be false. One must first show that the increase in registered cases in a specific region is not just a statistical fluctuation and calculate the probability that this correlation does not emerge by chance. If the ill health in that particular region presents itself only slightly above the average, and if the supposedly poisonous substance has only slightly higher concentrations than elsewhere, the probability that the latter is the cause of the former is low. One might suspect that there is a correlation but can't frame any reliable hypothesis. Then, consider that there are thousands of possible substances that might cause health issues and that there are several different physical disorders that each of them can cause. The probability that one correlates with the other by pure chance increases with the number of substances and disorders considered. This is because, in whatever region you live, you will always find a higher concentration of some specific substance X and frequency of sickness Y beyond the national average. This is potentially suggestive, but nonetheless, one still can't jump to conclusions and see this as final proof of a causal relationship between X and Y. Moreover, as in the case of the sunglasses and ice cream sales, one must rule out the eventuality that what causes the sickness Y isn't also the cause of the higher concentration of some substance X, because, if so, this would, again, be only an apparent correlation.

The larger and more complex a system is, the harder it becomes to convincingly detect, with certainty, the causes and effects. This is why every good scientist who deals with large data sets has been trained to cover up these false statistical correlations by learning tons of math, applying sophisticated tools of statistical analysis, and always working with a 'pinch of salt' before claiming whatever causal relationship.

The mind-body identification, which links the neural activity as the cause of conscious experience, is the almost unquestioned working premise based on an even subtler level of conflation between correlation and causation. It also escapes to well-trained scientists because it is not of a statistical nature; rather, it is of a more conceptual nature that can easily escape attention, especially if there is an ideological background. The serious and scientifically rigorous cognitive neuroscientist should be agnostic in this regard but, most scientists jump to conclusions and consider this causal relationship to be obvious. They are unaware of their *'confirmation bias'* –that is, their desire that some data or empiric evidence supports their beliefs and fits in their worldview (for example, as the case of Libet's experiments illustrated.)

Good science, however, did not progress by jumping to conclusions. It has always been based on a principle of exclusion of alternative explanations. Serious skepticism is not based on the parsimony of a hypothesis or assuming something to be true or false until proven otherwise, as most so-called 'skeptics' tend to do nowadays. Rather, it is based on ruling out that a specific fact or phenomenon can't be explained otherwise. The habit of correlating neural states and mental states, as the former being the cause of the latter, betrays an unreflective attitude that scientists would not allow in other contexts. It has always been an elementary codex of science that a theory is accepted only when other alternative explanations have been ruled out.

For example, if the Sun appears to move around the Earth, then the alternative theory – that the opposite might be true – should be considered as an alternative possibility until proven otherwise. If a signal measured by a measurement device should be accepted as evidence for a specific physical phenomenon, one must also convincingly rule out that the same signal may not have been produced by some other physical phenomenon. This is a procedural prescription in science that every good scientist knows well. Curiously, however, it is also a common scientific, intellectual misconduct to exempt oneself from this healthy skeptical attitude when one's metaphysical assumptions must be confirmed. Regrettably, one observes this attitude to be common also among post-materialists.

A 'bird's eye view' research of all the current theories of consciousness shows that, despite the fact that these provide utterly different explanations, each of them was able to gather empirical evidence to justify itself [71]. There is a tendency to construct the experimental setup to support one's own theory, rather than challenging it. That is, one opts for an experiment constructed in such a manner that it confirms one's ideological premises a priori. It is a sort of unaware built-in confirmation bias in scientific praxis.

So, what are the alternatives to the mind-body identification? Can we explain the strict relationship between neural correlates and mental states as the former being the cause of the latter? If so, consider following pertinent correlation-cause fallacies.

In what sense does this reasoning differ from concluding that the eyeglasses-acuity of vision correlation proves that eyeglasses produce vision and, since broken eyeglasses impair vision, then vision must be in the eyeglasses? If a display device, such as a computer monitor, is damaged and does not appropriately represent the data, must we then conclude that the data is in the monitor? Is the fact that, by cutting a tungsten filament of a light bulb leading to the interruption of the electric current flow, evidence for electric energy being generated by the filament? If light can be turned on and off by a light switch, must we conclude that the light is 'produced' by the light switch?

Going beyond these analogies, how does this concretely relate to the states of consciousness and our brain states? Let's take the correlation between the secretion of dopamine and the emotion of love. If we carefully think about it, this is no evidence for love being just a chemical reaction. Also, a flash of light on your eye retina triggers a chemical reaction in the visual cortex, but nobody would argue from that that light is just chemistry in the brain. One might then argue that the chemical origin of love is also documented by its external causation, namely, that by taking some drugs, typically cocaine, one can induce very similar emotions. But, again, with transcranial electrical stimulation of the visual cortex, one can also induce the perception of flashes of light (called *'phosphenes'*) [72]. It is even possible to induce the visual perception of luminous shapes, such as letter shapes in blind people [73]. And, after all, we all know well that we might be seeing stars or light flashes like lightning, just by pressuring or rubbing our eyes (an effect known as *'photopsia'*). But nobody would reduce the existence of real stars, lightning, or the letters you are reading only to chemical reactions in the brain. So, this line of reasoning does not prove or dismiss much. By applying a bit of critical analysis, it reveals how things are not as simple as the unreflective naturalistic neuro-philosopher would like us to believe. We can equally well sustain the contrary hypothesis: Love induces brain states that can also be induced by something else. More generally: Consciousness, mental states, and emotional states determine the physical state of a brain, not the other way around. And, if on particular occasions it seems to go the other way around, this is precisely because the brain is an interface that can attune to it. The brain might be a physical substrate *through which* these conscious states manifest.

The metaphor most idealists prefer is the *'radio metaphor'* or *'filter theory of consciousness,'* which dates back to an original idea of American psychologist William James, who stated: *"My thesis is now this: that, when we think of the law that thought is a function of the brain, we are not required to think of productive function only; **we are entitled also to consider permissive*** *or transmissive function. And the ordinary psycho-physiologist leaves this out of his account"* (emphasis in the original text) [74]. James thought of the brain and consciousness as the prism separating white light into colored beams, respectively. If a broken prism fails in its function to 'reveal' the colored light beams, this should not induce us into the logical fallacy that the prism 'produces' colored light. It is just an object with a transmissive function; it doesn't 'generate' anything.

Fig. 28 William James
(1842-1910)

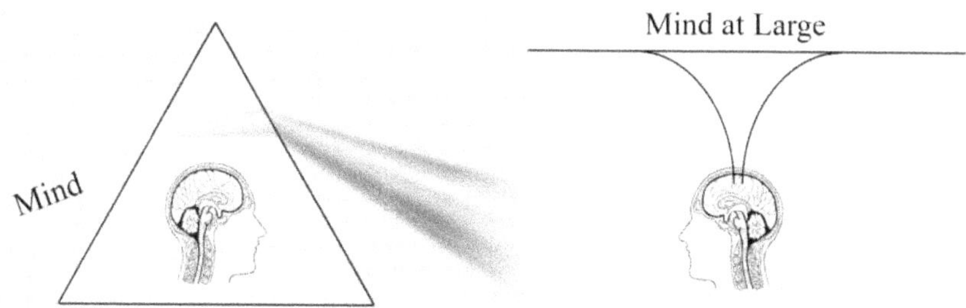

Fig. 29 W. James' and A. Huxley's metaphors for the mind-brain relationship.

The English writer and philosopher Aldous Huxley expressed a similar idea and proposed that the brain is a 'reducing valve' of what he called a *'Mind at Large,'* a universal or cosmic Mind that comprises all of reality with all ideas and all thoughts. According to Huxley our mind filters reality under normal conditions because, otherwise, we would be overwhelmed by the knowledge of this universal Mind. But psychedelic drugs can remove the filter and bring us into contact with the Mind at Large, leading to the experiences that several mystics describe. In his words: *"To make survival possible biologically, Mind at Large has to be funneled through the reducing valve of the brain and nervous system."* [75]

Fig. 30 Aldous Huxley
(1894-1963)

Thus, according to James' and Huxley's viewpoint, the brain is an instrument of consciousness, mind and something that goes beyond matter. It does not generate them. To express this thesis in its more modern version, let's first reason as follows:

If you mess around with the electronic circuits of a radio receiver, you won't be able to clearly receive the signal off a distant radio station, and will eventually shut down ('kill') the receiver entirely. Does this show that the information coming from the radio mediated by the (massless) electromagnetic signal was generated by the (material) receiver? Is the guy reading the news in the radio? If we knew nothing about biology, physics, and our world, we would find it acceptable to answer all these questions affirmatively. A baby, until the age of its first couple of years, probably believes this to be the case by listening to a radio or watching a TV. Therefore, should we conjecture about further alternative explanations? Let's refrain from multiplying entities and resort to Occam' razor by positing that, indeed, vision is produced by eyeglasses, the data is produced by the monitor, the light is in the filament or in the light switch, and the conscious being delivering us the news is in the radio. There is no logical reason to believe otherwise and to change our paradigm.

Admittedly, if this logic is pursued for a reasonable time, it has its motivations and merits, too. It is indeed a healthy scientific practice to try first to explain unexplained or apparently anomalous phenomena inside the dominant paradigm before jumping to other, potentially more complicated, or fancy alternative theories. It would be good scientific practice to first check whether some phenomena we might not understand at first can be explained in the frame of our present knowledge.

An example of this could be the supposed spontaneous generation of life. Before the 17th century, people wondered how life could emerge from nonliving things, such as flies from bovine manure or maggots from rotting meat. The conjecture was that a sort of *'spontaneous generation'* existed whereby nonliving things were capable of eliciting life. Aristotle thought that life arose from nonliving material if it contained a non-material *'pneuma,'* a *'vital heat.'* However, in 1668, the Italian scientist Francesco Redi, could disprove this by placing fresh meat into two different jars, one open and the other closed. A few days later, only the open jar contained maggots, and he could show that they must have come from fly eggs. Redi's experiment remained controversial, but later, similar experiments finally disconfirmed the spontaneous generation theory. This is an example showing that as long we can stick to a paradigm, we should not resort too soon to other, more or less contrived metaphysical explanations.[18]

However, after some time, if the present paradigm does not successfully describe the phenomena, especially if it reveals itself to be a no-progress quest after centuries of research, one might have to consider whether it isn't worth taking another look at the working premises and beginning to open one's mind to alternatives. Perhaps we have fallen into a correlation-causation logical fallacy, and the eyeglasses, the monitor, the light bulb's filament, the light switch, and the radio have no causative powers in themselves but are, so to speak, means of 'connections,' an 'interface,' a 'channel,' or 'relay stations' for something, and not the cause of the phenomenon itself.

Can we consider the brain a material 'connecting device', an 'interface' or 'relay station' between consciousness and mind to the material world? Can the brain eventually be a 'pre-processor' of sensory stimuli from the physical

---

[18] Paradoxically, however, Redi's experiment enforces the vitalist claim, namely that life can be generated only by other life. Something that has never been disconfirmed up to the present time.

world, a sort of 'physical mind' which computes these stimuli to be transferred and presented to mind and consciousness, but is not the mind and consciousness itself? What disproves the notion that the brain may not be just a sort of 'radio station' that tunes into consciousness and mind? A 'transceiver' that transmits information from the physical environment to the mind while receiving the mind's cognitive orders? In this view, the brain exists physically, as a highly complex physical system, responsible for pre-processing the most material and mechanical sensory information beforehand, but it plays no role in 'creating' consciousness.

Or should we dismiss it altogether by resorting to the magic wand of Occam's razor, which can always rescue our preferred worldview that there is no reason and no need to believe in such silly theories that multiply unnecessary entities and that one should always posit matter as fundamental without questioning and further thought? Other alternatives are not worthy of further scrutiny. This thinking is also reflected in the language and word choice one can observe in the specialized literature, in which one reads unjustified statements about how 'everybody knows' and 'it is an established fact' that the brain 'produces,' 'generates,' and 'creates' consciousness without feeling it necessary to substantiate such claims.

Of course, these analogies don't provide evidence for the opposite claim either, but they make it clear how careful we must be about jumping to metaphysical conclusions driven by our unaware assumptions, our potentially unaware ignorance, and especially our desire to read, in a few facts, an ultimate proof confirming our own ideological preconceptions and personal worldviews. This should hold not only for the materialist but for the idealist, post-materialist, and spiritualist (or whatever we would like to call them) as well.

Let us then discuss the several findings in neuroscience, including its weaknesses, which potentially indicate how the physical brain is not the source, origin, or cause of anything but is foremost a tool of perception and reception and a channel of expression. We will also argue that this is not only a viable hypothesis but even the most logical conclusion.

2. Am I my Brain? Where in my Brain?[19]

According to the traditional materialist viewpoint, the physical processes that lead to a mental state and subjective experiential awareness must be localized in one or a few areas of the brain responsible for its 'generation.' These views soon had to be proven wrong and had to give way to much more complicated conjectures. There is no evidence, not even indirect or

---

[19] The following chapter is an in-depth-analysis of the mind-brain identity theory that I published here: [394].

circumstantial, of a single little brain region, area, organ, anatomical feature, or Cartesian pineal gland that seems to take charge of this mysterious job of 'producing' or 'generating' consciousness. Most of the brain is busy processing sensory inputs, motor tasks, and automatic and sub- or unconscious physiological regulation.

However, according to current wisdom, since the *'prefrontal cortex'* (the front part of the front lobe — in short, the brain region behind your eyebrows) — is responsible for cognitively complex functions, such as thought, language, decision-making, and social behavior, it was considered the ideal candidate for the 'seat of consciousness.' In fact, one of the points which may not have received enough attention is the quite obvious fact that neural activity alone cannot be a sufficient condition to lead to a subjective phenomenal experience. Most of the brain's workings do not lead to qualitative experiences. Since the vast majority of things a brain does are unconscious (such as the heartbeat, breathing, the control of blood pressure and temperature, motor control, etc.), this raises the question: What distinguishes a neural process that leads to a conscious experience from that which does not?

For example, the *'cerebellum,'* the posterior area at the lower back of your brain, is almost exclusively dedicated to motor control functions, and its impairment leads to equilibrium and movement disorders. However, it does not affect one's state of consciousness. Its role in 'generating' experience seems to be marginal, if any. There are, of course, lots of anatomical and microbiological differences that distinguish the rest of the brain from the cerebellum, such as the size of the neurons, their biochemical properties, and electrical behavior. However, why this is supposed to lead to a conscious experience in one case

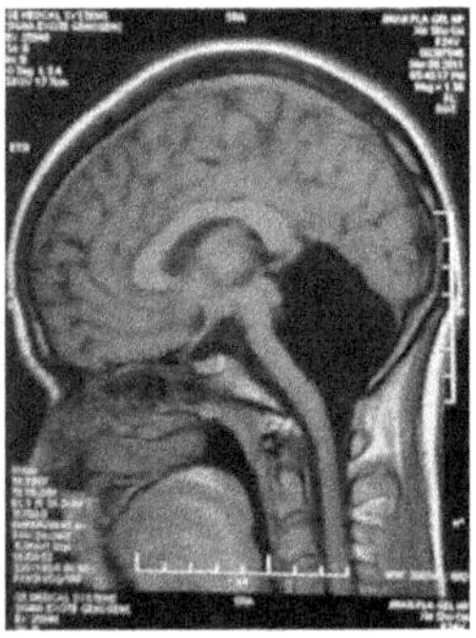

Fig. 31 Living (and walking) without the cerebellum. Credit: [76]

but not in the other remains a question to which science has no answer. There are also rare cases of people who live without cerebellum (a clinical condition known as *'cerebellar agenesis'*) and have only mild or moderate motor deficits or other types of disorders. These do not prevent them from living a relatively normal life and continuing to be a one and undivided self, what people call 'being oneself' [76].

There are no brain 'modules' dedicated solely to specific tasks. In fact, the same task can be performed with a variety of brain activation patterns. Strong activation in certain areas or networks can occur even when those regions are typically believed to decrease in activity during particular tasks. The same perceptual decision-making task can be achieved through multiple brain activation patterns [77].

The general idea of a "modular brain", with each module responsible for some particular function or cognitive skill, is furthermore contradicted by the fact that different brain areas can take over the same sensory or cognitive functions since birth. A dramatic example is that of an experiment that reconfigured newborn ferret's brains in such a way that the visual pathways were directed to the auditory cortical region. The ferrets developed visual functions in the auditory "module" [78]. These findings confirm the brain's proverbial *'neuro-plasticity,'* which we will see next through other extraordinary examples.

It may be worth recalling that the neuronal architecture in our bodies is not confined in the brain — that is, it goes far beyond our heads, through the brain stem, and down through the spinal cord. The *'central nervous system'* is made up of the brain and the spinal cord. The latter is responsible for the transmission of nerve signals from and to the motor cortex, and as is well known, injuries can result in paralysis. But, again, no cognitive deficit or state of consciousness is altered by impairments of the spinal cord. This leaves only one option: If there is a 'seat of consciousness,' it must be identified somewhere in the cerebral cortex or subcortical areas of the brain.

It is in this view that we understand the neurological diversity of an alien creature like an octopus. In octopuses, only 40% of the brain mass (only 10% centralized and 30% for visual processing) is arranged around the esophagus, with the remaining 60% distributed along their eight arms. The extension of their brains serves to coordinate the movement of their eight prehensile arms and the perception via their hundreds of suckers and instantly camouflage them against the background by activating thousands of color-changing cells (*chromatophores*). The octopus arms

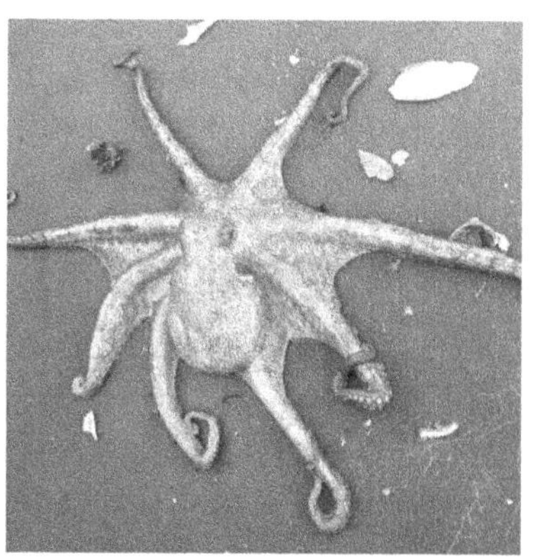

Fig. 32 The octopus.
1 head + 8 arms → 1+8 brains?

also have an independent processing power because if an arm is detached, it continues to crawl for several minutes — a curious and somewhat creepy fact showing that the central brain can give a high-level command but that the eight arms can execute autonomously. This is why people sometimes refer to octopuses as having 1+8=9 brains.

However, while octopuses are indeed weird creatures, we do not have to go so far as to conceive of them as being alien enough to have nine conscious personalities in one body. It would be more accurate to think of the nervous system in the arms of an octopus, not as independent brains (because they are extensions of and are connected to the brain), but as something akin to the cerebellum, brain stem, or spinal cord in humans, which perform independent and partially autonomous activities. It is unlikely that they constitute minds or centers of subjective experiences of their own.

Therefore, from the physicalist point of view, the old mind-body quest for localizing a phenomenal experience which 'generates' a mind with a self in a localized neural correlate of consciousness remains elusive. Identifying it in a gland, a cerebellum, or a spinal cord seems not to be a tenable hypothesis.

Another interesting example of how the correlation-causation fallacy conditions scientific and popular understanding of the mind-body problem can be illustrated by an interesting experimental finding in 2020 by a group of scientists from the department of psychology at the University of Wisconsin-Madison. They showed how stimulation of the *'thalamus,'* a small structure in the middle of the brain just above the brain stem, arouses macaques from stable anesthesia [79]. The awake, sleeping and anesthetized states could be aroused with the stimulation of the central lateral thalamus. The straightforward conclusion seemed to be clear. The ultimate origin and 'switch' for consciousness was discovered. If your consciousness 'depends' on the state of your thalamus, just by 'switching' it on and off by pushing a button, then the thalamus must be the 'seat of consciousness.' What else is there to think about?

This caused some excitement that didn't go unnoticed by the web and in the media. But, before falling into an unreflective confirmation bias, if one would like to put things in a proper perspective, the following should be considered:

First of all, as usual, the association of 'unconsciousness' exclusively with the absence of a waking state, assuming this to imply the impossibility of internal experience and awareness, is always implicitly and tacitly assumed in the premises and conclusions of these types of findings. We have already pointed out the evidence to the contrary in Ch. III and won't take this up here again. Even if we assume that there is no internal experience when we are anesthetized, the relevant question remains whether these sorts of experimental findings confirm that the thalamus is the 'seat of consciousness.' Is it a sort of modern replacement for Descartes' pineal gland in its materialistic monist version?

The thalamus is responsible for sensory information processing. It is known that its main job is to function as a relay and feedback station between sensory brain areas and the cerebral cortex. For example, it functions as a hub between the optical nerves that transport the visual information coming from our eye retinas to the visual cortex located in the posterior part of the brain. Even if one would remain conscious by turning down the functionality of the thalamus, one would no longer see anything because the neural pathways between the retina and the visual cortex are interrupted. From that, however, nobody would conclude that the thalamus is the seat of the visual experience for which the visual cortex is responsible.

It is, therefore, not surprising that there is a correlation between waking states and the physical state of the organ responsible for serving as a 'gate' for relaying almost all sensory impulses, perceptions, and sensations (except olfactory information) from the body to the cerebral cortex. Take away all the four senses and our subjective identification with our physical body, and then it is quite obvious that one must lose this identification. It is commonly called 'sleep,' which, in this case, was forcibly induced by anesthetic means and annulled by stimulating the macaques' thalamus.

If so, the natural question would be: Why should an organ with a relay and hub-like function be supposed to 'produce' consciousness? What is the causal relation between relaying, delivering, and broadcasting signals from one part of an information processing unit to another with the emergence of a conscious experience? Because a modulation, suppression, or disorder in information transfer leads to a 'modulation,' 'suppression,' or 'disorder' of consciousness, should we assume that information transfer in itself makes us conscious?

If so, should we conclude, then, that a data buffer — that is, a memory chip that temporarily stores information while it is sent from one place to another — becomes conscious and has experience? Of course, the thalamus seems to play a role in other functions, such as motor control and regulating sensory information. However, nothing gives us a clue as to what should be responsible for generating conscious experiences. It behaves much more as a switch that makes the sentience of the external world possible by turning it on and off, rather than being a 'source' of experience.

Other research, this time directly on humans, indicates yet another "gate of conscious awareness," namely the anterior insular cortex (the portion of the folded cerebral cortex) [80]. It seems to act as a type of gate between subconscious or subliminal and conscious sensory awareness. It acts as a filter that gates conscious access to the most important information. If it is active, external stimuli enter conscious awareness, and one will become aware of, say, an image or a sensory experience. Otherwise, one loses this awareness. That's why it is called the "gate for conscious awareness." But as the words say, 'gates' or 'filters' do transmit, route, seep through and allow consciousness to 'access to' something; they do not 'generate,' 'create' or 'produce' anything.

In the end, these findings don't tell us much. Also, it is easy for experienced and intelligent researchers to create theories and convincing scientific models of consciousness based on third-person empiric observational facts supported by an unaware correlation-causation fallacy and confirmation bias. These theories then convey an illusion of possessing an explanatory power that one finds they don't have once carefully scrutinized.

However, if there is not one single 'seat of consciousness,' could it be that the combination and activity of some or all the different brain areas do 'produce' the subjective experience?

In fact, one of the persisting myths is the idea that specific parts of the brain are dedicated to a particular psychological or organic function. But nowadays, we have sufficient evidence that must compel us to abandon this simplistic view of a compartmentalized brain. There is no brain region doing only one thing, and there are no neurons supposedly having only one function. Most neurons have several jobs, not a single purpose. It turns out that whenever we hear a sound, have a visual experience, have feelings or emotions, or perform some motoric task, the whole brain is involved. Pretty much everything our brain does is distributed over all its parts and can't be deduced by summing up the activity of these parts. Whatever happens outside or inside of us immediately triggers a wholistic brain-response, not the activation of one or a few cerebral centers. Even such an apparently highly specialized brain region as the primary visual cortex carries out information processes related to hearing, touch, and movement ([81], [82]). Another striking neuronal multitasking example is that of the *'anterior cingulate cortex,'* which is involved in decision-making, emotions, moral judgment, imagination, attention, empathy, and other higher cognitive tasks and yet also presides over much more basic necessities and functions such as hormone and immune system regulation. The reason why professionals nevertheless tend to associate specific brain regions with specific cognitive, sensorial, or motoric functions is that brain scans show only a temporal snapshot of the brain's most intense activity. We are seeing only a few 'tips of the iceberg' and missing the overall activity in the noise. When studies are conducted using less noisy but much more expensive and complicated detection methods, most of the brain's activity becomes visible [83]. On the top of that, it turns out that neuronal representations and functions gradually 'drift' over timescales spanning minutes to weeks. For example, if today a cell's function is dedicated to the detection of a specific odour, tomorrow it will detect a completely different smell ([84], [85]). This also entails that a specific brain region might change its job with time passing. This neuronal reorganization is known as *'representational drift'* and further complicates our understanding of the brain's activity.

Therefore, it would seem plausible that if consciousness arises from the activity of a complex aggregation of neurons, it could be something distributed among a wider brain region and may not be the epiphenomenal appearance of only one localized region. For example, one could conjecture that to 'generate' consciousness, at least some brain areas must work together in a unified whole via thalamic activity. From this perspective, the thalamus function is to 'integrate' the information flow of the several brain areas and, if disrupted, consequently lead to a 'loss' of consciousness.

This, however, would also suggest that if someone were to split your brain into two parts, and you survived, you would presumably feel somewhat less conscious and less 'yourself.' Right?

Some might be surprised to learn that this is not a thought experiment taken from a Frankenstein novel, but a very real surgical procedure performed since the 1940s: the *'corpus callosotomy'* (although only rarely used nowadays). It is an extrema ratio surgical procedure performed only to treat the worst cases of epilepsy (patients having up to 30 seizures a day) that did not respond to any medical treatment. In this

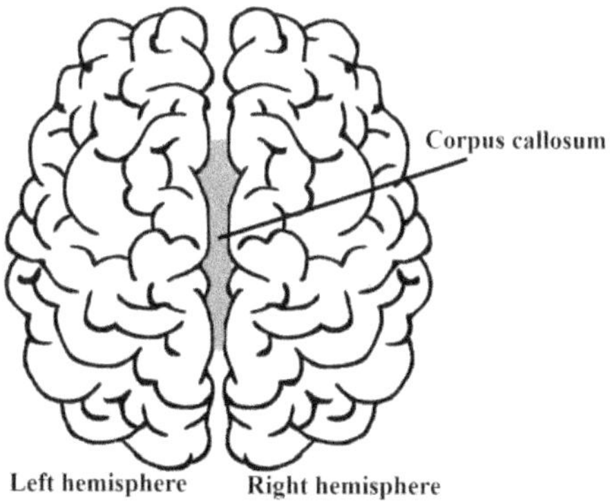

Fig. 33 Does brain-splitting cause 'self-splitting'?

procedure, the *'corpus callosum,'* the nerve tract connecting the left and right brain hemispheres, is severed (in part or, in some cases, entirely cut through), thereby avoiding the spread of epileptic activity between the two halves of the brain. Its natural function is to ensure communication between the two cerebral cortexes of the two hemispheres to integrate and coordinate motor, sensory, and cognitive functions, such as moving left and right limbs, the visual integration of the left and right sight, etc. Because most of the brain's activity is distributed on both hemispheres, with no indication of one or the other part being responsible for generating our sense of 'self' — the feeling of being a subject and individual, a sentient 'I' having experiences — one must wonder how the patients who have gone through such an acute surgical intervention feel. Do their split brains 'generate' a split personality? Does the patient claim to be two patients? Do they feel like another person entirely, or are there two "I's" in the same body?

Facts have shown that these patients do not have any symptoms of multiple personalities or display any signs of internal dissociation after surgery. When asked if they eventually feel the sensation of a 'divided self,' the answer is negative. Their self, mind, and conscious experience remain a unified whole of one subject and individuality. They deny being a different person from what they were before their brain splitting. Close relatives who knew the split-brain patients before and after surgery, didn't notice any change in personality. This surgical intervention eliminates or effectively diminishes the intensity of epileptic seizures and, quite surprisingly, preserves almost all of their cognitive and motor-sensory abilities.

Of course, there can be more or less severe drawbacks. Visual, auditory, and tactile information that was previously processed throughout the two brain hemispheres can now be processed separately in only one or the other part of the brain. This can (but must not) result temporarily or even permanently in difficulties in speech articulation and coherent motoric coordination between the left and right parts of the body or spatial orientation. In rare cases, strange disorders such as blindsight (see Pt.I-III.1) or the so-called *'alien-hand syndrome'* can take over, where one hand appears to have a mind of its own. This occasionally happens when the two hemispheres' representations of reality come into conflict, and one wants to override the other. In these rare instances, decision-making and volition between the two hemispheres clash. An example is the patient's struggle to overcome an antagonistic behavior, such as knowing what cloth they want to wear, while one of their hands takes control and reaches out for another cloth they don't want at all. However, this should not be confused with two personalities competing against each other (like in the case of *'dissociative identity disorders'*), as the split-brain patients identify with only one body and perceive their disobedient limb as being subjected to annoying motoric misbehavior; they do not report any sensation of some other internal personality taking control. Moreover, it confirms what we already saw: The brain — or, more precisely, our two brains — tell us two different stories, mostly fairy tales as interpretations of reality. It looks like the story of one hemisphere does not coincide with that of the other hemisphere but, in a non-split condition, we can integrate the information coming from both sides and make from it yet another coherent and unique story which we take as the 'truth' in order to evaluate what must be done and to make a decision. Not so for split-brain patients. They seem to identify with one of the stories — that is, consciously access one of its interpretations — and keep the other in a subconscious or subliminal awareness, sort of the blindsight case, what the American cognitive neuroscientist Michael Gazzaniga used to call the *'left-brain interpreter.'*

Gazzaniga and his colleague, Roger Sperry, became famous for their studies and experiments performed on several split-brain patients, which highlighted the role of the brain's left and right hemispheres. Noteworthy is how these patients get into trouble in verbalizing what they see in their visual field. The speech and language brain area resides in the left hemisphere. If someone having a split-brain is presented with a visual stimulus (say, an apple) on the right visual field — that is, the image is processed in the visual cortex of the brain's left hemisphere, the same hemisphere containing the language center — then the patient can correctly spell out what is seen: an apple. However, if an object is shown in the left visual field, which means that the visual stimulus is processed in the right hemisphere opposite the language center, then the split-brain patients are no longer able to say what they see. Nevertheless, if asked to write down what they saw, they perform the task accurately.

However, the fact remains that there is no 'self-splitting.' Split-brain patients return to normal life and are indistinguishable from 'normal brained people.' Most of their deficits are compensated for by visual, tactile, and auditory integration and become apparent only in laboratory conditions. Their consciousness and awareness are one undivided whole, and they refer to themselves always as "I" and "me." No personality dissociation appears, and the character, emotional stability, or intelligence remain unaffected.

Recent investigations also question the canonical textbook findings ([86], [87]). While it is confirmed that a corpus callosotomy splits the visual perception of the environment in two, several patients can nevertheless see them both and report it to the outside world — that is, they can access their language centers. Moreover, there is no evidence for memory loss. Their perceptions are unified in consciousness. There is no sign of two independent perceivers. If we don't confuse, again, mental states as being the origin or efficient cause for consciousness, then any apparent paradox dissipates. Split-brainers may have two (eventually even conflicting) hemispheric and motor-sensory mental states (something not entirely unusual in healthy subjects too) but, even if one argues and provides evidence for a 'two-minds' model, that wouldn't imply a split sense of identity or self-awareness.

This 'unity of consciousness' remains a deep mystery (fundamentally, a binding problem). Try to split your computer. If you saw your PC's motherboard into two pieces, you won't obtain the same result and can be sure that what you are left with is only good to be trashed once and forever.

These findings also challenge the theories of consciousness, which are actually highly rated among reductionists, such as the *'Global Workspace Theory'* (GWT) of American neuroscientist Bernard J. Baars [88] and the *'Integrated Information Theory'* (IIT) of consciousness of Italian neuroscientist Giulio Tononi ([89], [90]). For reviews of the landscape of modern theories of consciousness see also [91] [92], or, for a review of non-physicalist theories see [93]. GWT envisages a working memory that, once momentarily active, corresponds to a subjective experience. This memory is our inner domain in which we can bring the phenomenal event into our awareness. For a short time interval, the working memory contains what we are conscious of while all other processes that do not enter it are 'unconscious.' However, all the unconscious processes throughout the brain can be broadcasted and combined into the global workspace to form a unified perception of an identity experiencing the processes as a whole — that is, a unified consciousness. IIT is a theory that starts from the phenomenological point of view (instead of the other way around, as in most physicalist theories that posit matter as the source and the phenomenal experience as its outcome). Independently, from the assumed (material or non-material) substrate, it makes statements on how to measure consciousness, how it correlates with brain states, and how *"the loss and recovery of consciousness should be associated with the breakdown and recovery of information integration."* [89] A process of integrating the information coming from all the brain areas is supposedly the efficient cause of our experiential richness. The amount and integration of information determine the level of consciousness leading to a conscious entity. Both theories finally recognize that the old concept of a dedicated brain area for specific tasks was an overly simplistic viewpoint. The brain is an interconnected network in which no specific region carries out one function. Modern neuroscience thinks more in terms of network science, where several brain regions are highly interconnected and interdependent. This interconnectivity between parts of the brain enables the type of complex functions of which the brain is capable.

How far these conjectures align with reality is questionable. For both approaches, in severing the corpus callosum of a brain, one would expect a loss or at least a diminishing of conscious awareness because there would be a loss of working memory for the former or loss of information integration for the latter. But nothing like this happens.

To save the paradigm, the mainstream materialists like to point out that in not all documented cases was a complete transection of the corpus callosum performed. The truth, however, is that in several cases, the complete sectioning was performed and was even confirmed by MRI imaging or radiological means [94]. Then, they point out that a complete transection still leaves some residual subcortical structures (the brain structures below the surface cerebral cortex) intact, which allows for some communication between the two hemispheres, potentially maintaining the 'self' of the patients. Maybe, but this seems to be only a desperate strategic retreat that seeks to save appearances and doesn't want to confront the established facts: Cutting a brain almost completely in half leads to no change in conscious awareness and identity.

The hypothesis that consciousness emerges due to an integration of the activity of several or all brain regions is not convincing in the light of these findings. The hard problem of consciousness, the binding problem and the hypothetical existence of the so-called 'seat of consciousness', remain one of the deepest mysteries of science. While, if we look at these findings from the point of view of William James' filter theory or Aldous Huxley Mind at Large, they might make more sense. Consciousness must have a deeper source, origin and nature than just the activity of a bunch of neurons.

To further substantiate this, we could also mention that, to treat epilepsy, the most extreme surgical intervention is to remove an entire brain hemisphere, called *'hemispherectomy.'* Usually, this is done only in childhood because young brains can rewire themselves much more efficiently than older ones. The brain has such plasticity that, if one intervenes early enough, children who had a hemisphere removed will nevertheless recover from the momentary impairment because the brain will functionally reorganize and rewire itself. The remaining hemisphere resumes

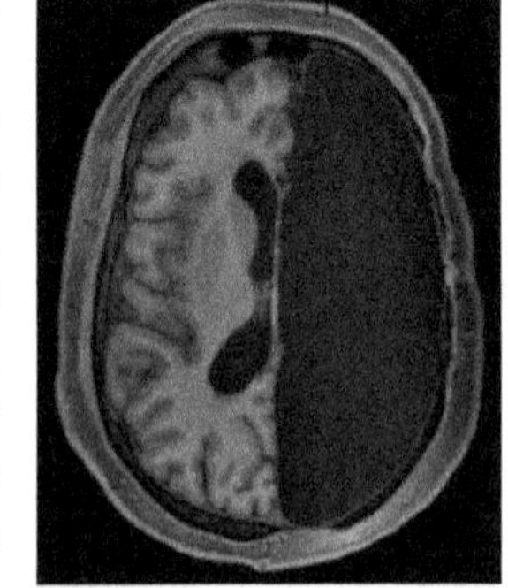

Fig. 34 Hemispherectomy. Living (quite well) with only half of the brain. Credit: [95]

efficiently all, or almost all, the functions previously taken up by the removed hemisphere. Fig. 34 shows a scan of the brain of an adult who had an entire hemisphere removed during childhood because of epilepsy [95].

So goes the theory. However, Nature seems to not take the left/right distinction and early plasticity hypothesis too seriously. That the left-right brain task distribution is not an inescapable neurological dogma is testified by people born with only one hemisphere. For example, while in healthy subjects the left visual field is represented in the right hemisphere and vice versa, someone born with only one hemisphere can develop maps of both visual fields in it [96]. Hemispherectomy on adults older than 18 years

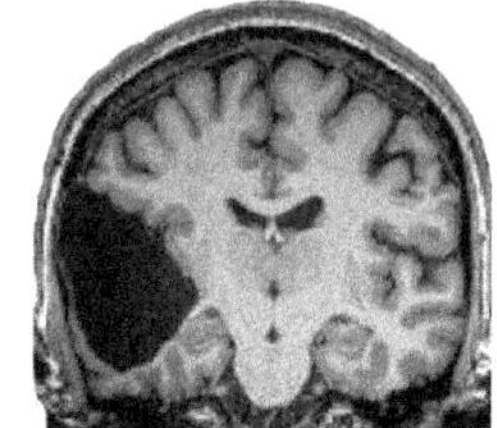

Fig. 35 Speaking without the brain's language centers. Credit: [100]

turns out to be safe and effective as in early childhood [97] and, even though sub-optimally, perform above 80% task accuracy on face and word recognition tests [98]. Even in the case of a left hemispherectomy, the language center — which in normal conditions, is in the left hemisphere — can be recovered by new reconnections in the right part of the brain [99]. Further evidence reports of subjects where the frontal lobe was missing from childhood without any measurable linguistic impairments, as shown by a case of a woman who grew up without her left temporal lobe (see Fig. 35), but speaks (and dreams) in English and Russian [100].

A possible explanation is that because these patients already had severe seizures originating in one of the hemispheres, the functional rewiring on the other hemisphere began before the surgery. But findings tend to disconfirm this easy way out. Though interconnectivity inside the brain networks increased, the interconnectivity between brain regions with the same function after hemispherectomy does not differ from that of two hemispheric control subjects [95]. That plasticity alone can explain this state of affairs is far from proven.

However it is, most patients become seizure-free, and their cognition is relatively unchanged after surgery (some motoric and cognitive functions decrease, but others improve). Overall, these patients appear to be 'normal.' Before and after surgery, there is no substantial change in IQ scores, and in everyday life, one could not tell the difference between humans having a brain or only half of one. No whatsoever 'half-self', 'half-awareness,' or 'half-consciousness' is observed or reported by the subjects.

Things can get even worse.

In 1980, the British pediatrician John Lorber reported that some adults cured of childhood hydrocephaly had no more than 5% volume of brain tissue and a cerebral cortex as thin as 1 mm [101]. While some had cognitive and perceptual disorders, and several developed epilepsy, others were surprisingly asymptomatic and even of above-average intelligence. Obviously, this didn't fit into the narrative of an established materialistic mindset and was simply ignored for almost three decades.[20] Then, in 2007, in Marseille, France, a

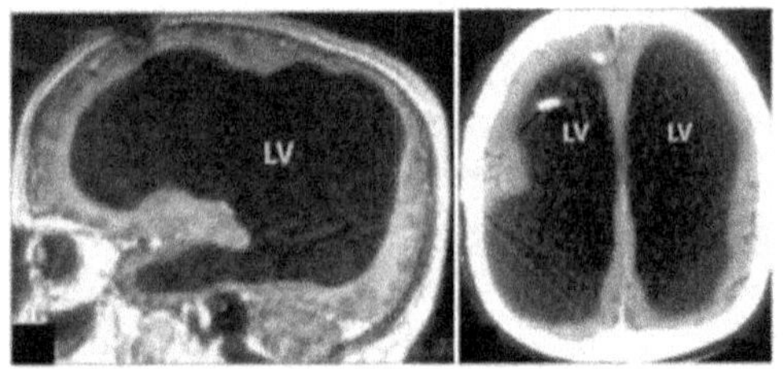

Fig. 36 MRI image of an almost missing brain. Credit: [102].

44-year-old man complaining of weakness in his left leg was submitted to an MRI brain scan [102]. As Fig. 36 eloquently shows, the skull is abnormally filled with cerebrospinal fluid, 99% of which is made of water, thus the name '*hydrocephalus*,' leaving only a thin sheet of actual brain tissue. As an infant, he had a shunt inserted into his head to drain the fluid, but it was removed at the age of 14. Evidently, the fluid build-up in his lateral ventricles (cavities containing the cerebrospinal fluid) didn't stop and ended up reducing the brain's size to 50-75% compared to its normal volume. One would expect this to be the brain of a dead person. Instead, this was an almost perfectly normal functioning brain. Though he had a below-average IQ (75/100), this man had a job, a family, and a normal life. His intellectual faculties were relatively intact despite the compressed but still existing cerebral cortex, which, however, had to take up all the functions of the missing subcortical parts.

This testifies again to the extraordinary plasticity of the brain — or, at least, seems to do so. It is astonishing how the brain can reconfigure itself and adapt to damage. The neuroscientific orthodoxy points out that some thin layer of white matter (underlying nerve cells connecting the cerebral matter lobe areas in normal conditions) is still present and could possibly connect up the upper grey matter, allowing it to function normally. Again, these are speculations aimed at saving the preestablished paradigm.

All these findings also confirm that brain size and the numbers of neurons in a brain do not (or, at least, do not necessarily) make up one's intelligence. Size matters for manipulative complexity, such as the more complex hand movements in primates, which humans can develop superbly (think of the hands of an expert musician playing piano) [103]. But a direct correlation between brain size and mental skills is not that straightforward. We like to believe that it is our brain size that makes us human but rarely do we question what one means by 'size'. The number of neurons? The weight of the brain? Its brain to body mass ratio? Or its volume? Humans don't have the largest

---

[20] Interested readers can convince themselves by finding, on the Internet, video interviews of people who have gone through the hemispherectomy or corpus callosotomy / 'split-brain' procedure or John Lorber's documentary on hydrocephalus.

brain size in any of the aforementioned senses. The human brain has about 90 billion neurons, weights ca. 1.1 to 1.4 kg and has a volume of 1300 cm^3. But the brain of an elephant has three times the number of neurons we have, and the weight and volume of the brain of a sperm whale measures six times as much and ants have a six times larger brain to body mass ratio. A bit of an extreme example that shows how cognitive skills and brain size are decoupled is the case of *'mouse lemurs'* that have a brain that is 1/200th the size of monkeys' but perform equally well on a primate intelligence test [104]. Therefore, brain size alone is not what makes up a more developed mind. Then what does?

Maybe it is its complexity. The more complex the network architecture of neurons is, the smarter it becomes. In fact, a system that has no complexity at all, such as a crystal, can hardly develop any form of cognition. Therefore, it is plausible that a certain degree of complexity is a mandatory factor for a brain or whatever material structure to display a form of intelligence and cognitive skills. The problem, however, is that it is unclear what kind of complexity this must be and what we mean by the 'complexity' of a material aggregate like a brain in the first place. For example, the orbits of stars in a galaxy can be quite complex, but nobody believes that this makes the galaxy a conscious being. Or the hydrodynamic flows in the atmosphere are complex enough to be computable only approximately by the fastest supercomputers, but no one believes this to be a reason why a bunch of air becomes something conscious. Therefore, if the neural complexity of a brain somehow scales with its IQ, it must be a special type of intricacy; it can't be just whatever contrived and complicated connectome. One could think of a measure of *'brain connectivity'* — that is, the number of wirings between neurons (through their axons, dendrites, and synapses) and the speed at which they transmit and receive signals — as an indicator of its complexity and see if it somehow scales with the cognitive functionality. However, MRI studies reveal that all mammals, including humans, share equal brain overall connectivity [105]. The efficiency of information transfer through the neural network in a human is comparable to that of a mouse. It is independent of the structure or size of the brain and does not vary from species to species. So, things can't be as easy as that. These findings are yet another reason to doubt the mind = brain dogma.

Let us now return to the idea that several neurologists or cognitive scientists hold, according to which consciousness — meaning, among other things, 'awareness' and 'attention' — resides in the cerebral cortex, the outer gray layer of the brain. The previous cases of split-brain, hemispherectomy, or hydrocephalus, no matter how extreme the surgical intervention or the brain malformation, appear to be in line with this hypothesis. The cerebral cortex was halved or compressed and ill-formed by the cerebrospinal fluid but was largely present.

This belief isn't unproblematic. First of all, because the *'neocortex'* — that is, the layers of the cerebral cortex — exists only in humans and other mammals, one must conclude that birds, fish, octopuses, amphibians, and reptiles are, per definition, all 'unconscious' and incapable of having some, more or less elementary form, of conscious subjective experience. There is no sentience; they don't feel pain, fear, or pleasure or have whatever feeling. These are only outer appearances of automatic reflexes, sort of 'animal p-zombies.' This is reminiscent of

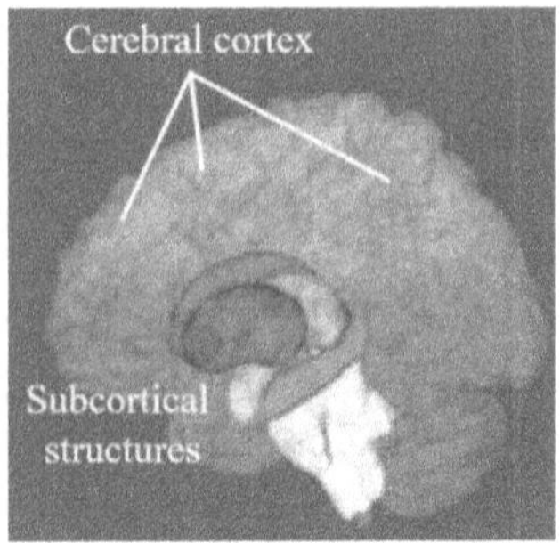

Fig. 37 The cortical and subcortical structures of the brain.

the old Cartesian doctrine that considered animals only as automatic and reflexive organisms operating like clockwork or automatons. It is a narrative that warns about anthropomorphizing animals' behavior but then falls into a form of semi-anthropocentric worldview itself by denying conscious experience to other forms of life as supposedly proper only to some animal classes, obviously with humans at the top of the pyramid. This attitude denying that some animals are conscious, sentient beings because they lack a specific organ may also come from the scientists' psychological necessity to cleanse their own conscience. We shouldn't forget how modern neuroscience relies heavily on the massive use of animal experimentation.

At any rate, as expected, evidence is beginning to emerge that, for example, the neural correlate patterns of sensory consciousness in a corvid bird aren't substantially different from the neural correlates in humans having a similar sensory conscious subjective experience [106]. It is sort of the two-stage process of unconscious vs. conscious vision — that is, the no-report and reporting of conscious experiences — where the activity correlates with conscious experience delayed relative to a stimulus onset signal that we discussed in Pt. I-II.3 and III.1,4. Moreover, one wonders how some birds can also perform amazing cognitive feats despite their forebrains consisting of lumps of grey cells. It turns out that cortex-like circuits in avian birds exist that are reminiscent of the mammalian forebrains, and the idea that advanced cognitive skills are possible only because of the evolution of the highly complex cerebral cortex in mammals is becoming less plausible [107]. Of course, if you try hard enough in finding neural features that could explain all this, you will always find something that confirms your worldview [108]. There is sufficiently strong evidence to conclude that both cephalopods and crustaceans are sentient [109]. After all, this is unsurprising: Common sense doesn't really need any scientific proof to accept that ravens, crows, octopuses or lobsters are sentient beings.

However, the worst thing is that modern neurology does not recognize that some 'class of humans' has the right to be treated as conscious and sentient beings able to suffer, despite all evidence to the contrary. A good example that illustrates this is the cases of children in a *'developmental vegetative state'* — that is, what is officially considered by the American Academy of Neurology (as declared in its guideline report in 1995 and confirmed in 2018) as being a neurovegetative state in which there is *"no evidence of purposeful behavior suggesting awareness of self or environment."* [110] In other words: a universal rule that reduces them to unconscious children who cannot suffer because this supposedly requires a functioning cerebral cortex. Something that our intuition naturally suggests that even reptiles and birds must have is not allowed for humans because of an inability to respond to the environment as other sentient beings do.

Nevertheless, only one case that shows the contrary should be sufficient to disprove a universal rule. Four such cases were brought to light in 1999 by a group of pediatric neurologists at the University of California, led by D. Alan Shewmon [111]. They studied the states of awareness in *'congenitally decorticate children'* — that is, the cases of four children who were almost completely lacking the cortical tissue and were neurologically certified as being in a vegetative state. Yet, the loving care of their mothers (or of someone who adopted them and bonded with them via dedicated full-time caring) could gradually 'awaken' in them a conscious awareness. From an initially unresponsive state, they showed clear signs of having developed auditory perception and visual awareness (despite the total absence of the occipital lobe that, in normal conditions, hosts the visual areas). For example, they tracked faces and toys, looked after persons they recognized and could distinguish from their mothers or caretakers, listened to music for which they manifested preferences with their facial expressions, including smiling and crying, and, at least in one case, gave clear indications of self-recognition in a mirror. One can't think of better evidence for a subjective experience and awareness of the environment and the self. Shewmon notes:

"Nevertheless, biologists (not to mention animal rights' activists) speak meaningfully of 'consciousness' in animals, where only the behavioral, operationally definable, nonreflective dimension is implied. Unarguably, such 'consciousness' is just as properly attributed to the decorticate children described here. Were they not humans studied by clinicians but rather animals studied by ethologists, no one would object to attributing to them 'consciousness' (or ability to 'experience' pain or suffering) based on their evident adaptive interaction with the environment. This alone is surely remarkable. Even prescinding from the question of self-awareness, the possession by decorticate children of even animal-type 'consciousness' thoroughly contradicts prevailing permanent vegetative state orthodoxy, which predicts that they should be precisely vegetative, not sentient and intentionally behaving."

These cases seem to contradict the prevailing theory according to which the cerebral cortex generates consciousness in humans. However, the mainstream proponents point out (as they did in the case of the still incomplete disconnection between the split-brain hemispheres or the possible gap-filling white matter in hydrocephalus brains) that the children were not completely decorticated, as some cortical tissue was still left (for, example, the bottom left image of Fig. 38 shows that a remnant of the frontal lobe is still present).

First and foremost, if we believe the cerebral cortex being responsible for consciousness, one wonders why they are considered nonetheless in an unconscious vegetative state in the first place? Secondly, it is unlikely that the few fragments of the cortex lobe could alone account for the totality of their conscious behaviors. Everything indicates that a subcortical mediation of consciousness exists, and it must play a role in conscious awareness.

Despite this compelling evidence, the conviction that consciousness is 'produced,' 'generated,' or 'created' by the cerebral cortex and that someone having only subcortical brain tissue must be 'unconscious' remains the prevailing dogma. This is because physicalists must necessarily find the brain area, the 'locus' and 'seat of consciousness' that allows them to pinpoint, or at least determine, some spatial-material relationship for the 'emergence' of consciousness in a physical structure.

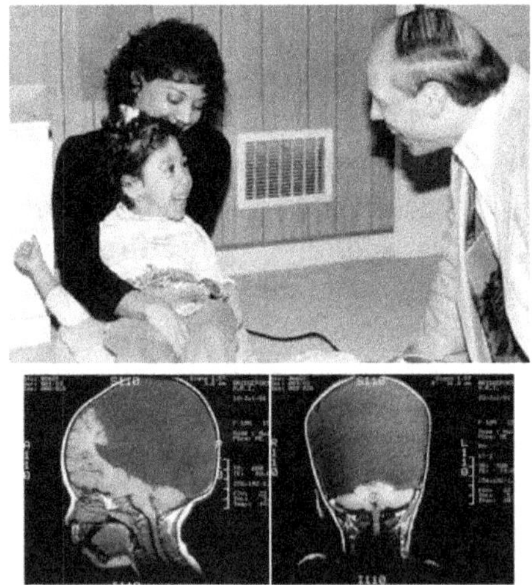

Fig. 38 Top: Congenitally decorticate children socially interacting with Dr. D.A. Shewmon. Bottom: MRI brain scan (midline sagittal and posterior coronal plane). Credit: [111]

Shewmon labeled the *"developmental vegetative states as a self-fulfilling prophecy."* Those children whose parents uncritically accepted the physician's dehumanizing verdict that they were 'vegetables' or 'reptiles' turned out to behave exactly as such. Those parents, however, who ignored the prognosis and "showered their children with loving stimulation and affection" could teach mainstream science something about the neurophysiology of consciousness.

Of course, in the case of decorticated children, we are speaking of cognitively and motorically severely disabled cases that cannot compare to the neurologically normal 'split-brainers' or hydrocephalic patients. The cerebral cortex seems to be essential to gaining awareness of the physical external world and at least part of the higher cognitive functions. But the point in question for us here is: What is the source of the subjective experience? It can't be localized in a specific area, not as a whole-brain integrative physical process, and not even in the cerebral cortex alone.

Though many still struggle to accept this state of affairs, the fact that consciousness does not require the cerebral cortex was later confirmed by other findings [112]. Now the physicalists have been forced to retire to the last cerebral bastion for the seat of consciousness: the brainstem [113]. Indeed, its stimulation can also trigger intense emotions and feelings. But one wonders what mysterious property a neural circuitry dedicated to the most physical and basal control of cardiac, respiratory, and temperature regulations containing mainly neurons for motor and sensory functions is also able to give rise to such an apparently immaterial and completely different and unrelated 'function' or 'property' as a conscious experience. Also, these aspects add to the line of evidence that the brain does not generate experience but, rather, mediates it. We guess that it is only a matter of time before this stronghold will crumble like all the others. Mounting evidence indicates how conscious behavior exists in life forms that have no cerebral tissue in the first place. But, before stepping towards the world of plants and cells, let us first overview some aspects that emerged for the higher cognitive functions we have seen so far.

Overall, with the exception of congenitally decorticate children, the cases mentioned above of people who have undergone corpus callosotomy or hemispherectomy, or people suffering from hydrocephalus, cerebellar agenesis, or several other types of brain damage, show how surprisingly intact their higher cognitive function remains. One would expect that the first victims of such invasive neurological changes or surgical interventions would be those complex and high-demanding cognitive functions so characteristic of the mind, such as intellectual skills, abstract thinking, decision-making, reason, logically and willfully planning actions, and so on. Instead, it turns out that even if large brain masses are injured or absent, the cognitive skills of the subject remain substantially unaltered.

All these findings require an explanation from the physicalist viewpoint, which identifies the mind with the brain. Of course, as usual, one could resort to the conjecture that neural plasticity explains all things. In principle, it can always be invoked, like a magic wand, to save appearances. But how far are we willing to go with that? How far can the brain's plasticity intervene to restore mental functions? Until 50% or 30% or 10% of the brain is left? Can the stability of cognitive functions under severe structural damage of the brain be ascribed to neuroplasticity alone, or is there more? Thus, some caution would be appropriate. For example, a recent study challenges the idea of adaptive circuit plasticity according to which the brain recruits existing neurons to take over for those that are lost from stroke. Undamaged neurons do not change their function after a stroke to compensate for damaged ones, as the conventional re-mapping hypothesis believed [114]. Another study finds that our brains are not able to 'rewire' themselves, despite what most scientists believe [115].

The question is also why sensory or motoric functions are much more prone to be disrupted than mental ones. The IQ and abstract thinking of these subjects seem not to be affected (for a review of the discrepancy between cerebral structure and cognitive functioning, see [116].) These recent findings confirm what was already known from the studies of the American neurosurgeon Wilder Penfield. His surgical specialty was the mapping of seizure foci by stimulating the brain regions of locally anesthetized but awake patients. Observing the patient's response, he was able to show how different brain electrical stimulations would cause a seizure or evoke a sensation, a perception, a movement of muscles, a memory, or even a vivid emotion but, interestingly, never evoked or inhibited thinking [117]. The normal reasoning functions, those of mind, intellectual skills, and rational analytic thought, were never affected by whatever stimulation. From a third-person perspective, thought can seemingly be switched off entirely by triggering a seizure or by anesthesia. However, it can't be weakened or enforced by weakening or enforcing the activity of any brain area. Penfield also noticed that his patients were perfectly able to distinguish their own sensations and experiences from those triggered by the stimulation of the electrodes in their brains. They retained a third-person perspective on cortical stimulation — that is, they perceived how the evoked percept by the brain stimulation was done to them, but not by them.

Furthermore, if the human's analytic and rational functions are a mere cerebral product, one would expect to find some observable difference between the ordinary brain of someone with a low IQ and that of a genius like Einstein. In fact, Einstein's brain was removed after his death and has been conserved until nowadays for analysis to find some cerebral signature that could account for his extraordinary intellectual achievements. Yet, nothing relevant was found, and Einstein's genius remains a mystery. More recently, a study evidenced that his corpus callosum was thicker than average, indicating that the connectivity between the two hemispheres of Einstein's brain was generally enhanced compared to other 'normal' brains [118]. But, if the connectivity between brain hemispheres accounts for someone's intelligence, this immediately raises the question, again, in light of what we have discussed above: Why does a hemispherectomy not lead to a loss of IQ or no substantial psychological and behavioral change? No evident correlation between intelligence and brain structures has been found so far.

This is also confirmed by the fact that, despite decades of research with brain imaging techniques, the correlation between brain anatomy and personality traits or mental pathologies remains elusive. The neurobiological variation matching up symptoms of mental disorders such as ADHD, autism, or schizophrenia is weak, and its significance is questionable. The lack of this correlation is a bit at odds with the wide held belief that the mind is only a cerebral epiphenomenon. Inside a materialistic paradigm where the mind-brain identity theory is the foundation of one's way of seeing the relationship between mind and brain, one would expect to see some correspondence between mental disorders and brain disorders. Yet, there is no consensus among neurologist whether there is such a correlation at all. It is not even clear whether mental disorders should be understood to be brain disorders and what conditions need to be met for a disorder to be rightly described as a brain disorder [119] [120]. An illuminating discussion among neuroscientists about this surprising aspect can be found here [121].

One might also question if, besides the spatial distribution or localization of the neural correlates of consciousness, its intensity might also play a role in generating a conscious experience. For example, it is well known how drugs can change our brain chemistry and give rise to subjective psychedelic experiences. What will a brain scan show in these conditions?

Research has been conducted with *'psilocybin,'* a psychoactive drug obtained from hallucinogenic fungi, so-called *'magic mushrooms,'* which, apart from its therapeutic applications, produces intense subjective effects such as an altered state of consciousness similar to that of LSD, alterations in perception, mood, and thought, enhanced ability for introspection, mystical experiences, changes in time and space perception, and changes in the perception of self. These are rarely

Fig. 39 The psilocybin mushroom.

experienced except in dreams, deep meditative states, religious exaltation, or acute psychosis.

From a physicalist's perspective, which equates mind and brain as being one and the same thing, one assumes that the intensity of 'mind-expanding' psychedelics must be directly proportional to an increase in neural activity and connectivity. However, the contrary turned out to be the case. A BOLD-fMRI (Blood-oxygen-level-dependent fMRI) study reported a significant decrease of brain activity — that is, a decreased blood flow and venous oxygenation— as being inversely proportional to the intensity of the subjective experience reported by the test subjects [122]. The authors of this research pointed out how this finding is consistent with Aldous Huxley's 'reducing valve' metaphor in the brain that acts to limit our perceptions in an ordinary state of consciousness. These findings were later confirmed by further studies with other hallucinogenic drugs such as LSD and ayahuasca ([123], [124], [125]). While not conclusive and not necessarily indicative of a particular metaphysical mind-body ontology, this is yet another piece of the puzzle that seems not to fit as expected in the mainstream materialistic narrative.

Without sliding into a metaphysical discourse, an interesting clue that comes from the purely physiological domain is that, wherever and however consciousness manifests, it is subjected to the effects of organic activity in several bodily regions, not only in the brain. By 'organic activity,' we mean not only neuronal activity but, in general, any biological process. In fact, mounting evidence shows how our consciousness is, if not 'generated,' at least 'modulated' by other parts of the body.

For example, gut microbes (*'microbiota'*) influence our cognition, emotional state, and memory. There is a relationship between them and pathological states, such as anxiety, mood disorders, or developmental disorders such as autism. The amount and diversity of microorganisms inhabiting our intestines affect how we think, perceive, and experience the

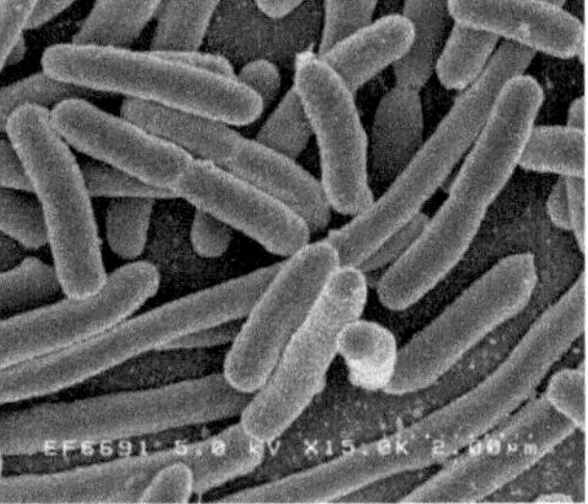

Fig. 40 Human gastrointestinal microbiota.

world [126]. This is physiological evidence that suggests how our emotional states cannot be reduced to only a brain-centric view but arise due to a complex interaction between the brain and the rest of the body [127]. There seems to be intuitive wisdom when we say that we have a 'gut feeling' about something.

If we can state that all our thoughts, emotions, and cognitive functions are 'produced' by our brains, only on the factual basis that a physical alteration of the brain leads to an altered state of consciousness, why then are we not also allowed to state that the very same functions are a 'product' of our intestines instead? If we assume that emotions are 'generated' by the limbic system because, for example, changes in the structure of the amygdala 'elicit' mood and anxiety disorders, why can't we say the same for our gut, as its microbiota 'generate' similar effects? Moreover, it is a widely accepted fact that a sort of *'enteric brain'* exists — that is, a nervous system of neurons that governs the function of the gastrointestinal tract independent of the brain, the spinal cord, and the brain stem. Maybe our 'seat of consciousness' is not in our brains but, instead, in our guts?

The only reason why nobody takes this seriously is that the physicalist posits a priori 'brain-centrism' as dogma, and frankly, we all would like to avoid identifying ourselves and our personalities with our intestines!

Of course, we can stick with the brain-centric model and reshuffle everything into a *'gut-brain axis'* according to which the gut microbiome influences the brain, with the brain remaining the ultimate source of conscious experience. In fact, communication pathways between the gut microbiome and the brain exist, reassuring the dominant paradigm.

This, however, can become an even more complicated endeavor. It could imply that we inadvertently reduce the whole to one of its parts and, paradoxically, consider that part as the ultimate cause and sufficiency for the process of the whole — say, for example, like regarding digestion as being driven entirely by the processes taking place only in the stomach, considering it the cause and sufficiency for digestion itself.[21] It might sound ridiculous, but this is essentially what the reductionist and physicalist mindsets are trying to do when they consider the brain as the ultimate causal and sufficient center for the processes of mind and consciousness.

And, after all, one might also question: What is the cause and what is the effect? We take it for granted that some biochemical states of an organ 'elicit' a mood or an emotion but rarely question if it might be, at least on some occasions, also the other way around — namely, that the mood or emotion affects that biochemical state that we misinterpreted as its cause.

As science advances, it becomes increasingly difficult to distinguish between causes and effects while maintaining the brain as a separate central master-organ, with all other organs and physiological processes subservient. This view is challenged by recent findings which suggest taking a perspective that considers the body as a whole — that is, a vastly more complex ecosystem of molecules, microbes, and neurons distributed throughout the entire body rather than in only a 1.5 kg clump of wet, soft, gray and white matter in our skulls ([127], [128]). Psychological conditions are sustained by complex interactivity between the brain and bodily processes. Other findings show how mental and cognitive processes in the brain work in tandem with other type of cells (for example, immune cellular processing) operating across the entire body, challenging the idea that only the neuronal cells in the brain have the exclusive ability to learn or cognize [129].

The tendency to reduce all mental, emotional, and conscious states exclusively to the state of the brain is misleading. Most scientists tend to agree with this view, as long as generic metabolic functions are considered. However, not when they consider consciousness. It is posited a priori as an axiom that it *must* be generated somewhere and somehow in and by the brain.

Aside from phenomenal consciousness, there remain other aspects to be explained which escape the physicalist's paradigm with a strikingly similar pattern. One example is the neural correlates of memory. Also, in this case, one thing is sure: Memory is not stored in a specific brain area like it is on a digital computer. Nevertheless, more than a century of research for the *'engram cells'* — that is, the group of neurons supposedly responsible for the

---

[21] An analogy first suggested by French philosopher Henri Bergson.

physical representation of memory — has not led to tangible results providing convincing evidence that such cells really exist.

On the one hand, it appears that some brain regions are more involved in memory consolidation than others. For example, the *'hippocampus,'* a part of the limbic system, seems to play a significant role in memorization. In the amygdala, emotional memories play a role, such as the remembrance of fearful events. Motor events are stored in the cerebellum, while the prefrontal cortex is responsible for higher-cognitive memorization tasks, such as storing words and semantic content. Neurotransmitters must also be fundamental in memory formation because they are responsible for the communication among neurons determining its synaptic strength.

On the other hand, what cognitive neuroscience can observe with modern brain imaging technologies is that memory storage and retrieval appear to happen throughout different brain regions in a complicated manner depending on which kind of information (sensorial, experiential, conceptual, etc.) is processed and if it is encoded or retrieved in short-term memory or long-term memory. Overall, the real physical, cerebral memorization mechanism continues to remain elusive, and there is still no credible theory explaining where and how experiences are memorized at a neural level.

Again, one may argue that facts show how damaging the hippocampus leads to a short-memory loss, especially in object and facial recognition tasks, but this does not affect other types of memories. Moreover, the fact is that, as many have unfortunately experienced in their families or themselves, a brain stroke or dementia can lead to consistent memory loss. Therefore, everything seems to indicate that memory is in the brain. What else?

This logical memory-brain identification has similar difficulties in terms of standing up to scrutiny as the presumed consciousness-brain identity and is highly prone to becoming another correlation-causation fallacy. What makes us so certain that brain damage causing memory deficits is caused by physical impairment of the physical memory content? The only thing that we know is that a brain injury can lead to the impossibility of retrieving the old memory. A failure to retrieve a memory does not necessarily imply a memory loss. The inability to recall something does not imply the absence of it. Using an analogy: Damaging a computer's CPU will render it useless, but the memory in the hard drive is not necessarily affected.

Indeed, it is well known that when a brain injury occurs, causing some form of amnesia, what was thought to be lost forever may remerge into awareness, sometimes after years. Those whose loved ones suffered from dementia know well how memory and clarity of thought can suddenly and quite surprisingly reappear in a brief moment of lucidity, called *'paradoxical lucidity,'* or, even *'terminal lucidity.'* There can be bursts of mental clarity that sometimes occurs shortly before people die. Credible reports documented cases in which people with dementia, advanced Alzheimer's, schizophrenia, and even severe brain damage, suddenly return briefly to a normal cognitive state [130].

This suggests that the disease-specific permanent structural brain changes can't be associated with the repository of memory. Eventually, they could be neurological routes through which, in normal conditions, memories pass but are not the seat of the stored

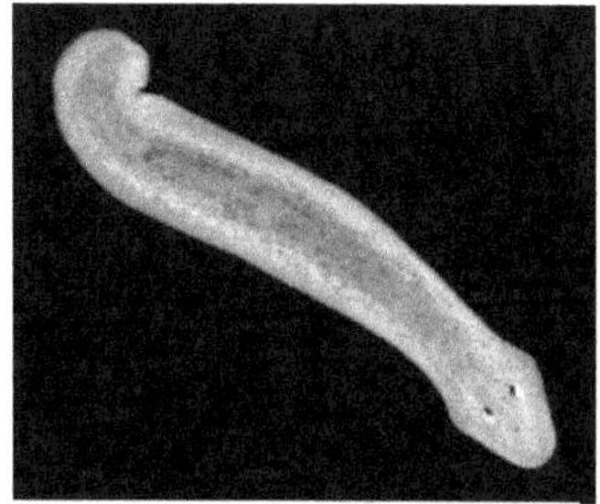

Fig. 41 A planarian flatworm.

information itself and can, ultimately, be bypassed to retrieve memories later by other means and mechanisms.

What makes things even trickier are the above-mentioned cases, especially those of hydrocephalic subjects who do not present substantial memory loss. This seems to be counterintuitive. If memory is not encoded in a specific brain region but overall through its neural networks, one would expect a relationship between the amount of cerebral tissue and its storage capacity. Even for an artificial information processing system based on complex neural network circuitry, the number of neurons determines (linearly or non-linearly) its storage capacity. From the materialist's perspective, the information content should somehow scale with the brain size. But nothing like this is observed in hydrocephalic individuals. In most cases, their cognitive skills and their mnemonic functions continue to work remarkably well and are comparable to those of healthy subjects. The same could be said for those patients who have undergone hemispherectomy. How can it be that someone without half of the entire brain has no memory loss? Obviously, we can, as usual, explain this away by resorting to the plasticity of the brain or the functions of residual brain tissues that might jump in to save the paradigm. Or, we could conjecture that memory is stored in both hemispheres; therefore, if one hemisphere goes lost, the other remains nevertheless unimpaired (a hypothesis which could also fit well with supposed evolutionary advantages). Or because it is always the diseased hemisphere that is removed in all these cases, Nature might have provided for a mechanism that transfers the memories to the healthy hemisphere before surgery. However, we should be aware that these are conjectures, hypotheses, and speculations, not scientifically established truths. Memory storage and retrieval in biological brains remains a largely unexplained mechanism, and no conclusive evidence exists that proves it to be of physical nature.

Other research that might suggest how and where memories are stored in brains comes from experiments performed on freshwater flatworms called *'planaria.'* These creatures can be trained to associate an electric shock with a flash of light — that is, they can learn by association that whenever they perceive a shock, a light flash will appear shortly thereafter. That they remember and associate the electric impulse with the light flash can be shown by the fact that once trained, they will curl their bodies whenever the light is flashed, even without the discharge. Therefore, it is reasonable to assume that they must have encoded the experience in their brains.

However, things are not as easy as that. Planarians have an incredible self-regeneration ability. If this worm is cut in half, each amputated body part regenerates as two new fully formed flatworms. Not only does the part with the head form a new tail, but the remaining tail also forms a new head with a brain and eyes. In 1959, James McConnel, a professor at The University of Michigan, showed that the newly-formed planaria with a new brain also maintained its conditioned behavior. The new-formed living being never received the electric shock and light flash of the training phase, and yet it reacted as if it still had a memory of the training it had never received. How could an initially headless worm acquire a memory that is supposed to be stored only in the brain of the other worm?

McConnel cut the worms many times more and observed that all the worms retained their memory and learned response. This suggests that memories, if physical, may not be stored only in the brain but throughout the body, in non-neuronal tissue. He went further by training worms, killing them, and then feeding them to other worms. Surprisingly, the cannibal worms that ate their siblings and that were not trained, when presented with the task of reacting to the flash of light, curled their bodies and did so even faster than the original worms did.

McConnel's findings suggested a memory transfer phenomenon. His idea was that RNA molecules (the cell's messengers that carry the protein-making information from the DNA) could transfer memory from one planarian to another as a "memory molecule." Motivated by this idea, he then injected worms with RNA taken from those trained and reported that the training had been transferred. This seemed to support the idea that memories are encoded in the RNA structure.

However, further research could not reproduce McConnel's experiments convincingly. Even though experiments using rats showed that memory transfer is possible, after a period of notoriety, McConnel's experiments were dismissed and forgotten by the scientific community.

Then, in 2013, Michael Levin and Tal Shomrat of Tufts University vindicated McConnel's first experiments by using a computerized training of planarians, replacing manual procedures that caused previous test attempts to fail [131]. In 2018, Alexis Bédécarrats from the group led by David Glanzman, of the Department of Integrative Biology and Physiology of the University of California, resurrected McConnel's idea, showing how the extracted RNA from a long-term trained sea slug, the *aplysia,* can induce sensitization in an untrained aplysia [132]. This is taken as evidence for the existence of engrams and the hypothesis that RNA-induced epigenetic changes (changes that switch the genes on or off but do not alter the DNA sequence) lead to the protein synthesis required to consolidate or inhibit memory. These local translations into synaptic proteins determining the neural structure of memory are actually the mainstream engram model.

Furthermore, researchers at Coleen Murphy's lab published their findings on worms sharing memories with others by swapping RNA [133]. In fact, Caenorhabditis elegans, a transparent roundworm, not only learns to stay clear of poisoning food but genetically embeds the threat of skanky meals into its kids to force them to stay clear as well. These worms absorb strands of RNA from their toxic meal through their intestines and cause specific genes to be switched on allowing it to remember and avoid a particular pathogen. This is passed on not only in their offspring but at least for the next four generations. The fact that a parent's physiology could be imprinted on their offspring was considered impossible and is another indication that RNA might play a role in memory consolidation.

However, the problem with this hypothesis is that the fastest protein synthesis causes cellular changes on timescales of minutes. This raises the question: How could it possibly be responsible for our ability to store and recall memories almost instantaneously?

Moreover, Glanzman's group challenged the idea of memory mapped as synaptic connectivity in the brain in previous research, in which they showed that it is possible to erase synaptic connections while maintaining the same conditioned behavior in the mollusks. Long-term memory and synaptic changes can, at least in some cases, be dissociated [134]. It has also been shown that the brain tissue turns over at a rate of 3–4% per day, which implies a complete renewal of the brain tissue proteins within 4–5 weeks [135]. How then can the memory supposedly internalized in that tissue remain intact?

Similar challenging evidence comes from hibernation. Since animal brains undergoing hibernation are subjected to severe changes in structure, one would expect equally severe memory losses. Nevertheless, while ground squirrels tend to forget their conditioned tasks after hibernation, they retain their social memory [136]. Memory is retained during hibernation in Alpine marmots [137]. If memory is encoded in neural networks, bats seem to benefit from an as yet unknown neuroprotective mechanism to prevent memory loss after hibernation of their brain [138]. Memory seems also to be immune from seasonal skull and brain size changes, the so called 'Dehnel's phenomenon' [139]. Memories formed in the earliest embryonic states of frogs survive extensive remodeling of their brains and bodies [140]. Most impressive is how moths can remember what they learned as a caterpillar (the moth larvae) despite a complete metamorphosis of its brain [141]. During the period of this metamorphosis from larvae to moth, most of its brain tissues is literally dissolved into a messy soup–that is, into their constituent proteins through a process called 'histolysis'–and later reconstructed into that of a moth. Yet memory remains unaffected (for an overview on the stability of memories during brain remodeling see [142]. How can animals whose brains have been so drastically remodelled still recall their past experiences?

Also the already mentioned representational drift ([84], [85]), which causes neuronal representations inside the brain to change with time, challenges classical notions of engrams. Classical models of memory consider the stability of the engram as the basis for the persistence of memory. If all mnemonic cerebral configurations change over a time scale of few days how can a memory persist over much longer periods?

Last but not least, the search for engrams resorts to the correlation between the memory evaluation based on fear conditioning behavioral tasks of rodents and its presumed associated neural changes. It turns out that almost all findings are vitiated by inadequate sample sizes, leading to questionable statistical correlations, which represent only weak evidence but then, contrary to the math, are presented as strong evidence for the engram cells hypothesis tested [143]). This, again, shows how, even among trained scientists, the correlation-causation fallacy is always lurking around the corner, especially when one isn't aware of one's own confirmation bias and needs to publish as much as possible to promote one's career, and how sensational hype in the media should be taken with care.

However it is, these findings are still controversial and must go a long way in terms of being reproduced and confirmed before they become an established scientific fact. The debate on the 'seat of memory' isn't new. It dates back to Henri Bergson opposition against a reductionist understanding of memory [144]. Bergson considered memory to be of an immaterial and spiritual nature, rather than being stored in the brain. But the search for the neural engram will continue, as what else could we look for? Most scientists believe that memory *must* be somehow encoded into neural states, as from the physicalist perspective, it can't be otherwise. The physicalists will never consider alternatives such as an 'extracorporeal information storage' (for a review of this point, see [145]) and never look for any evidence for it a priori.

The cases we studied so far, if singled out separately from the context, might still find a systematization inside the dominant paradigm. Some physicalist theory, in particular neuroplasticity or some contrived neural hypothesis, might still be a viable speculative option that could save the orthodox worldview. But, if we look at the mass of evidence in its entirety, as a whole, with a bird's-eye view that can apprehend them all together, they don't look very credible. While, a filter theory a la W. James, or a Mind at large theory a la A. Huxley, that see the brain as an instrument, rather than a 'generator', are a much more convincing hypothesis.

The fact that McConnel's findings were ignored for more than half a century should tell us something. Similarly, despite the potentially interesting implications for consciousness studies, the groundbreaking findings of Lorber regarding hydrocephalic patients in the 1980s or those of Shewmon regarding congenitally decorticate children around the turn of the millennium remained isolated initiatives that did not lead to further investigations. The reason why these studies and findings have met with little interest might be because they were focused on extreme pathologies, which are relatively rare and difficult to study and, therefore, represent only a small niche of the medical literature. Of course, neurologists, surgeons, and physicians are busy treating their patients and may not be interested in existential and philosophical questions such as the mind-body problem, the search for the 'seat of consciousness' or in the cannibalistic tendency of flatworms.

Nonetheless, I can't escape the impression that this doesn't explain the lack of interest in further pursuing these lines of research. This is demonstrated by the vast amount of literature which, instead, is almost exclusively focused on finding the neural correlates of consciousness. My feeling is that the real reason for this omission is not so much of a practical nature but is rooted in an unaware state of denial. Any finding that does not confirm the mechanistic bias is systematically ignored.

At any rate, we will not conduct a historical review of the rise and demise of one or the other theory or philosophical approach but, rather, will point out further interesting aspects which, while not being conclusive, are at least indicative and should be considered seriously for any sound statement for or against a theory and worldview. Some aspects aren't usually discussed in academics but are nonetheless well-established scientific facts relevant for a serious examination of the connections between matter, consciousness, and Nature. That consciousness and mind may precede the brain's functions, and even any nervous system, is suggested further by the recent discoveries in plant and cell biology. Let us inspect some of these findings in the following chapter.

3. The Cognizant Plant

It was once believed, and it is still the prevailing opinion, that plants are just multicellular organisms which we call 'alive' only because they are systems that are composed of 'living' cells, and that undergo metabolism, grow, and can reproduce, but we don't think of them as having any form of cognition or ability to learn or make decisions, let alone have a subjective conscious experience. Sooner or later, everyone has discovered that plants move, adapt to the environment, and grow and lean towards the Sun, but considering them 'intelligent,' or having a 'mind' and being 'conscious,' whatever that might mean, is something that most of us, especially those more scientifically minded, would consider a too farfetched idea. We can't allow ourselves to contemplate the eventuality that something which does not have a brain may nevertheless possess such attributes, even if only in a more or less primitive or involved state. It is, however, not at all a new idea and has always fascinated the collective imagination.

Already in 1867, the Italian botanist Federico Delfino concluded from his studies that denying intelligence to plants is *"a serious mistake, born of a superficial appreciation of the facts."* [146] In 1889, no less than Charles Darwin and his son studied the movements of plants [147] and compared the plant's roots to some sort of primitive brain, an idea that has become known as the *'root-brain hypothesis.'* However, Darwin's authority wasn't enough to pave the way for further investigations and, because the strictly mechanistic and molecular-based biology would soon have become the main trend, his insights were almost forgotten.

The Indian scientist Jagadish Chandra Bose, in 1926, wrote about *"The Nervous Mechanism of Plants"* and was the first to conduct experiments which led him to conjecture about similarities between the nervous systems of animals and the signaling paths in plants. Bose studied the responsive mechanisms of plants and reported how *"the excitatory polar action of an electric current and its transmission to a distance, proved that the conduction of the excitation of the plant is fundamentally the same as in the nerve of the animal."* [148] Bose openly spoke of 'plant-nerves' and the scientific community valued his discoveries, but the times were not ripe to allow him to go too far with the neurobiological analogy.

A brief popular interlude took place in 1973 when Peter Tomkins and Christopher Bird published their bestselling book *'The Secret Life of Plants'* [149] and evidently struck a chord in the popular conscience. Tomkins' and Bird's book, however, was more of a collection of speculative research sliding into the paranormal (such as plants able to read human minds and enjoying classical music) rather than a rigorous scientific account. Their claims were dismissed and nowadays are considered to be unproven 'New-Age pseudo-science'.

Nevertheless, about three decades later, science began to consider plant intelligence a serious topic of research. Speaking of plants that 'learn', 'process information', 'remember', or 'develop strategies' is no longer taboo. Mounting scientific evidence from plant biology demonstrates how vegetal life shows elements of intelligent behavior that were not suspected or just considered impossible.

At the turn of the millennium, terms like *'plant neurobiology'* appeared in the scientific literature very much in line with Bose's understanding, drawing parallels between the complex information processing and signaling system in plants with the animal's neuronal activity [150]. Because, as is well known, plants do not possess neurons, this implicitly self-contradictory terminology sparked some controversy. The articles of some scientists published in serious peer-reviewed journals suggesting that plants may have some sort of cognitive abilities soon met with resistance. While the dominant mechanistic materialism can no longer deny the experimental evidence, it does not consider these phenomena as true evidence for what is commonly meant by 'intelligence,' 'cognition,' or even less a 'mind' or 'consciousness' in plants. After all, one can imagine any behavior being nothing more than the result of a complex adaptive internal signaling process leading to mechanical non-cognitive reactions of a biological system. It is just instinctive, hardwired feedback to stimulation coming from the environment as a thermostat reacts to temperature. Therefore, so the story goes, we should not anthropomorphize plants. The assumption of intelligent behavior in vegetal life forms is an unnecessary hypothesis. A concerned appeal against the 'rationale of this concept,' calling for more 'intellectual rigor' that refrains from 'superficial analogies' and 'questionable extrapolations,' was signed by 33 scientists in 2007 [151].

However, setting aside debates about the nomenclature, it is quite clear that what prompted this concern was not motivated so much by a desire for scientific terminological rigor but, rather, the fact that speaking of plant neurobiology implicitly suggests the existence of an intelligent behavior where there are no neurons. Obviously, this flies in the face of the brain-centric belief system that there can't be any intelligence without a brain.

Fortunately, this did not prevent this new branch of science from growing in the following decade, with some also proposing a new 'philosophy of plant neurobiology.' [152] Let us study some examples of recent research in the field of plant intelligence.

One finding that met with considerable attention was the discovery that plants learn by association. Associative learning is the ability to understand that there is a relationship between two stimuli. A classic example dates back to the famous *'Pavlov's dog experiment'*. A dog was trained to associate the sound of a bell with the receipt of food. Even before the dog saw the food, it began to salivate once the bell rings.

A similar experiment was done with plants by a group led by senior research fellow in plant behavior at the University of Sydney, Monica Gagliano. In this case, the bell and food were replaced by a fan's airflow and blue light, respectively. The Australian group claims that garden pea seedlings (Pisum sativum) change their foraging behavior — that is, their direction of growth — if trained to associate a running fan with a light source shining an hour later after the fan's operation. First, the pea seedlings were trained by exposure to the fan and light. During this training session, the fan was switched on for 60 minutes before the light was switched on as well — that is, the plants were 'alerted' that light was coming in about an hour. Notoriously, plants tend to avoid wind but grow towards the light. How will they behave if, after the training session, in a testing session, they are presented with the fan stimulus without the light?

It turns out that if in the previous training session, they were presented with a fan + light stimulus in the same direction, in the testing session in which no light appears, most will nevertheless grow towards the fan, contrary to their innate instinct to avoid air currents. The experiment was a bit more complicated than that (the interested reader can read the article [153]) but the bottom line is that Gagliano alleges[22] that plants are capable of associative learning: They associated the fan and light stimuli and, once trained, 'learned' to 'predict' that the fan was announcing that a light source would have been turned on in a specific location. Like Pavlov's dog, which associated the bell with the arrival of food, plants also associate the fan with the appearance of light.

Another example that raises important questions about the predictive abilities of plants and how they perceive the environment was an experiment that analyzed the goal-directed movements of the same pea plants that Gagliano used. Research led by Silvia Guerra and Umberto Castiello of the University of Padua, Italy, showed that the climbing plant searching for a support to attach to exhibited an anticipatory prehensile mechanism and that it was able to plan its movements *before* having any physical contact with the support [154]. The stimulus was a wooden pole (see Fig. 42) which the plant could use as a support, grabbing it with the tip of its tendrils (the stem specialized for attaching to supports). The temporal evolution of the spatial trajectories of the two tendrils' tips (in blue and red) when there was no stimulus — that is, when the wooden pole was missing — is depicted in Fig. 43 (a). A limited oscillatory movement

Fig. 42 Climbing pea plant with wooden pole support. Credit: [154]

occurs, also called *'circumnutation'* (well known since the time of Darwin's observations), but there is no evidence for an 'intention' to grab for something.

In the case of the presence of a stimulus — that is, the wooden pole — this back and forth circumnutation substantially increases, as shown in Fig. 43 (b), and once the plant's tendrils come in the vicinity of the 3 cm thick wooden pole (but still before a tactile contact), they show purposeful anticipatory behavior by changes in the aperture of the tendrils preparing to grab the support. Just as we humans do with our hands when reaching and grabbing for an object, the pea plant reaches and grabs for the support first by a progressive

---

[22] We should point out that this experiment is contested. At the time of this writing one attempt to reproduce it failed [384].

opening, followed by a gradual closure until it matches the object to be grasped. Everything indicates the presence of motor planning and an 'approach-to-grasp' behavior. Also, the average speed of the tendrils was significantly higher in the presence of the stimulus.

The experimenters also tested how smart the plant was in detecting a 'fake stimulus' by replacing the real wooden pole with a 2D stimulus — that is, a photograph of it. Then, a similar behavior as that of the no-stimulus situation was observed, as shown in Fig. 43 (c).

This experiment shows the plant's ability to purposefully plan its behavior by grasping and opening its tendrils before having any physical contact with the support. It does not flutter randomly throughout the environment until it bumps into something on which to grab. The plant seems to literally 'see' the support! Otherwise, how could it coordinate such action in advance? The authors could only speculate.

A similar conclusion is suggested by another experiment that showed how the Boquila trifoliolata, a plant native to temperate forests of central and southern Chile and Argentina. is capable of flexible leaf mimicry [155]. It mimics the form of the leaves of a nearby artificial vine plant. Indeed, plants also possess photo-receptors called 'ocelli' which, however, do not form an image. What kind of mechanism stands behind this 'plant vision' remains unanswered.

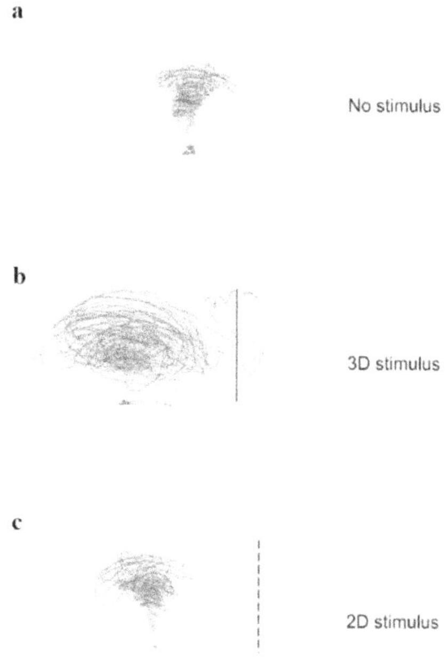

Fig. 43 Spatial trajectories of the plant. [154]

Another possibility is that plants use chemosensory perception or even acoustic perception. In fact, evidence is mounting that plants communicate with each other via these sensory means. For example, it is now recognized that the huge network of the roost of the trees in a forest builds a signaling and communication network. Mycorrhizal networks facilitate tree communication, learning, and memory. Mycorrhizal are almost microscopic fungi that enter a symbiotic relationship with the plant's roots. Mycorrhizal networks extend far beyond the roots of a single plant; they can link the roots of all the trees of a forest, similar to a neural network and the biochemical signals resembling those of neurotransmitters. These allow for a communication and recognition system among trees with cognitive qualities such as perception, learning, and memory. To the dismay of rationalist physicalism, some scientists have begun to speak of a 'forest intelligence' that *"may contribute to a more holistic approach to studying ecosystems and a greater human empathy and caring for the health of our forests."* [156]

Plants also respond to sounds. They are especially sensitive to auditory signals of about 200–300 Hz, which is likely due to the frequency range of running water. The roots of the plant grow in the direction of a water sound source. Moreover, a study suggests that when plants are stressed, they emit ultrasonic sounds, which could possibly be used as a communication system to conveys information to other plants or animals [157]. Sounds in the 20–150 kHz range were recorded when tomato or tobacco plants were under drought stress or by cutting the stem. About 30 ultrasound 'clicks' per minute were recorded. It was even possible to show that the different types of stress led to varying types of emitted sounds.

A possible explanation for the generation of these sounds is *'cavitation'* — that is, a mechanism whereby air bubbles form and explode in the *'xylem'* (vascular water and nutrient transport tissue which extends from roots to stems and leaves). It is conjectured that other plants or animals may use these sounds to inform themselves about the state of the plant and behave accordingly (for example, a bat could prey on caterpillars attacking a plant).

This is a completely new area of research that has yet to establish itself, but it is very suggestive of the fact that we might have missed something about the plant kingdom, which was considered silent until then. Nevertheless, it is interesting to note how this possibility did not entirely escape humans' understanding and might have been perceived subliminally (triggering subconscious fears) because this is also reminiscent of a myth according to which Mandrakes, also known as Mandragoras, are sentient and magical plants whose roots, whenever unearthed, scream and eventually can kill a person who hears it. A myth that also appeared in several Harry Potter movies.

These were only some examples of the recent findings. There is extensive literature now that, especially in the last decade, has consistently shown how plants change behavior and adapt, respond predictively, possess some form of memory, resort to air and underground communication systems based on chemical, visual, and acoustic signals, have learning abilities and can evaluate their surroundings, make decisions, and even have a social life and cooperate (for a not-too-long review, see [158].)

This is what people mean by 'plant intelligence' that should, of course, not be confused with animal or human intelligence. But it is not inappropriate to speak openly of a minimal or 'proto-cognition' of plants, what scientists call a *'basal cognition'* (for a modern review of this elusive concept see [159].)

Anyway, this is no longer popular romanticism or pseudo-scientific woo but, rather, facts reported by well-established empiric findings published in peer-reviewed scientific journals.

Nevertheless, despite all this evidence, the fierce resistance to any notion of plant 'intelligence' continues unabated. In 2019, another group of scientists declared that "Plants Neither Possess nor Require Consciousness." [160] They drew this conclusion from a comparative evolutionary study that surveyed brain anatomy and functional complexity during the evolution of consciousness in simple and complex animal brains. Therefrom, the core statement is that consciousness emerges as a result of many special and diverse neurobiological features which are numerous and amazingly complex, meaning that a specific level of organizational complexity of the brain is required for subjective experience. We shouldn't speak of plant intelligence but, rather, of complex adaptive mechanisms to the environment, and we shouldn't anthropomorphize plants. But this doesn't tell us anything as regards the explanatory gap in terms of why phenomenal consciousness exists in the first place, confuses a no-progress quest for a progressive-quest, and of course, posits (more or less subconsciously) in the premises the conclusion which, after some contrived argumentation, obviously comes to the desired proclamation: Because consciousness requires a brain, there can't be any form of consciousness in plants, no matter how much evidence there is to the contrary.

As an off topic, but contextually relevant case, it is also worth mentioning a case of cognition without the brain, not in plants, but in jellyfish. This tiny creature is no bigger than a few centimeters and has a nervous system, but lacks of a brain. Scientists trained Caribbean box jellyfish to dodge obstacles, showcasing the animal's capability for associative learning. This discovery challenges the belief that complex learning requires a centralized brain and raises questions about the evolutionary origin of learning. [161]

The question, then, is: Where is the threshold between adaptive behavior and an intelligent one? If one does not specify any boundary, then we could also downplay every human behavior as 'adaptive.' After all, we can see every of our own human actions and thoughts as a series of 'adaptations' to gain some advantage, satisfy some desire, or avoid displeasure. Of course, we shouldn't anthropomorphize animal and plant behavior, but this argument can be reversed: Refusing to assign sentience to other forms of life is an anthropocentric tendency as well.

For example, we tend to associate plants with a totally unconscious organism because plants live in an entirely different time dimension that our human temporal experience can't grasp. Only with time-lapse video recording techniques does the plants' movement appear in all its detail and subtleties. It is in this time dimension so alien to us that the plants no longer appear to be inert and unconscious beings. If we had not been conditioned by our human-centric point of view and perceived the plants' behavior in this different time dimension, with all of its associative and predictive learning skills, its cognitive abilities, and its social, competitive, or collaborative behavior, we would become much more open to the idea of plant cognition. But, as usual, in our everyday life, as in science, we tend to see reality from our anthropocentric point of view. We tend to label everything that does not express our human waking state of consciousness as unconscious. Equally, we consider plants as unconscious only because they don't live in our own temporal and cognitive dimension.

One might say that we should not compare these plant abilities literally to human or animal behavior. In fact, we are speaking of relatively basic cognitive functions as compared to human cognition. The plant's cognition is only an elementary form of 'intelligence.' However, the decisive point is that facts show how plants display complex behaviors and decision-making abilities that were previously thought to be impossible for an organism without a brain. A growing body of literature makes it increasingly difficult to maintain the orthodox belief system that would like to reduce organisms without a central nervous system to biochemical machines dominated solely by purely mechanistic reflexive reactions. Meanwhile, it is no longer implausible to conjecture that plants might have some sort of primitive sentience and rudimentary cognitive processes. Mind, or at least a simple or 'basal' form of mind, might be fundamental, inherent in life, an intrinsic property of matter, and that could exist also independently from brains, nervous systems, and neurons.

It is interesting to note how popular intuition preceded science. The wide popular acceptance of the idea of plants having a more or less developed 'inner life' was an intuition which people could not rationalize but relied upon a feeling and ancient wisdom that turned out to be, at least to some degree, a correct 'precognition.' Later, we will argue that this is not just a historical and sociological freak in which archaic superstitions and science coincidentally correspond; to the contrary, it is the manifestation of an inner connectedness of the collective with subliminal dimensions that transcend mind and matter and which, on the surface, though at times confusingly and irrationally, manifest concretely in trends, thoughts, and eventually fashions that nevertheless, deep down, reflect an inner perception of natural truth.

4. The Mind of the Cells - Part I

But how far can we go in search of the origin of cognitive behaviors in living organisms? If we can't identify the whereabouts of mind in the brain or the 'seat of consciousness' in some particular neural correlate and instead find elementary forms of cognition in plants that have no brain at all, can we go even further and look for some form of cognizant behavior in single cells?

Though this is also not a new question, it was not in the minds of most biologists and cognitive neuroscientists until recently. It was at about the turn of the millennium when a renewed interest in this field gained momentum, especially due to an increasing amount of evidence that is slowly but steadily transforming our understanding of how mentality emerges in living organisms and even questions the very notion of 'mind' itself.

The question for which some scientists have pursued an answer is: Can the Pavlovian conditioned behavior that is observed in plants (and obviously in animals and humans as well) also be observed in single cells? After all, one might think that plant intelligence could emerge as the result of a complex interaction of the plant cells, as the neural network in a brain does (which is why people speak of 'plant neurobiology' even though plants don't possess neurons). From this perspective, though purely speculative, one might still save the monist brain-mind identity: Instead of claiming that mind and consciousness are nothing other than an epiphenomenon of the brain, one might refine this hypothesis by considering that mind, consciousness, complex mental tasks, and cognition are emergent phenomena from a complex web of interactions of an aggregate of cells, though these cells must not necessarily be neurons. However, if a conditioned behavior can be demonstrated in a single cell, this extrapolation would not hold. Indeed, it turns out that this is the case. Several experiments with unicellular creatures have made it clear that conditioned behavior in single cells exists and is comparable in its complexity to that of plants.

An example of an interesting finding in this sense could be the evidence of conditioned behavior in amoebae. A Spanish group analyzed the motility pattern of the *'Amoeba proteus'* under the influence of the two stimuli and consistent with associative conditioned behavior [162].

The Amoeba proteus is a particular species of unicellular *'eukaryotic'* organism — that is, a *'protozoan'* (single-celled eukaryotes, unlike prokaryotes, are cells having a nucleus within a membrane) which can change its shape and move by extending and retracting its *'pseudopods'* (the arm-like projection of the cell's membrane — see Fig. 44 bottom-left.)

This amoeba is an interesting experimental subject because it exhibits *'galvanotaxis'* — that is, it perceives the presence of an electric field and moves towards the cathode (the negatively charged electric pole), which makes it amenable to stimuli with standard electronic devices. On the other hand, the A. proteus also exhibits *'chemotaxis,'* which means that it moves in the direction of the presence of a chemical stimulus, a *'chemoattractant'* (here an nFMLP peptide), more commonly known as 'food.' Fig. 44 shows the movement of the amoebae in the experimental chamber in different conditions (the '-' symbol indicates the cathode, the '+' symbol the anode, and 'p' the 'peptide').

Fig. 44a shows the locomotion of amoebas without any stimulus. As expected, their movement is almost random; no preferred direction is visible. Fig. 44b shows the dynamics of the cells under galvanotaxis — that is, under the influence of the electric field — with the cathode on the right. Clearly, a preferential movement is visible. Almost all amoebas move towards the negatively charged electric pole.

Fig. 44c illustrates the case for chemotaxis only where the peptide is placed on the left. This time, there is no electric stimulus. Then, though with a less pronounced tendency than in the galvanotaxic case (a minority seems to be less attracted), most of the amoebas clearly show the tendency to move towards their food in the left direction. Next came a thirty-minute induction phase process. In this case, the unicellular organisms are subjected to both of the previous stimuli simultaneously and are left free to choose their preference, namely, if they prefer the chemical nourishment or the 'electric food.' Fig. 44d shows that an almost equal amount of amoebas opted for one or the other possibility, with a slight majority of 53% for the peptide.

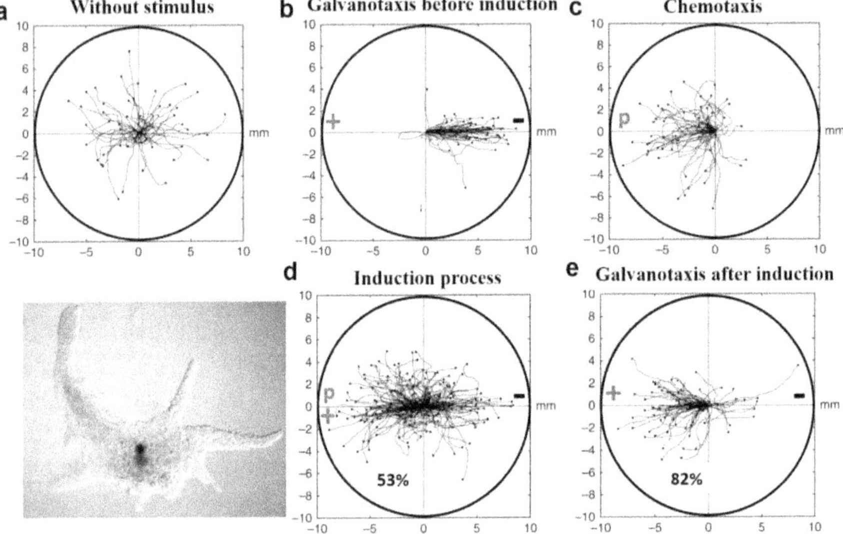

Fig. 44 Evidence of conditioned behavior in amoebae. Credit: [162]

Finally, those amoebas that went towards the peptide during the induction process were manually extracted and, after five minutes of isolation with no stimuli, were placed into another galvanotaxis phase. As Fig. 44e shows (compare this with Fig. 44b), despite the presence of an electric field that should attract them towards the cathode, they nevertheless move towards the anode.

Everything indicates that they remembered the peptide's location and therefore decided to follow the electric field in the direction opposite what they followed previously in the induction process. Not all amoebas seemed to have been conditioned, as about 18% returned to the old behavior (or recognized the trick). However, once they felt the presence of the electric field, most of them (about 82%) nevertheless moved towards the anode, evidently associating it with the presence of food, as they remembered from the conditioning phase of Fig. 44d which, however, in this case, is absent.

This suggests that single cells can also associate stimuli. Here, it was the anode's positive electric polarity with the presence of food, as in the pea plant the fan with the light or with Pavlov's dog the bell announcing the arrival of food. This shows how simple unicellular organisms can link past events and act according to what they remember. It is a modification of behavior against an innate tendency. To put it bluntly: The amoebas 'changed their minds.'

Another quite surprising behavior was (re-)discovered recently in another protozoan. In 1906, the American zoologist Herbert Spencer Jennings noted how the *'Stentor roeselii'* could escalate actions to avoid an irritant stimulus. Stentor roeselii is a free-living protozoan, a trumpet-shaped ciliate species — with hair-like organelles called *'cilia'* which allow it to swim and sweep food into its 'mouth.' Jennings'

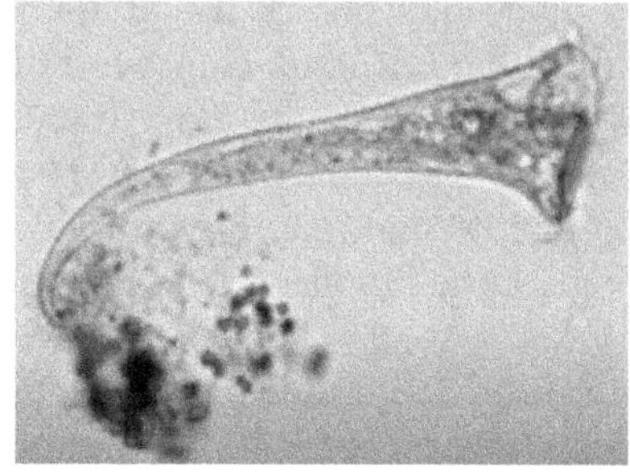

Fig. 45 *Microscopy image of a stentor roeselii.*

observations are illustrated in Fig. 46. The S. roeselii bends away its body if exposed to an irritant powder. If the irritant stimulus persists, its next strategy is to put its cilia in motion in an attempt to expel the powder particles. If this doesn't help, it contracts down to its holdfast (a root-like structure anchoring it to the substrate). Finally, if still nothing helps, it detaches from the holdfast and swims away.

This is not just a mechanical and reactive behavior. It is a complex hierarchy of avoidance behaviors in which the protozoan first enacts a strategy, sees if it works, and if not, resorts to another strategy in a series of attempts to solve an irritating problem. It is a learning and decision-making process that people could hardly believe to be possible for a single cell.

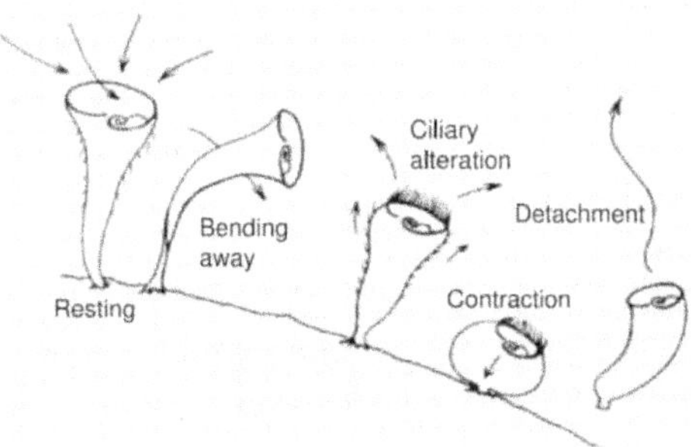

Fig. 46 *Hierarchy of responses to repeated stimulations. Credit: [166]*

Jennings expressed this as follows: *"The writer is thoroughly convinced, after long study of the behavior of this organism, that if Amoeba were a large animal, so as to come within the everyday experience of human beings, its behavior would at once call forth the attribution to it of states of pleasure and pain, of hunger, desire, and the like, on precisely the same basis as we attribute these things to the dog."* [163, p. 334]

However, science wasn't ready, and Jennings' observations were ignored until 1967 when someone tried to replicate it, which was unsuccessful because another species was used that did not display the same behavior as S. roeselii. For the dominant reductionist and physicalist mindset, that was enough. It is pointless to search for a cognitive behavior in single cells: The dogma was that only in an organism with a brain or at least a nervous system made of a neural network can such a cognitive behavior arise. Conjecturing that cells might have some sort of elementary mental skills is an unnecessary hypothesis, as Occam neatly told us. When we do not believe something to be possible, we usually don't search for it. Then we tell ourselves that there is no evidence for it without realizing that nobody even seriously looked for that evidence (for a short historical account, see [164]). In fact, only recently historical research brought to the light that the evidence for learning in single cells was already there, buried in the archives, ignored and forgotten by conventional academia [165].

Fortunately, times are changing, as are people's minds. One hundred and thirteen years later, in 2019, Jennings' observations were finally confirmed by a group at Harvard Medical School [166]. Indeed, this unicellular organism can modulate its behavior about how to respond to the environment in an escalation of actions that, to date, represent the most complex behavior known for a single (brainless) cell.

A less elegant (and slimy) but quite remarkable example one could present about cellular cognition is that of the abilities of the *'Physarum polycephalum,'* a large amoeba-like slime mold *'plasmodium'* — a fungal cytoplasm containing several nuclei but enclosed in a single membrane — that can be considered as a single giant cell. It is a creature reminiscent of

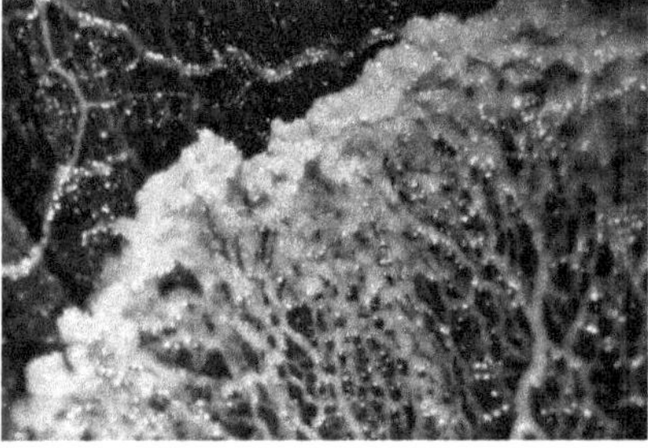

Fig. 47 The Physarum polycephalum slime mold.

the 'The Blob,' the alien amoeboid being in the popular sci-fi and horror film in 1958. It changes its shape as it crawls in search of food as a yellow network of tube-like structures that grow a few centimeters per hour and whose movement can be captured via time-lapse recordings.

This slime mold has several skills and behavioral patterns that could be labeled as 'proto-intelligent' and that one would hardly associate with such a primitive creature. For example, it can find the minimum length between two points in a labyrinth. When P. polycephalum is grown in a maze with nutrients at two spots, it first invades the entire labyrinth until it finds its food. Then, it retracts its pseudopods, leaving only those corresponding to the shortest path connecting the two food sources intact (see Fig. 48). [167]

Further research showed that P. polycephalum could minimize the network path and complexity between multiple food sources [168]. Conditioned behavior was shown as well. When this plasmodium is exposed to a sequence of three drier and cooler life-threatening conditions at constant time intervals, it reduces its speed of growth or stops entirely for a while before starting to grow again after each pulse. Once conditioned, it learns to anticipate the arrival of the second and third shocks even if only the first pulse is administered and the other ones are not. This shows that this slime mold can memorize and learn to anticipate periodic events [169].

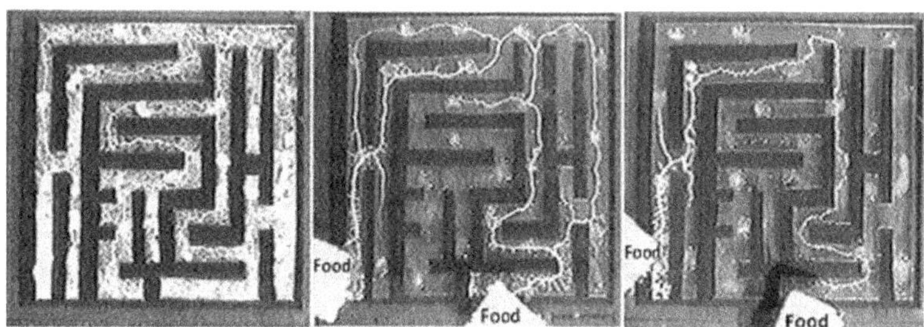

Fig. 48 P. polycephalum finds the shortest path in the maze connecting two agar spots.
Credit: [167]

P. polycephalum is also able to adjust to unfavorable circumstances [170]. If one forces it to cross an agar bridge with caffeine or quinine at toxic but not lethal concentrations, it first slows down. However, after repeated attempts, it nevertheless crosses the bridge at the same speed in the absence of the irritant substances. It was believed that this was something only neural networks could do — learning to ignore repeated negative stimuli. That is, a learning process of habituation took place.

Slime molds also remember food location, as they have a memory about the environment. Research indicates that this memory is encoded in the tube's diameters of its network-like body [171]. The slime mold's tubes grow and shrink in diameter in response to the nutrient's location.

These brainless organisms not only possess remarkable associative memory and problem-solving skills but also exhibit some form of self-awareness. When it encounters a structure, such as a wall it has to go around, the two arms of the slime mold branch out around the wall; once they come together and touch one another, one of the arms retracts. This suggests that the slime mold knows itself to be the same unity — that is, it is self-aware of itself as a single organism (for an in-depth review on slime molds see also [172] .)

Basal cognitive skills beyond neural tissues have been shown also in fungi. This went unnoticed until recently because we tend to reduce these rather complex creatures to the surface appearance of the familiar mushroom and consider them to be part of the plant kingdom. However, fungi make part of a separate kingdom that includes also yeasts and molds, are genetically more closely related to animals than to plants and, most importantly, express their real life with the growth of their *'mycelium'* and microscopic interconnected network of *'hyphae'* that are only partially visible to the naked eye. The mycelium is the vegetative part of a fungus bacterial colony, forming a mass of branching hyphae–that is, its long branching filamentous structure. When fungi are observed as a whole in time lapse

Fig. 49 Example of fungal mycelium.

and under cognitively challenging conditions it turns out that mycelial networks possess memory and engage in relocation decision-making, such as responding to quantity and location of new resources, retain a memory of growth direction, express a sensitivity to a changing environment, respond to restrictions in their physical space becoming narrower and branching less frequently, alter their developmental patterns in response to interactions with other organisms and are capable of spatial recognition and learning coupled with a short-term memory ([173], [174], [175]). Forest floor-dwelling fungi can send one another electrical signals to form word-like clusters, according to another research [176]. Whether that represents something akin to language isn't clear. Anyway, clear is that fungi exhibit behavioral and at least some sort of primitive communication skills.

But, after all, are these complicated experiments really necessary to convince us that these unicellular or multicellular beings possess some form of cognition? When we look at an amoeba, a Stentor roeselii, or a paramecium under a microscope or plants in a time-lapse video, what do we see? Dead automatons behaving mechanically? If one looks carefully, one can see that all the behaviors described above were already manifest and evident in the microscopic or time-lapse videos.[23] Is a tiny unicellular creature that swims through a fluid, hunts for its prey, avoids obstacles, has a memory, and can even predict events in advance just a simple machine to which we shouldn't ascribe at least some form of primitive cognizance, a basal cognition, and eventually even some elementary form of sentience? Is a climbing plant that nervously flutters its tendrils throughout space, analyzes the environment, grows towards a support it apparently 'sees,' and begins to grab for it before even touching it only a machine driven by a chemical reaction?

It might be worth mentioning, in this regard, the case of *'warnwoiid dinoflagellates,'* a unicellular organism that has an eye-like *'ocelloid'* consisting of subcellular analogs to a cornea, lens, iris, and retina. It can point its ocelloid in different directions, and it probably uses it to detect the prey or predator, or the ocelloid served an evolutionary purpose in the past. This is not to say that this cell literally 'sees' as we do. In fact, the lens does not produce

---

[23] Online, one can find several videos illustrating this, such as [365].

a real image on the retina; it only concentrates incoming light to increase sensitivity, which allows it to 'see' only flashes of light. Nevertheless, the ocelloid consists of subcellular analogues to a cornea, lens, iris, and retina [177]. However, to 'see' these flashes, it does not need a brain, and we can't escape the suggestive feeling that the preconceived idea of single-cellular organisms as being mere automatons that do not know, feel, or see anything might be misplaced. We don't know what it is like to be a dinoflagellate and if there is something that experiences anything at all. However, the doubt that these tiny creatures are more than a complicated photodiode and may possess some form of primitive and elementary phenomenal consciousness is legitimate.

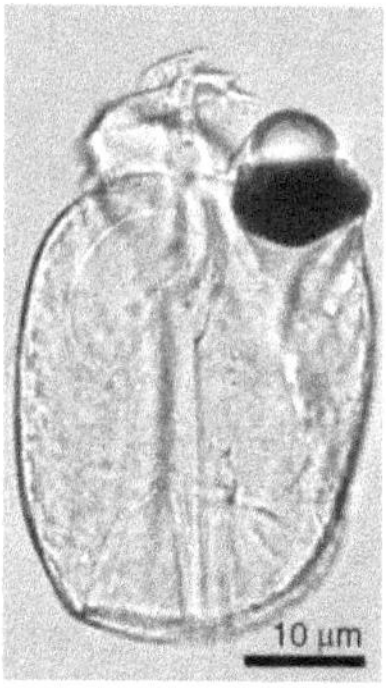

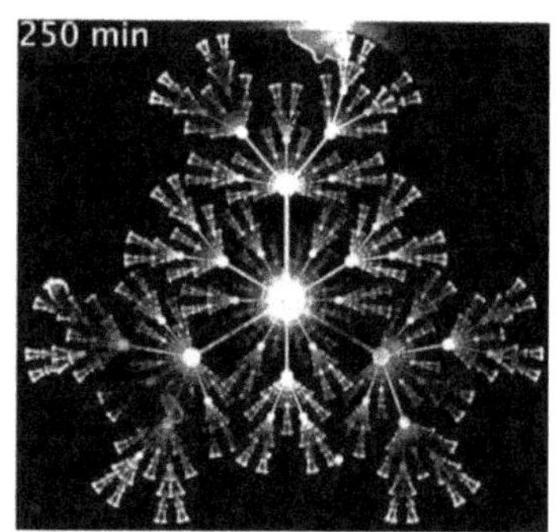

Fig. 50 The warnwoiid dinoflagellate with its forward looking ocelloid. Credit: [391]

What about the simplest life form, namely, bacteria? Bacteria are considered the most elementary form of life because they are prokaryotic cells — that is, without a nucleus but with the DNA freely located within the cytoplasm — and are thought to have arisen before eukaryotes. Nevertheless, it has been shown that they can sense the environment, actively move within it, target food, avoid toxic substances, and meaningfully change their swimming direction. Most astonishing is their behavior when they come together and form a bacterial community which shows surprising problem-solving

Fig. 51 Bacteria navigating inside a tree-like fractals maze. Credit [178]

abilities. Bacteria communicate with each other and coordinate gene expression, which determines the collective behavior of the entire community. This leads to a collaboration that allows the community to achieve a common goal. An example of such a complex bacterial community we all know well is the 'biofilm' plaque on our teeth. Together with viruses and fungi, bacteria are part of the gut microbiome and are associated with the bacterial communities in soil mycorrhizal fungi, as already mentioned in the case of tree communication. They are found everywhere and can be beneficial, as in fermentation or decomposition, or harmful, as in infections.

These communities elaborate functional structures to determine if the microbes nearby are enemies and eventually secrete antibiotic compounds toxic to other species except their own, anchor in a place or stay motile, divide for the growth of the community, release spores (resistant structures used for survival under unfavorable conditions), etc. This allows them to work together to survive in stressful environments. Analogous to P. polycephalum, bacteria's collective intelligence becomes evident when they are confronted with relatively complex task-solving problems such as route-finding in mazes and fractals. A group of biologists from Princeton University, led by Trung V. Phan, created a maze on a silicon chip flooded with food chemicals and with no dead ends, where wrong paths merged into other wrong paths, which led to infinite loops. They observed if and how the bacteria successfully navigated the labyrinth by chemotaxis [178]. The same experiment was performed with tree-like fractals. Chemosensitive motile Escherichia coli bacteria were placed at the beginning of the maze (or center of the fractal) and allowed to explore the entire maze (or fractal). They collectively escaped from both of these topologies by eating through and reproducing themselves. At first, this might not sound surprising; they might have just randomly expanded until they found their way out. The point, however, is that they can do this in times much shorter than the random walk for which their growth speed would allow. First, they got trapped in the dead ends, but then they built up in clumps and collectively launched in waves out of it. This is a wave behavior that appears to be the result of inter-bacterial communication informing the collectivity which way not to explore anymore. Everything indicates that bacteria developed collaborative problem-solving abilities to evade the maze.

These were only a few examples of cognitive cellular behavior. Other examples could be cells sensing a shortcut using self-generated gradient and selecting a new minimal route (see [179] or, for a review of bacteria's behavior, see also [180]). Further studies showed that the emergence of conditioned behavior is not confined in some individual type of cells. In recent experiments 2000 individual cells belonging to three different species were shown to be able to acquire new migratory patterns modifying their response to a specific stimulus by associative conditioning [181].

Everything indicates that cognition is a fundamental property of the living realm that predates neural networks.

Can we dig even deeper into the cell structure and still recognize some cognitive and/or collective behavior? It seems to be the case, indeed.

Studies of proto-cognitive capacities demonstrate that even subcellular biochemical dynamics—such as gene regulatory circuits and protein pathways—can exhibit various forms of memory that remain stable over long periods and are resilient to noise. In fact, in many instances, noise can actually enhance memory potential. However, no specific network property reliably indicates memory, suggesting that this phenomenon cannot be fully explained by network dynamics alone [182].

Moreover, it turns out that organelles–that is, the subunits within the cell itself–also display collective and cooperative cognitive behaviors: Mitochondria appear to communicate with one another and act in coordinated networks. Mitochondria–the 'powerhouses' of the cells–are the organelles responsible for generating biochemical energy to power their internal reactions. Like ants in a colony, they collaborate, form groups, divide up tasks and synchronize their activities. Mitochondria communicate with signaling networks within and between cells. They even merge and subsequently split again, sharing chemical signals, proteins and a genetic code. They process information individually

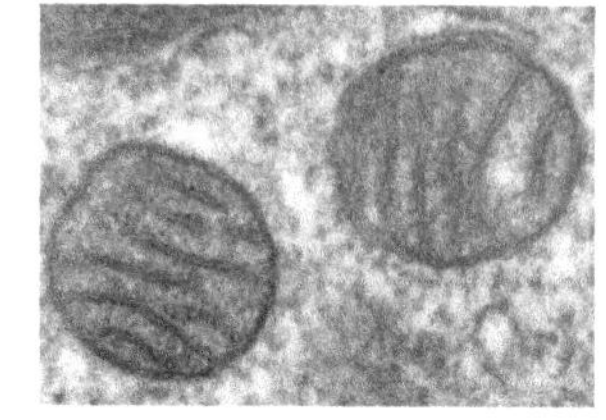

Fig. 52 Mitochondria with their matrix and membranes as shown by electron microscopy.

but then operate jointly. Their collective behavior, interconnectivity and cooperativity led some to speak openly of mitochondria as being 'social organelles' that form 'social networks' [183].

It is hard to escape a feeling of 'déjà vu' reminiscent of the collective behavior and cognitive skills observed only in organisms with a brain. Just an innate anthropomorphizing tendency of humans? Of course, one can resort to the eternal progress-quest attitude, claiming that it is only a matter of time and everything will be systemized inside a (probably monstrously complicated) mechanistic model.

Nonetheless, it is a fact that among living beings without a brain, an intimation of collective intelligence arises when several units or 'nodes' — that is, unicellular organisms — connect and form an informational network. Once several of these nodes signal to each other in a complex connectome, a new order and level of functional skills arise that the single-cell doesn't have. The single-cell already seems to have some elementary cognition, but something new and qualitatively different emerges once these little entities connect in a complicated communication web.

This higher-level organization cannot be analyzed and understood in terms of a system described as the sum of its parts. It is the mutual connection, information transfer, and interaction among the single units that define the properties of the entire web. It can't be reduced to a simple sum of its parts because it is also the dynamical interaction of its units that decides what kind of skills, functions, and qualitative properties it has (for a review of cellular intelligence see [184]; another review with the author's emphasis on the drawbacks of reductionism, see [185].)

The most striking feature is that there seems to be something reminiscent of the neurons in our brains: The single neuron can do almost nothing by itself; only by intercommunication with thousands of other neurons can it form an incredible perceptual and cognitive machine. Precisely this coincidence should make us wonder if the brain isn't just one among the several cellular networks that allow cognition? One doesn't necessarily need a network of neurons to support some form of cognition, thought, or consciousness. Our assumption that only brains — that is, a network of neurons — can cognize, think, and perceive, while other forms of connectomes or aggregations of different types of cells can't, is an unwarranted and unsubstantiated inference. If, instead, we suppose that there is a cellular 'proto-mind,' a basal cognition, however elementary and primitive, connecting with other cellular 'proto-minds,' forming a higher-level mind, this can furnish us with a more complete, interesting, and effective perspective on the neuroscience of mind and consciousness. Further findings on the learning of slime molds not mentioned here suggest that different information-processing can be realized on different architectures, not just in brains [186].

Scientists now begin to suspect that all living cells are cognitive [187]. Understanding cellular computation is becoming a central goal of research on cognitive evolution. A synthesis of cell biology and neuroscience in computational and metabolic terms could lead to new deep insights and set the stage for a paradigm shift [188]. We might even go so far as to speculate that cognition does not 'emerge' at all from a cellular network; rather, cognition is fundamental and takes the physical substrates as a means of expression–cells as physical support for an elementary but immaterial mind (sort of a 'monad' a la Leibniz — see Pt.II-II.2k), not the origin of mind itself, and which, once properly organized, form a physical hierarchical structure that can support the action of a higher-functional immaterial mind, but is not its causal origin. Science is not there yet, but the idea of cognition as something being elemental and that transcends biology is emerging in the literature (see [189], and also my paper [190].)

However it is, neurosciences, molecular biology, and genetics have allowed us to gain enormous insight into the cell's physiology and genetics but have failed in furnishing us with a consistent theory about the emergence of mental abilities not only of organisms but, to a large degree, about the cognitive behaviors of the single-cell itself. A purely reductionist approach that explains the whole as an intricate system of interacting tiny entities was supposed to lead us to a deeper (eventually ultimate) understanding of the cell's workings. Instead, it turns out that the more we dig into the cell structure, the more its overall functioning becomes complex and increasingly difficult to overlook.

Most people vastly underestimate the complexity of a cell. Sometimes one also hears biologists comparing the complexity of a cell to that of an airplane. Many professional molecular biologists recurringly show an inclination to oversimplify the complexity of cells and organisms. This shows a typical weakness of the linear-thinking analytic human mind which always tends to play down the complexity of things of which it has no grasp. You can be a very knowledgeable scientist and still have no idea how complex some of the phenomena or natural structures that you are studying are. The complexity of a cell is far beyond that of an airplane. Even on the most basic molecular level, things are much more complex than that.

For example, we still don't understand how the amino acid sequence specified in a gene folds into the three-dimensional functional protein shapes. For decades, scientists have been trying to figure out the protein folding mechanism with very little success (more on this later). And airplanes aren't able to make an exact copy of themselves, as they don't reproduce. The mechanism of cell division remains one of the most complicated biological processes. We know that there are particular reproduction stages, called 'mitosis,' but there is no understanding of why and how these phases occur. Contrary to popular belief nowadays, we know that the reproduction mechanisms cannot be reduced to a simple DNA instruction set. DNA does not contain any instructions for making cells or any specification or blueprint of a cell. Apart from a tiny fragment of the overall picture, and despite all the hype and exaggeration that come from the research in the field, we have no understanding of the molecular factors that cause a cell containing millions of ribosomes and thousands of mitochondria to undergo reproduction. It is an incredibly complex process that we are far from having elucidated. Cells are shown to have a high degree of chronological, structural, and sequential coordination, which allows them to assemble their parts and make them fit together in a proper arrangement. They can remodel and repair themselves. We are far from having any decent understanding of how they manage to do that. The fantastic complexity of multicellular life is millions of times more complicated than that. The tendency to conceptualize the living cell as an intricate piece of machinery that can be explained reductionistically and deterministically, different to a man-made machine only in terms of its superior complexity, is no longer tenable. The recent introduction of novel experimental techniques capable of tracking individual molecules within cells in real time reveals the cell's dynamic, self-organizing nature of its constitution, the fluidity and plasticity of its components, and the stochasticity and non-linearity of its underlying processes [191]. Networks of cell signaling are complicated enough, and this still can't explain how biological organisms form. Stem cells differentiate into various tissues by an enormously complex chemical and mechanical signaling process which remains largely unknown. The scientific reason is a too-low-level form of cognition to comprehend the complexity of things and which it must instinctively downgrade to simplistic toy models. The microscopic and molecular analysis of the structure and dynamics of a tiny thing like a living cell has revealed a lot of insights to us but, at the same time, made it clear how naïve our belief was that soon the mystery of life and its workings would be disclosed. We are nowhere near having even a superficial understanding of what is really going on inside a cell.

5. Neurocentrism, Hypes and the Illusion of Knowledge

We are now assisting in a sort of renaissance of experiments that suggest how many intelligent actions can be performed without neuronal activity. Of course, one can interpret all these behaviors according to one's own preferences and belief systems. Humans can condition and blind themselves to go so far as to deny, even other humans, the right to be free or to live because they have 'no souls' or are considered an 'Untermensch' (a 'subhuman'). Therefore, it is unsurprising that we are not willing to grant any consciousness, sentience, and dignity to any other life form if we struggle to allow it even to ourselves. The materialist's typical objection is that, as scientists, we should maintain an emotionless and unemphatic objective attitude refraining from anthropomorphizing plants or cells. However, with this overly cautious attitude, one ends up replacing anthropomorphism with anthropocentrism. This almost complete and pervasive attitude of pretending that cognition, consciousness, sentience, feelings, and emotions are exclusively human traits or, at best, something we should ascribe only to organisms with the prefrontal cortex, denying it to other living creatures, is a disguised form of anthropocentrism as well. Scientific anthropocentrism is subtle nowadays and dressed up more elegantly than in the past but is alive and well and expresses itself precisely by accusing others of anthropomorphism. This is not something new; it already happened in the history of science towards non-human primates. As American anthropologist, Barbara J. King observed: *"For much of the 20th century, it was common practice for ethologists to resist acknowledging the profound emotions expressed by these animals* [monkeys and apes]. *Anthropologists and zoologists who deviated from tradition and described animal grief and other emotions such as joy were accused of anthropomorphism, the projection of human capabilities onto other species. However, the tides began to turn as ever-more research in the field and in captivity showed unmistakable evidence of animals deeply feeling what happens to them. More than ever before, researchers now recognize that grief and love are not emotions experienced only by us humans."* [192]

Moreover, notice how there seems to be a common trait in many of these experiments. Several authors involved in plant and cell intelligence recognized how many of these findings were already made decades ago but have been ignored for a long time and have been re-discovered only recently. This shows that what prevented the discovery of these facts was not a technological limit but mostly due to a preconceived ideological mindset that excluded a priori that cognition could exist without the substrate of a neural network. If you don't believe that something can exist, you won't look for it and, therefore, you won't find it. Even if you find it, you might well be blind to it and continue to point at the supposed lack of evidence that confirms your bias. This is the main reason why some fundamental discoveries about the cell's basal cognition have only recently resurfaced from the pages of history books and gained recognition in the 21st century. The last century biology (as many other sciences) fell so in love with its exclusively reductionistic, molecular, and genetic point of view that it did not realize how it was slowing down further progress in other fundamental domains. The idea of a single-cell being capable of performing complex behaviors such as learning associatively or solving problems was considered at best a curiosity that sooner or later had to be recast into that very same reductionist paradigm. Science may have lost a century of progress in the field of plant and cell cognition. Finally, it seems that times are changing. Hopefully.

To sum up, we saw how science searched for the seat of consciousness or the mind and found no reliable locus, brain region, or neural correlate that could neither shed light on its whereabouts nor tell us how it comes into being. We saw how surprisingly plastic the brain must be to recover its cognitive and mental functions despite extreme damage, surgical intervention, or deformation and wondered if the relatively intact mind and consciousness of these altered brains could be explained away only by the magic wand of neuronal plasticity. Paradoxically, we found that plants — organisms without neurons — show behaviors that display an elementary associative intelligence, a form of 'proto-awareness.' Plants have no neurons. Yet, they are capable of actions that we previously thought were possible only in an organism with a central nervous system. We went further and discovered that even single-cells must possess some basal cognition. This is a property that non-neuronal cells, such as bacteria, reinforced by creating collaborative networks. All this looks as if the mind, at least an elementary form of mind, is already present in a 'pre-neuronal' form. The bottom line is that the deeper we dig into the microscopic biological realm, the more we find mental signatures still present and active.

At this point, the question we would like to ponder first is: Given the above findings and new insights, is the dominant physicalist science in its present format, which conceives of mind and consciousness only as an epiphenomenon of the brain — that is, a science based on an exclusively neuronal paradigm — still tenable? Is the 'you are your brain' dogma still a 'fact' we shouldn't doubt?

We want to maintain a conservative attitude and refrain from replying with a too-self-assured negative answer. We consider the evidence from experiments with plants and cells that refutes mental phenomena as the sole result of neuronal activity to be compelling. However, we also understand that these findings may be legitimately considered as not completely conclusive and require further research. After all, the fact remains that any higher-level cognitive function is always associated with neural activity.

However, we believe that it is time to question the exclusively reductionist approach that cognitive neuroscience has taken, especially in the last decades. There is no doubt that neuroscience, especially in the last three or four decades, has made enormous progress. From medicine, neurology, neuropsychology, and cognitive sciences to computational modeling or AI research, these have had a — and may have an even more — dramatic impact on our knowledge in the coming decades. Further technological advances in neuroimaging will undoubtedly continue to shed light on the brain's activity. The importance and usefulness of these neuroscientific advancements, especially as diagnostic tools, is undeniable. Due to this success, neurosciences experienced a revival that was to be expected and is partially justifiable. We ourselves made ample reference to neuroscientific findings.[24] There is a takeaway from the advances in this important field, which tells us something about how we perceive the world or, more precisely, how we construct the world.

However, rather than resorting to an ever-increasing magnification of the details and failing to notice what the whole is about, we must finally recognize its limit. One point in question is what these empirical data are supposed to explain, suggest, or imply from an ontological standpoint. The uncritical acceptance of the (more or less unaware and metaphysical) neuronal premise, particularly that of the relationship between neural activity and mind or consciousness being an undeniably given datum, should be questioned.

This 'neurocentric' attitude is not only ideological; it also has its roots in the political and financial structure of the academic and research system. Nowadays, neuroscientific achievements must be overemphasized because a research project on something 'neuronal' has more chances to attain funding than projects that do not. These ample financial concessions are, however, not due to past scientific findings. The willingness of governments, foundations, and society to finance costly brain research projects based on a purely neuro-reductionist paradigm is not triggered by previous results which, in most cases, did not go much further than correlational evidence between what happens in the brain and what one subjectively reports feeling, seeing, thinking, and perceiving. The neurosciences did not transform traditional psychology and therapeutic approaches as was expected and didn't disclose the mysteries of human consciousness that could close the explanatory gap. In many respects, especially when it comes to potential applications for neurodegenerative diseases, such as Alzheimer's or Parkinson's, neuroscience did not meet the expectations, either. Technologies such as MRI, fMRI, MEG, EEG, PET, and other non-invasive scanning techniques are valuable diagnostic tools and have given us important insights into the anatomy, physiology, and function of the brain and other organs, allowing for the study of memory, learning, drug effects, and diseases such as schizophrenia, autism, or clinical depression, or can be vital for preparing surgical interventions. However, despite 40 years of hopes and promises, these did not lead to tangible progress in the treatment of mental disorders or other practical applications. The failure of a European

---

[24] The author's animal rights bent struggled with this. I would love to see less animal experimentation and more alternative forms of investigation.

mammoth project such as the Human Brain Project only further testified to this. We are waiting to see how a similar US-based '*BRAIN initiative*' that foresees "*fundamental new discoveries about the brain*" by 2025 [193], or the Chinese '*China Brain Project*' and the Japanese '*Brain/MINDS*' mapping program, are going to develop. They may gain some insights into the information processing of the human brain, but any claim that they will allow us to manufacture drugs to treat neurodegenerative diseases, let alone contribute to explaining and reproducing conscious and mental functions in any form, is unfounded.

Nowadays, we hear people talking about artificial neural networks implemented on supercomputers that simulate millions of neurons. What we are rarely told is that an artificial neuron is an extremely simplified version of a real biological neuron. The simulated units are sum-up-inputs and spit-out-spike boxes. To simulate a single neuron, one would need a far more complex functional object. Strictly speaking, a single neuron must be simulated by a neural network itself. All the billions of neurons in our brains are not just neural networks. The brain is a neural network of neural networks (for a more in-depth analysis of this, see [194], [195]). Making things even worse is the fact that the connection between neurons, the dendritic arms of some human neurons, can perform logic operations that require whole neural networks [196]. This means that a realistic simulation of a biological neural network should be simulated by tens of billions of neural network of neural networks, each connected to tens of thousands of other neural networks. Even if the most optimistic predictions in the progress of supercomputer technology for the next hundred years came true, we would come nowhere near to such computing power.

The effort and enthusiastic drive that stands behind this 'neuromania' is neither scientific nor based on research findings or justified by its potential pragmatic applications. It is based mostly on a hegemonic ideological and cultural belief system anchored in the physicalist, reductionist, and material monist ideology, also not rarely backed by financial interests and policies that are prone to the same collective suggestion. However, that does not prevent society from continuing to support and fund the same rational and scientific reductionist and neuro-centric paradigm. For example, according to the American psychiatrist Allen Frances, former chairman of the Department of Psychiatry at Duke University School of Medicine in North Carolina and member of the task force that produced the fourth edition of the Diagnostic and Statistical Manual of Mental Disorders, or DSM-IV, since 1990, "*the National Institute of Mental Health (NIMH) suddenly and radically switched course, embarking on what it tellingly named the 'Decade of the Brain.' Ever since, the NIMH has increasingly narrowed its focus almost exclusively to brain biology — leaving out everything else that makes us human, both in sickness and in health. Having largely lost interest in the plight of real people, the*

NIMH could now more accurately be renamed the 'National Institute of Brain Research.'" [197] While, as of 2022, a review showed that *"...despite three decades of intense neuroimaging research, we still lack a neurobiological account for ANY psychiatric condition"*(emphasis mine) [198].

This state of affairs is also reflected in popular media and the collective culture. If something is explained in neuronal terms or brain areas subjected to some bio-physical reaction such as the cause of an ailment or disorder, or supposedly explaining anything about our human nature, it is much more likely to be accepted by the popular audience as final scientific proof. The prefix 'neuro' is second only to that of 'quantum.' Everything that is labeled with some 'neuro'-specification is instinctively considered to be a source of ultimate scientific truth that explains everything, obviously from a purely reductionist and physicalist perspective. While in physics, the media outlets' hyperbolic announcements contribute to a vast 'quantum woo' movement, we fail to notice how much 'neuro woo' is pervading it as well. A title that slaps the prefix 'neuro,' 'genetic,' or 'molecular' onto a finding or account of things creates a sense of increased explanatory power and 'truth-value,' which almost subconsciously lures us into the illusion which transforms a correlation into a causal claim.

Moreover, it has been shown that overestimates of effect size and low reproducibility of results in neuroscience are much more common than assumed [199]. Because of a *'publish or perish'* policy which is nowadays the almost exclusive directive and evaluation parameter for academic career success, researchers must publish as soon as possible in order to succeed, which leads to a hasty overestimation of their own assumptions and claims with statistically non-significant results reflecting effects that are only apparently true. This is not specific to the neurosciences but generally to medical research. Linguistic spin — that is, reporting strategies aimed at convincing the reader that the beneficial effects and efficacy of the experimental interventions are greater than those shown by the results — was found in more than half of the abstracts of medical peer-reviewed papers published in scientific journals [200]. On top of that, overstatements or even misinformation proliferate on internet-based health news, using causal language and suggesting a strength of inference in academic and media articles where there is none [201]. It turns out that 58% of media articles had inaccurately reported the results of the academic studies.

However impressive and thrilling the techniques of neuroimaging are, when it comes to the question of the mind-body problem and the nature and origin of consciousness, especially to the question of why neural firings become phenomenal experiences, no progress at all has been made. The findings of the last decades didn't yield any understanding, and the brute force identity between mind and matter is not supported by facts but has always been posited as an a priori metaphysical premise.

The beautiful images of the brain tell us something about the neural activity but nothing about the mind in itself. The fMRI images have now reached a resolution of about a few millimeters, which is indeed an impressive technological achievement, but this pixelated volume (a so-called *'voxel'* — that is, a cubic volume of a few mm^3) still contains millions of neurons and billions of synapses and doesn't tell us how these neurons and synapses are doing their job. It is like reducing the map of a city with a million inhabitants into a pixel and trying to deduce from its flickering the social behavior of its population. These increasingly precise mappings do not directly measure the neural electric activity. Rather, they are pictures of the fluctuations in the metabolic changes of the neurons involved in a task. MRI or fMRI imaging determined real progress in enabling more accurate neurosurgery, the mapping of internal injuries, or the detection of tumors. These images, however, did not provide any insight into what mental functions do, how they do it, or what they are. For a specific motor-sensory action or cognitive task, it is great to see how a specific brain area 'lights up.' However, this, per se, does not tell us much about how these tasks are achieved.

fMRI brain scans are also vitiated by a low signal-to-noise ratio, meaning that they display a signal affected by high noise levels and only 1% change from the baseline. Picking the statistical threshold, which determines what counts as a significant signal, can be a matter of preference. Each scan produces enormous amounts of data, which processing aimed at sorting out signal from noise is left to the discretion of the individual researcher who is already stressed by the pressure to publish as much as possible, and potentially leading to misapplied statistics. In fact, it was shown that several studies misidentified random noise as signals where there isn't really one, correlating brain activation with emotion, personality, and social cognition measures, portraying implausibly high correlations [202]. Even in a dead salmon's brain, neuroimaging voxels can exhibit activity if not corrected for multiple comparisons [203]. But repeating MRI experiments is expensive and time-intensive. This is a reason more that forces researchers to play around with analysis parameters until they see a signal where there is none.

Recent findings have shown that things might be even worse than that.

When 70 independent teams were asked to analyze identical brain images, no two teams chose the same approach, and their conclusions were highly variable [204]. Human error and subjective methodological decisions lead to serious issues with scientific reproducibility.

Moreover, the supposed correlation between brain activity and specific tasks identified by fMRI brain scan imaging turns out to be much less reliable than previously thought. A US research group from the Department of Psychology and Neuroscience at Duke University has demonstrated that common task-fMRI measures are unlikely to indicate the real locus of brain activity associated with some specific cognitive task [205]. It turns out that the regional brain activity for a given task, such as reading a text, observing an object, or performing some specific motoric task, not only changes from subject to subject but changes over time in the same subject performing it. If one performs an fMRI of the brain of someone who is, say, reading aloud a meaningful series of sentences, the fMRI images taken only a couple of hours apart while the person is performing exactly the same task will look completely different from each other.

Moreover, it turns out that neuroimaging studies are massively affected by replication failures and are hyped by inflating small effects with dubious statistical significance. [206] Reproducibility requires samples from thousands of individuals. Something that is usually much too expensive to do. In other words, what you read about brain-scan research (beyond its diagnosing purposes) is much too often nonsense. When all things are considered, the frustrating verdict is that *"We can count on less than a hand the number of these studies that have held up under scrutiny and are really driving treatment,"* as postdoctoral fellow at Harvard Medical B. Tervo-Clemmens said. [207]

The bottom line is that brain scanning images perform very poorly in terms of identifying the centers of brain activity responsible for a particular cognitive process. Despite all the extraordinary progress of neuroscience, there remains an enormous gap dividing us from connecting a clump of neurons acting like a flaming marvel of electrical discharges to the comprehension of how and why they are supposed to forge a mental and emotional life inside a subjective phenomenal experience.

At any rate, even if we could find a way out from these technical oversights, cognitive neuroscience is mainly about the relationship between our higher cognitive functions (such as attention, memory, response inhibition, etc.) and the structural characterization, localization, activation, and temporal changes of specific brain areas seemingly responsible for it. It tells us how some regions of the circuitry underlying the performance of some executive functions can be described by neuronal short- or long-range connectivity and changes, local patterns of activation, or changes in anatomical structures in the brain, eventually including other genetic, physical, physiological, and developmental factors. Nonetheless, all these fantastically complex phenomena still don't tell us much about how and what the brain is really doing and, ultimately, what or who we are.

In hindsight, that shouldn't come as a surprise. It is like knowing everything about the electronics of a computer and being able to map all the internal electrical current flows in the CPU, the memory, and inside all the circuitry in its chips down to the microscopic level — but that will not tell you much about what the computer is doing. Trying to reconstruct a complicated AI compiled software running on a computer only by observing where the current flows in its integrated circuits and seeing which group of transistors and logic gates flicker up is an almost impossible task. While, recreating the exact physical structure of a computer would not lead us to a machine making anything meaningful or useful. The software–that is, a running code written by an intelligent external agent– is needed. Here also a computer is only an instrument, a means of expression for a cognitive entity, not its origin or source. In fact, studying a microprocessor with the same criteria employed by modern neuroscience, trying to reverse-engineer its functions, would fail: We would not be able to explain how it works, let alone reveal anything about the running code, and which is the real 'agent' that causes the behavior of the machine [208].

The lesson is that knowing everything about the hardware won't tell you much about the software running on it. Whatever functionalists might tell us, even if we were able to trace back our 'mental software' from the brain's extremely complex activity, that would furnish us with only a functional description of what is going on. Why should a sequence of internal physical states, however complicated, complex, intricate, convoluted, or multilayer-networked, give rise to a subjective experience? If you love the technical jargon so in fashion nowadays, which has only a seemingly explanatory power and adds nothing to a real understanding, choose labels such as 'recursive feedback loops,' 'self-modeling,' 'resonances,' 'information integration,' 'synchrony,' 'phase-locking,' 're-entrant circuits,' 'self-referential processes', or whatever. That didn't lead to any new insight regarding the question of how phenomenal consciousness and its qualia come into being, either. Anything that loops could be a vessel, but not a producer. The heart 'loops' blood throughout the body, but doesn't produce it. The reason for the blood to go back to the heart has nothing to do with the 'production' of blood. There is no rationale in believing things to be different with the 'production' of consciousness.

With the gigantic progress of microscopy, we could build microscopes that can almost see molecules reacting inside a living cell. We could look into the new world of molecular biology and make astonishing discoveries. And yet, by going deeper into the elementary structures of our biological existence, we found how things became increasingly complex. It is a complexity that grew exponentially with the degree of magnification and of which we frequently have no understanding, let alone control. The reductionist assumption that everything can be explained by a bottom-up approach that dissects and maps everything in the minutest details has delivered a false sense of 'knowledge' which is not, in fact, knowledge but, rather, a superficial representation at the level of raw data–only the tip of an iceberg that we misinterpret for the whole. Knowing everything down to the neuronal, cellular, or even molecular level of a brain function or the molecular description of the physiology of a cell no doubt tells us something. But it is only a partial understanding which becomes an illusionary knowledge when it is mistaken for the whole of knowledge. It delivers a false sense of assuredness based on the assumption that once a territory is mapped, it is also known. The idea that knowing the map is knowledge is an illusion hard to eradicate. It is like analyzing every molecule and elementary particle of the printer's ink on a piece of paper without realizing it to be Shakespeare's sonnets.

That's why the assumption that dominated especially the 1980-90s, according to which an increasing knowledge of the neural correlates of phenomenal consciousness would have explained it, elicited a sort of 'gold rush' that revealed itself to be delusional. In this sense, neuroscience did not progress much further than where Descartes was, other than telling us that the mind is not in the pineal gland. The rebuttal of dualism, which replaces it with a mind-brain monism, has no evidence whatsoever supporting it. The fact that world-known high-ranking hardcore physicalists claim otherwise, stating this to be something self-evident that no longer needs any explanation, doesn't make it true. The mind-body problem and the hard problem of consciousness remain a controversial issue more than ever. On the contrary, instead of leading us nearer to the solution, modern neuroscience only deepened the mystery.

This is also the reason why some scientists resort to extreme attempts to save, in my opinion, an untenable position proclaiming the inexistence of phenomenal experience in the first place as a meaningless concept and wrong construct and going so far as to deny their own experience and phenomenal consciousness as a 'category mistake.' There is a modern movement in the philosophy of mind that clings, at any cost, to the functionalist or physicalist paradigm, the so-called *'eliminative materialism'* or *'illusionism'* or just *'eliminativism.'*[25] According to this school of thought, consciousness is 'eliminated' from the outset as an illusionary representation of the brain. Only matter, the laws of physics, and the material brain are real. The impressions of having qualitative, subjective experiences such as feelings, sensations, perceptions of pleasure, or pain — that is, qualia — do not really exist. These are only 'false beliefs', a 'wrong representation' of our brains, just illusions.

But even if we assume that consciousness is only an illusion, the obvious and natural counterargument is that illusions are also a phenomenal experience. There is still someone or something having such an illusion. A hallucination might be a complete misrepresentation of reality, but that doesn't make the illusionary experience disappear or less experiential.

You will then be told that this is a misinterpretation of eliminativism. This theory states that phenomenal consciousness is a 'disposition.' Consciousness

---

[25] For reasons of personal mental hygiene, I won't furnish a precise account and literature of eliminativism. The interested reader can search online names like Daniel Dennett, Paul and Patricia Churchland, Susan Blackmore. The following is just a personal summary inspired by an online debate over an article by Keith Frankish [369], and an article by Michael S. A. Graziano [385].

is just a 'disposition that puts you in contact with the world.' It is an 'introspective mechanism that misrepresents the patterns of brain activity as phenomenal properties.' It is the self-monitoring of the brain that leads to a 'pseudo-conscious' experience, but these experiences are misportrayals of real brain states, a sort of self-delusional fiction of the brain. Therefore, so goes the eliminativist narrative, we are conscious, but we should reject a certain qualitative conception of what consciousness is. Qualia seem to exist but don't really. Consciousness is a complex of informational and reactive self-modeling processes of computational information conservation which we erroneously interpret as an introspective awareness and misrepresent as a simple quality, say, the pure feel of yellowness. It is a belief of having an intangible experience we call 'consciousness', and coming from an inaccurate information model in the brain that depicts us having it. To have the illusion of qualia is to be inclined to believe that one has them. In essence eliminativism tells us that our subjective experience does not exist, it is just a delusion, it is only a false 'self-representation', we are just a bunch of matter that thinks to think, but doesn't.

But why should a misrepresentation or a false belief translate into a phenomenal property? Why should a 'misrepresented informational reactive process' translate into a subjective experience, or the 'inclination to believe in something' lead to a phenomenally conscious perception? However illusionary and misrepresenting the process might be, the hard problem of consciousness remains a mystery. This line of reasoning is unconvincing. It amounts at saying that we don't exist, our existence is only due to a false belief.

With all due respect, the author's feeling is that these are only pseudo-philosophical reasonings that resort to obscure and nonsensical babble. That wouldn't be worrisome if these theories were not taken very seriously by many academics of high-ranking universities and if they were not considered the state-of-the-art of the modern philosophy of mind. It is the easiest shortcut: If you can't explain something that doesn't fit into your ideological belief system, instead of reconsidering your metaphysical premises, you simply deny the existence of the thing that must be explained in the first place. This is what we see with climate change denialism and conspiracy theorists. Eliminative materialism is intellectually no better than a flat Earth theory of consciousness. It is the last desperate attempt of the new Ptolemaic who don't know themselves and try to save appearances and a crumbling anthropomorphic paradigm at any cost. It is to expect that the attempt of physicalism to explain consciousness will go into the history of science like logical positivism in philosophy, or behaviorism in psychology, or like alchemy tried to transmute base metals into gold. To date, these are still the mainstream approaches that attract grant money in ways that the more serious theories cannot. One can only wonder how deep the contemporary philosophy of mind has fallen and it even prides itself for that.

That might, however, lead to an unexpected positive outcome. An increasing number of scientists are realizing how the attempt to answer the questions about mind and consciousness from an exclusive standpoint is leading to untenable, sometimes even ridiculous, lines of thought. Many are now becoming aware of how the vast amount of data and functional insights on brain activity, interpreted only through the purely reductionist and physicalist neuroscientific glasses, delivered a false feeling of certainty and confidence: an illusion of knowledge. Becoming aware of an illusion is the first step to getting rid of it.

Part II
The Higher-Mind Seeing and a First Step towards a Post-Material Science

I.   Philosophical Idealism

The enigma of consciousness remains an unsolved issue. Nobody can claim to have a definite answer that can be applied as a final resolution to all the aspects and questions that this elusive 'thing' we call 'consciousness' posits. However, after having clarified what the mind-body problem is about in the light of the modern findings and its experimental and observational evidence, we can fairly say that the simplistic physicalist point of view is less and less tenable.

So far, the aim and function of what has been discussed were to soften our deeply ingrained but unaware and naïve form of realism. The neuroscientific insights and the first-person perspective on how our perceptions, mind, and consciousness work help us become aware that, contrary to our instinctive idea, what we see, perceive, and think of the world is not a one-to-one correspondence with reality.

We have learned that our senses can deceive us, that our minds build up fictions, that the very notions of mind and consciousness are not at all interchangeable, that consciousness has something intrinsically mysterious with no seat and origin and that mind, or at least some elementary cognitive functions we could label as 'mental', seem to be intrinsic in Nature without requiring a brain. This laid the groundwork that will now allow us to deconstruct naïve scientific realism, seeing the functions and powers of reason as well as its flaws, and let us take a step beyond in search of a new paradigm, a new working hypothesis, and point of view.

In the following, we will show how this alternative has its roots in an ancient and well-known philosophical current that humanity nurtures and revives throughout the centuries: *'philosophical idealism'*. It had only partial and, at times, alternate fortunes, but the very significant aspect is that it never died despite many attempts to falsify and dismiss it. When this happens, we must seriously consider whether this might not be a sign, an invitation to think further and look beyond the present premises, because the apparent philosophical immortality of idealism may be due to the fact that it captures some essential core truth which, despite all the repeated attempts to ignore it, we will sooner or later have to take more seriously and confront.

The takeaway message of the idealistic philosophy is that we should never forget how we are beings subjected to the illusion of the senses and the mind. While the Enlightenment, which intellectuals also like to call the Age of Reason, especially with its outgrowth in modern science, recognized that we must take care of the illusion of the senses, it failed to recognize the illusion of the mind and reason itself.

What we mean precisely by that is the topic of the following chapters. However, we will do so by taking a different and more pedagogical approach than what one finds in conventional expositions. We will defer the description of the philosophical thoughts of the classic philosophers such as Plato or Plotinus and the alternative theories that opposed the materialistic realism of the Enlightenment, like in Berkeley or Schopenhauer, to a later discussion. At times, these are difficult to understand, and sometimes their thoughts are encapsulated into a cryptic language. Instead, we would like to take a somewhat easier route first, a sort of self-made conceptual tour, a more modern version, even at the cost of a less rigorous philosophical language and conceptual framework. This, in turn, can help and guide us in understanding the writings of idealism later and, hopefully, be a more comfortable and accessible entry to the deep questions it raises.

Before tackling philosophical idealism as a worldview with some of its historical and conceptual foundations, we would like to approach it by pondering our notion of 'reality' based on psychological and scientific facts we rarely ponder on. The author believes this to be the most suggestive introduction to idealism reformulated in modern language. We will then review idealism in its main concepts. This will only be a brief and admittedly superficial self-made introduction, still without referring to the great minds from antiquity to modernity whom we will mention later on in Pt.II-II. Finally, we will review and discuss the findings of the last century in the field of quantum physics in the context of idealism. Its strange paradoxes and counterintuitive reality may afterward appear less weird and unnatural than they seem to be when we give up a reductionist, deterministic, and local form of realism and replace it with the idealistic standpoint. Of course, this won't solve the paradoxes of quantum physics and lead us to a new theory of everything, but it can help us to later bridge the enormous gap that separates mind from matter in a more natural form.

1. A First Step into Idealism

From the previous discussions in Pt.I, we may slowly become aware of where the root of the problem lies with consciousness and what the logical error is of conceiving consciousness as an emergent property of matter. There must be something wrong with the materialistic idea of consciousness in the sense that we are unconsciously trying to explain something (consciousness) with something else (matter), where the latter, as we conceive it, is already somehow itself indirectly the product *in us* of that which it wants to explain.

Yes, it sounds contrived, and it is. But precisely for this reason, we have such a hard time capturing the problem and seeing the fallacy of physicalism. The question that physicalists must answer is: How is the emergence of consciousness from matter possible if the notion of matter is already an 'object of meaning' in the first place? We know matter only as an experiential phenomenon but then posit it to be devoid of any subjective and experiential content. The unaware assumption on which science works, and the idea that empiricism is objective knowledge of the world, is logically circular. Mental knowledge is an interpretation of sensory inputs. The very notion of 'matter', that conceptual entity physicalism posits as fundamental and from which it wants to derive everything, makes no exception: Matter is a 'condensation of meaning' as well.

We are not used to considering the solid stuff that we call 'matter' as a figment of our conscious perception because we have been accustomed since birth to see the world just in that way and never give it further thought. We live in a world based on the fallacy arising due to the unaware separation between the perceived object and the subject's juxtapositions. However, there is only a sensorial experience at the bottom and a perception of meaning already far removed from the experienced and mentalized object itself. Science extends this perception of meaning to measurement devices or abstract mathematical concepts but, ultimately, works with the same fallacy of our everyday cognition. It is based on what we called the *'sense-mind'*—that is, a mind that categorizes, organizes, and conceptualizes sensory perceptions in forms, numbers, and symbols and with which we *'re-present'* things but that can't *present* us with the inherent nature of things. We will learn how this fallacy was already known in its various forms in the time of Plato, but that nevertheless remains the foundational bedrock of science and analytic philosophy today.

When we speak of 'perceptions,' we must intend this in a broader sense. We have not only subjective perceptions through our five sensory means (seeing, touching, smelling, hearing, and tasting) and of emotional states (joy, sadness, fear love, anger, etc.) but also of objects of mental cognition (meanings, names, forms, numbers, etc.). Ideas,

Mental perceptions
(Names, forms, numbers, symbols, meanings, etc.)

Emotional perceptions
(Joy, sadness, excitement, fear, love, anger, etc.)

Physical perceptions
(Sight, touch, sound, smell and taste, etc.)

Fig. 53 Also, thoughts are subjective phenomenal experiences in us.

mental images, or abstract notions are also objects of perception. The only difference is that we are dealing with phenomenal mental content instead of sensory stimuli coming from the physical senses. This recognition resists acceptance if we look at our experiences from the scientific third-person perspective but becomes obvious if we take the first-person perspective. It is not too difficult to become aware of the fact that ideas are subjective perceptions as well. The appearance of thought and its relative semantic perception is, of course, of a very different kind than a qualia such as a color or the sensation of pleasure and pain, but it is fundamentally not so dissimilar in its nature. We have the sensation of the appearance of semantic content in our field of conscious awareness. What unites all these perceptions is their subjective experiential reality.

In a certain sense, our experience of the world is as in the case of the duck/rabbit Gestalt figure. We have seen that our representation of the 'outer world' is many-sided. However, these representations are in us, not out of us. There is no reason and no necessity to believe that when I look, say, at a table, a different cognitive process is at work, as in the case of looking at a Gestalt figure. When I say, "This is a table," I'm making a statement that, from the ontological point of view, is as real or unreal and correct or incorrect as claiming, while looking at the figure of the duck/rabbit, "There is a duck" and not a rabbit. The same can be said about any notion of matter in terms of particles, waves, or solid objects with a given form. These objects of the material world are not inherently existent in themselves, just as the figure of the rabbit is not more inherently existent as the figure of the duck. Everything depends on the 'switch of meaning' that we choose to turn on or off.

We must conclude that the material realm and every inner or outer realm we perceive must be manifold, has not a single and unique aspect, appearance, property, and feature that we are accustomed to thinking of. It has, in itself, no inherent features, properties, and qualities at all in its intrinsic nature. All these are juxtapositions and labels that come into our field of conscious awareness as a subsequent stage of perception.

We will never sufficiently emphasize that what we are hinting at is not some form of solipsism or empty nihilism. We still have no reason to believe that what we see does not exist in some sense. After all, it is tremendously hard to think that everything we perceive, be it mental, physical, or emotional, should be simply nonexistent. If I look at a real rabbit, I cannot 'switch' it to a duck or eliminate it from existence. But recall what the solution to the Molyneux problem was: The congenitally blind who, after treatment, could see again, yet didn't see a thing. The pure observation without any subjective and mental juxtaposition, which is so cherished by science, does not allow you to see things. The latter is an internal cognitive process that arises in us at a later stage and must be learned. There is somehow a realm that exists and lends itself to some of the many possible interpretations; it exists 'out there' but not in the sense that it possesses the inherent properties that we construct with our minds. Resorting again to the Gestalt metaphor, we could say that neither the duck nor the rabbit are 'real' and 'exist' in itself, but rather represent a conceptual construct that emerges in us long after the perceptual experience, still the figure on the piece of paper is 'real' and 'exists'. Something that produces these impressions in us is 'out there' in some form or another.

The very notion of what it means for something to 'exist' must be reanalyzed again from scratch. The notion of something 'existing' because it has some specific properties and features that we can see, perceive, measure, and calculate and that we bind together in a meaningful semantic object of cognition appearing in our conscious awareness is no longer tenable. We must leave behind a naive scientific realism that rests on the assumption according to which only that which is independent of our observations is real. Such a reality is an impossibility and a logical contradiction in the first place. At this point, the only thing we can say is that we live in a world of perceptions, a perceptual field of sensory events, an incredible web of mental contents arising in us and that we call 'the universe' which appears inevitably and always to our consciousness and through our constructs. Science is not 'objective' in the sense that it describes the material world 'as it really is', independent from our subjective perceptions and mental representations, as it claims to be able to do. Also, the thing we call 'matter'—that is, all material objects such as rocks, mountains, tables, chairs, molecules, atoms, particles and bodies, neurons, and brains—is already a 'condensation of meaning' because what is seen is meaning already and we are mistaking meaning for the real thing.

What we call 'matter' is intrinsically related to our conscious experience. In a specific state of consciousness, as in our normal waking state, in which the experiential cognition of meaning occurs in a fraction of a second without us being aware of the process, matter appears as a self-evident independent entity, but the notion of a material object has no reality in itself if there is no inner conscious act of binding and 'condensing meaning'. There is an intrinsic and indissoluble connection and interrelation between the material world and consciousness that we can simply not get rid of, despite any attempt.

Moreover, our perception of the emergence of meaning and how the properties of the world are bound together, or fail to be bound, as in the case of people affected by cognitive disorders such as Balint's syndrome, make it clear how the qualitative properties of the whole cannot be explained simply by knowing the parts that compound it. However, if our act of 'condensing meaning' is an operation that appends a label of meaning to any perception, then the 'true world out there' stands before such action. What is supposed to be the 'reality' must be something without labels such as wetness, colors, forms, etc., that our subjective experience suggests.

So, what would it be like if we were really able to suspend meaning, as is still the case in formerly congenitally blind children immediately after surgery or as mystics claim to be able to do? The material world would not disappear, as the supporter of extreme relativism would like to believe. Instead, a reality would be unveiled in which neither duck nor rabbits appear. The material world where we recognize properties with definite quantitative meanings (shapes, forms, extensions, etc.) and qualities (such as colors, feelings, etc.) then becomes a random patchwork of perceptions. The significance of the outer material world is switched off from a conceptual existence. What remains is a reality that stands prior to its subsequent reconstruction through a perception-act of meaning in, of, from, or through our consciousness but without mental conceptualizations: A featureless realm.

What we call 'reality' is inherently property-less in its intrinsic nature and without qualities. The qualities and properties of matter are already emergent structured appearances in our consciousness and not the basic elements, the stuff that the world is made of and from which, commonly, consciousness is supposed to emerge. The properties of matter are not intrinsic elements inherent in the observed object but, rather, are subjective emergences in the consciousness of the observer.

Notice also how our innate resistance to accepting reality as an illusion derives from the unaware and subliminal necessity of taking consciousness itself into our logical account of things. We might also be open to the idea that our entire experience of the world is an illusionary misrepresentation of it, but we have a hard time eliminating with it also the conscious experience itself and that we are willing to believe the duck, rabbit, giraffe, or person have. I might also be willing to admit that my perception of a person's body, voice, and smell is just an illusionary reconstruction in and of my mind, but it feels a bit too far-fetched to eliminate the person as a conscious and sentient being, denying even its emotional, mental, and spiritual reality. This is why many feel uneasy about abandoning their naïve physicalist realism. Because, if one declares matter an illusion, one might feel compelled to abandon also consciousness as being inexistent. Deep down, we subliminally know this to be untrue. Once we give up the physical monism, only then do things appear in their natural context.

2. The Unwarranted Distinction Between Primary and Secondary Qualities

If we take physics as the most fundamental science that is supposed to describe all of reality, we can reduce everything to three elementary 'objects': space, time, and matter (or force as matter can be reduced to the manifestation of a force field). These three ingredients are considered to be the three objective fundamental primitives in the sense that they are not further reducible to something else and from which all of our physical existence is supposed to emerge. Concepts like space, time, and matter were posited as givens, as objectively 'true' and self-explanatory entities.[26]

Primary qualities	Secondary qualities
Solidity	*Colors*
Extension (size)	*Sounds*
Position in space	*Tastes*
Figure (form, shape)	*Smells*
Motion or rest	*Heat-cold perception*
Number	

---

[26] It might be useful to point out that this changed recently among physicists because there are good reasons to believe that the failure to find a theory of quantum gravity is related to a too-naïve conception of space, time, and matter.

This is why one distinguishes between so-called *'primary qualities'*, such as the notion of form, extension, measure, etc., which are supposed to be inherent features of the objects of the world out of us, and *'secondary qualities'*, namely, colors, tastes, smells, etc., which we tend to consider as unreal or not inherent features of physical objects.

The distinction between primary and secondary qualities can be traced back to Galileo Galilei and John Locke in an attempt to demarcate between perceptual illusions and the supposed objective real properties inherent in material objects. Galileo already expressed this view in 1683 in his work *'The Assayer'* (Il Saggiatore), also called the *'Galilean gap'*, according to which matter or any 'corporeal substance' has figure, size, a location in space, motion, contact, and number, while color, taste, sound, and smell are not 'conditions' residing in the material objects and are only pure names existing solely in our' sensorial bodies'. If we remove ourselves from the latter—that is, our senses from these objects of cognition—these corporeal substances would nevertheless retain the former properties.

A few years later, in 1689, Locke argued as follows [7].

Primary qualities of bodies.

"[...] Take a grain of wheat, divide it into two parts; each part still has solidity, extension, figure, and mobility: divide it again, and it still retains the same qualities; and so divide it on, till the parts become insensible; they still retain all those qualities. For division (which is all that a mill, or pestle, or any other body, does upon another, in reducing it to insensible parts) can never take away either solidity, extension, figure, or mobility from anybody, but only makes two or more distinct separate masses of matter, of that which was but one before; all which distinct masses, reckoned as so many distinct bodies, after division, make a certain number. These I call original or primary qualities of body, which I think we may observe to produce simple ideas in us, viz. solidity, extension, figure, motion or rest, and number.

Fig. 54 John Locke (1632-1704)

Secondary qualities of bodies.

"Secondly, such qualities which in truth are nothing in the objects themselves but the power to produce various sensations in us by their primary qualities, i.e., by the bulk, figure, texture, and motion of their insensible parts, as colors, sounds, tastes, etc. These I call secondary qualities."

On this distinction of primary and secondary qualities, the whole scientific human intellectual adventure has been based. Though you might not find the distinction in modern textbooks, it is clear that all modern science rests on this deeply ingrained assumption: Primary qualities should be considered the real and objective properties of the world, while the secondary qualities are merely subjective illusions, nothing more than experiences in our brain, with no inherent reality.

After centuries of an almost uncritical acceptance of such an idea (with only a few notable exceptions, such as philosophers David Hume and George Berkeley), for cultural and historical reasons, it has now become extremely difficult to doubt this division, especially after the undeniable success of science that based its foundations on this premise. The belief in the 'objective existence' of primary qualities against the delusionary non-existence of secondary qualities is so fixed and deep-seated in our reasoning that it might be perceived as an insult to our intelligence to think otherwise.

Nevertheless, when pressed to take a closer look at how our cognition and perceptions work, we might gain a glimpse into this fallacy. For instance, if people are asked if the taste of a tomato is an inherent property *in it* and independent *of us*, or if the perception of taste is a phenomenon occurring in our consciousness that has no reality in the tomato, assuming that they understand what we are talking about, most will agree that the second case has to be regarded as true. However, if we ask if the quality of redness of that tomato is an inherent property of it or, again, something that appears in our conscious experience, most begin to hesitate a while before giving a response. Instinctively, we are guided to consider these qualities as inherent properties of the objects that surround us. However, after a little reflection, our intellectual strength suggests otherwise, and most will still agree that colors are a qualitative construct in us as well, an experience of and in phenomenal consciousness, but not something objectively real 'out there'. Redness is a qualia, a subjective perceptual experience, a transcription in our consciousness of a particular frequency of an electromagnetic wave and not a quality independent from the observer. You may not be entirely convinced, but at any rate, that's what modern science tells us.

By the way, color vision is a complex science telling us that our color perception is determined by many more parameters than a wavelength. For example, what we really see is not the wavelength but, rather, an integration of the reflectance of an object with the power distribution of the illuminating source on different wavelengths (you may have noticed that, in a badly illuminated environment, colors tend to appear somewhat differently). For the sake of briefness and simplicity, we will not deepen this technical aspect other than pointing out how it further supports the thesis that colors cannot be an inherent property of the world.

Now, if we ask further whether the shape, form, solidity, and position in space of this same red tomato is an inherent property of it, or if it is, again, something appearing to our consciousness, then the strength of our cultural belief systems prevails and betrays itself. The instinctive urge to reflect a little further on this seems to be much weaker, and, in most cases, agreeing with

Galileo and Locke, we will answer that these are objective inherent properties of the objects of the world, independent from our perception.

If so, we should ask ourselves: What grants, to the extension and form of a material object, a higher degree of truthfulness compared to its color? Why should we accept as an objective reality the shape with which we measure matter in space while rejecting its color as only a subjective experience? If we carefully and honestly examine the situation, we will not easily find such a precise and clear-cut reason that should lead us to such a belief. What seems to be for someone an absurd wordplay or the empty sophistry of philosophers like Berkeley may, instead, hide a deep secret.

A reason for the human tendency to believe primary qualities as superior properties that are inherently contained in the bodies of the world might be that spatial information of shape and size can be double-checked via visual and tactile information, while smells, tastes, and colors cannot. When we open our eyes while having an object in our hands, the visual sense is a powerful ally and testimony to our touch sense. It confirms our mental 'space projection'. The converse is also true: We can construct such notions as shape, size, texture, etc., from the visual information alone and afterward control them through our touch experience. Smells, tastes, and colors do not have the possibility of such a double-check and are therefore regarded as less reliable in the so-called 'discovery of objective truth'. In this sense, spatial properties seem to have a greater element of reality.

To further investigate this state of affairs, it is necessary to trace back the source of our firm conviction that conceives of space as something having an objective and inherent reality in itself. The fact that our brains must learn to recognize two- and three-dimensional objects in space is no novelty. Indeed, infants have some innate capacities to recognize shapes and depth, but as already mentioned, most of the spatial capacities must be acquired through learning and experience. Parents know well how babies have that annoying impulse to grab any object they can find and immediately throw it to the ground with full force. This is not the manifestation of a fancy or rude character. On the contrary, it is a necessary exercise and entertainment by which their brains investigate the correspondence between what they see and the 3D depth of the environment. We never see anything 3D because all light is projected onto a 2D surface: our eye retina. Because newborn babies see only an unintelligible 2D image without significance, the brain circuitry of their visual cortex must still learn to reconstruct a meaningful 3D world from a 2D image. The observation of the falling objects gives their brains information about the 3D structure of the environment. Our brains take advantage of movement to reconstruct what we perceive as depth. Only once we perceive depth can we also reconstruct a 3D depth where there is none at all, for example, in a photograph or painting. Our visual space is, first of all, a reconstruction, not an image of the space as we believe to be in itself.

But is it a reconstruction of something real and independent of our seeing and our cognitive processes? What about M. C. Escher's world-famous impossible figures? Fig. 55 illustrates the *'impossible planks'* that show how our inherited logical elaboration of visual space information can be deceiving.

Fig. 55 The impossible planks.

Our perception of space requires very complex information processing that exploits many features of the world. Normally, this process is so fast that we aren't aware of it and tend to underestimate its complexity. For our daily experience, space perception is such a normal and given fact that only rarely do we think about how it comes into existence; we take it for granted as an obvious and indubitable truth. Space is perceived as something very 'real out there', and it would be very hard to convince us that it may contain a subjective element, too.

Of course, we know that there are a lot of spatial perceptual illusions, but the question is also about whether the notion of space, with or without objects contained in it or containing it, has an objective reality in itself. Should we regard the fundamental notion of space characterized by height, width, and depth, with all its related ideas that we append to it, such as shape, extensions, and quantitative measures, as a real object of the world or a mere conceptual entity that we construct in our subjective inner experience? Is space effectively 'out there' or a fiction that our consciousness constructs to convey a logical and coherent perception necessary to survive in some indefinable realm? Has space any meaning beyond consciousness? In other words: Is space something objectively existent out of consciousness, or is it, again, a subjective construction in us, a symbol that appears to our consciousness? Is a third option plausible, such as that space somehow exists but has nothing to do with the concept we have of it in our minds?

What we can say for sure is that a cognitive and semantic construction is necessary in order to have the perception of space in its depth and extension. Space certainly does not present itself as it is to our mind without going through a complex information processing and reconstruction, as is the case for other percepts. One example is that of optic ataxia in Balint's patients. Due to lesions of the posterior parietal cortex, they lose the capacity of space depth perception that places an object in relation to the overall environment. This affects the ability to interact with objects presented in the visual field and its spatial location. Patients with optic ataxia fail to accurately reach for objects because the target for gaze and limb movement are decoupled. They reach in front of them randomly through space until their hand or arm bumps up against the object they see as if space 'collapses' in front of them. Space is not represented; only the shape of the object is perceived without the awareness of its location. It becomes difficult to tell whether an object is large or small because its distance remains unknown to them.

Another example of spatial representation deficit is the *'unilateral spatial neglect'* (also called *'hemineglect'*) when someone, due to an injury on the left or right parietal lobe, can see only the right or left spatial-visual field, respectively. Any spatial information of one of the two visual fields is simply neglected. Patients with this neurological condition behave as if part or all of the left or right side of their world (eventually including the left or right side of their own body) does not exist. This could include not eating off one side of their plate, not shaving or putting makeup on one side of their face, not reading the left or right pages of a book, and not acknowledging anything or anyone on one side of a room. Some will even deny that their left arm and leg are theirs. The following figures drawn by these patients of a flower and a clock might clarify this.

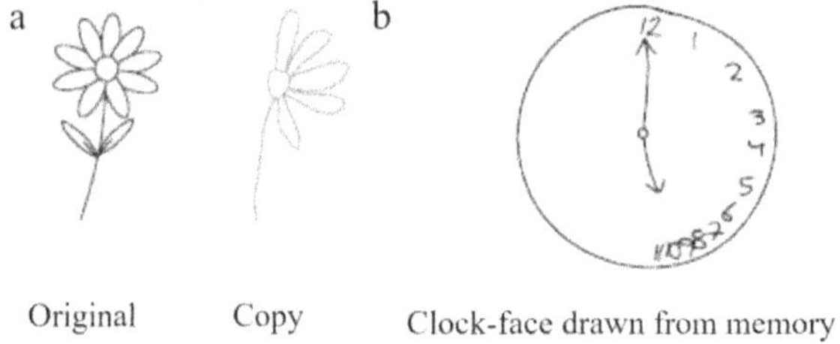

<table>
<tr><td>Original</td><td>Copy</td><td>Clock-face drawn from memory</td></tr>
</table>

Fig. 56 Drawings from patients with unilateral spatial neglect syndrome. Credit: [209]

When asked to draw a copy of a flower, a patient with unilateral spatial neglect syndrome only draws what is perceived on the right side of the visual field, thereby producing only a half-flower (Fig. 56 a+b). When asked to draw a clock-face, the patient tries to fit the numbers and clock-hands into the right side of the clock. This shows that neglect patients have an awareness of only the right side of their perceptual space, though they perceive the entire space and even store the information of the whole picture in their minds.

Among the primary qualities, there is also motion. Motion perception is also a reconstruction, a binding act of our conscious perception. This becomes quite clear in a rare neurological disorder resulting from an impairment of the visual cortex areas V1 or V5, called *'akinetopsia'* or *'motion blindness'*, which prevents patients from perceiving motion. What they perceive are only stationary objects or a series of still images of the world. Motion is perceived as a cinema reel or a multiple exposure photograph—that is, a superimposition of several exposures perceived as a single image. Other visual functions such as depth, shape, and face recognition are still intact, but due to motion blindness, the patients nevertheless get in trouble if they are asked to reach for objects, cross a street, drive a car, or engage in other tasks that require a change in motion and for the movement of people or objects to be assessed. For example, patients affected by motion blindness describe the pouring of a cup of water as a fluid appearing frozen like a glacier and being unable to know when to stop pouring because the movement of the rising fluid is absent.

This shows that not only space but also motion is not a given. As the phi phenomenon or the flash-lag illusion already eloquently demonstrated, motion blindness shows again that our perception of movement in the world is not an absolute that we discover as it is 'out there' but, rather, is a phenomenon that we construct in ourselves, binding it together with another mysterious 'object' of cognition, namely, time. There can be no motion and space without a concept of time. Space, time, objects, and motion are deeply intertwined conceptual entities whose ultimate nature has continued to elude the best scientific and philosophic minds through today.

Locke was unconsciously using what we call 'common sense'. We construct motion or rest from birth on, and it is really difficult for any one of us to understand that motion is a concept arising from the conscious experience in us and might be a feature of the world no more than color, taste, and smell are. There is no real reason to think that motion belongs to a special quality of a body. Not to mention that physics tells us that motion is a relative quantity—that is, an object might appear at rest to someone and in motion to someone else. The relativity of motion was one of the main insights of Galileo and became the bedrock of Einstein's theory of special and general relativity.[27] After all, motion, as well as velocity and momentum, are abstract concepts through which we describe a dynamical relation between bodies, not a property in or of a body. The speed of a car is not an inherent property of the car, like its mass or size.

These neurological cases don't necessarily tell us something particularly deep or philosophically relevant, but they might help us remember that space, with its accompanying concepts of extension, figure, depth, left and right side, etc., is a reconstruction in our conscious perception. If we would take the pure positivist approach, which insists on the application of Occam's razor, we could conclude that the existence of space as something independent from our inner consciousness is an unnecessary hypothesis.

Berkeley didn't need such findings to support a similar thesis. He reasoned about secondary qualities first and then argued in the same way with the primary ones. One of his famous examples is as follows.

---

[27] However, this does *not* imply that the theory of relativity states that motion depends on any personal subjective preferences.

The quality of heat is a sensation that occurs in us and is not a sensible object in the outer world. This becomes clear when we place our hands in water for a while—one hand in cold water, the other in warm water. If afterward, we put both hands in water at room temperature, it is possible to perceive it with one hand as warm, whereas with the other one, it will be felt as cold. So, he argued, the common qualitative notion of heat and temperature, which we instinctively associate with objects outside of us, is not intrinsic to them but rather a subjective perception and experience *in us*.

At first, such reasoning may appear trivial. What is novel in this? But further reflection shows that all this has much deeper consequences than we might first suppose. Berkeley did not center himself in space taking a third-person perspective but adopted a first-person subjective approach. This makes a subtle but decisive difference. He realized how our cognitive processes are an acquisition of meaning in us through a process of association with particular experiences that are still in us. Locke instead was inclined to regard meanings out of us and of a different nature than a subjective experience without realizing that meaning is already a subjective perception. There is no reason to believe that our conscious awareness of phenomenal events related to material objects with primary qualities like solidity, size, and position in space is working differently.

Thus, is there any reason to believe that touch is something that gives us information about things that should be regarded as more realistic than the optic-visual experience? We have also seen that, besides the well-known visual illusions, there are illusions occurring on our skinned body surface, such as the saltation effect discussed in Pt.I-II.3. Nevertheless, there is somehow a strong tendency to believe that the touch sense conveys more information about 'reality as it is' than any other senses.

Imagine the following situation: You see an object and try to reach for it. Suddenly, you realize that nothing is there. Which of your senses—vision or touch—would you doubt? Conversely, imagine bumping suddenly into something you perceive with the touch sense, but you don't see anything. Again, which of the two senses would you doubt? Probably you would take touch as the ultimate arbiter (it is also interesting to see how our language betrays this common assumption when we say that we "lose touch with reality".) The reason why we believe so strongly in our touch experience is probably because anything that does not interfere with our touch sense can be ignored and is regarded as not decisive for survival in our daily lives. It is, perhaps, rooted in an evolutionary advantage. At any rate, ultimately, the perception of solidity and the 'discovery' of material objects is always a qualitative experience of forces acting on our skin.

What is solidity after all? "Well, bump your head running against a wall and you will know!" you might say. But by doing so, you already point to a painful experience, the most subjective possible perception of a physical attribute. How do we know, if not through a subjective touch experience, if an object is solid, liquid, or gaseous? Is the physical attribute of solidity something extrinsic to our subjective experience? It is clearly, again, nothing other than a mental transcription of a subjective experience, like color, smell, etc. Why, then, is solidity an attribute we ascribe, say, to a hard piece of nougat chocolate, as being something very 'real', whereas the notion of its sweet taste is something 'unreal' and subjective? Of course, something such as solidity can be measured objectively, while the taste of nougat or chocolate cannot.

It might also be worth pointing out how Locke's idea that the repeated division of a grain of wheat *"till the parts become insensible"* would *"never take away either solidity, extension, figure, or mobility"* is an assumption contradicted by modern science. As we will see, when dealing with the microscopic objects of the quantum realm, even the notions of position, momentum, and solidity–and, thereby, of extension and figure as well–become quite uncertain qualities and no longer lend themselves to be defined by the conventional conceptions of classical physics. Molecules, atoms, and even more so elementary particles are subjected to laws and quantum phenomena in which these primary qualities no longer retain their aspect as we are accustomed to in our everyday experience of the macroscopic world. Of course, Locke couldn't know anything about quantum physics, but this historical detail shows how, also, bright minds frame ideas, hypotheses, and conjectures about things and their inherent nature upon unaware and a priori unquestioned assumptions.

So, let us see what solidity is according to modern physics. Why can't we walk through a solid wall? Physicists know very well that this apparently naive question is not at all an obvious fact. There are electric forces and quantum principles at the atomic level that prevent us from going through walls. And why do particles collide? Because they exert forces mutually (electromagnetic or of another nuclear type of force on a subatomic level). Otherwise, it would be perfectly possible to go through walls and see particles and objects flying through each other. As we will see later, in modern quantum physics, there is no concept of 'collision' in the classical sense we think of, such as solid marbles kicking each other around, and there are no objects with a defined extension in space as we imagine them in our still Aristotelian or, at best, Newtonian minds. Shape, extension, and solidity are only different aspects of a subjective experiential phenomenon with which we represent an abstraction that stands for an observed set of forces acting in time on our senses at a macroscopic level. At a microscopic level, in the domain of particle physics, notions such as the extension, shape, or solidity of a particle become vague and are instead intimately correlated to their force action. We can even say that the notion of spatial extension comes into being in us through a complex relational network of force events evolving in time. What we call 'matter' is something that exerts a force on our skin. Solidity and matter are an aspect, a mental abstraction we construct from the internal subjective experience of the actions of physical forces (or measure through mechanical devices). In Pt.II-I.5&6 we will also learn how, in quantum field theories, there isn't anything such as 'matter', as conceived in classical physics and which we conceive of in our daily lives at macroscopic human scales. This idea is much too naive and was definitely rejected by modern physical science of the last century. The only things that 'exist really' for science are space, time, and force fields.[28]

If matter is basically only a force-expression, what, then, is independent of our own mental ideas, a force? Physics defines it as the change of momentum–the quantity of motion–per unit of time. But the momentum of what? Of a given mass–that is, of a chunk of matter, of course!

So, we got trapped in a recursive and self-referential logical loop. It arises because we, at the bottom, can't speak of anything outside our phenomenal experience of the events that transcend these 'experiential objects of meaning'. Modern science is ultimately based on phenomenal events in our conscious experience that have been enthroned to 'semantic segments' supposedly having an inherent and independent existence in themselves, which, however, don't have beyond a conscious act of perception and an internal 'condensation of meaning'. We construct abstract notions such as matter, labeling them with equally abstract attributes such as solidity, volume, density, mass, etc., from a subjective experience of a relationally bound set of force events, registered and organized appropriately, associating them afterward with a meaning that finally emerges in our minds. But there is no reason to believe that this 'outer world' we construct contains something even remotely resembling such a thing in the realistic sense of which we are accustomed to thinking. The attributes of the solid physical world, such as solidity, mass, etc., which science considers as objectively real, emerge from our conscious force-perception, just as the meaning 'duck' or 'rabbit' emerges in our conscious visual light-perception. There is substantially no difference in tactile perception. What we finally really do is that of building in us a unified representation of conceptual or geometrical objects of meaning from a stream of information coming from our touch sense. The touch receptors and the involved brain areas may also be very different than the receptors we have in our eye retina (cones and rods) and the visual processing brain areas, but the processes in our minds occurring due to the binding process—that is, by binding and constructing, as in the case of the Gestalt unification, the emergence of meaning—are substantially the same. Everything we said for the visual sense also holds for the physical 'force-sense' of touch.

Therefore, the notion of force, and its emergent properties in the form of solidity or impenetrability, are a perception of meaning as well. Force is a

---

[28] For a more in-depth analysis of this aspect from the point of view of modern quantum physics, see also [36] VII.1,2.

subjective perception qualitatively on the same level as sound, color, and taste. There is no true reason to behold it as something that has a superior truth with respect to any other secondary quality.[29] This is something that modern physics begins to intuit but still must process further until it comes into its awareness.

Now let us reread Locke's statement. It seems that Locke believes that secondary qualities *"in truth are nothing in the objects themselves"* but primary qualities instead are, in truth, something in the objects themselves and are *"utterly inseparable"* from them.

What does it mean that a primary quality is inseparable from an object? This statement would have a meaning only if we could conceive of an 'object' without its shape, form, or solidity. But the attempt to define this strange entity obviously fails because we posit in the first place that primary qualities define what we mean by an 'object'—again, a hidden unaware logical circularity that arises from the third-person perspective which tries to eliminate the first-person one and thereby falls into a recursive self-referentiality.

Are secondary qualities such as taste, smell, and color properties separable from 'objects'–that is, from primary qualities? One does not—perceive the redness of a tomato or the sweetness of a piece of sugar in one place and the tomato or the sugar shape and size in another place. Interestingly enough, we saw, however, that something on this line seems to occur in our brains. Recall Pt.I-II.3, where we pointed out that, in our brains, the properties of the color, texture, and shape of the things we perceive are processed in different brain areas (see Fig. 9). It is also known that there exist neurons dedicated to the task of identifying the motion, position, and visual angle of the *same* object *separately*. However, no center and method of integration have ever been discovered, and nobody has the slightest clue how this information is suddenly perceived in a single integrated whole that we call an 'object', with all its primary and secondary qualities bound together.[30] Moreover, there are people able to separate primary from secondary qualities consciously: Recall Balint's patients affected by visual simultagnosia who can separate colors from objects (see Fig. 19).

At any rate, it is this binding of primary and secondary qualities that allows things to emerge as meaningful objects in our conscious awareness—that is, to allow us to say that they are 'real' and that they 'exist'. In fact, what kind of properties are we thinking of when we speak of the 'existence' or 'reality' of

---

[29] We might also consider force as a fundamental primitive provided we recognize how we never perceive a force directly, but rather feel or measure only its effects. More on that in Pt.III-II.3.

[30] There are some purely speculative theories such as this one [373] (see also references therein). Note how scientists, although they stick to the physicalist perspective, feel pressured to shift from the materiality of neural networks to theories of non-material electromagnetic fields. Wrong theories also might be a small step in the right direction.

a material object? It is not difficult to understand that, perhaps without being directly aware of this, we are thinking prevalently of its primary qualities, such as its form, extension, dimension, solidity, and position in space. In fact, nobody needs to look at a material object to ascertain that it exists. We can declare its existence simply by the use of touch, moving our bodies in space, and trying to reconstruct its shape mentally as blind people are shown to be able to do, who, however, must have a very different understanding of shapes than normal seeing people have. The use of the sense of touch associated with the mental image of a solid geometrical object in our minds is the very essence of what we believe to be, per definition, 'the existence of a material object'. What is not immediately realized is the fact that it is the mental image of it that we believe to be real—again, something appearing in our minds and which, from an ontological point of view, isn't much different from color perception.

Moreover, the touch experience is not a direct contact that reveals the meaning of the object because, as we discussed earlier, meaning is not an intrinsic attribute in things. Touch is a perceived 'force pattern' in magnitude, time, and frequency that is perceived first; only later, through an organizing idea, do we associate it with a meaning. Touch, like vision, does not disclose a truth that we discover. Touch also is only a random patch of sensory signals as long we do not code them mentally in ourselves, as babies know much better than grownups or the adults who were later treated for congenital cataracts had to realize to their dismay once opening their eyes.

Therefore, there is no qualitative ontological superiority between the subjective feeling of the solidity of a rock and its temperature perception, the perceived colors of the petals of a rose and its forms, or the sweet taste of a piece of chocolate and the form or size of its mold.

It is not clear on what logical grounds we should consider primary qualities as inherent properties of the world, while secondary ones are only fictitious appearances. Primary qualities have been conferred with supremacy that they do not deserve. We can only state that primary and secondary qualities are both qualitative experiences appearing in us. More than that, we can't say. The mystery of consciousness and its relationship with the world and the very notions of 'reality', 'objectivity', or 'existence' are far from clear. Nevertheless, there is no logical or scientific reason to posit primary qualities on top of 'truthfulness', 'reality', or 'existence', whatever these words might mean.

Yet, there is something very strong in us that refuses to accept that things aren't somehow 'out there' independent of someone looking at them. If everything is simply so illusory, why at all does this universe in its immense extension appear? If I have to take a flight from New York to Paris, I clearly perceive that there is in between a great spatial extension 'out there' that must be overcome. We cannot, through a simple act of imagination or 'meaning switch', beam ourselves instantly from one point of the globe to another like in a Star Trek movie! Let us see where the problem comes from.

Albert Einstein was a realist, someone who was convinced that things exist independent of our observations. One of the many anecdotes about Einstein says that one day he was walking with his colleagues, discussing the strange and counterintuitive features of quantum mechanics. A. Pais recounts Einstein's reflections as follows:

"We often discussed his notions on objective reality. I recall that during one walk Einstein suddenly stopped, turned to me and asked whether I really believed that the moon exists only when I look at it. The rest of this walk was devoted to a discussion of what a physicist should mean by the term 'to exist.'" [210]

From where did Einstein's problem arise? What do we *really* see when observing the moon? The problem arises because we are unaware of the fact that we are misinterpreting a first-person subjective experience for an 'objective' observation and are still using highly subjective concepts like 'existence' or 'reality'. To state that something is 'real' when it continues to exist without someone looking at it is founded on the unaware assumption that things exist *as* we perceive them. We have already seen that it is not valid to state something about the true and objective reality of an object independent of our subjective experience.

Einstein was deceived by the ontological distinction between primary and secondary qualities. We accept the fact that colors are qualia arising from a complex electrochemical process that, for whatever reason, then appear in our conscious awareness. It makes no sense to question whether a color exists when we are not looking at the colored object—simply because colors, as qualities, always exist in us and appear in our brains but are never outside of us as something inherently existent in the object we are examining, not even when we are observing it. In other words, an object reflects or emits electromagnetic waves but does not 'possess' a color; it only *appears* to our conscious perception as colored. This is the subtle but fundamental point we always forget. Of course, this holds for any secondary quality, such as the taste, smell, etc., of any object. Does the secondary quality of grayness of the moon—that is, its white-grey color—exist independent of our conscious experience of the quality of colors? We would negatively answer when referring to colors but not to the size, spherical shape, and distance of the moon from Earth

Now, the point is to precisely understand the difference between the distinction of the subjective world in us and what we call 'the real world out of us'. For this purpose, we may ask in what sense does the question "Does a material object exist when we are not looking at it?" differ from the question "Does the color of a material object exist when we are not looking at it?" or "Does the sweetness of a piece of sugar exist when nobody tastes it?"

There is no reason to believe that primary qualities are objective, whereas secondary qualities are subjective only because of their different qualitative natures. Primary qualities are subjective in this sense exactly as any quality too. From the ontological point of view, what we call 'the distance of the moon from the Earth' has no more or less reality than its color. Not even that: The smell or taste of a lunar rock is as real or unreal, true or untrue as its shape, solidity, and distance from the Earth. Everything comes down to a mental subjective inner reconstruction of perceptions. And yet, there is no reason to reject the existence of an objective reality (whatever 'objective' might mean).

We could answer Einstein by saying that the moon isn't there even when one looks at it. What the observer sees and describes as the 'moon' doesn't exist in the first place, as it is only an internal subjective qualitative abstract representational experience in us. In this sense, we can say that the moon does not exist *as we perceive it*, also when we are observing it. What we are seeing is a reconstruction of our minds, not something really 'existing out there'

So, why did this distinction between primary and secondary qualities come into being in the first place? Just a sociological, historical, and cultural fashion?

We don't feel it necessary to go so far. Closer scrutiny of the distinction between the two types of qualities from a first-person perspective makes it clear that the real difference between primary and secondary qualities comes from the fact that the former can be made easily *'inter-subjective'*—that is, it is possible to make a precise measurement of quantities (say, for example, the length of a table) on which anyone can agree with. Secondary qualities, instead, are *'only subjective'* and don't always lend themselves to such an agreement so rigorously. What distinguishes them is not their ontology but the fact that primary qualities lend themselves to quantification, in numbers and a common language that is inter-subjective, while secondary qualities are less or not at all amenable to this. Primary qualities are measurable, which means that we can agree on a number that makes something of it as inter-subjective.

But, at the bottom, the inter-subjective primary qualities are also always subjective. The belief that what is inter-subjective deserves a higher ontological status has its roots in its practicality: Inter-subjective qualities are a means of formalization of the world that can be communicated to others. Nonetheless, while primary qualities lend themselves to quantification on which everyone can agree, they are subjective experiences of phenomenal qualities as well, no more and no less than secondary qualities are.

Why should the quantification of an inter-subjective experience confer it with more 'reality' than taste? Locke's primary qualities are subjective as secondary ones, with the difference being that they lend themselves to an inter-subjective quantification. We might take this per definition as the demarcating criterion between 'real' and 'unreal' entities for pragmatic reasons. Obviously, a carpenter that needs to measure the length of a table couldn't care less about our philosophical ruminations. Also, since the time of Galileo, science has done the same. But such a demarcation can't be a good ontological characterization from a philosophical point of view that looks for the intrinsic and ultimate inherent reality of things. This would not bring us further in the explanation of the problem of consciousness and how experience comes into existence through these primary qualities, which are already an inter-subjective perception themselves.

Therefore, if we deny reality to secondary qualities, it follows that primary ones are 'unreal' as well. This

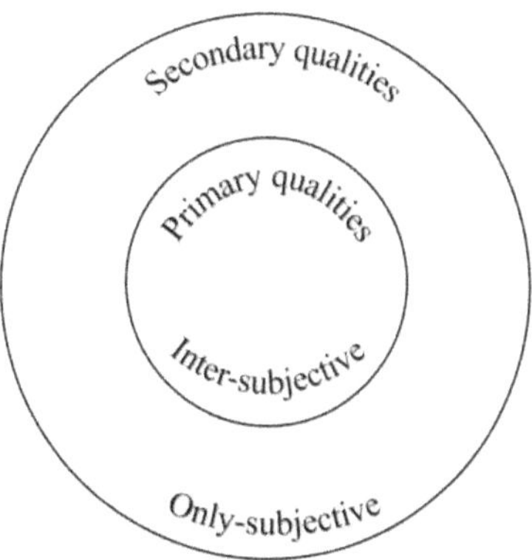

Fig. 57 Primary and secondary qualities are both subjective.

means that, for example, if we consider the sweetness of a sugar cube as only a subjective perception in us that has no reality in the object itself, then its cubical form is also 'unreal' and subjective, not an inherent property independent of our perceptions. Or, to put it the other way around, if we insist on granting primary qualities with objective truth, then we have to grant secondary qualities the same ontological status of 'truthfulness'. The sweetness of a sugar cube should then be considered as real 'out there', independent of us tasting it, as is its form. This is a philosophical form of realism that most of us are probably not prone to endorse.

Again, we are not advocating for a form of solipsism that considers everything mental or nihilism that denies existence to everything. We can maintain a form of realism in which we conceive of 'something' that 'exists', and that 'something' causes qualitative experiences—that is, qualia—to come into existence in us, and that replaces reality with the experience of reality. What we can say for sure is that the universe we observe lends itself to a quantitative and qualitative description but is beyond quantity and quality. Quantity and quality manifest to our consciousness and are not inherently real properties that inform us about the true ultimate nature of the world. If we take this to its extremes regarding the primary quality of extension size and shape, we could say that, in some sense, the universe itself collapses into a unique singularity in which the notion of extension loses significance. Space is also only an experiential object that has no meaning beyond conscious perception. Its objective truth must be atemporal and beyond primary as secondary qualities–that is, beyond any intellectual reason. We will take up this point later.

Because the form and size of an object is a derived mental representation, a 'mental qualia' so to speak, coming from sight and/or tactile sensory stimuli, one might ask, as T. Nagel famously did, "What is it like to be a bat?" Even though, contrary to common belief, bats can see with their eyes (but are almost color-blind), they rely mostly on echolocation (sonar)–that is, they emit ultrasound waves that, once reflected by an object or potential prey, are interpreted by their brains. In other words, here, Nature replaced electromagnetic waves with sound waves. There is no question that bats can fly through a complex 3D environment and that they possess a fine spatial orientation and localization that allows them to quickly find their way. The question is whether bats see properties of objects and have a subjective perception of space in terms of depth, length, and height, as we have. It is reasonable to assume that their 3D representation may not differ much from ours. But some interesting clues might come from the fact that bats identify not only the shape and size of objects but also their density (ultrasound waves are absorbed and reflected differently from objects having the same shape and size, but with different material densities). How does a bat 'see' the density of an object? Like we perceive colors? Unless you can literally 'incarnate' your mind in that of a bat, you will never know, but for sure, there is 'something like to be in' experiencing the density of objects, perhaps like we experience its color or taste.

A much more striking example from the animal kingdom that highlights the dichotomy between experienced qualities misinterpreted as objective properties of things and the features of the world comes from several studies pioneered by H. W. Lissmann in the 1950s-60s ([211], [212], see also more recent ones [213], [214]), of a Nile fish, the Gymnarchus Niloticus. The philosophical implications of these studies were first pointed out in 1991 by William Seager [215]. The Gymnarchus is an electric fish and is almost blind; it distinguishes only day from night. Its main sense is a pulsed 300 Hz electric field, running from tail to head, which it produces with an electric organ in its tail.

Fig. 58 The Gymnarchus Niloticus and its electrolocation of objects by electric field lines.

It perceives the intensity and variations of its own electric field distorted by objects in the surroundings with an electrolocation system based on epidermal electroreceptors. Any object having a different electric capacity and conductivity than that of water distorts the structure–that is, the gradient and contours of the electric potential distribution around the fish–which, therefore, perceives the surroundings by analyzing the 'electric images' of the field projected on its electroreceptive skin surface. It explores the environment in the dark and literally maps it onto its own electrosensitive skin surface all over its body as a replacement for an eye retina. When its electric organ discharges, it sends an electric impulse throughout the environment, whose (radar or sonar-like) reflection it receives back on its sensitive skin surface.

Behavioral experiments have shown that, by categorizing the amplitude and waveform of these reflected impulses, the Gymnarchus electrolocation system distinguishes the form, location, and conductivity of nearby objects possessing 3D depth and even the shape perception of objects independent of their size (distance), rotation, and material. Distance location is, in itself, already a remarkable ability because it lacks stereoscopic 'vision', as we humans are accustomed to, thanks to our binocular sight. As Lissmann showed, the perception of conductivity of objects allows this fish to 'see electrically' *through* objects, for example, by discriminating an outwardly identical-looking container holding water from another containing a glass rod. The Gymnarchus discriminates objects of the same form and location by their conductivity, those with the same form and conductivity by their location, and, finally, those with identical location and conductivity by their form.

Therefore, what this electric fish 'sees' is an amalgam of the object's location, form, and conductivity. The spatial location and properties of material objects that a Gymnarchus perceives are not its form, location, or conductivity taken separately, but to us, a completely mysterious combination of all of these. What is it like to be a Gymnarchus? It must perceive the world in a manner totally alien to humans.

The Gymnarchus is a strange being in which two primary qualities of location in the space of an object and the form of the same object are united in a unique and condensed meaning that contains them both but without perceiving them separately, as we do, because it lacks a stereoscopic vision and tactile grip experience. This would associate a completely new property to the perceived objects of the world that we cannot even imagine—a world where objects have not a form and a location in space but only 'formlocation' and, on top of that, are distinguishable from their electric conductivity.

If humans would have a different structure of sensory organs and were to perceive the world as this Nile fish does, John Locke would have had a very different appearance of the world and a different opinion on the distinction between primary and secondary qualities. Perhaps we would have come up with another complete categorization of qualities. This shows further that there is no logical reason to believe that primary qualities are qualities outside our conscious perception and inherent properties of the world, at least not more than secondary qualities are. This distinction can be purely arbitrary and serve instrumental and survival purposes, but we must acknowledge that, upon closer scrutiny, the sort of realism that Galileo and Locke suggested does not hold.

To conclude, let us reread Locke's definition. We get to know more about the idea of primary qualities as inherent properties of bodies. He says that secondary qualities *"are nothing in the objects themselves, but powers to produce various sensation[s] in us by their primary qualities..."* In other words, secondary qualities must be explained in terms of primary ones. On what is this belief grounded? The attempt to explain an experience of color, the perception of taste or smell, etc. beginning from size, bulk, and figure amounts substantially to the same epistemological operation of doing the contrary: Why not explain the perception of depth and three-dimensional objects from secondary qualities such as taste and smell?

This sounds ridiculous, and it is, but the physicalist approach that tries to explain consciousness in terms of primary qualities does precisely this. The hard problem of consciousness has one of its roots in such uncritical acceptance: the idea that we can explain secondary qualities from primary ones. It is an intellectual exercise that has the same prospects of success as that of trying to prove a mathematical theorem in terms of perfume, aromas, and milkshake tastes.

But, after all, the refusal to discuss again such issues is comprehensible: If we reject Locke's distinction of primary versus secondary qualities, on which the whole scientific endeavor has based itself, over 300 years of supposed scientific realism would collapse into a human inter-subjective construct and would force us to the painful conclusion that all that has been known so far is a toy model with no inherent reality that will forever remain incapable of telling us what the physical universe really is.

Perhaps the time has come to take this step and have the intellectual courage to revise some of our assumptions, to go beyond our naive forms of realism with which daily science unconsciously works.

A new science is needed—one that is capable of extending its vision beyond these misleading and unaware assumptions. Many of the strange paradoxes and problems that are apparently without explanation in the frame of the classical scientific reason can find a new and unexpected resolution if we place primary qualities at the same level as secondary ones. The day we will not give for granted such a dogma, we might take a step further towards a new science of consciousness and a more conscious science.

3. What is Reality?

> *"Imagine a painter who, having painted a self-portrait, points at it and declares himself to be the portrait. This, in essence, is what physicalism does".* Bernardo Kastrup [216]

In this chapter, we would like to make it clearer, through a little 'experiential-philosophical' experiment and other conceptual thought experiments, how the perception of what we call 'the world' does not provide us with information about what we think the 'world out there' is made of. There is an intrinsic relation between a subjective experience and what we think the 'objective world' is supposed to be, and that cannot be avoided, including in science. We would also like to show how the education and cultural environment we are accustomed to is heavily based on a very naive form of realism. Perhaps it might not be easy to understand where the real problem lies at first because we must revise our assumptions from their very roots and learn to look inside. After all, every change of paradigm unavoidably requires a revision of our belief systems, and at times, this can become a psychological challenge. Looking inside ourselves can be a difficult journey that challenges our ideological and sometimes even religious preconceived and unexamined assumptions, which we are unwilling to give up. However, such an exercise may allow us to proceed and take a step further in understanding the world and, consequently, ourselves.

An introduction to exploring how we perceive and conceive of the world could be that of following British philosopher and author Owen Barfield, who once invited us to do and ask ourselves the following: "Look at a rainbow. Is it really there?" Of course, we shouldn't regard a rainbow as an object having any concrete spatial and material existence. It is an intangible appearance about which we can't say that it is "really there". We believe that only the raindrops and the Sun are "really there" and that the rainbow is not.

Nonetheless, we can't state that a rainbow is a mere hallucination either, as others also see it. This fact, however, doesn't imply that the rainbow is actually there; rather, it is a shared or collective appearance or representation—that is, an inter-subjective experience—and not an objective entity.

Therefore, by logical extension, one can also argue that everything else is only an inter-subjective appearance, not something "really there". A tree, a mountain, and the moon are "really there" only in the sense that the rainbow is "really there". As the appearance of the rainbow comes into existence due to the interaction of the sunlight with the rain droplets, so does the appearance of a tree, a mountain, or the moon come into existence due to the interaction of the sunlight with their molecules and atoms. One can press the argument even further by noting how molecules and atoms are themselves, yet another appearance realized due to the interaction of a measurement device onto the elementary particles of which they are made. This leads us into the realm of quantum physics where, instead of reaching a final 'objective entity' that is "really there" (whatever that might mean) and from which we could build, by a bottom-up process, all that there is, Nature answers us with even more elusive and ephemeral phenomena further detached from any of our sensory means and conceptual constructs, as we will discuss in Pt.II-I.5&6. At the bottom, everything boils down to representations of our sense-mind based on our sense perceptions. Using Barfield's words

"...the familiar world which we see and know around us--the blue sky with white clouds in it, the noise of a waterfall or a motor bus, the shapes of flowers and their scent, the gesture and utterance of animals and the faces of our friends–the world too, which (apart from the special inquiry of physics) experts

of all kinds methodically investigate–is a system of collective representations. The time comes when we must either accept this as the truth about the world or reject the theories of physics as an elaborate delusion. We cannot have it both ways." [217]

We will study Barfield's ideas further in Pt.II-II.2dd. Meanwhile, another approach that clarifies the same insight from a more neuroscientific standpoint could be Don Salmon's, an American psychologist, who introduces us to an exercise that explores the relationship of mind to matter from a somewhat different perspective from what you might find in textbooks on neuroscience [218] [219].

"Take a simple object; something right at hand: a pen. Hold it in your hand; look at it; tap it on the desktop, and listen to the sound. Now, consider your experience of this 'sound'. What do scientists tell us about it? By coming into contact with the desk, the pen causes a series of acoustic vibrations in the atmosphere, in the space between the desk and your ear. This process, i.e., the compression and rarefaction of air (basically, contraction and expansion), continues at a certain frequency, and these waves of compressed and expanded air eventually reach your ear. (Edgard Varese, a modern French composer, when asked what he did for a living, once replied, "I am a disturber of the atmosphere".)

Do you hear a sound when these movements of air reach your ear? No, not yet. First, the atmospheric movements have to set in motion an immensely complex set of other vibrations. The vibrations go from the eardrum to the hammer and anvil, then to the cochlea where the cilia start vibrating (like the strings of a 'kindred harp'). Then the vibration of these cilia gets converted into tiny electrical currents which race along corresponding auditory nerve fibers to the brain.

You still haven't heard a thing!

These electrical vibrations are sent to the temporal lobe of the cortex, and are then processed in several other centers of the brain. And not a single neurophysiologist has yet been able to tell us how, but somewhere, someway, unbeknownst to anybody, all of this physiological activity is converted to the experience we call 'sound'. So, where does this 'sound' exist? Until it reaches your brain and is converted into an experience, there is no 'sound'. There are vibrations of the atmosphere, pressure waves in a gas, but no sound; the word 'sound' has no meaning apart from the experiencer.

Now, look at the pen: look carefully; pick it up in your hand, roll it around, observing the color, the shape, the subtle shading, the play of light against the pen. Let us again hear what the scientists have to say. You don't actually 'see' the pen, not directly anyway. Light is reflected off the pen, and you only see that reflected light. It may be interesting to pause for a moment and ponder the fact that according to science, you never see, in a direct way, the physical world at all. Your entire visual experience is of reflected light. It is an interesting exercise in itself to take some time for a leisurely walk, holding in mind the fact that all that you see is nothing but reflected light of the objects, not the objects in themselves. Now, you don't actually see the light either. The light is reflected off the pen, and it reaches your eyes.

You haven't seen anything yet!

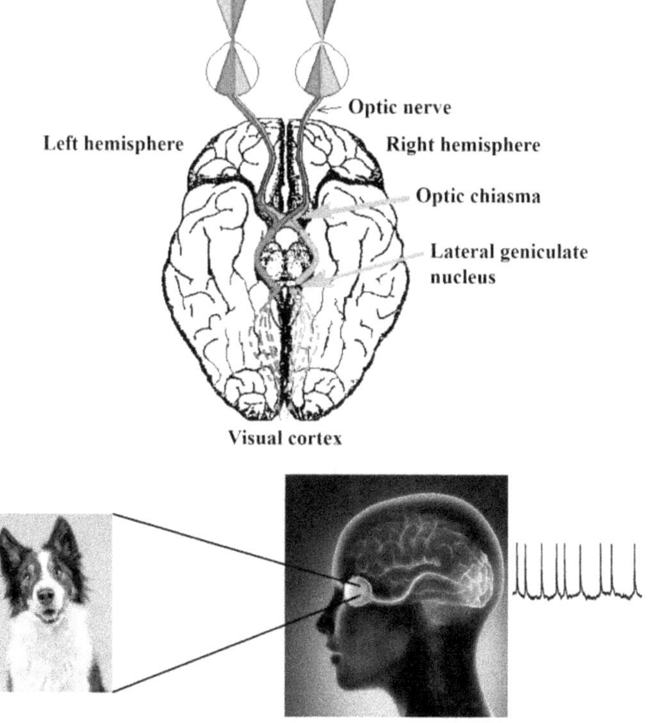

Fig. 59 The visual pathways in the brain and the neural spikes.

As with the vibrations of the atmosphere with regard to sound, the light has to enter the eyes, and through an even more complex process than with hearing, it is converted to electrochemical energy and travels along the optic nerve back to the occipital lobe of the cortex.

And you still haven't seen anything!

Once again, there has never yet been a neurophysiologist who can tell us how, after this extraordinarily complex analysis of this electrical energy by the brain, it suddenly becomes a visual experience. But they tell us it is only after this cortical analysis that you have a visual experience.

Without going into the other senses, suffice it to say that regarding the ordinary experience of the 'external' world, what you experience is not direct physical contact but the reaction of your nervous system to stimuli impinging on your body. What those stimuli are in themselves are not what you are experiencing: In other words, you don't directly experience light, atmospheric movements, or the variations in heat, mass, etc., which scientists say make up the physical world. Rather, you experience the reaction of your nervous system to these stimuli, which your brain analyzes and then projects out into space. These are not metaphysical or philosophical speculations; they are the commonly accepted findings of physiological and perceptual psychology.

When presenting these findings to audiences, experimental psychologist Charles Tart, compares our situation to the experience of a pilot training in a flight simulator. The flight simulator is designed so well that the novice pilot is unable to experience the difference between flying a real plane and working with the flight simulator. Similarly, you are living inside what he calls a 'world simulator', experiencing only the reactions of your nervous system to something whose actual nature you do not have direct access to in your present state of consciousness.

Now, take some time for another leisurely walk. Really take some time to feel this—it's much more effective if you don't think it out; rather, try in your bones to get a feel for this—that what you are experiencing is a construction of your brain, it is not at all a direct contact with an external independently existing physical world. To get a feel for this, as you are walking, it may help to switch your attention between visual, auditory, and tactile sensations. Focus on one kind of sensation at a time, and consider that your experience is of the brain's transcription of external stimuli, not of direct contact with these stimuli."

Thus, you never see the image projected on your retina. The image is disassembled in the brain and there is no region where you will find a representation of it. What you ultimately perceive are only and exclusively 'neural spike trains' (see Fig. 59)–that is, the electric activity of neurons made of fast short lived and sharp electric impulses. What we really perceive are only sequences of neuronal action potentials firing, and which temporal pattern encodes the information. Every single perception and thought in us is nothing else than the perception of these neural spikes. When you look at a dog, what you 'see' is nothing that retains the structure or representation of the dog. You become aware only of these electric impulses. Even in this moment where you are reading these lines, and in your whole life, you have never perceived anything else, other than this millisecond duration electrical activity within your brain.

In some sense, we live in a virtual reality without knowing it (sort of like in the film, *The Matrix*) and can't access the 'real reality'. This should further make it clear how, also from the strictly scientific and neurological point of view, everything we perceive is something that happens in us; it is never outside of us. This neurological analogy further clarifies how we tend to believe that a sound is something happening outside of us while our thoughts appear in us. However, the sound appears in the same field of awareness as the thought; it is a qualia experienced by a conscious experiencer. The idea that sound is outside of us comes into being because of a (quite fast) mental projection that imagines it to be outside

Let us try to answer the following well-known philosophical question: "Does a tree falling in a forest where no one is present still make noise?"

Some answer positively, grounding their argument in the fact that we have no reason to think that it should be otherwise. The noise of the falling tree is a fact completely independent of the observer. Others simply say that it is useless to question something that we will never know because there is no way to test this, yet they subtly imply that, still, the tree may make noise or perhaps also not.

Hopefully, we have made it clear enough where the whole misunderstanding lies. Both answers show the unconscious assumption we are working with. Both assume that the subjective experience we call 'sound' is something objectively present in the outer world. However, what is acting in the outer world is not sound, and there is no noise. Rather, out in the forest, we have pressure waves—that is, vibrations of air molecules. These become noise, sounds, or experiences that convey some sort of qualitative information—that is, qualia—only after a very complex data analysis of and in our brains, where something very different from the vibrations of air molecules has been processed. The experience of sound is not a quality of and in the world or a property of an object but is first and foremost something occurring in our heads, an experience which our minds then project outside of us. The tree produces pressure waves in the atmosphere, but sounds are only qualitative experiences in our consciousness. As with Einstein's question on the existence of the moon, in this case, the best answer could be: "The noise is only in our head, the tree doesn't make any noise at all, even in the case someone is present hearing it falling."

To better understand how our brains construct *a* world and not *the* world, we will go on further with what I shall call the *'monitor metaphor'*—of course, not even a thought experiment but, at best, a metaphor that, however, serves to outline the logic standing behind the reasoning we propose. [31]

Suppose your monitor is conscious(!) What would a conscious monitor believe to see? Say it is conscious of what appears on its screen, which is a sequence of images that it believes to be the real world. Suppose on the screen appears a film that shows a beautiful landscape. Our conscious monitor would then claim to see an objective reality: "I see a beautiful landscape." But where do these images come from? They are nothing other than an electronic

---

[31] The 'monitor metaphor' is similar to the 'user interface theory' of Donald Hoffman [363]. We feel it misleading to invoke a principle of evolutionary advantage to explain why our cognitive system misinterprets reality. With or without an evolutionary advantage, mind, reason, and our sensory apparatus would do so anyway. Therefore, we will present a modified version of an analogous metaphor here. While the metaphor might be an introduction to Max Velmans' *'reflexive monism'* [247] [246].

reconstruction, a 'translation' of a data set stored in a memory chip—that is, in millions or billions of microscopic cells that can attain an electrostatic on-off voltage state corresponding to ones and zeroes.

This digital information results from a complicated manipulation of software that receives digital signals from a stream of light impulses propagating through fiber optic cables or received via a satellite dish and modulated with some given standard. Previously, this information was encoded in electromagnetic waves propagating through air and space before reaching the antenna. There is no landscape 'out there'. What is really out there are initially only electromagnetic waves conveying some sort of information to the system but do not—not even slightly—resemble the landscape and have nothing to do with it at all.

What the conscious monitor 'sees really' is not the landscape but just an internal subjective symbolic representation of the electromagnetic waves in the chip memory states, which is displayed on the pattern of pixels of the monitor and finally lights up with different light intensities and colors. Nothing of the electromagnetic impulses remains, let alone the original object they encode.

If my monitor were conscious, it would erroneously misinterpret the complex encoding in computer memory with the transmitted image of the landscape as the objective reality. But at this stage, these pictures are nothing more than the representation on a horizontal per vertical matrix of pixels of the internal physical, electronic memory states. The 'monitor's brain' perceives it as its own subjective experience and, by attaching a meaning to it, would swear that what it is seeing is an objective and undeniable reality. The images it perceives are still only abstract symbols, an apparent 'objectification' into a structured spatial organization of pixels that are apprehended and bound as a whole still with a subjective conceptualization and an emergence of meaning.

Passing from this metaphor to the everyday reality that we perceive in our brains, we may ask: In what sense is our situation different?

Looking at a landscape with our eyes, we, too, receive information through electromagnetic waves of higher frequency, which we simply call 'light.' These light impulses, which fall on our retinas, are first converted into electrochemical impulses and then must be transmitted to the respective brain areas where a tremendously complex manipulation occurs. Only after this process has been completed do we begin to see something. But, when this happens, nothing of the original electromagnetic waves falling on our retina remains, let alone of the objects from which they were reflected.

In other words, what we see is not the landscape or even its reflected light, but an abstracted, codified, and deformed inner electrochemical reaction inside our skulls that no longer has anything to do with any 'objective reality' and is not even remotely a representation of it. The 'reality out there' that we subjectively observe stands to the 'objective reality' as it is, as the image on the monitor stands to the electromagnetic wave or the electronic, physical states of the chip that abstractly encode it.

We inevitably must conclude that we experience projections of a reality that never existed and that will never be accessible to us anyway, at least not with our actual sensory means and present cognitive functions. The world we perceive is a subjective symbolic projection of some indefinite realm in our consciousness. What we see of our world is a symbol. We do not even see a representation; we see only abstract symbols, icons.

There is a fundamental difference between a representation and an encoding symbol. For example, the representation of a table could be a picture, a sketch that one might draw with pencil and paper and that still captures some of its reality, such as the shape, form, proportions, color, etc.[32] A symbol for the table would simply be the word 'table'. The ink on paper that encodes an object in letters is completely removed from the object itself and retains absolutely nothing of its inherent truth and reality.

One might argue that the analogy does not hold because we do not experience words, numbers, or abstract objects appearing in our field of awareness, but sensations, objects of experience that are mental, sensory, and emotional qualia. However, these do not represent reality as they are; they are experiences impressing onto our phenomenal consciousness and therefore become encoded as 'subjective units of mental and sensory phenomenal events', what we shall call the *'experiential symbols'*. But they remain just that: still symbols. Similarly, we do not represent the world in our brains. Rather, we construct it with these experiential symbols that encode it. It is a subtle but decisive difference. What we perceive as tables, rocks, bodies, or through observational devices, atoms, cells, stars, or galaxies, are not even representations but a subjective symbolic encoding of phenomenal experiences, and certainly not the objects in their inherent existence. There is no real separation between the subject and the object.

Let us bring the consequences of this metaphor a bit further.

Suppose that our conscious and intelligent monitor begins to question its own nature, constitution, and consciousness and tries to understand how a subjective experience of the scenes it perceives on the screen, which it regards as the real objective world, comes into being.

Then the desire to explain its own consciousness experience will drive it to analyze this symbolic 'picture world', leading it to discover that it is made up of parts and sub-images to which it associates a meaning with new symbols, such as trees, rivers, stones, molecules, atoms, neurons, brains, etc., which invariably appear on its 'subjective screen of perceptions' as objects of which

---

[32] In a sense, this sentence contradicts our previous reasoning on first and secondary qualities as being equally subjective, but we are coming to a point where the limitation of language becomes apparent.

the world is made. But this world does not exist; what really exists is, again, only the outer electromagnetic signal that is encoded in the chip memory and gets displayed on a matrix of pixels on the monitor surface. What is 'objective' and 'real' in this metaphor is the display made of pixels—say, the light-modulated liquid crystal display (LCD) and the pixel's light turning on and off with different intensities and colors.

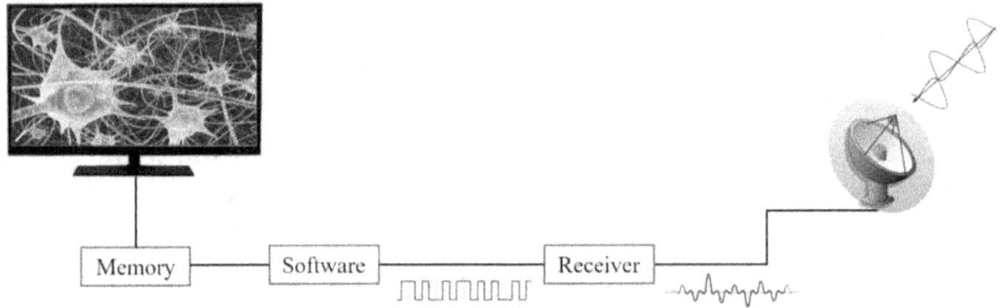

Fig. 60 The 'conscious monitor' metaphor.

Say the monitor concludes that its conscious experience must be explained by an image on the screen: A display of a neural network. This also has no reality in itself, no more and no less than the word "table" stands for the real table. However, with this conscious monitor circuitry being a die-hard 'picture-realist', it will start from this image—that is, the percept it binds into an image of a neural network—to explain how its phenomenal consciousness comes into being from the neuronal visual concept itself. It will imagine a complicated mechanism of neuronal interactions that are supposed to explain the emergence of its own consciousness starting from the neuron images—that is, from the projection of abstract symbols represented by the LCD—and failing to notice that this image is already an epiphenomenon determined by the memory chip and, before that, by the modulation of an electromagnetic signal.

Obviously, it can only fail. Things become recursive, for the simple reason that it falls into a trap: It is trying to explain its display technology, the microelectronics, the software that encodes the electric signal coming from the satellite dish starting from the images on the screen they produce–that is, in terms of the symbols appearing on such a screen itself. This is obviously a hopeless and meaningless attempt.

After a long struggle, it will conclude that the attempt to explain the emergence of its consciousness, by constructing some mechanism starting from the pictures it perceives, will lead nowhere. It tries to understand the emergence of its experience by arranging and combining symbols without being aware of how these symbols are already elements of the same experience it wants to explain. Failing to do so, it discovers the hard problem of consciousness.

Coming back to our daily reality, we should ask again: In what sense does our subjective experience of the world differ from this metaphoric monitor consciousness? And in what sense is the attempt to explain consciousness in a pure physicalist manner less naive than the monitor's attempt?

When we refer to matter, stones, molecules, and neurons, we are not aware of the fact that we are already speaking of symbols and subjective phenomenal events that are apprehended in our inner 'screen of consciousness'. These objects that we perceive as having a concrete reality are already objects of subjective experience–that is, experiential symbols; they are events in consciousness. To explain consciousness in terms of events, perceptions, and images appearing on our 'display' of phenomenal consciousness is like the monitor trying to explain the function of its LCD in terms of the images appearing on the very same display. We unknowingly still try to explain the inherent reality of an object by something that is inherently subjective. This is not only hopeless but also logically inconsistent.

When the materialists posit neurons—or, more generally, matter—to be fundamental and consciousness to be an emergent epiphenomenon, a byproduct of a complex arrangement of matter, they seem to not be aware that they fall into the same fallacy: They try to get rid of any subjectivity by positing, as a foundation, an experiential symbol, starting from subjective units of phenomenal events in the first place.

Expressing this in more general terms with a diagram, as in Fig. 61, we can represent this as the lack of *physical causal closure* of a theory. Physical causal closure means that everything has only physical causes. It is a principle and an a priori assumption that is at the basis of the physicalist ideology.

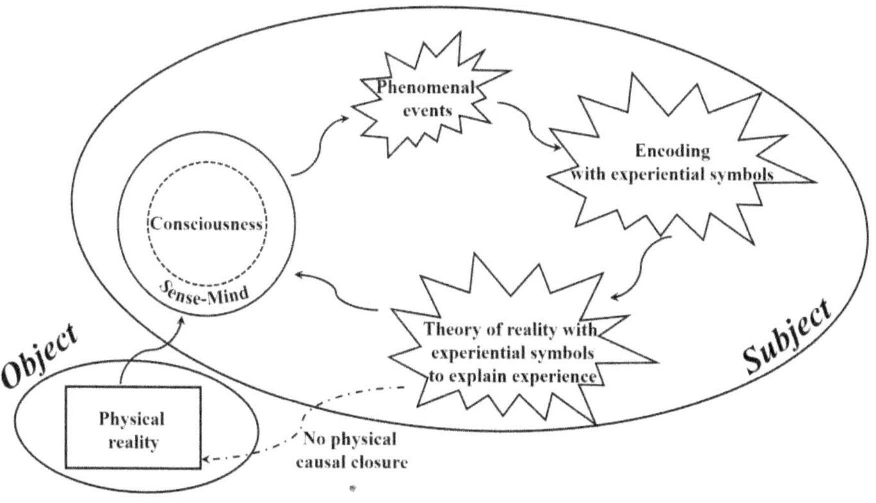

Fig. 61 No physical causal closure for physicalism.

We experience the physical world through the impression of physical interactions onto our sensory perceptions of our bodies and a mind that receives these perceptions, and that we simply called sense-mind.

This sense-mind will experience phenomenal events that it will almost instantly actualize in a scheme encoding reality in semantic units, the experiential symbols. If our minds try to build a theory of reality with experiential symbols to explain experience itself, then they are still working with the same experiential stuff they wanted to get rid of. But they can't get rid of the subjective element to describe the objective one. At the bottom, there is still no distinction between subject and object because the latter remains out of the domain of the subjective experience, which means that it fails to realize a physical causal closure.

If consciousness emerges from an aggregation of matter and its interactions, then we must expect that in reducing that aggregate to its components, nothing that can be related to a phenomenally conscious experience will be left. In general, the attempt to make a property emerge as an epiphenomenon from a complex aggregate assumes that, once we deconstrue it, the property should cease to exist. For example, the wetness of water emerges at a macroscopic scale as an interaction among a huge number of microscopic H_2O molecules. The wetness property ceases to exist when we look at the fundamental constituents that make it come into being. It would make no sense to state that a water molecule is wet. Nevertheless, putting together two atoms of hydrogen with one of oxygen is not a simple operation of adding individual entities. A new quality emerges when a huge number of these molecules interact and constitute a whole—that is, wetness—and that has an existence not in itself but only in us as a perception of a qualitative construct in our consciousness of its viscosity (recall Pt.I-II.2.) In other words, water is not wet; it becomes wet only through our perception. Wetness is like colors or smells in that it has no meaning in itself at all if not for our subjective experience. The existence of wetness in water is as real or unreal as the existence of the perception itself. Wetness in water exists, as does sweetness in sugar. It makes no sense to speak of these qualities as independent of a subject.

This example makes it clear how the idea of explaining qualities from a reductive perspective is misplaced or, at least, much more intricate and subtle than we may assume in the first instance.

At first, this seems to be the case when we look at a realm that we perceive as inanimate matter: Dissecting the brain, in every single neuron, molecule, and particle, that property we call 'consciousness' ceases to exist. However, the problem with this is that the notion of matter, neurons, molecules, and particles itself is still an experiential symbol. The claim that consciousness will be explained starting from matter assumes that consciousness will find a description in terms of something unconscious. But matter, its properties, and the physical processes that govern it are already conceptualizations in the mind or events in phenomenal consciousness itself. It is a sort of metaphysical materialism that posits something abstract, which it mistakes for independent of mind, and then tries to explain mind in terms of the very same abstraction of mind itself. It doesn't recognize that matter is a way of seeing, not something that is known and seen in itself. It wants to reduce the qualities of experience to non-experiential and non-phenomenal events without realizing that these are irreducible. It amounts to the attempt to explain the emergence of something with a construct made of the same thing and then wonder why, by doing so, the thing doesn't generate other than itself. Trying to explain consciousness with matter is like trying to deduce the properties of water with wetness. Or, to put it bluntly, it is like trying to explain the nature of chocolate with bits of chocolate and then wondering why one isn't able to explain it in terms of something that is not chocolate. At the bottom, it wants to generate something out of magic.

Consciousness is beyond the symbols with which we would like to explain it. It apprehends qualia, mental events, thoughts, and the experiential symbols we call 'matter', 'neurons', 'neural networks', 'molecules', etc., as the witness that stands beyond any of these experiential or phenomenal symbols. Consciousness is the 'screen', the 'monitor' on which these events appear, and cannot be explained by the images on the monitor itself. One can't describe the nature of light in terms of the movie characters it projects on a cinema screen. Conflating the movie content with the light, declaring the latter as an epiphenomenon of the former, is the fallacy of physicalism. It is as absurd as trying to explain the structure and function of a table as a complex arrangement of the symbols "t", "a", "b", "l", and "e". That's what physicalism does.

The experienced material world is, ontologically speaking, no different from a written text impressing our conscious subjective experience we associate with a meaning, not an objective realm existing on its own, separated from experience. What we hear, see, smell, touch, and taste appear to our consciousness as a code and translation into experiential symbols that have entirely lost any similarity to the inherent nature and existence of the things they represent. It is the translation of the mind through the senses, a 'sense-mind translation'. What science is doing is not a description of things as they are; rather, its attempt consists of finding a 'sense-mind language' that is the best possible translation of nature's language. However, the objective truth of things–that is, a truth that does not depend on our subjective way of cognition– remains inaccessible to us. We run into trouble only when we believe that our 'sense-mind symbols' that appear to our consciousness are the reality–as it is in itself–and that it is independent of mental constructs. Then, obviously, a missing link between matter and consciousness appears because we miss how our abstract constructs have already crept in.

This circling becomes apparent when science would like to explain something inherently subjective by trying to objectify it—say, for example, the sensation of being a subject, the first-person feeling of 'I-ness'. At this point, reason would like to shift towards a third-person perspective but finds itself trapped in a loop in which it is trying to exceed itself and inevitably falls back onto itself again, as an eternally switching of meaning in a Gestalt figure. Scientists and philosophers of mind indeed intuit this logical circularity. That's why, throughout the literature, there is a tendency to resort to explanations that involve 'recursive neural mechanisms', 'feedback loops', 'self-instantiating reflexive self-directedness', 'resonances', 'phase-locking', 're-entrant circuits', 'internal monitoring', etc., which supposedly have an explanatory power that they do not have, not even in principle.

It makes things in no way better if we delegate our perceptions to a mechanical measuring device. In what way does a subjective experience of the perception of, say, a chair through our senses, with the subsequent symbolic representation in our minds of it, differ from the instrumental measurement, with some device, of a particle position, which is also mapped in our minds through some other symbolic representation? There are obviously technical differences here, but finally, we are always left with an abstract mental reconstruction that appears to our consciousness. In this sense, the outcome of a measurement device—say, in the form of a number that measures a physical magnitude—is no less real or unreal than the symbol appearing in our minds when we perceive the light reflected from an object such as a chair. What we continuously forget is that we never see the objects as they are; we perceive only the mind's construct of it and in us through our conscious experience.

Someone may protest, saying that this argumentation amounts to placing, say, a chair on the same ontological footing as, for instance, a number such as π. The answer is[33]: The chair in itself is not an abstraction like π, but our mental perception and experience of it is. What appears in our minds and what we see when we are looking at a chair has no objective and ontological superiority in comparison to the 'conceptual appearance' of π in our mental screen. Obviously, the difference is that we can forget about π but not about the chair; otherwise, we will stumble upon it. There is something in physicality that is more concrete than just a number or thought. Nevertheless, this does not elevate the mental concept of it created by the phenomenal experience of the chair in us to a higher status; it remains a symbol that has no more reality than any other symbol. It is a symbol that stands for the real thing but retains nothing of the thing itself. It is, in principle, also conceivably a conscious being with a different nervous system and brain which, when looking at a chair, may perceive the appearance of the notion and meaning of π. It is a being that would probably have a very short life because such a world perception wouldn't grant any evolutionary advantage. However, the point we want to make here is that we have to avoid the deceiving distinction between our own world perception and the perception of symbols. Otherwise, if we forget how we create our inner mental phenomenal constructs, taking it as the ultimate reality, problems

---

[33] Unless one embraces a Platonic form of idealism where also numbers are assumed to have a concrete reality.

inevitably arise because we unconsciously mistake what we perceive for what is perceived.

At the bottom, this is why there is no way, not even in principle, for us to derive consciousness from matter, as matter is an appearance, a figment, a projection, or simply an experiential symbol appearing to consciousness itself. Because consciousness has this irreducible and primary character, it would be much more interesting to know how this appearance we call 'matter' is derived from consciousness.

The possibility that, nevertheless, something exists 'out there' is still an open question. To realize that everything is a construction forwarded to our consciousness does not imply that everything is unreal or that nothing exists. Things exist, but first, we must acknowledge that this idea of objects having qualitative or quantitative properties independent of our consciousness and our sensory means has no logical ground. Secondly, it is the choice of words, such as the meaning of 'existence', that should be thought all over again.

What we perceive to be 'real' is a reification in us. *'Reification'* is the assumption that our senses or a system of measurement informs us about the intrinsic nature of the measured or perceived object. The entities that we reify exist (whatever 'existence' might mean). However, their essence is beyond any of our experiential, conceptual, or verbal designations.

This doesn't mean that we are soul-less algorithmic machines crunching symbols. Observing the world, we see and perceive the beauty of a flower, the aesthetics of a landscape, we feel a sense of awe at observing the power of a natural phenomenon, the unity of and in Nature, we capture an undefinable vibration in the manifestation of things. All that can hardly be reduced to a dry sequence of conceptualizations. Precisely this has motivated the whole work you are reading. There is a Spirit that *shines through* our figments and mental constructions. But it is not a Soul that can be derived by our abstractions. Rather, it is an immanent harmony in things and Nature that we learn to perceive *despite* all our filtering perceptions and reductive conceptions. As we will see next, there is an archetypal truth in things we can capture.

For the time being, this should make it clear how misplaced the present belief is that science studies an objective reality independent of our subjective experience and mental designations. The belief that scientific theories are objective, one-to-one representations of the universe, as it exists independent of our experience, is a metaphysical assumption underlying modern science that, upon closer scrutiny, is revealed to be wishful thinking. Physics was originally designed to explore the essential nature of reality, but from this perspective, instead of progressively reaching this goal, paradoxically, it has found itself removed further and further away from it as time passes.

Most scientists are not at all interested in these philosophical and existential questions. Such concerns are deemed uninteresting, for experiments can be carried out, and mathematical theories can be formulated without any reference to ontological questions. However, such an indifference veils a naïve and unreflective realism. A disinterest in spirituality does not lead to an objective assessment of how things are in themselves but rather results in an unconscious and unintelligent adoption of a particular form of metaphysics, usually a simplistic, if not grossly ignorant, form of realism. The assumption that science is describing an objective world independent of our consciousness and our conceptual constructs is a deeply engrained metaphysical belief that most scientists are unwilling to reconsider, for it is regarded, often implicitly, as a priory truth. It is a form of realism assuming that if our representations and models can predict phenomena, they reproduce the physical constitution of external objects as they are. This metaphysical view also remains the undeclared and unaware dogma that saturates most of our institutional education in science today. These unaware assumptions that permeate a naïve scientific realism are rarely addressed in classrooms. It is simply taken for granted and absorbed subconsciously and uncritically from generation to generation and exerts a powerful influence on the thoughts and attitudes of our society as a whole.

And yet, while we can accept that matter is not as it appears to us, it is hard to believe that the stuff we perceive with our sense isn't somehow present outside of our brains in the world 'out there'. Someone may wonder if, despite all this, a correct mental construction of reality could exist. Is there, after all, a true correspondence between a subjective experience and an objective reality? We could still argue that perhaps we don't see things as they really are, but it is, at least in principle, conceivable to think that some way of looking at them may convey to us the correct way of seeing them as they really and objectively are. In other words, can there exist a one-to-one correspondence between 'reality as it is' and a representational model in our minds?

For instance, if I look at a wave, I can be intellectually aware of the fact that it is a transcription, a figment in my brain (just as, for example, the abstract mathematical sine or cosine function is an intellectual transcription of a wave as well), and that fails to produce such one-to-one correspondence. But, after all, there is 'something out there' that causes that transcription or mental figment to present itself to our awareness. It is something we call an 'object,' and we can hardly believe that it is only a dream or a hallucination. May I ever become conscious of the real ontological nature of the ultimate reality of things?

We can say that this is not possible, not even in principle, by resorting to our human sensory means and the mind alone, at least as they are actually developed in the species we call 'homo sapiens'. Our sense-mind, the rational intellect, and reason alone can give us only a superficial and distorted sense-mind image of things. On the one hand, we must become aware of the limits of how the materialistic knowledge of the world which is supported by a low-level cognitive apparatus such as our mind, reason, and senses, can't be the final and highest forms of cognition, as we would like to believe, and as science tacitly assumes. On the other hand, this leads to the question of whether we can suppose that some superior form of cognition can deliver us such knowledge—say, a sort of 'knowledge by identity' where object and subject identify and become one and the same, revealing to us the inherent essence of things as they are in themselves.

We don't have a crystal ball and can't look into the future evolution of mankind. Nevertheless, it is plausible to think of higher forms of cognition beyond reason and the sense-mind thought processes, capable of presenting to us the thing-in-itself and all truth about it. A knowledge by direct contact and an unfiltered experience is, in principle, conceivable. Something that goes beyond the sensory and mental vision not only is plausible but seems almost inevitable if we accept evolution: It must come as the result of further evolution of the development of our cognitive faculties. This is an aspect we will take up again later in Pt.III.

4. Kant's Noumenon, Plato's Cave and Non-duality

Here, we will recast the findings of the preceding chapters in the parlance of the school of philosophical idealism, especially from the perspective of physics. We stress in advance that we don't believe idealism to be the ultimate philosophical point of view that correctly describes the relationship between our mind and the world. Idealism is only a transitional step in the evolution of humanity's thought and the understanding of reality. There are still higher forms of knowledge that contain but transcend idealism. Nevertheless, idealism is a first step in the right direction that can lead us a bit further than the present anthropocentric physicalist view. Idealism is a connection, a bridge that can help us to step beyond the naïve materialism still plaguing the conceptual foundations of science towards a more developed perspective that will be elucidated in Pt.III. We will first recall some basics of philosophical idealism and, in the next section, a few aspects of quantum physics (QP).[34]

The doctrine of philosophical idealism dates back to the time of Plato, reemerged in different forms over and over again throughout the centuries and continents, and continues to influence philosophical inquiry to the present day. There are so many versions of idealism and subtleties to be considered that it is difficult to define them in a few sentences, and we won't be able to go through all of the details here. However, what they all have in common, and that will be relevant later for the philosophical implications of QP, is their recognition that mere empiricism—that is, the act of measurement of a physical quantity alone and which characterizes all the exact sciences—does

---

[34] This and the next section are a less rigorous and modified and reduced version of the two chapters that the author published in his books on quantum physics. [221] [36]

not lead to the revelation of an objective reality, if by 'objective reality' we mean a model that faithfully represents things independent from our mind.

The philosophers of ancient Greece or of the 18th–19th centuries did not know anything about neurobiology, but they were acutely aware of the fact that our senses and our minds have no direct access to 'reality as it is'. This led several of them to take sides regarding a conception of reality in which we can't abstract from human sense-cognition.

Broadly speaking, one can distinguish at least two categories of philosophical idealism. One is *'ontological (or subjective or metaphysical) idealism'* which holds that all reality is a mental construct and, at its foundation, regards mind (instead of matter) as the fundamental constituent of reality. One of its most notorious supporters was George Berkeley. Because QP does not (or does not necessarily) support the claim that all of reality is a mental construct, we will focus our attention on some formulations of the second type of idealism, that of *'epistemological (or objective or formal) idealism'*' expressed by 18th- and 19th- century German philosophers such as Immanuel Kant, Friedrich Schelling, and Arthur Schopenhauer and that, later, grew into a philosophical study known as *'phenomenology'* in the 20th-century and is represented by philosophers like Edmund Husserl, Alfred North Whitehead, Martin Heidegger, and several others. These are just some names that may furnish a direction for further reading and that we will discuss a bit more in detail in Pt.II-II. In the next chapter, we will try to illustrate how a similar worldview emerges from QP if we give up our naïve realism and embrace a larger concept of reality that questions whether science is the ultimate tool for investigating ultimate truths. Without having the pretension of exposing a rigorous definition and explanation of epistemological idealism, we try to characterize it as follows.

Epistemological idealism posits that all our knowledge about reality, being inevitably mediated through our senses, is only a formal representation emerging in our minds and that these forms can't be taken as representative of things as they are 'really'. Relating to an object should not be mistaken for knowledge of the *'thing-in-itself'*, as Kant used to call it, being undistorted and independent from any mental activity. Every insight that our knowledge reflects is a structured thought in our minds where we conflate knowledge with the object of knowledge. Our minds process sense data by organizing it into abstract propositional knowledge that refers to quantitative and qualitative properties which the things-in-themselves do not have. Kant, in his magnum opus *"Critique of Pure Reason"* and in a second edition (1783), in which he challenged Berkeley's ontological idealism by adding a *"Refutation of Idealism"*, introduced the form of idealism known as *'transcendental idealism'*. Idealism, says Kant, does not concern the existence of things independent from our perceptions, the things-in-themselves, and that he notoriously called the *'noumena'*, but is only about our modes of representation of the perceived '*phenomena*.' Also, entities that we consider to be self-evident and fundamental (such as space and time) are not noumena but, rather, are emerging phenomena in our minds. Whatever we know, we can't avoid knowing it by filtering it through a sense-mind perspective. This doesn't imply that there is nothing at all or that we must surrender to some form of relativism or nihilism. There exists a mind-independent reality in which something else must be the ultimate foundation of all things. However, that ultimate reality of things cannot be accessed by our senses and our minds, which can present us with only an abstraction of it. There might well be a subject-independent world, with things 'having a being' without us looking at it, but this is beyond the possibility of our normal cognition. One may argue for or against an 'absolute truth' but, if any, also in science, we are allowed to know it only through a fiction based on subjective experiences.

Therefore, Kant's transcendental idealism remains agnostic as to the constitution of the ultimate reality of things. If by 'realism' we mean that objects exist independently of us, then idealism does not contrast with realism. Though we know only the mind-content structure of perceptions inside of us without any possible direct contact with things outside of us, we have no reason to deny the existence of the external objects that trigger these sensations, concepts, and apprehensions. The human mind can't have a knowledge by identity with things. Our knowledge of the world begins with a sensorial experience which translates itself into representations, pictures, figures, figments, and reflections leading to a state of mind.

Kant was by no means the first one in pointing out the distinction between the phenomenal and noumenal world. Plato used his '*cave allegory*' already a couple of millennia earlier coming to similar conclusions. We are like those who have lived their entire lives chained inside a cave, forced to stare at a blank wall without having access to the outer world other than by seeing, on the wall, the projected shadows of people passing in front of a fire at the cave's entrance. We misinterpret the shadows for real persons because we know nothing else. Unless we become aware that we know only projected shadows of reality, we will forever live in a world of delusion without even being aware of it. Even if we become aware that we are living in a virtual reality, we cannot avoid reifying inside of us those phenomenal appearances in our minds. There is, indeed, some sort of underlying reality. The universe is not just fiction in our mind, but that reality has nothing in common with the constructs and appearances it triggers inside our mind, no more than a shadow of a human body telling us something about the nature and complexity of a living body.

Indeed, a typical claim of the mystic is that the world we perceive and that we mistake for real is, instead, nothing other than a 'shadow of the Spirit.' We see only the mental shadows of the outer world projected into us and mistake the shadows for the real things. Shadows are something intrinsically inexistent that can nevertheless cause perceptual events. Matter as we perceive it is a shadow. We exchange for objective reality the image of a shadow for the object that projects it.

Fig. 62 The Plato's cave allegory.

Thus. the claim of the philosophical idealist is that Plato's cave allegory is more than that, it represents our perception of the actual world: What we call 'the material world' is also a shadow that lacks inherent existence, and which science instead exchanges for a self-evident fact and bases as a foundation of everything.

Let us expand this allegory. Imagine living on a 2D surface, a sort of 'flatland world.' All of reality and its objects, however, are placed in a 3D universe, while all their shadows are projected on the 2D world we live in. We are 2D beings chained on this surface, and never ever, not even in principle, can go beyond this 2D surface. We can perceive only the shadows of the 3D things. Using this analogy, the real 3D objects are the things in themselves, Kant's noumenon, while the projected light represents our act of sensorial observation (our senses 'reach out'), and the shadows represent our ordinary perception of things, Kant's phenomena. We mistake these shadows for being the real things as they are, without realizing that they are only projections, figments, or 2D 'icons' that have no inherent reality and are stripped of almost all the real properties of the 3D objects projecting them.

And what is the cave, or the screen on which these shadows are projected and become visible? In this metaphor, the screen is consciousness. This 'screen of consciousness' is what all of reality is projected onto. What is real is the screen–that is, consciousness–not the shadows or the images projected on the screen. If the light that projects the shadows is turned off, the shadows disappear, but not the objects that previously projected those shadows and not even the screen.

In this sense, consciousness is fundamental while matter, or at least our perceptions and conceptions of matter, is an illusion. Consequently, trying to explain the emergence of phenomenal consciousness by means of brain activity is like trying to explain the existence of the screen by means of the shadows and images projected on the screen. This is a senseless attempt that only highlights our ignorance of how we perceive the world and, ultimately, of what and who we are.

If we want to progress toward a deeper understanding of what consciousness is, we must become aware, at least intellectually, that we create in ourselves, on an internal screen of consciousness, like the projections of the shadows on the wall of Plato's cave, sensory representations of things which, however, do not reveal to us anything about the things themselves. What physics calls 'matter,' 'space,' 'time,' and 'force' are the reflection of a cosmos that we experience as shadows in our minds. The true nature of things independent of ourselves is forever precluded to our human and scientific sense-mind knowledge. This is because, to use Plato's allegory again, cutting the chains and leaving the cave so that we can know things as they truly are means that we must first eliminate the sensory perception and mental construction of it. This amounts to an attempt to apprehend the world without interacting with it–a quite difficult task. For this reason, the belief that science represents an objective truth independent from any mental abstraction which, ultimately, is based on events in phenomenal consciousness, is misplaced. This belief assumes that a *'view from nowhere,'* as T. Nagel used to call it, is possible. It is yet another unaware form of an anthropomorphic faith pretending that we are capable of looking at the world from an absolutely objective frame, a sort of 'God's eye view.'

We live in a world given to us as 'being', without allowing us to gain direct contact with that state of being—at least not with our present abilities of cognition. Having existence of being' is all there is. Everything else is derivative. At least in the form of epistemological idealism, idealism is not an anti-realist conception but rather a different form of realism that does not posit matter, space-time, forces, or any other phenomenal and even not mental experiences as the fundamental 'stuff' of reality.

Also, the most empiric, exact, rational, and analytic science, as physics could be considered, must resort to experiments composed of mechanical sensors and measurement devices which replace, extend, and are delegates for our human sensory abilities but, at the end of the experimental chain, it must always and inevitably convey its result in the form of numbers, data, or pictures that the human mind can interpret. The character of things in themselves cannot be inferred from any amount of empirical data. Science can only speak about the appearances of cognition but can't go far enough to tell us anything about the inherent essentiality of the things it cognizes. Also, our experiments in the laboratory, as precise and 'objective' and independent of human intervention as we might consider them, must be reduced to a cognition of appearances— that is, 'shadows'—rather than an inherently objective knowledge of things. Scientists should be aware (something which, unfortunately, most of the time they are not) that empirical facts are not the revelation of ultimate truths and, at least when making philosophical claims, should distinguish between appearances and things-in-themselves.

The fact that things like stones and mountains are permanent objects of perception and are subjected to what we call the 'laws of physics' is by no means a guarantee of objective knowledge if, by 'objective', we mean the inherent essence of things as they are, independent of our perceptions and modes of cognition. The perceived world cannot be considered 'real' only because of its permanency, while we consider dreams as 'unreal' because we find ourselves in a different dreamworld every time we fall asleep. We instinctively consider the world we perceive in the waking state as 'real' because every time we wake up, we find ourselves in the same world as we were in when we fell asleep. However, pointing out this permanency is not a sufficient argument to make our everyday experience the ultimate designation of reality. Because also the shadow of a stone or a mountain on the cave's wall is permanent and subject to the laws of physics; however, this does not allow us to replace the shadow with Kant's noumenon, the real thing-in-itself.

Kant thought there must be something that exists independently of us being the cause of our representations of the world that appears to us. Otherwise, these representations and appearances of the world we perceive would arise out of nothing. That's why we are forced to conclude that the things-in-themselves which are the cause of our representations exist, but they are mind-independent or 'transcendental objects'. There is a substratum underlying the phenomenal world, the noumenon, but we can't investigate the true inherent nature of the things-in-themselves and have objective knowledge of it because we always start from the representations—that is, from the phenomenon, the symbol.

All this does not necessarily imply that reality is ultimately mental (the claim of Berkeley). If we are left with nothing other than mentation in investigating the outer world, this does not imply that the world is mental—or, at least, not only mental. The fact that we perceive the world with 'mental glasses' does not make it mental, nor more and no less than the fact that perceiving only things with rosy glasses makes the world pink.

One can't fail to notice how reminiscent this ontology is of some Eastern philosophical traditions—in particular, the Indian spiritual doctrine of *'non-duality'*, which was developed between the 7th and 8th centuries by the Advaita Vedanta Hindu school of thought and whose roots can be traced back to the ancient Vedic texts.

According to non-duality, and as seen from higher states of consciousness, reality is an undifferentiated 'One without a second' ultimate

Fig. 63 The Hindu Om symbol for the ultimate Reality.

Reality—the *'Brahman'*. The nature and source of everything is this Brahman, whose essential reality is a featureless Unity in which duality, separation, and fragmentation are only illusions in an essentially undivided Whole. Only when the appearances of polarity deceive our senses and mind will we be caught up in the the cosmic '*illusion of Maya*', also called the '*veil of Avidya*'(Ignorance). A fragmented universe appearing to our senses and mind in the form of particles, atoms, molecules, things, and beings does not reflect its underlying essential reality. Ultimately, everything is the manifestation of essential non-dual Oneness; however, our sense-mind, which is by nature a dividing tool of cognition that distinguishes, separates, and has the tendency to reduce everything to sub-components, can only know the world by contrasting and comparing perceptions and sensations, making it appear to be a reality of irreconcilable polarities. This lures us into the false belief that we live a separate individual existence, identifying ourselves in a mental 'ego', caught in the play of dualities, the source of all suffering and worldly despair, where the knower and the known object seem separate. The illusion arising from our perceptual and mental representations of reality is summarized in the Vedic parable of the rope and the snake. The 'unawakened' live in fear because they mistake their mental world for real, similar to someone who mistakes a rope for a poisonous snake. The existence of the rope is not denied, but the image of it that we project onto our internal 'mental screen' is a fantasy that has nothing to do with reality. Meanwhile, those who are 'liberated' ('moksha') recognize the falsehood of Maya through an inner realization (usually effectuated by a long psycho-physiological preparation through practices such as meditation and yoga), and 'awaken' to a state of consciousness in which one lives in the transcendent, blissful, time-less, featureless, non-dual, undivided, and unborn Brahman. In hindsight it makes sense. Anything that is without form, qualities, or properties, can't come in or out of existence.

The difference between Western idealism and Eastern non-dualism is that the latter not only is a philosophical school of thought but also, in a sense, distinguishes itself by going exactly in the opposite direction. The Eastern experiential approach dethrones the mind and transcends the philosophical thought, rooting its knowledge in a mystical experience which, however, once framed into words, must inevitably take on a philosophical mask. However, the conclusion they arrive at, if not the same, has at least strong similarities. Something along the lines of an absolute monism can also be found in other Indian or Buddhist teachings (for a modern Buddhist view of physics and the mind, see [220]), as well as in some Christian mystical traditions—above all, in the 14th-century mystic and philosopher Meister Eckhart.

And the overall common conclusion they all share more or less implicitly is that consciousness is primary, the fundamental root of all things, the ultimate fundamental stuff of the universe. Taking a more modern evolutionary view, this amounts to saying that it is not evolution that creates consciousness; rather, it is consciousness that emerges in and through matter. Consciousness did not evolve by material processes building upon chunks of dead and unconscious matter. On the contrary, consciousness is the ground and primary force that impels the evolutionary process already from the times of the beginning of the universe.

5. Quantum Physics: Facts About a Weird Reality

The interested reader who is looking for a more detailed and technical introduction to QP can resort to the author's two volumes ([221], [36]). Herewith, we will take a quick glance at those facts, experiments, and aspects of QP that will be relevant for further thoughts supportive of an idealistic conception of reality.

Now, why should we regard QP as a source of idealistic considerations? After all, this is a philosophical school of thought that existed long before the inception of QP. Since the times of ancient Greece, philosophers have questioned what reality is and the fundamental essence of things. Philosophical idealism can also be discussed and defended inside a purely Newtonian paradigm, and one does not need to resort to quantum mechanics (QM).[35]

The point, however, is that QM led us a step (or many more steps) further by confronting us with facts that clearly evidence how mental abstractions and constructions are tools for practical purposes, not representations of an inherent reality. Previously, we could still delude ourselves into regarding the appearances of the permanence and solidity of a deterministic and predictable clockwork universe as the ultimately 'true' properties that build up an inherent reality independent of our subjective perceptions and conceptions. Newtonian mechanics and Einstein's relativity were still in line with these unaware assumptions, and idealism seemed to be a distant philosophical remnant or romanticism. However, if QM is taken seriously (that is, not just as a calculation tool or a theory that we would like to see return to a naïve Newtonian or Einsteinian deterministic realism as many do by resorting to a plethora of so-called 'interpretations of QM'), philosophical idealism, in one form or another, is inescapable.

It is precisely in the quantum domain that the value and inescapable logical conclusion that points to an idealistic interpretation of reality presents itself most pressingly. If we look at quantum phenomena from the perspective of the philosophical idealist, several aspects that previously seemed utterly at odds with a naïve sense-mind-based form of realism now acquire a different significance.

For example, we realize how the notion of a particle's rotational intrinsic angular momentum is already a heavily sensorial and mental-laden representation borrowed from the everyday macroscopical realm. In physics,

---

[35] Quantum mechanics refers to the dynamical aspect of quantum physics. For sake of simplicity we will use both as synonymous interchangeably.

by the intrinsic angular momentum of a particle, also called *'spin'*, one means the rotational quantity of motion that a particle carries with it.

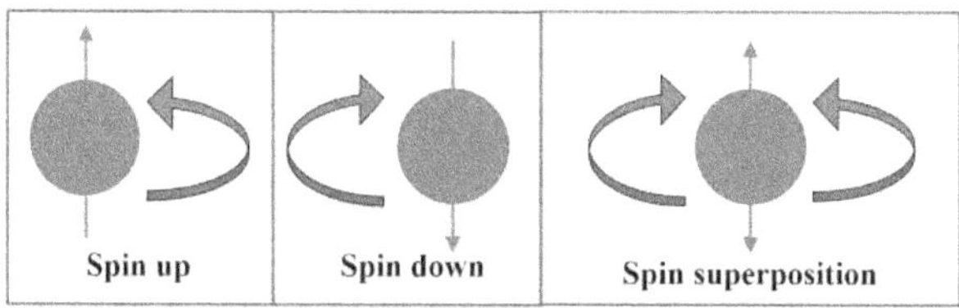

Fig. 64 Intuitive concept of the spin of a particle and its quantum superposition.

Or, more intuitively, imagine how fast the particle rotates around a given axis. By convention, we can speak of a 'spin-up' when a particle spins counterclockwise and a 'spin-down' when it spins clockwise. In QM, things become much less intuitive when we realize that, once the spinning direction is fixed, there can be only these two spin states with a fixed amount of angular momentum and nothing in between; you won't find a particle spinning half as much. Moreover, it turns out that quantum particles, contrary to the macroscopic objects we know from our everyday experience, can possess both spins at the same time—that is, one speaks of a *'quantum states superposition'*.

This weird property defies our natural understanding of the properties of the world to which we are accustomed. This aspect of QM becomes so weird because we unconsciously equate the very elementary concept of what we call a 'physical state' of an object to a configuration of properties that this object is supposed to inherently possess without being aware that these properties are our own juxtaposed sense-mind constructs. We imagine (which means we project a shadow on our internal mental screen) an extended body with some form and properties changing its appearance periodically in space and time. However, the very notion of 'rotation' is a concept we reconstruct from a series of subsequent appearances presenting themselves as stimuli in our brains and that we bind together, as one does with the sequence of photograms of a film, into a mental movie. When it comes to the question of whether this macroscopic notion known as 'angular momentum' makes any sense at the fundamental level of the inherent reality of things 'as they are really', we must realize that we are speaking of useful inter-subjective qualities that, however, have nothing to do with properties of an ontological reality. In this sense, the spin of a particle stands to an objective reality like the subjective experience of the color of an object stands to the object itself.

Quantum superposition is not limited to spins. The same can be said about every 'physical state' of a particle. For example, when we say that a particle 'has' a specific position in space, we are already working with an internal representation of it which mistakes mental projections like space and point-like entities for 'real' properties of something 'out there'. By doing so, we have already projected into ourselves a fiction of reality that is detached from 'reality as it is'. In this sense, the position of a particle in space and time is no more and no less 'objectively real' than its taste, smell, or color. In fact, in QM, particles behave as if they could be in a spatial superposition—that is, in two or more places at the same time.

The typical case where this can be observed is the double-slit experiment, which eloquently illustrates the famous *'wave-particle duality'*. When an elementary particle—say, a photon or a material particle such as an electron—passes through sufficiently small and nearby slits, it behaves like a wave by being diffracted at the slits (think of water waves or light diffraction at an edge). It appears to go through both slits at the same time.

However, once it reaches the screen (say, a photographic plate or a CCD camera), it is always and only detected as a particle—that is, a spot, a single interaction in a precise place. Which precise position turns out to be the observed one is a purely quantum random event. However, if many particles are 'shot' through the double slit, interference fringes appear, which are the sure sign of waves interacting with each other. In other words, a particle like an electron behaves like a wave but is seen only as a particle, then again distributes itself along the detecting screen according to a statistical law that constructs an interference pattern typical of a wave.

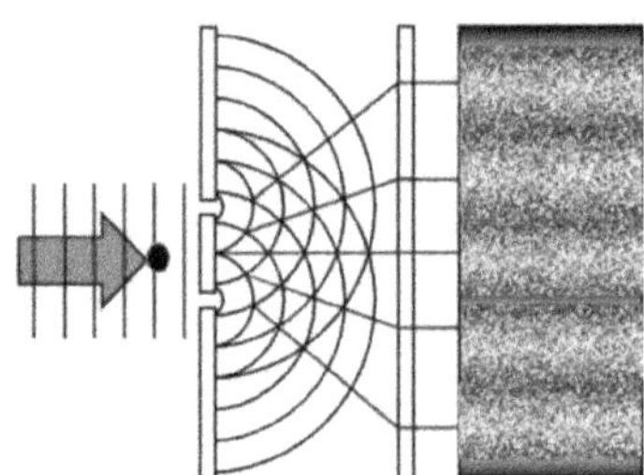

Fig. 65 The double slit experiment and wave-particle duality.

So, what is true? Is there a particle or a wave or something else? It looks like what we call 'elementary particles' in QM have a specific 'wavy-physical state' described by a *'wave function'* as long as they are on their way to a detector and then 'collapse' into a particle state once measured—the famous *'collapse of the wave function'* (also called *'state reduction'* and *'quantum projection'*.)

These strange quantum properties have created a headache for great minds such as Einstein and the Austrian physicist Erwin Schrödinger. We can't relate these phenomena to our everyday experience and understanding of the world. It turns out that we can't say with our limited sense-mind abilities what a 'physical state' of an object really is in itself, other than formulating it as a 'state-of-being'. Quantum states are states-of-being which we would like to project into a conception of a distinct and unique spatio-temporal relation of properties typical of the macroworld and that we are accustomed to mapping into our minds. However, a map does not say much about the territory. The superposition of quantum states is simply another state-of-being of something that finds no correspondence in our cognitive map.

To use the Platonic allegory again, we could compare it to the projection of the shadow of a cylinder on a 2D surface. The measurement of an object in quantum superposition and its related wave function collapse corresponds, in the analogy, to the projection of the cylinder into the figure of a disk or a square (see Fig. 66). At each measurement, we randomly get one or the other result, and we cannot understand how that could be considering that, from the quantum mechanical point of view, the initial physical state of the object is always the same.

In this analogy the cylinder represents the real thing-in-itself, the object in its state-of-being, in its essence, as it really is, while the two possible 2D figures stand for the 'shadowy' scientific sense-mind perspective that modifies and alters a projected experience into an abstract internal representation which is inside of us and which no longer has anything of the essentiality proper to the measured object. The random appearance of one or the other shape does not imply that the cylinder has anything to do with squares or disks. Despite the fact that we 'objectively'–that is, inter-subjectively–all agree to observe squares and disks, nobody could explain with it a world made of cylinders. Unfortunately, that's precisely what several interpretations of QM try to do

Fig. 66 The cylinder metaphor for quantum superposition.

.'Heisenberg's uncertainty principle' reflects the quantum randomness that determines where a particle hits the screen. One can reformulate it by saying that the precise position and momentum (the quantity of motion given by the product of the mass and velocity) of a particle can never both be known at the same time with arbitrary precision. One can eventually determine with high precision the location of the particle, but then one will have only an uncertain outcome once measuring its momentum. The opposite is true if we determine, with precision, its momentum; then the position will be uncertain. This is the infamous indeterminism of QM. Heisenberg's uncertainty principle sets limits on how precisely we can know things. Beyond a certain limit, everything becomes unpredictable, indeterminate–fuzzy, so to speak.

This flies in the face of the deterministic Laplacian conception, to which, in particular, the physicalist's realism clings. In my opinion, an anthropomorphic understanding of the world in which everything must be mechanical and automatic—like clockwork—and where, for any cause, there is only one possible effect and outcome. A conception of phenomena taking place outside strict cause-and-effect mechanics, an 'acausal' reality that is at odds with a deterministic conception of a mechanistic universe of the pre-quantum classical physics dominated by point-like particles that kick around each other like tiny marbles. Quantum indeterminism might hide a secret, a power, a meaning that we have not discovered and possibly won't ever know if we don't allow ourselves to know it by sticking to the 19th-century naïve realism.

What about *'quantum entanglement'*? This is yet another strange property of quantum particles that challenges our understanding of a local realism positing space as a self-explanatory fundamental datum. However, Nature reminds us that it isn't. If two particles interact with each other, they will *not* form an object made of two parts, like gluing together two bricks. The two particles become intrinsically indistinguishable because they become the same quantum field and will form a unique and undifferentiated whole. This undifferentiated whole is expressed again by a wave function that spreads out into space and collapses back into the two particles only when measured with a macroscopic measuring device.

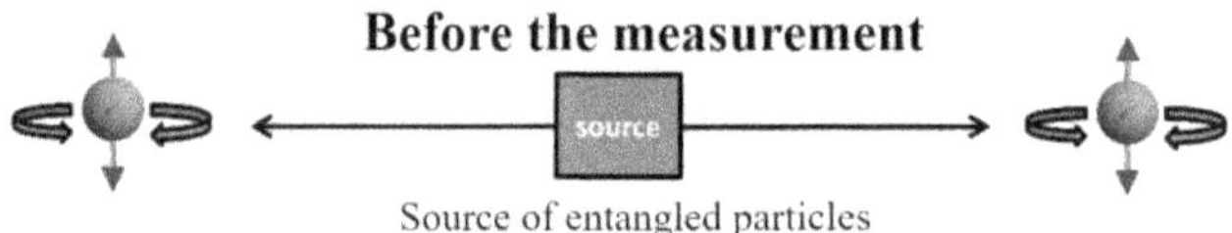

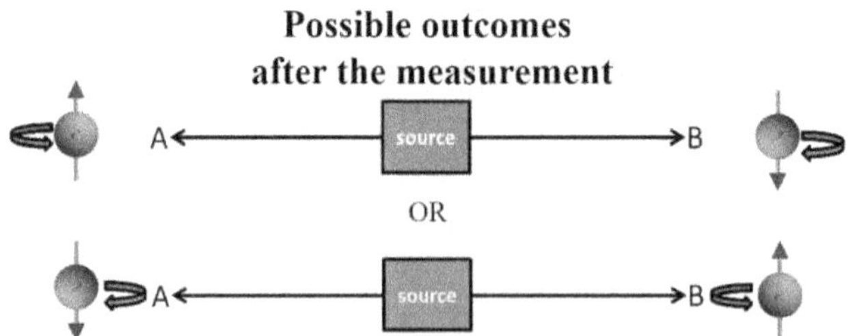

Fig. 67 The quantum entanglement of two particles.

Until then, they can be separated light-years away and still form an undifferentiated whole. At the instant of measurement of one particle, the other particle, eventually light-years away, will collapse to its identity with, for example, anti-correlated properties such as showing up with the opposite spins. If, on one side, someone measures the particle in the spin-up state, someone else, on the other side of the universe, will measure the spin-down state on the remaining particle. This seemingly 'spooky action at a distance', as Einstein called it, is, again, contrary to all our inborn perceptions and mental representations of what space, the extension of bodies, and the local character of reality are supposed to be.

QM tells us that reality is non-local. That is, a measurement performed on a particle in one place will immediately influence the state of its entangled particle, even if they are separated by a distance that cannot be bridged by the speed of light in the time interval during which the two measurements take place. This apparently also defies the theory of relativity of Einstein, but doesn't really. Quantum entanglement tells us that our notion of space and time is a misunderstanding. There is an inherent holistic aspect in reality—and that, obviously, is impalpable to the materialist reductionist mind.

Quantum entanglement appears somewhat less mysterious from the standpoint of the philosophical idealist. Our difficulty in accepting a state-of-being of something singular with no parts and distinctions can be traced back to the mind's intrinsic disposition and natural cognitive function to distinguish, particularize, dissect, divide, and fragment in space and time for analysis and interpretation. A quantum-entangled object, without differentiation and distinguishable parts, is in a state of being that is inherently other and opposite to what the mind's function is able to apprehend. It is like trying to understand the meaning of a word by analysing it in terms of phonemes. If one is cognitively unable to bind sounds together into a unique and undifferentiated perception, the real meaning that the word represents will forever remain precluded. A quantum-entangled system is in a state-of-being of inherent unity in plurality—something that to the mind is self-contradictory, and that can't be reconciled.

The '*principle of quantum indistinguishability*' is an inherent state-of-being of electrons. With its instinctive and natural function of separation and distinction, it is the mind that can't follow. At least for those who embrace epistemological idealism and who do not regard mental constructs as representations of an inherent reality, it is not entanglement that constitutes a conceptual and ontological problem—quite the contrary. Once we accept that mental conceptions of space and time are mere projections in our conscious internal experience and that they have no substantial reality, the question arises: Why does a fragmented universe appear to us at all? Why isn't everything permanently in an entangled and undifferentiated state? Why does this appearance of fragmentation emerge in the first place? This is a much more mysterious fact than the question which asks about things the other way around.

From the point of view of idealism, it becomes clear why the pervading and almost irresistible desire to reduce everything to deterministic paths of point-particles is simply another sense-mind reification in us and does not represent an inherent aspect of reality as it is. The very notion of a particle is a conceptual entity that arises due to a cognitive mental act of distinction and separation that points dots on a mental screen of consciousness due to the interpretation of a sensorial data stream. However, this is already a fiction created by a state-of-mind representation arising in our conscious experience and must not be mistaken for the inherent reality of things. Also, in the so-called 'particle physics', we never observe particles. What is measured is always and only a localized interaction of something with something else. We subsequently reify–that is, imagine–the effects of this 'something' as a mathematical point-like object surrounded by a force field, but no one has ever seen such a thing directly. All experiments in particle physics are about the measurement of the effects onto a finite-sized body, such as a pixel on a CCD camera, a photodiode, a spot on a photographic plate, etc., none of which are points. Quantum uncertainty is the natural state-of-being of things that become defined and determinate only at the instant of the experimental measure, which is always and inevitably a mapping into a mental representation based on distinction, reduction, division, and separation.

Just as we cannot describe the physiology of a person's body by looking at the shadow cast by that body, neither can physics describe the inherent truth of things like a spin, a path in space and time, the position of a particle, or the notion of a particle itself, as they are independent of our sensorial perceptions and mental constructions, which are built upon abstractions that are, in turn, based on other sensory data and cognitive constructs. The question is no longer why microscopic things appear to be uncertain and resist being cast into a reductionist worldview, but rather why an intrinsically undifferentiated whole allows itself to be reduced to what we perceive at the macroscopic scale as being an extremely fragmentated and definite reality.

And yet, the fact is that what we experience is made of parts, 'chunks of matter', which can be explained with molecules and elementary particles subjected to the laws of physics and chemistry. They seem first to be in state superposition but, when measured, appear in one specific state, which in quantum mechanical parlance is called the *'eigenstate'* —that is, the state of a system which, once measured, will remain the same after several measurements. However, this 'collapse' occurs only once a measurement device 'observes' it.[36]

This raises the question of why we don't see these things happening if they are commonplace in the microscopic realm. This question is commonly framed as the *'measurement problem'* of QM, and it still hasn't found a complete resolution. The issue with the quantum collapse of the wave function, together with the weird aspects of quantum superposition, entrained many bright minds.

---

[36] Notice that in physics, by 'observation', one means a measurement performed by some mechanical device, such as a video camera or a photodiode. It has nothing to do with the mind or consciousness of the physicists making the measurement.

Well-known is the thought experiment of Schrödinger with his famous *'Schrödinger's cat paradox'*.

Schrödinger imagined a radioactive atomic nucleus in a decayed and non-decayed quantum superposition—that is, an atom that is going to emit a particle but maybe still hasn't. If the radioactive decay takes place, the emitted particle will trigger a mechanism that will poison a cat inside a box. Extrapolating this to the cat, one could therefore argue that the cat must also be in a dead and alive superposition state and that it will remain in this rather ambiguous state until we open the box and, by collapsing the wave function, see which of the two potential alternatives is realized.

All this sounds quite absurd and alien to our common sense. One can hardly imagine this to be true—namely, that the poor cat can be dead and alive at the same time—until we look at it. What Schrödinger wanted to illustrate with this strange thought experiment is that a paradox emerges—namely, that there is, in principle, no logical reason to believe that the superposition of states is only a microscopic property; it should also be observed at our scales. The theory posits no boundary between the micro and macrocosm, contrary to our everyday experience.

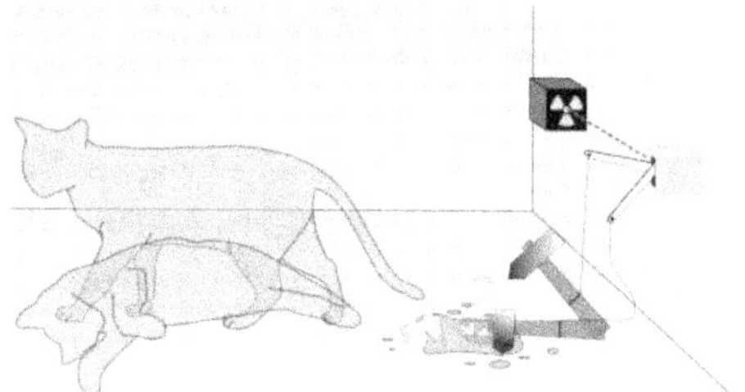

Fig. 68 Schrödinger's cat experiment.

The Schrödinger's cat paradox can be partially resolved by a phenomenon called *'quantum decoherence'* which, to put it bluntly, explains the loss of coherence—that is, the state of superposition or entanglement—due to the thermal interactions of the environment with the gazillions of molecules of the cat's body and that induces the collapse. Closer inspection showed, however, that this can't be the whole story because it leads to formal contradictions in the mathematical foundations of QM[37]. Therefore, at the bottom, the measurement problem is still an open question.

Some resort to the idealist perspective and claim that, after all, it is consciousness that collapses the wave function. The collapse can occur anywhere in the chain between the interactions with the measured microscopic object and the macroscopic measurement device, as well as the physical

---

[37] For the technically inclined reader: the violation of unitarity and the linearity of Schrödinger's equation.

phenomena taking place, which make a macroscopic device indicate a result (the so-called *'Von Neumann chain'*). Because, in physics, no reason or law prescribes where we have to place this boundary, one might conjecture that the consciousness (or the mind) of the observing experimenter looking at the device is the cause for the quantum collapse. This interpretation has sparked many controversies and passionate discussions, especially among non-materialists who think mental causation to be the source of the quantum collapse and as an argument for the existence of consciousness as an unmaterial entity.

However, while it certainly did not escape the reader's attention that who is writing is very open to idealistic and non-material explanations of reality, this interpretation leads, even in me, to some eyebrow-raising. This, indeed, would imply that the cat is effectively dead and alive at the same time until we open the box and consciously perceive it as being in one or the other state. Poor cat! This time, I play the part of the physicalists and prefer to believe that the measurement problem will find a resolution inside a more conventional but extended theory (possibly a theory of quantum gravity) resolving the issue in another way.

At any rate, in the following, we will continue to show that whatever the resolution of the measurement problem is, we don't need to resort to such extreme interpretations of reality. Orthodox QP alone, without any need to resort to contrived theories about the supposed 'role of the observer in QM', is more than sufficient to argue that the reductionist local realism can't stand scrutiny and how this leads necessarily and unavoidably to some form of idealism.

We already pointed out how human perceptions and mental models of matter portrayed as a solid and impenetrable entity is an illusion. From atomic physics, we know that over 99% of the material making up an atom is concentrated in the nucleus, which is about 100,000 smaller than the whole atom's boundaries (defined by the wave function). Moreover, modern quantum field theories tell us that, again, another 99% of the nucleus mass is determined by the binding energy that 'glues' its constituents together—namely, the protons, neutrons, and quarks of which they are made—and that we cannot imagine it as being built up by solid particles, such as mass-like marbles, as our imagination would like to have it. At the bottom, there is no 'matter' at all but, rather, an energy foam of forces and interactions that don't have anything to do with the mental models with which we perceive the world at a macroscopic human level.

A similar situation we encounter in the quantum domain is the so-called *'virtual particle'*. These are other strange quantum objects emerging as a direct consequence of the laws of quantum field theories (QFT) and according to which particles seem to literally pop into and out of existence from nothing. In QFT, which is an extension of QP to Einstein's theory of special relativity, empty space—that is, vacuum—is not really empty but, rather, is envisaged as a continuously fluctuating foam of energy in which things can come into existence for a fleeting interval of time. The good old law of energy conservation that forbids energy from being created or destroyed from or into nothingness is saved considering that these fluctuations come and go at time intervals shorter than a billionth of a billionth of a

Fig. 69 The vacuum ‚quantum foam'.

second. Because, according to modern physics, there is an equivalence between matter and energy expressed by Einstein's famous formula $E = mc^2$ (with E the energy, m the mass of the particle, and c the speed of light), one can also interpret these energy quantum fluctuations as the coming and going in and out of existence, even of material particles.

This is, of course, only our mental model and interpretation of what the theory tells us. In QFT, one speaks of 'fluctuations', as these serve only as a mental image for our too-naïve understanding of the real phenomena that occur at quantum scales. More correctly, one speaks of the expectation value of the vacuum ground state—that is, a potentiality of empty space in a special place and at a particular time to manifest an energy content and an exchange of forces among particles during a scattering process. We will never see a particle come and go from nothing because its existence is so short that any measurement would necessarily have to act faster than light—something disallowed by Einstein's theory of relativity and violating several fundamental laws of physics. Nevertheless, this is how we project into our minds a model of the quantum vacuum and the interaction of forces. Most (but not all) physicists understand that this is a fiction and a naive interpretation of what is taking place in reality, but they can indulge in these simplistic models because, when it comes to the calculations and the predictions of the theory on which it is based, all works fine. The standard model of particle physics, which is the brainchild of QFT, is one of the most successful and experimentally best-tested theories in the history of science.

Another property of a world ruled by the laws of quantum mechanics is *'contextuality'*. In our everyday experience ruled by classical physics, in which we look at an object several times in different experimental conditions, we don't observe that the object's physical properties change. For example, the speed of a particle—that is, its momentum—is independent of the measurement set up as long as the measurement device does not interact too much with it. Whether one measures the particle's momentum or speed, and also its position, makes no difference. In classical physics, the momentum and position of particles are considered two independent intrinsic physical properties that do not depend on the context in which one measures it. One says that classical mechanics is a non-contextual theory in which properties—such as the position of a particle in space and time, its speed, its angular momentum, etc.—'exist out there' independent of how they are subjected to measurement.

Not so in QP. Here, things become trickier. For example, in the double-slit experiment, as long as one does not try to determine through which slit the particle went, it will always hit one of the interference fringes on the detection screen—that is, it behaves like a probability wave diffracted and refracted by the slits (see Fig. 70 left).

If, however, one attempts to discern which way it went, it will behave like a point-particle hitting two limited areas of the screen corresponding to the two slits without the appearances of the interference fringes (see Fig. 70 center-right). This is the simplest and most-cited *'which-way'* experiment that illustrates how the outcome of a measurement in QP is contextual —that is, it depends on the context of the measurement or, to put it more colloquially, on how we ask the question to Nature.

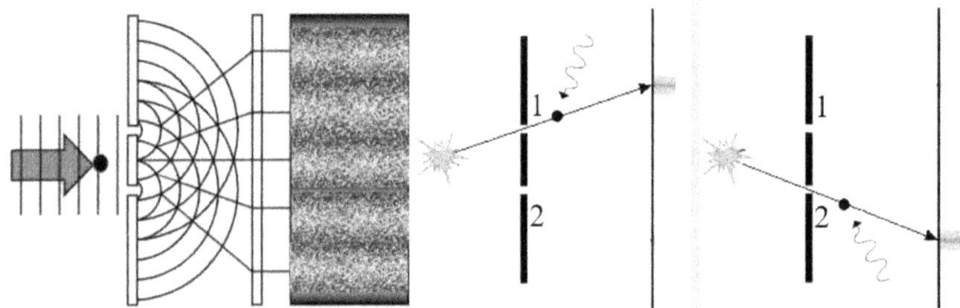

Fig. 70 Left: no information on which-way. Center-right: the which-way experiment.

Despite the particles always having the same initial conditions, they behave differently according to whether we try to gain information about their whereabouts.

One important aspect to keep in mind is that, contrary to popular belief (and, unfortunately, also due to misrepresentations of physicists themselves when they explain their findings to a popular audience), this is not because measuring the particle's position implies an interaction with it. It is not the interaction that destroys the interference pattern appearing on the screen but, rather, only the experimental context. With modern quantum optics technologies, it was possible to build more complicated but equivalent experiments that imply no or almost no interaction between the measurement apparatus and the particle, but nevertheless, the outcome remains the same. It is the information on the which-way that determines how things turn out, however small the interaction may be.

This suggests, again, that the things' out there' making up the physical world do not possess intrinsic properties in themselves and independent of how we observe them. These properties, such as the specific position of a particle, come into existence at the instant of measurement, not before.

The most impressive and weird results concerning the model of reality we get from modern experiments conducted with quantum optic devices in the last three decades all confirm this. These could effectively test thought experiments like the which-way information retrieval or other famous thought experiments that were also realized in a laboratory setting, such as the *'delayed-choice'* and/or *'quantum eraser'* experiments (we will no longer analyze these here, for that would go beyond the extent and scope of the present document; see [221], [36]) and that, like it or not, confirmed how reality is non-local and/or non-deterministic, and contextual. As difficult as this might be to accept from the reductionist perspective, it becomes almost easy to recognize once we realize that what we call a 'physical property' is already a mental reconstruction, a perceived quality in consciousness, a qualia appearing inside us.

People try to save appearances by resorting to conceivable possible 'loopholes' of QP, especially noting that the present structure of the theory still allows one to conceive of some extreme versions of a deterministic non-local or non-deterministic local theory. We consider these to be desperate attempts to promote the survival of a crumbling and dying paradigm that hopes to save an anthropomorphic conception of reality.

Notice, however, that also contrary to some assertions coming from the opposite camp of some idealists, this does not (or does not necessarily) imply that the world is only mental. Quantum contextuality only makes clear that the notion of a 'physical property' and its related primary qualities are figments coming into existence with the act of measurement and the readout of the macroscopic devices that our minds translate into a representation of its own, that have no reality and meaning in the microscopic universe. If we see things only *through* the mind, this does not imply that everything is mind.

Another fundamental aspect of QP, which highlights how a holistic character must be inherent to physical reality, is the so-called *'non-additivity'* of entropy. In statistical physics, the notion of entropy can be associated with the degree of disorder of a system or, more precisely, with a state function proportional to the number of possible *'microstates'* a system can acquire. A simple example that we all know from our ordinary perception of reality is the possible number of states (or 'configurations' or 'degrees of freedom') of a single die—namely, six. A collection of two dice can acquire a number of W = 6 x 6 = 36 possible states. With three dice, one has W = 6 x 6 x 6 = 216 configurations, and, generally, by tossing N dice, one can obtain 6^N different possible states.

Something similar can be said for a system of particles, say a gas, in which molecules or atoms can be individually in many physical states, say, the different energy levels, and which, taken together, allow the gas as a whole to acquire an enormous number of possible configurations or microstates. This number of possible microstates (more precisely, a quantity proportional to the logarithm of this number) defines the entropy of the system. The higher

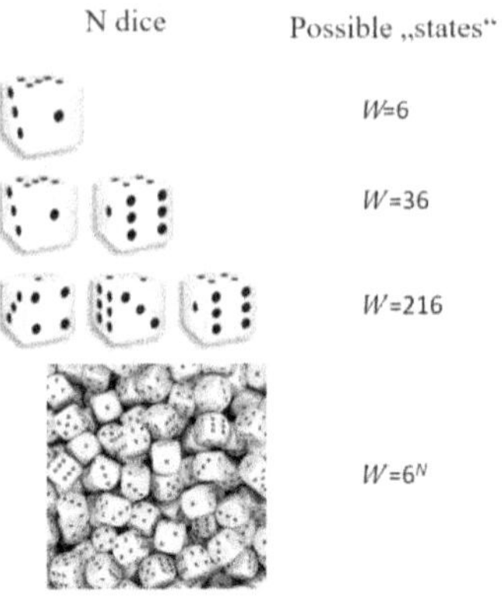

Fig. 71 Number of possible states of N dice.

the entropy, the higher the disorder of the system because, microscopically, it can acquire a huge number of possible microstates but without the system changing macroscopically. For example, each second a gas can go through gazillions of different microstates but without varying its macroscopic state parameters, such as temperature, volume, or pressure. In a certain sense, the entropy is a measure of our ignorance about the actual microscopic state of each of its constituents: We know the gas can be in one of the many possible microstates, but we don't know which one.

So far, entropy is in line with the classical concepts of physics. In particular, it is an additive measure, which means that the entropy of a system grows with the growing number of its subsystems.[38] For example, the number of possible states of a two-dice system is larger than that of a single die (36 > 6, obviously). This is in line with our intuitive notion that the whole is the sum of its parts.

However, let us extend this to QP, with the so-called quantum analog of entropy called the 'Von Neuman entropy', to the case of two particles, each having two possible spin states. In classical physics, if the number of possible particle states is described separately by two state functions (a spin-up *or* spin-down state associated with each particle), the number of possible states of a two-particle system is 2*2 = 4 possible microstates. This is still in line with our classical notions because we can describe each particle with its own state-function. However, the analogy breaks down for a system of quantum-entangled particles. A system of two entangled particles has only one state because, in QM, it is described by only one state-function (a spin-up *and* spin-

---

[38] More precisely, for those acquainted with a bit of math, it is the sum of the entropies of its subsystems because the entropy is expressed by a logarithm. That is, if a and b are the number of possible microstates of two subsystems, their entropies are defined as $S_a = \log(a)$ and $S_b = \log(b)$, and their total number of microstates is a*b. Then the total entropy of the system is $S = S_a + S_b$ because $\log(a*b) = \log(a)+\log(b)$.

down state associated with both particles), meaning that it has to be considered as one and unique whole without subdivisions. Before the measurement that determines the collapse of the wave function to two separate particles, each with its own 'state of being', an entangled system has only one unique state. Paradoxically, in QP, the number of microstates of two entangled particles turns out to be smaller than the number of microstates of one particle—that is, smaller than the number of microstates of its components. This contradicts our intuitive notion of how the number of microstates of a whole system comes into being in the first place. It is as if we would obtain fewer degrees of freedom—that is, possible outcomes—by tossing two dice instead of only one. This can be extended to many particles. The weird thing is that if several particles—say, even thousands or millions of them–are entangled, the number of possible microstates is still one!

This is a less theoretical possibility than we might think, as such a strange state of matter has also been realized in the laboratory: the *'Bose-Einstein condensates'*. These are special and very peculiar states of matter that can be realized by freezing a gas down to almost absolute zero temperature until its atoms coalesce into a single quantum entangled whole.

This strange property, which, however, is intrinsic and almost natural in a quantum realm, makes sense only if we accept that at microscopic levels, it is futile to think of an entangled system as two (or many) particles somehow interacting with each other. There are no longer particles or subdivisions; the system becomes a holistic and undivided single whole–a whole that is not the sum of its parts because there are no longer parts in the first place.

It may be useful to add to this section another quantum effect, the *'quantum Zeno effect'*, and that did not attract the same amount of attention as other weird quantum effects, but it is worth mentioning because it will acquire significance in the frame of an integral cosmology. The quantum Zeno effect is a phenomenon in which repeated and fast measurements slow down a quantum system's time evolution. It is a peculiar quantum phenomenon and applies to every system and situation in which any kind of transition occurs in time. In order to understand this, it might be useful to compare it first to our everyday experience.

As is well-known, if you sufficiently heat a pot of water, it will sooner or later start to boil. That is, the water pot is a system undergoing an energetic thermodynamic transition from non-boiling water at a low temperature to boiling water at a higher temperature (100° C at sea level). It goes without saying that this is supposed to happen regardless of whether or not you are watching the pot.

Not so in QM. In this case, we are dealing with a microscopic and quantum system at the atomic level or not much more than that. The Zeno effect describes the dynamics in the quantum regime whereby the transition probability from a quantum state to another is modified because of the measurement—that is, simply because it is 'observed'.

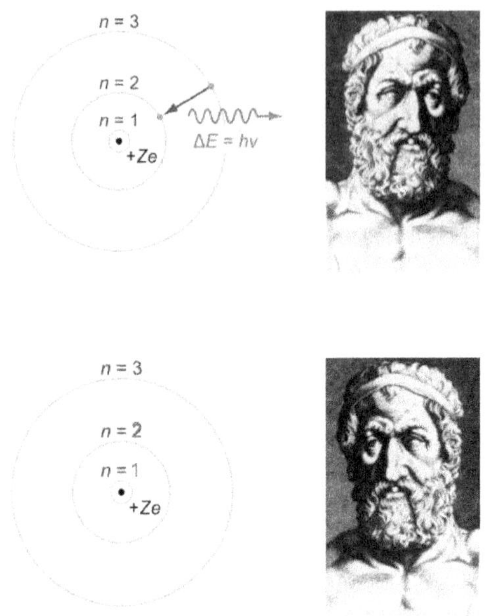

Fig. 72 The Zeno effect:
Observing an atom constantly will avoid its radiating energy.

For example, take an unstable atom that has some probability of decaying next in time. The radioactive *'half-life time'* of a specific type of atom is the average time after which 50% of a large number of atoms of the same type has decayed. This half-life time for each atom is not something we can engineer. It is a constant given by the laws of Nature. However, if measured continuously, the same atom will never decay. This implies that the probability that an atom will undergo an energy transition emitting a photon is lowered if the system is subjected to frequent ideal measurements.

Consider an atom that is excited by a photon and that undergoes an energy transition to a higher excited state. It will, sooner or later, relax to the ground state and emit a photon. The probability that this occurs within a specific time interval is entirely dictated by quantum laws. However, if the atom is subjected to fast and repeated measurements, the probability that this transition occurs will become lower and lower as the frequency of these measurements increases.

Ideally, if the measurements are so fast that we can consider the observation as continuous, the transition probability tends to zero, and we can consider the system almost frozen in its eigenstate. In other words, measurements can suppress the time evolution of a quantum system. This is because each measurement projects the *'state vector'*–that is, the quantum state of the system before the measurement–back to the eigenstate. To put it in other words again, it causes the wave function to collapse to the same eigenstate from which the system wanted to evolve away.

Of course, the Zeno effect owes its name to one of the famous Zeno paradoxes, the arrow paradox, of the Greek philosopher Zeno of Elea, who tried to argue that motion is an illusion. He imagined time as a line made of timeless segments. Zeno states that for a motion to occur (for instance, an arrow in flight), it must change the position it occupies, and in any one (duration-less) instant of time, the arrow is neither moving to where it is nor to where it is still not. At every instant of time, no motion is occurring. If everything is motionless at every instant, and if time is composed entirely of instants, one reaches the paradoxical conclusion that motion is impossible. The similarity to this in the quantum Zeno effect is that if the time intervals of our observations become shorter, nearing the 'timeless time unit', the system will no longer be able to move from where it is. The continuous observation will prevent motion.

Notice again how, by 'measurement' or 'observation', one must intend every sort of interaction that causes the projection of the system onto the eigenstate, not a human observation (even though, for the sake of simplicity, this is what Fig. 72 might suggest with Zeno's severe look, but let us not take things too seriously!)

The quantum Zeno effect is not just a speculation or a theoretical conjecture but has been experimentally shown to exists [222]. So, at least in QM, there is much truth in the good old saying that "a watched pot never boils"!

There are many more aspects of QP that would be worth notice and that seriously challenge our ordinary sensory and mental understanding of reality. However, we consider this quick review of QP to be sufficient to highlight how illusionary our perception of reality is.

More than anything, it is the holistic, non-local, and non-deterministic character of QM that arouses so much resistance in reductionist and deterministic thinking, especially in the materialist conception of reality. The reductionist mind is, per definition, based on the idea that reality must be explained in terms of a bottom-up paradigm[39] and, obviously, it believes that all this is too weird and impossible. The belief is that sooner or later, with further experimental evidence and theoretical research, everything will be recast into some more or less deterministic and/or reductionist local realism. This is its original and inherent belief system, in which a composite ontology and foundation of all reality is about particles having a precise position and momenta moving in space and time and interacting through the exchange of other particles. So goes the assumption and premise in almost every modern scientific compartment. This model worked fine and is still at the base of

---

[39] In Pt.III we will show how this is a mysterious fact in itself.

modern particle physics, but this form of realism is limited and naïve. From the ontological standpoint, it has been replaced by QP, and the resistance to giving it up is neither scientific nor rational but, rather, an emotional attachment to the human's way of seeing the world—no more and no less than the attachment to anthropocentric sense-mind-based worldviews like geocentrism or flat-earth theories.

On the other hand, it is no coincidence that several other scientists open to alternative views have conjectured that brain functions might also be determined by quantum phenomena. For example, one might ask if the binding problem we discussed earlier may find some resolution inside a quantum paradigm of consciousness. One can draw a metaphorical analogy between how the mind binds its perceptions and thoughts in a unified whole with the physically non-composite undivided quantum whole of entangled particles. The uniqueness we experience, for example, when our minds bind the form, color, and size of an object into a single-featured undivided conceptual and sensory whole is, at least metaphorically, somewhat reminiscent of the coalescence of a many- particled aggregate into an object described by a unique wave function that can't be reduced to the sum of its parts, like the entropy of a Bose-Einstein condensate.

Only metaphors, or is there more? So far, attempts to connect consciousness studies with QP have not been very successful, apart from some insights of the so-called *'quantum biology'*, which, however, at the time of this writing, did not go much further than showing how quantum phenomena, indeed, play a role in some chemical reactions in life's functions, such as photosynthesis. However, this idea of a 'quantum consciousness' that trickled so deeply into the collective, leading to much pseudo-science, still lacks robust experimental evidence. Nonetheless, as the popular belief inwardly felt truth that it couldn't rationalize with plant's intelligence and that the science of the time disregarded, it might well happen again with the relationship between consciousness and QP. It is something that most scientists deem unlikely but that they might sooner or later have to revise. Only time will tell.

6. Quantum Idealism

"[...] it is the general recognition that we are not yet in contact with the ultimate reality. We are still imprisoned in our cave, with our backs to the light, and can only watch the shadows on the wall."
Sir James Jeans, The Mysterious Universe

Idealism is rarely addressed by scientists or philosophers debating the philosophical implications of QP. This is probably because physicists have no background in philosophy, and philosophers have a poor understanding of the foundations of QM. Another reason may be that we live in a time and culture strongly dominated by a Laplacian physicalist mindset. Most scientists and philosophers of science hold an unshakable belief in a purely deterministic and local material conception of reality in which matter is fundamental and all there is, while everything else must be derivative from it. This belief system is so deeply ingrained that its adherents never allow themselves to ponder alternative thoughts. That's also the main reason why so many are busy sometimes elaborating bizarre interpretations of QP, such as the 'M*any World Interpretation*', the *'pilot wave theory'*, or *'superdeterminism'*, which are ontologies that try to rescue the naive human-centric sense-mind local and deterministic perception and understanding of the world. The author believes that these are yet another chapter in the history of humankind in which we tried to save appearances at any cost, as we did with geocentrism and creationism.

An idealistic monistic philosophy might not be the ultimate truths either, as these are also human-made attempts to explain the world of appearances in which we live. In fact, in Pt.III we will go beyond that. However, they furnish a platform that is much more in line with the non-local and holistic character of QM. The which-way experiments, quantum non-locality, Von Neuman's subadditive quantum entropy, the Bose-Einstein's condensates, as well as the quantum delayed choice/eraser experiments we didn't mention here—all indicate an intrinsic reality which, beyond and independent of our sense-mind representations, refuses to be conceived of as a differentiated system in which the sum of the parts equals the whole.

However, idealism should not be considered another interpretation of QM since it does not offer a new ontology. Rather, it is a suggestion for a different approach to the subject and an attempt to point out how our 'human-centric' unaware assumptions regarding the nature of 'reality' and 'objectivity' can deeply mislead us into adopting an anthropomorphic understanding of the world—a misunderstanding to which science is not immune. It invites us to look forward towards a new understanding and paradigm of reality instead of trying to save and rescue the old worldview, however successful and progressive it might have been in the past. Philosophical idealism may be useful for making us aware of our unconscious assumptions, heightening the philosophical standards that can eventually reflect themselves in the 'modus operandi' with which we pursue science in practice and, only then, eventually, be integrated into new interpretations of QM.

Sooner or later, we will have to come to terms with these advanced experimental tests of quantum ontology which put us in front of a scientifically established fact: Like it or not, there are no such things as 'particles' traveling one or another path. Rather, only a Kantian thing-in-itself displaying an undifferentiated whole reminiscent of a Vedantic non-duality that we can't describe with anything better than an abstract mathematical concept like the wave function. The author can hardly avoid noticing the similarity between these unitarian, monistic, non-dual worldviews, which repeatedly emerged in such different historical and cultural contexts, and the non-local entangled character of quantum particles in the microscopic domain. To me, this suggests that some aspect of fundamental truth stands behind these philosophical thoughts or mystic experiences—an aspect that our still-too-naïve sensorial and intellectual reductionist material scientific conceptions can only grasp superficially.

Fact is also that the contextual character of QM, which arises when we attempt to gain knowledge about the which-path, and where the quantum system answers, so to speak, according to the question asked, makes it clear how our naïve realism, which intrinsically resorts to a thought process based on *'counterfactual definiteness'*, is terribly at odds with the facts. Counterfactual definiteness is a philosophical term with which philosophers of science indicate how a theory assumes the existence of the properties of objects, even when they have not been measured. It is about the definiteness— that is, the existence—of the properties of an object that are not yet measured but derived and assumed from what we know of that object. For example, if one knows the direction of motion and the speed of a particle, one can calculate its position in the future or where it was in the past without measuring it in that past or future instant. In QP, one can do that only by making probabilistic statements, and not because of our ignorance about the particle's position but because its whereabouts in space and at an instant in time come into being with the measurement. The same can be said about the spin of entangled particles. It is the act of measurement at the very instant at which it occurs where the property of spin comes into being by a projection of the entangled state 'collapsing' it to a state of two separate particles. From the idealistic point of view, this apparent mysterious behavior becomes almost naturally understandable. Because a measurement does not reveal any intrinsic property of a thing-in-itself; rather, it is a projection process into our mind that, obviously, can take place only at the instant it is performed. The image of two spinning particles is derived from a thought process based on a counterfactual definiteness. It is something we reconstruct a posteriori by imagining how an unmeasured object was or will be. This, however, is, again, a mental projection in the past or future and that we should not elevate to an ontological truth.

With his colleagues Nathan Rosen and Boris Podolsky, Einstein defined realism as follows: *"If we can predict the value of a physical quantity with certainty [...], then there is an element of physical reality corresponding to this physical quantity."* [223] This statement inherently assumes counterfactual definiteness. From the idealist's point of view, this is not necessarily an incorrect statement, provided that we realize that the 'value of a physical quantity' tells us nothing about what that 'element of reality' is.

Moreover, if we inspect more closely what classical and quantum information is really about in terms of sensorial and mental cognition, we will realize that it is first and foremost a physical quantity that measures our ability to distinguish. The statistical notion of entropy as a state function which counts the (logarithmic) number of microstates of a system does precisely this: It measures our cognitive ability to distinguish, discern, and separate something from something else (here, the sum of all the physical states we can distinguish from one another). By doing so, the idealist would say that we are already outside an objective cognition of things in themselves. We are reconstructing the world within ourselves through an abstract representation of distinctive features in our minds, starting from a series of sense-data which are the results of an impression on a measurement device whose only function is to distinguish, separate, and count, leaving us with a number, such as an entropy value, that is useful for practical purposes but doesn't tell us much about the ontology of what is being measured. Everything in science, and especially in physics, is about distinction. However, this distinction is not already present in a reality a priori to that act of distinction. Rather, the cognitive act that we perform at the instant of measurement is what subsequently reifies it in our sense-mind as the 'reality out there'.[40]

That strange collapse of the wave function and the measurement problem might arise due to our minds' unconscious but quite stubborn habit of reducing, collapsing, and determining quantitatively spatio-temporal relations and properties, without realizing how these relations and properties are epiphenomenal events in the form of the emergent phenomena of subjective qualities appearing in us and not to us.

Kant's idealistic standpoint regarding what space and time are may also be relevant to modern theoretical physics. He downgraded Newton's space and time from a fundamental and self-given objective inherent feature of reality to an a priori mental construct derived from our subjective cognitive experiences. Space and time are 'a priori representations' or 'pure intuitions' of the phenomenological subjective appearances of objects, which allows us to form images of things in advance. Space and time appear as necessary representations of things and events but without being features of the universe and of things as they are in themselves. Substance and material objects have properties and attributes that stand in spatio-temporal relations, which, taken together, are what appear to us as the law-governed whole we call Nature. Matter, energy, force, space, time, or space-time are relational categories.

What QM tries to tell us is how the demarcation line between objective and subjective becomes questionable—not in the sense of the 'role of the observer in QM', but in the sense of embracing an idealistic view of reality. Plato believed that because we cannot gain access to reality as it is, we must posit, as the ultimate constituent of things, its forms, and numbers—a Platonic

---

[40] For the more technically inclined reader, we may also note the curious fact that a Lagrangian density is a local measure of information, called 'Fisher information', which can be interpreted as the ability to distinguish between two nearby probability density functions [386].

Absolute that he called *'hyperuranion'*, which is supposed to be a perfect realm of forms and ideas of things. Plato's idealism conceived of numbers, forms, and eternal mathematical truths as the ultimate tools for describing the essential nature of all things and reality.

In particular, theoretical physics was extremely successful in embracing this philosophy, which is nowadays neatly summarized by the 'shut up and calculate' attitude. Everyone who studies the mathematical structure of QP, Einstein's relativity, and the standard model of particle physics must admit that purely quantitative science stands in front of us in all its splendor. However, that came at a heavy price. Reducing the search for truth and reality to primary qualities has, indeed, realized Plato's, Galileo's, and Locke's dreams insofar that we have built a giant intellectual construct of reality that, empirically, works fine. Yet, contrary to their original intention, paradoxically, what we are left with is not a description of reality as it is. From the philosophical perspective, we find ourselves in a quantum quagmire that reflects an abstract platonic hyperuranion that nobody knows how to interpret and, let alone state which kind of ontology it fundamentally describes.

After the beginning of the 20th century, the fortunes of Western philosophical idealism quickly faded. Scientific and technological breakthroughs, the exponential growth of industrialism (due in large part to two world wars), the success of a reductionist and atomistic physicalist approach in physics, biology, and medicine, the paradigm shifts in physics ignited by QP and the theory of relativity, and the great hopes invested into the (now-failed) philosophical project of empiricism, and logical positivism made philosophical idealism appear to be the old-fashioned, almost mystical, nonsensical gibberish of a few nostalgic and backward philosophers.

It was in this spirit that the 20th-century British philosopher Bertrand Russel revolted against idealism, trying to deconstrue it. However, this didn't lead him where he expected. It seems almost ironical that Russel, who tried to dismiss and avoid any form of idealism, finally had to admit: *"We know nothing about the intrinsic quality of physical events except when these are mental events that we directly experience."* And: *"Physics is mathematical not because we know so much about the physical world but because we know so little: it is only its mathematical properties we can discover."* [224]

Fig. 73 Bertrand Russel. (1872-1970)

This led him to frame what nowadays goes under the name of *'Russelian monism,'* which maintains that there must be two properties of reality–generally named *'quiddities'* (a term borrowed from the scholastics, see Thomas Aquinas): the most basic entities of physics (mass, charge, spin, etc.) and others that must underly phenomenal consciousness. More than his theory,

I find Russel's intellectual development sociologically interesting. Individually, he went through an intellectual and psychological process that science has gone through in the last century and is now slowly becoming aware of. First, you try hard to dismiss something. Then, at the end, you find yourself left with precisely that which you initially wanted to avoid.

Indeed, about a century later, we have harvested an incredible amount of knowledge and progress in the sciences and technology. However, deep down, the sensation that, along the way, we may have missed something that could answer our innate thirst for deeper knowledge and a more comprehensive theory of meaning about the world and existence is resurfacing. Despite all the attempts to negate it by resorting to a purely rational and still-too-Newtonian and Laplacian worldview in scientific thinking characterized by an "inborn and childish realism", as Schopenhauer used to call it, there remains a feeling of dissatisfaction. The desire is mounting for more.

This was already manifested in the thoughts of some of the founding fathers of QM. Max Planck once stated:

"I regard consciousness as fundamental. I regard matter as derivative from consciousness. We cannot get behind consciousness." [225]

"As a physicist who has devoted his whole life to down-to-earth science, to the study of matter, I think I can safely claim to be above any suspicion of irrational thinking. And so, after my research into the atom, I say this: There is no matter in itself. All matter is created and exists only by a force that makes the atomic particles vibrate and holds them together to form the tiniest solar system in the universe.

Fig. 74 Max Planck
(1858-1947)

However, as there is neither an intelligent force nor an eternal force in the whole universe, mankind has not succeeded in inventing the much longed-for perpetual motion machine—so we must assume a conscious, intelligent spirit behind this force. This spirit is the source of all matter." [226]

Erwin Schrödinger made no mystery of his adherence to the non-dual Advaita Vedanta philosophy:

"[...] The plurality that we perceive is only an appearance; it is not real. Vedantic philosophy, in which this is a fundamental dogma, has sought to clarify it by a number of analogies, one of the most attractive being the many-faceted crystal which, while showing hundreds of little pictures of what is, in reality, a single existent object, does not really multiply that object. We intellectuals of today are not accustomed to admit a pictorial analogy as a philosophical insight; we insist on logical deduction." [227]

Fig. 75 Erwin Schrödinger
(1887-1962)

Werner Heisenberg, who was acquainted with Kant's idealism and Aristotle's philosophy, wrote:

"Natural science does not simply describe and explain nature; it is part of the interplay between nature and ourselves; it describes nature as exposed to our nature of questioning." [228]

"I think that modern physics has definitely decided in favor of Plato. In fact, the smallest units of matter are not physical objects in the ordinary sense; they are forms, ideas which can be expressed unambiguously only in mathematical language." [229]

Fig. 76 *Werner Heisenberg (1901-1976)*

"[...] atoms and the elementary particles themselves are not as real; they form a world of potentialities or possibilities rather than one of things or facts. The probability wave [expresses] the tendency for something. It's a quantitative version of the old concept of potentia from Aristotle's philosophy. It introduces something standing in the middle between the idea of an event and the actual event, a strange kind of physical reality just in the middle between possibility and reality." [228, p. 160]

John von Neumann, the mathematician whose name will be forever associated with the mathematical foundations of QM, acknowledged:

"The sciences do not try to explain, they hardly even try to interpret, they mainly make models. By a model, a mathematical construct is meant which describes observed phenomena with the addition of certain verbal interpretations. The justification of such a mathematical construct is solely and precisely that it is expected to work—that is, correctly to describe phenomena from a reasonably wide area. Furthermore, it must satisfy certain esthetic criteria - that is, in relation to how much it describes, it must be rather simple."

Fig. 77 *John von Neumann (1903-1957)*

David Bohm, in his well known "Wholeness and the Implicate Order", states: *"In principle, this reality is one unbroken whole, including the entire universe with all its 'fields' and 'particles'"* and *"To begin with undivided wholeness means, however, that we must drop the mechanistic order."* [230] Bohm formulated a holistic view of the universe, which he divided into an *'implicate order'* and an *'explicate order'*. The implicate order connects everything with everything else, where any individual element could reveal *"detailed information about every other element in the universe"* in an *"unbroken wholeness of the totality of existence as an undivided flowing movement without borders."* Meanwhile, the

Fig. 78 *David Bohm (1917-1992)*

explicate order is the manifest world. It is secondary and derivative, and it flows out from the laws of the implicate order.

It is a curious fact that these philosophical positions of several founding fathers of QP, who were already known worldwide, and some of whom were Nobel prize-winning scientists, were completely ignored. World War II might have played a detrimental role. Planck's son was executed by the Nazi regime. Heisenberg's fate was only slightly better. After Adolf Hitler came to power in 1933, he was branded a 'white Jew' by a pseudo-intellectual movement that wanted to establish a new 'Aryan Physics'. Schrödinger was at grips with the rising power of Nazism as well, while, on the other side of the Atlantic Ocean, the fear that Hitler could possibly win the race in building nuclear weapons convinced von Neumann to participate in the Manhattan Project, which led to the first atom bomb. They were times when everyone had to be careful not to upset the regimes in power and when blank survival preoccupations became much more urgent than any philosophical musings about the philosophy of physics or questions of meaning.

However, this attitude remained crystallized after World War II and was the almost completely pervading attitude until the turn of the new millennium. A strictly reductionist, deterministic, materialist, and naively realist worldview dominated almost the entirety of physics-related academia—and, to a large degree, continues to do so today. However, things are changing; fortunately.

In fact, it is science itself which, if it becomes aware of its own fallacies and recognizes that it is a limited tool that does not unveil the ultimate truths of things, can point out its own limitations and make us realize that the world, as we perceive it, is an illusion.

The problem is that we humans do not know ourselves. We are like the slaves in Plato's cave who misinterpret shadows for the real thing. In this sense, scientists are no better than anyone else. The dissatisfaction with the 'shut up and calculate' approach and the lack of a clear quantum ontology manifested in the innumerable interpretations of QM are symptoms and indications that human nature will not allow itself to stop there. It was precisely this refusal of science to nurture that innate human thirst for deeper truths and meanings and to go beyond that naïve and unreflected realism, supported by a skeptical physicalist and militant atheist movement, that, paradoxically, did foster the new age pseudo-scientific theories and quantum woo that now flood the popular media. These are expressions of (a more or less unconscious and confused) revolt of human nature against the almost dogmatic closure of a purely pragmatic, utilitarian, reductionist, and physicalist analytic reason that does not allow for an inquiry that seeks to answer deeper questions about meaning and essence. Fortunately, this tension is becoming increasingly apparent and acknowledged, thereby sparking renewed interest in philosophical and scientific conceptions aimed at transcending the dominant neo-anthropocentric worldview. It is very likely that some sort of philosophical idealism—not only in physics but generally in all sciences—will return. And, indeed, it is.

II. Higher-mind Philosophy Reloaded

1. Motivation

The previous chapters were meant as a preparation for this coming brief summary of the idealistic and spiritual insights of Western philosophers. It is a sort of recap using the original words, concepts, and theories of Western philosophy, from ancient Greece to modernity. This anticipation had a pedagogical motivation aimed at preparing the grounds for identifying ourselves better in some aspects that nowadays are part of a continental philosophy which, however, is perceived at times as being obscure and unintelligible if one resorts directly to the original writings without preliminary preparation and reflection.

Modern philosophy is rediscovering its historical roots. It is the failure of modern neuroscience, or the weird aspects of quantum physics and all the related emergent philosophical issues, that led some academics to reconsider these past doctrines. If idealism and its more or less marked mystic ramifications were considered relics of the past, or, at best, musings for a few nostalgic romantics of the early 20th century, the questions about the reality of matter, the unification of fundamental forces, or the nature of space-time in physics, but especially the questions about the nature of consciousness in light of the new neuroscientific discoveries, are leading several philosophers and even scientists to reconsider some old teachings that might not have been so off-track in their understanding of reality.

The purely materialistic and intellectual conception of reality that emerged from the Enlightenment was a necessary purge for humankind as a collective; it had to go through this evolutionary process to refine its mental and intellectual skills. Mind has still not completely conquered humankind, as eloquently indicated by the spread of pseudo-scientific woo, anti-scientific denialism, and the pervasive post-truth conceptions. Reason has still not settled itself completely in the human race, and an almost exclusive concentration on analytic skills was necessary to initiate a mental crystallization in the collective. The 'trick of Nature' to counteract the irrational and obscurant tendencies was—and more than ever remains to the present—to resort to extremes. The trick consists of luring us into an extreme exteriorization of our consciousness towards physicality, material values, science, and technology, fostered by an industrial revolution that is still ongoing and that is backed by a financial system based on consumerism. It is a process and cycle that has not come to its end because we are still, to a large degree, instinctive and emotional beings.

However, as is usually the case when things are taken to their extremes, while the collective consciousness is focused on the development of one of its inherent powers, this also distracts it from realizing and further developing other potentialities. This historically recent, almost exclusive concentration of homo sapiens on the exterior materiality of things[41], as also the inebriating successes of a dry analytic and impersonal reasoning, led it to a detachment from inner spiritual domains to such a degree that it now refuses to recognize its own inner nature, declaring it an evanescent mirage, an old remnant of ancient superstitions. This lost ability of most humans to look inside and recognize other dimensions of reality has led it to an equally irrational form of spiritual denialism. In other words, we not only threw out the baby with the bathwater but forgot, and even denied, that there was a baby in the first place.

This led to a sort of forgetfulness of one's own philosophical traditions tied to a spiritual understanding of reality that did not mistake one's inner intuitive perceptions for an illusion. In particular, 20th-century philosophy was captured by the hypnotic lure of the scientific and analytic sense-mind. Analytic philosophy became, and remains to a large degree, the dominant tradition. The almost exclusive emphasis on formal logic, mathematics, language, natural sciences, and empiricism became its main trait together with its intent to accurately expunge any reference to metaphysical, let alone theological, innuendos.

Logical positivism was one of its major brainchildren. It believed that philosophy should be concerned only with propositions verifiable by formal logic or with facts that could be experienced by a third-person human-centered sense-mind cognition, while any other speculations, such as metaphysics or an ontology based on first-person accounts, were branded as illusions, meaningless nonsense and dead disciplines that had to be confined, at best, to history books, and not regarded as possible sources of knowledge. Great minds like Gottlob Frege, Rudolf Carnap, Bertrand Russell, and Ludwig Wittgenstein, to mention just some, tried hard to pursue an intellectual program capable of reducing all of reality into the context of pure logical and empirically verifiable propositions.

Logically, logical positivism exhausted itself and died out pretty quickly. It is nowadays widely considered a failed project by the vast majority of philosophers. Positivism has itself become a relic of the history of human thought because it tried to anthropomorphize the universe through the lens of ordinary human reason itself. This is not to say that the spirit that stood behind it has died out. Several other, more or less openly declared, forms of scientism dominate the philosophical and scientific academic environment today. At least an adherence to strictly analytic and third-person observational approaches with the empiric quantitative measurement paradigm of physics is

---

[41] Part of the indigenous cultures make an exception and it is to expect that they will play a role in helping the collective to recover a lost spiritual and environmental wisdom. More on this in Pt.II-III.2e.

still considered, by the majority, to be the only meaningful way to investigate the world. However, the limited power of logical, analytic cognition will sooner or later have to recognize how its knowledge is bounded inside a domain much narrower than it likes to admit.

Teachings that we deemed as archaic remnants of superstitious minds or, at best, as outdated and no longer necessary speculations are taught in schools and to undergraduates as historical heritage, but not as worldviews having a value of their own. Western metaphysics and idealism were, and are still, perceived by most as philosophically outdated doctrines that might be studied and debated but only from the perspective of their historical, anthropological, and sociological influence on contemporary thinking, as necessary appendix that complete one's educational philosophical curriculum but that have been surpassed, or that are even considered blatantly wrong models of reality. Instead, it would be a step forward in Western philosophy to recognize how the attempt of logical positivism to get rid of any metaphysical, idealistic, and spiritual speculations by devaluing any first-person report as nonsensical per definition and, more generally, by placing the analytic tradition at the center of the philosophical universe as something representing all of philosophy as a whole [42] was an evolutionary necessity that is now hampering further progress.

It should come as no surprise, then, if some metaphysical speculations nowadays experience a comeback. All this makes a lot of sense if seen from the evolutionary perspective. Reason limited to the sense-mind has taken itself to its own extremes because of an evolutionary necessity: First to completely develop its own skills and second to explore its domain and see how far it could go. But, more or less subliminally, people begin to inwardly perceive that we are reaching the limits of this domain and begin to feel it somewhat dry, exhausted, and narrow-minded. This state of affairs may not have come into full awareness, but a non-negligible part of humanity feels inwardly an uneasiness and dissatisfaction with the purely rational and materialistic account of what we are supposed to call 'reality'. An increasing part of humanity feels that the image of the meaningless and spiritually empty world that reason and its physicalist allies offer us can't be the last word. That's at least one of the reasons why scientism has lost its appeal. Though there is still no clear understanding of what is supposed to replace it, the time has come to look beyond.

The viral spread of pseudo-sciences and forms of pseudo-mysticism is not a regression to the past but a still confused and immature reaction against the present. One of the main factors which led to this treatise is the desire to avoid the great danger of reverting and de-evolving to the past instead of evolving and taking a step towards the future. In fact, those who are still struggling to

---

[42] One might object that continental philosophy has also played a role. However, its influence on professional academia remained limited.

develop and apply in its full potentiality the discriminating mind usually resort to simpler forms of spirituality which might be branded by the ordinary materialistic mind as backward religions and rituals that survive the rational age because of unconscious fears and desires, such as the fear of death or the desire for redemption. But also, this judgmental ordinary mind is only the one-eyed child of its time, which is equally unable to look beyond its narrow limits, which mistakes it for the whole of reality.

On the other hand, there is now a growing part of humanity in which the rigors of reason are fully established and yet are nevertheless increasingly felt as insufficient and superficial—not because of a lack of analytic faculties but, on the contrary, precisely because the full possession and application of the rational, materialistic sense-mind, while disclosing the richness and complexity of the physical universe, reveals its own limitation and narrowness. Sooner or later, the call from within is not only a natural reaction but also the inevitable destiny of the collective.

From this perspective, some of the old metaphysical and spiritual traditions acquire a new meaning and fresh vigor. It is no coincidence that so many points of contact exist between the spiritual traditions and idealistic philosophies throughout the times and throughout cultures. They must have gotten into contact with a more fundamental truth of things that we still struggle to capture. Therefore, it might be appropriate to consider the eventuality that at least some of the spiritual, intuitive and idealistic constructs of the past aren't just the product of a mere analytic dialectic applied to metaphysical, theological, or teleological quests, as most assume, but the first intimations of the rise of the ordinary mind towards higher domains of cognition. We should reconsider in a new light ancient philosophies or non-religious mystic traditions which the ordinary mind doesn't and can't understand because of its inherent limited nature, but that we can now appreciate fully by rising above the ordinary thinking patterns limited to a pure analytic sense-mind cognition.

The resurgence of metaphysics and idealism in the West is palpable but still in a preliminary and undeveloped phase of reawakening.

Our present aim is threefold.

First, in this second part, we seek to refresh our memory with some of the historic metaphysical, idealistic, and spiritual Western-philosophical traditions. Of course, this is not something that can be done in a chapter of a few pages. It is not our intention to do justice to the vast amount of wisdom and knowledge that spans Western metaphysics, as this would imply a full reconsideration, analysis, and rediscovery of a huge treasure that collects the gems over a period lasting 25 centuries. Nevertheless, we would like to pick out some of these gems—that is, the representative personalities of this philosophical, spiritual, and mystic movement—making a quick review, just as an act of remembrance of a past and ancient wisdom that might be worthwhile to rediscover and go back to study. We provide a few pixels that hopefully encourage the reader to long for a much larger picture than that which could be drawn here.

Secondly, we will convey a feeling of how these past philosophers were not dealing with infra-rational or subconscious domains but, on the contrary, were gaining access to something trans-rational that stands beyond a strictly analytic thinking mind and that could, at least in part, transcend itself towards what we will call, from now on, a *'higher-mind'*[43] cognition of reality. This is a different form of cognition that is essentially still mental but from which our ordinary intellectual processes are only a derivation and appendix. We will describe the qualities of the higher mind in further detail in Pt.III-I.4f. For the time being, it suffices to say that the higher mind is clarity of a mind that still works with thought-forms but where logic can, but must not be, the primary lens with which one sees things. It is a different form of cognition that captures by a single view the whole with no need for inferences. It gets more intimately in contact with the truth of things that the analytic mind alone can't seize. This doesn't imply that the philosophers we are going to consider were not analytical thinking personalities. Quite the contrary; they mostly still dwelled on the level of the conventional mind, yet there was something in the background that was silently trying to express itself—the baby that analytic philosophy threw out with the bathwater.

Thirdly, the rediscovery of what we will call the 'higher-mind philosophy', together with the new ways of seeing that could potentially lead to a new form of science we will discuss in Pt.II-III, shall prepare us for a greater leap and synthesis that will go far beyond the present Western materialistic naturalism or Eastern spiritual views. The opposition and incompatibility between the analytic and spiritual thought and also the apparent contradictions between several schools of thought inside the Western higher-mind philosophy itself and which sometimes appear as mutually exclusive, will no longer be seen as an assemblage of speculations in competition with each other but as the different facets of the very same world as it appears to a cognition expressing itself on different levels of consciousness. From this perspective, also, idealism will be realized as only one step along the path and will be transcended. An 'integral cosmology' which sees from above by a grand vision the different interpretations of reality which emerged throughout history, either in its analytic, idealistic, or mystic expressions, will reveal how the extremes and apparently irreconcilable poles appearing therein are, instead, a natural understanding of the world from different cognitive levels of consciousness.

---

[43] The adjective 'higher' should not be regarded as designating any moral or ethical superiority, let alone any implicit judgement of any kind, no more and no less than software engineers speak of 'higher programing languages', mathematicians of 'higher-order logic', pedagogues of 'higher education', or neuroscientists of the 'higher cognitive functions' of the prefrontal cortex. It only indicates a functional advantage.

2. Rediscovering the Western Spiritual Philosophy

The ancient Greek philosophers were not only on the verge of discovering the scientific method, which Galileo and Newton introduced about 15 centuries later. They were already sensing themselves through a path that could have led them towards a more mature form of spirituality, higher-mind seeing, and unification with the scientific method. This is consistent with the logic of a natural evolutionary process trying to elevate a strictly rational analytic tradition to something which began to rise beyond. However, history—that is, Nature—took a longer and more painful path. The pragmatic and militaristic Roman empire conquered ancient Greece, and slowly but steadily, the Hellenistic tradition faded away. After its decline, not much was left behind, and Europe plunged into the Dark/Middle Ages, followed by the materialistic Enlightenment of which we are culturally still an expression. Today we look at the history of Western philosophy through a rational, analytical, and scientific-minded lens, also failing to appreciate its trans-rational component. It is predictable that sooner or later, a 'higher-mind philosophy' will have to establish a lost and forgotten balance as a natural process of the evolution of consciousness. By focusing exclusively on the rationalistic and naturalistic side, modern philosophy and science forced themselves into a cave of shadows and even prided themselves on doing so.

A heavy price was paid for this shortsightedness. A very one-sided naturalism forgetful of its own roots and past achievements has become prone to reinventing the wheel over and over again without being aware of it. It forgot Newton's advice that, if we want to see far away, we can do so only by standing on the shoulders of the giants of the past. Therefore, before looking toward the future, let us engage in a quick (rather fragmented and patchy) recap from Heraclitus to our modern times.

a. Heraclitus of Ephesus

Heraclitus was a native of the city of Ephesus, nowadays a province of southwest Turkey. He was a self-taught thinker also known as "the obscure". Heraclitus declared the unity between being and becoming—that is, the One is the multiplicity and its change itself. It is the One Divine that is and contains all the changing appearances of Nature. Famous is his statement "Panta rhei", meaning "everything flows"—an understanding of life and reality reminiscent of the

Fig. 79 Heraclitus of Ephesus
(535-475 BC)

Buddhist doctrine of impermanence. Nature is seen as a never-ending cycling process at the origin of which stands the 'Logos', the ultimate mind, the rational force permeating the universe, that is the originating and organizing principle of the world symbolized by the Sun. Also, the opposites are all expressions of the same Logos in its plurality which contains them all. There is a unity of being and becoming that manifests in the changing phenomena of Nature. We perceive movement as a change in the multiplicity and plurality from being to non-being, but that does not affect the Being itself. It is the opposites of being to/from non-being that we perceive as a temporal manifestation. This change is never-ending and eternal, and Heraclitus summed it up by saying: *"From all things one and from one all things."* Moreover, the duty of every human is to bring into correspondence the 'inner Logos' with the 'outer Logos', meaning that our thoughts and actions must be in harmony with the eternal natural cycle of metamorphosis which reflects the essence of a 'divine Logos'. Heraclitus could be considered the precursor to—if not even the founder of—pantheism, equating God with Nature. This cosmology will be taken up again only much later by Spinoza in the 17th century.

b. Parmenides of Elea

Parmenides, a contemporary of Heraclitus, argued that one couldn't get behind thinking by thinking. One must go beyond the world of illusions created by our structures of perception and realize the 'Being' as the ultimate Reality. Only the uncreated, unchanging atemporal, and neither bound nor unbounded whole Being exists. The things we perceive are only non-actual appearances. The phenomenal appearances of the material world are like the mirage of a lake over a hot desert area when there is nothing

Fig. 80 Parmenides of Elea (515-460/55 BC)

there. The experience of change and motion is an illusion of the senses themselves. Every change implies the appearance of something coming in and going out of existence. Think of an apple. We say that it is an apple—that is, there is something that exists, and we call it an 'apple'. What happens once one eats it? Has the apple gone out of existence? To avoid absurdities, we must conclude that change is only an appearance. Therefore, Parmenides contends that ultimately everything is the non-actual appearance of a non-divisible 'Being' and the apparent differentiation and its derivative change being non-actual. Also, time is illusory, and everything must be simultaneous in the now. What we perceive are different facets of the same Being, a perception of the Oneness from different perspectives which appears as the many but still is only the One. Finally, we are left with the One, which is Being where there is no no-being. His slogan was "all is One". A monistic philosophy in which all reality is fundamentally the One. A conception that is very similar to the Indian non-dual teachings of the Brahman which is the 'One without a second'. In fact, Parmenides was not only a rational thinker but also a mystic. He described his journey into the heavens where he met the goddess who revealed to him the 'way of truth' and how the 'daughters of the Sun' escorted him to a temple where day and night meet—a place where all opposites meet into a greater Oneness of reality. These mystical insights disclosing an ultimate reality and logical argumentations made Parmenides a good example of a philosopher who could unite the rational with the trans-rational.

c. Plato

Shortly after Parmenides and Heraclitus, Plato introduced a form of idealism in which he conceived of geometric forms as real objects existing in some higher realm beyond the world of senses. Plato contended that forms such as a square, a triangle, a circle, etc., are not the creations of our minds; rather, they are existing ideas that our mind can access. At first, this might sound bizarre. After all, we can carve out regular and geometrical forms which make up any material object. For example, we might believe that someone who draws a 2D square on a blackboard needs not to get into

Fig. 81 Plato (~425-348 BC)

contact with some undefined ideal and immaterial hyperuranion to be able to conceive that square. However, Plato would object that what has been drawn on the blackboard is not a square but only a discontinuous line of chalk on a rough surface, which has a thickness (it is not 2D but a 3D track of calcium carbonate). It certainly has some irregularity, the lines are never perfectly parallel, the edges are not perfectly at 90° to each other, etc. In other words, we have never seen a real square but only approximate reconstructions of it. We can imagine it but never find it in the real world. The square on the blackboard represents an idea that we could not draw if we would not have it already in mind—the same fits for every geometric object.

The question is, then, where does our mind get the idea of these objects if we never observe them in the physical realm? Plato argues that there must be more to the world than the physical reality we are aware of with our human limited sensory means. There must be a realm of *'ideal forms'* that is invisible to the senses but can be seized by the mind and, only after that can be cast into an approximate physical representation. Platonic forms are intelligible but not sensible. Without having the concept first, one could neither see nor create its representation. Plato goes so far as to claim that every 'substance'—that is, matter itself—is only the dim shadow of that realm of ideal forms (therefrom the allegory of the cave) in which it participates and, to some degree, imitates but replicates only imperfectly. Our senses perceive differences, but the mind can rise above and apprehend the ideal form. All the variety of material objects of a specific class that we observe in the natural world (say, for example, the different types of trees) are only variations of one unique and single ideal form that they represent, the *'Universal'* (say, the universal we generically call 'tree', but will never perceive with the senses). For Plato, 'learning' means recalling a knowledge that is already latent elsewhere, namely, in the realm of forms.

Therefore, in the frame of Platonic idealism, only the objects of thought have a true reality. This holds not only for geometrical objects that could be described by quantitative mathematical descriptions, but also for 'qualitative forms' like beauty, truth, courage, justice, etc. At the top of the hierarchy of all these ideal forms stands the Logos of Heraclitus.

The same could be said about our bodies and souls. The body reflects (shadowy) the form of our soul. When the former dies, the latter continues to exist in the realm of these ideal forms. Similar to the Eastern teachings of reincarnation, Plato is famously known to have believed in the theory of transmigration of the soul, which comes and returns from and into this realm of forms.

d. Aristotle

Aristotle was Plato's scholar. It is impossible to summarize his thoughts and achievements (as well as his failures) in a few paragraphs. It may only be said that while Plato speaks with spiritual intuition, Aristotle will throw the seeds for a natural philosophy that will become, much later, about 18 centuries later, the springboard towards the modern scientific method. Aristotle had, compared to Plato, a more proto-scientific attitude towards Nature and reality. He can be considered the founder of the empiric and observational method,

Fig. 82 Aristotle (385-323 BC)

together with logic and classification, which, as well known, until nowadays, are the pillars of modern scientific inquiry. He objectively tried to observe, catalogue, and classify things in the domains we nowadays call physics, biology, physiology, and meteorology. More than furnishing answers, his real gift to humanity was offering us a methodology.

Nevertheless, Aristotle also had a somewhat less known metaphysical personality. In his treatise *"Metaphysics"*, a term he coined meaning 'that which is beyond the physical', the empiricist and logician Aristotle showed his spiritual-minded side. He conceived of the human soul as being an organizing first principle, the beginning of all things, the *'arché'*, and the essence of an organism. As for Plato, the living body was the form of the soul. The latter was the actuality (the *'first energeia'*) while the former the expression in potentiality (the *'second energeia'*).

Aristotle distinguished four causes of change or movement:

- The *'material causes'* (depending on the material composition of a body),
- The *'formal causes'* (depending on the form of a body),

- The *'efficient cause'* (the agent of change, such as the action of forces), and

- The *'final causes'* (the purpose or goal of the change or motion).

While modern science recognizes the first three causes, it denies any reality to the latter. This is because final causes seem to suggest a teleological character inherent in Nature—that is, they seem to imply an inner and hidden source of intelligent *'agency'* that always stands behind the natural phenomenal appearances. This is anathema in modern thinking. But Aristotle argued that we must look beyond external appearances and intuitively feel how every living being is moved by an inner urge to become greater than what it is. Everything is more or less manifestly evident to our sensorial experience, moved to fulfill a *'télos'* —that is, a purpose and specific goal or aim. Even illnesses, 'mistakes' of Nature, or what modern Darwinian evolutionary concepts would call a genetic 'disorder' caused by causal accidental random changes in the DNA , are in the eyes of an Aristotelian the effects of the resistance and inertia of matter to a forming and purposeful force. But for Aristotle, human activities like those of an artist painting a beautiful natural landscape are the attempt of the human soul to represent an inward significance and beauty that isn't immediately visible to the senses. Ultimately, everything is guided from within by its *'entelechy'* —that is, that which has a purpose within, some sort of vital and self-organizing principle which aims to realize a specific design or purpose. Of course, God is the Source and Origin that determines this *'télos'* and thereby is the ultimate cause of all changes and movements in Nature.

It is interesting to note how Aristotle distinguished between a *'passive mind'* or *'passive intellect'* and an *'active mind'* or *'active intellect'* (nous poiêtikos). The perishable passive mind is that which is continuously modified by our sense-experience and the interactions with matter. The separate, unaffected deathless, and everlasting active mind illuminates the passive mind to lighten it up like light does in making colors actual. It is the active mind which recognizes the forms as 'profiles' of the One. Thus, there is at least an active mind he doesn't identify with the brain or any bodily organ.

The precise interpretation of Aristotle's philosophy of mind is controversial, but this distinction suggests that he recognized the difference between what we have so far called the 'sense-mind' and higher forms of cognition—an understanding of our thought processes, intuitions, and cognitive function which transcends the ordinary mind concept and emerges throughout the cultures and ages.

A detail worth mentioning is Aristotle's concept of *'hylomorphism,'* a philosophical view that conceives of every object as composed of two essential components: a material essence we call matter ('hyle') and an immaterial form ('morphe'). It is a metaphysical theory of substance that sees form as an essence giving every material object and living body its organization, structure, properties, and identity–that is, the 'whatness' and essence of a thing. This distinction implies that matter is brought into existence only if form is already present and active in it as a causative principle.

Hylomorphism is also closely related to Aristotle's idea that potentiality exists for the sake of actuality. Similarly, matter exists for the sake of receiving its form. Matter represents the potentiality of an object, while form represents its actuality activating its potency.

e. Hermes Trismegistus

The hermetic spirituality of Hermes Trismegistus, a syncretic combination of the Greek god Hermes and Egyptian god Thoth, and purported author of the Corpus Hermeticum, a collection of 17 Greek writings. These series of sacred texts are the basis of Hermeticism, which is often interpreted as a mere theological, philosophical or even a magical treatise. This, however, is a narrow understanding of a deeper ancient wisdom. Hermeticism is primarily concerned with experiential practices intended to transcend the ordinary limited state of consciousness rising to higher forms of spiritual knowledge known as 'gnosis', to realize the true nature of reality behind the hallucinatory veil of appearances of

Fig. 83 Hermes Trismegistus (~3rd-century BC)

the mind. Its aim is to realize cosmic consciousness and union with the divine (for a modern introduction, see [231]). Gnosis is a Greek word meaning 'knowledge.' However, in the hermetic context (and the present treatise) it is not intellectual or logical knowledge but is rather a spiritual and mystic knowledge that transcends words and any representation of the analytic mind. The aim of gnosis is to realize one's own true nature and obtain union with the Divine that is all, in all, and non-dual (a theme that, as we will see, is also central especially in non-dual Eastern philosophies.) In Hermeticism[44] the descent of the soul, a spark of God's soul, into a body obnubilates its inherent wisdom and perfection. That's why in man is made in the image of God and can unite with God. However, in its incarnation the soul lives in a state of hallucination, subjected to the passions of the body and life, created by a *'demiurge'* that has created the physical universe and its false appearances, behind which reigns the true, the good and the beautiful. Yet, the universe is described as beautiful and perfect. Our goal in life is not to escape the world to reach heaven but to bring heaven to Earth. We should not reach the Gods but allow them to descend in us. By cultivating rituals and offerings, but especially practicing piety and the admiration of the Gods we can obtain

---

[44] Hermeticism should not be confused with Gnosticism, the Jewish and Christian mystical doctrine founded later about the 1st-century AD.

salvation. The material world is permeated by the light of God, but the sensible reality is wrapped into several material spheres (Moon, Mercury, Venus, Sun, Mars, Jupiter, Saturn, and the spheres of stars). While the heavenly three spheres are the 'Ogdoad' (the realm of the souls), the 'Ennead' (the ream of divine Light and gnosis, the Nous), and the 'One' (the Source, or God.) The soul once descended into body and now must ascend back through all the seven planetary spheres to obtain purification. Once purified, it will access the last three spheres beyond space and time rejoining with its Source and acquiring supreme knowledge. To do so, however, it needs to free itself from the 'torments' of ignorance, sorrow, lust, anger, etc. and defeat the powers of darkness. Then it will obtain the powers of light, such as self-control, joy, truth, generosity, etc.

We will see how the hermetic spiritual practice is in many respects comparable to that of Eastern philosophies, especially the practice of yoga. It represents one of the many examples that show how these mystic experiences were not particular to any culture or historical period but are the common trait of human experience throughout all ages and geographical or cultural contexts. Something that can't be considered a coincidence and without deeper significance. These mystics saw something that makes part of a vaster Earthly and cosmic reality that escapes the ordinary states of consciousness.

f. Plotinus

Another interesting figure was Plotinus, a Hellenistic philosopher and one of the major figures of neoplatonism. Plotinus was neither an intuitive-minded man like Plato nor a rational empiricist like Aristotle. Rather, he was a mystic who translated his experiences in philosophical language.

According to Plotinus, focusing our senses and mind too much on external physical things leads to the false notion (the false *arché*). Detaching from our superficial

Fig. 84 Plotinus (205-270 AD)

sensory perceptions, we can get into contact with supra-sensible realms and realize the three *'Hypostasis'*—that is, the three fundamental substances which underly the whole of Reality: the One (the Good, God, that which is beyond being itself), the rational intellect (the *'nous'*, the mind, the intellect, the being and the realm of Platonic forms), and the irrational soul (endowed with desire, entangled in matter, having sensations, etc.). The human's mission is to climb this ladder of existence in order to know the One in an ascesis he coined as "returning home". In fact, for Plotinus, the lower levels are only images of the levels above. Matter is considered the ultimate separation from the One, while every material form, as any Platonic form, is still an abstraction of the very same One. Everything flows down from the One, meaning that matter, sensations, and concepts that our perceptible world is made of, are successive 'emanations' or fragmentations of it. But the One is uncaused or self-caused[45], containing only itself. At the level of the One, there is no separation; everything is in the One and is the One itself. Plotinus also states that, in order to obtain this insight, it is not sufficient to train the intellect or exercise virtues which master the irrational soul, but that alone does not lead us to the realization of the One. It is by an interior life, detached from material desires and emotions and through the contemplation of the One, that we can soar to the identification with it.

Again, it is difficult to avoid noticing the similarity, if not the full identity, with the Eastern non-dual teachings, the Buddhist tradition, and the Hindu yoga practices. Some scholars contend that Hellenic philosophers might have had some contact with the Indian peninsula. Indeed, it should be noted that Alexander the Great, whose imperial expansion reached northwestern India, was a scholar of Aristotle. It might well be that he established a scholarly transmission between Eastern and Western traditions, though no historical document or event confirms this. At any rate, even though such a cross-fertilization might have taken place, these striking similarities between the two mystic traditions have to be found much more in human nature than in the exterior historical context. Sooner or later, the human being finds its way inward. And, in essence, what it finds there is always the same undifferentiated and eternal oneness, the monistic idea that all of reality can be derived from a single principle, the One.

g. Thomas Aquinas

The medieval Italian friar and saint Thomas Aquinas tried to reconcile religion with reason, proposing a school of thought of natural philosophy known as '*Thomism*,' which took inspiration from Aristotle and built something new. It is impossible to do justice to his philosophy in a few paragraphs. Here, we will pick out only a few points of his doctrine that tries to unite natural philosophy, science, mind, intuition, Nature, and the Spirit and that represents a pre-scientific attempt to create the kind of extended synthesis we will develop in the third part.

Fig. 85 Thomas Aquinas (1225-1274)

For Aquinas, faith and reason are not opposites but, rather, complementary. God reveals himself through both rational thinking and the study of nature. Truth can be known through natural revelation (reason and faith) and supernatural revelation by the inspiration of the Holy Spirit. As did Aristotle, Aquinas posited that life came from non-living matter, and he was one of the earliest thinkers (after Empedocles) to have the intuition that species mutate

---

[45] Plotinus first introduced the important concept of 'self-causation', and that, as we shall see later, will be taken up by Spinoza calling it '*causa sui*'.

by directed evolution. He framed five arguments for the existence of God, the *'Five Ways.'* He distinguished between *'natural laws,'* those captured by reason, and *'eternal laws'* or *'divine laws.'*

Aquinas took up Aristotle's hylomorphism with the idea that form and matter are two distinct categories. Matter cannot be without form, but form can preexist matter. It is the form that causes matter to be. There are *'substantial forms'* that determine the essential and immutable properties of a material object and *'accidental forms'* that determine the mutable non-essential properties. Moved by this idea, he postulated that the human soul is an immaterial substantial form, while the body reflects it as a material embodiment. This reflects Aquinas' existentialism, which maintains that all things have transcendental properties of being: At the base of everything is the 'being' of things. This 'being' is a principle of and in things that is made of a *'quiddity'* ('whatness')–that is, the essence, form, or nature of things. In line with the Aristotelian doctrine, all things have final causes, a purpose, and an aim.

Relevant to the philosophy of mind is Aquinas' perception of the human intellect as being secretly moved by God, though he admits that reason can be self-sufficient to a certain degree. Humans can share the divine wisdom. The mind and will are the highest power of the human soul; they are not to be found in a bodily organ. The mind's and will's natural determinations of the good ultimately originate in God, but humans have free choice in accepting or rejecting it by determining themselves. The best definition Aquinas proposes for the notion of 'freedom' is to be found in an entity being *'causa sui'*–that is, cause of one's own and self-determined. By free choice, he meant that we are the cause of our own actions and desires, for our own sake. God must be self-determined because there is no one else beyond it that could determine it.

h. Meister Eckhart

After Plotinus, Neoplatonism had a considerable influence on the history of Western philosophy. Its monistic idea that the essence of all reality can ultimately be traced to the One is its distinctive character. The German mystic, philosopher, theologian, and Dominican friar, Meister Eckhart could also be cited as someone who was part of a group of neo-platonic philosophical mystics with clear parallels to the Buddhist and Hindu traditions.

Fig. 86 Meister Eckhart (1260–1328)

His statement on the nature of God was clear but also dangerous: God is in all beings, a 'footprint' in the soul of all creatures. There is no real difference between our soul and God. Our soul is God. And yet, God is 'nothing', meaning that no adjective, word, or concept could express its real nature, which is beyond words. Only if our self (*"das Ich"*, the

"I") disappears can the soul awaken and become one with God. Eckert distinguished between a *'Nature that is created'*, and a *'Nature that creates but is not created'*. *"All things are God itself"* was his message. It is a worldview that, in modern terms, we would call 'panentheistic'. *'Panentheism'* (from the Greek pan = "all", en = "in" and Theos = "God") conceives of everything as being in God—that is, God pervades and interpenetrates every part of the universe and, nonetheless, extends beyond matter, space and time. God is not a distant entity but, rather, a divine spark that we can find in us without the need for external authorities. This is an unheard-of and unacceptable heresy for a religious establishment that considered itself the sole mediator between the human and divine.

But Eckhart's recognition of God in ourselves does not have to be misinterpreted as an attempt at human self-aggrandizement. Rather, to the contrary, it is an invitation to an act of humility that should subject our outer nature to the inner Soul. Eckhart was a master of mystic intuition, experience, and inner realization of God as Oneness in everybody's soul that we can find by means of contemplative practices. In fact, if we want to discover our inner God, we must practice a discipline of 'detachment' (*'Abgeschiedenheit'*) from our desires, emotions, thoughts, inclinations, longings and learn to connect with our inner soul and rest in it. It is about detaching from our outer sensorial and mental nature and coming in contact with the inner one—a practice of equanimity, an unmoving detachment aimed at calming the mind and emotions, not too dissimilar from the Eastern psychological practices of meditation and yoga. We must leave everything behind us and surrender to the Divine in us. Once we have emptied ourselves like a cup, the same cup will be filled with God's bliss and joy.

Eckhart died before the inquisition could harm him. But after his death, several sentences were eliminated from what was left of his written texts. He was light years ahead of his time, and his teachings remain, until today, underestimated not only among Christians but also by the contemporary philosophy that, at best, focuses only on his rational writings and is unaware of how he was one of the major figures of an underground current of the history of Western culture interpenetrated by a mystic philosophy that the analytic mind fails to understand. Eckhart's theology was not based on an abstract reasoning but indicated a practical spirituality that allows everybody to find their own way to an inner divinity and a trans-rational knowledge. If embraced, this could have formed a whole new spirituality in Western civilization, as it did in parts of the Asian continent. This is something dangerous for a dogmatic religious thinking and non-sensical for the analytic sense-mind-based intellect of the Age of Reason that had yet to come.

But the human spirit does not grow and evolve linearly. Stagnation or even relapse into a previous stage of development is possible, and most importantly, it might have to forget itself to focus on other aspects of existence.

i. Giordano Bruno

The Italian natural philosopher, former Dominican friar, and cosmologist Giordano Bruno is considered, even today, the icon par excellence of the free thinker, the martyr who stands until the last for truth and who doesn't bow down to the established religious, political, and cultural establishment.

Fig. 87 Giordano Bruno
(1548-1600)

Bruno soon laid down his Dominican robe and became a staunch critic of the whole Christian doctrine. One of his most fundamental teachings was the infinity of the material universe in an infinite ether. If God is infinite, and if all Creation is its manifestation, then the latter can't be limited; it must be infinite as well. Our world is not the center of this infinite universe. Man is only a 'deep shadow', a detail in the infinite, just an infinitesimal speck. We are neither light nor truth.

Contrary to the established belief of a divine but static and immutable universal order, a Heraclitan 'panta rhei' resonated in Bruno when he declared that everything is change and the world is in an eternal metamorphosis. Matter, being infinite and permanently in movement and transformation, is a unity expressed in action, form, and power. It is a living matter that can be spiritualized because the universal matter is the expression of the same universal World-Soul residing in every living being.

Bruno became well known for his conjecture that stars are just other suns similar to our Sun and that other planetary systems must also exist beyond ours. But for him, all celestial objects were not just rocky planets or a hot burning bunch of gas; rather, they all were 'larger beings' one could call 'Gods' that 'dance' in the cosmic harmony and, at the same time, harbor other forms of life not destined to death.

Though Bruno was not a mystic, like Meister Eckhart, he can be considered a panentheist. He posited an immanent transcendence (or transcendent immanence) in all things. That is, God is in all natural things and, at the same time, is not identical to the World. Everything is 'pulsed through' by the 'World-Soul', the 'Anima Mundi'. Here, again, we find the dual aspect of Nature: the One immanent that is and becomes everything, and its Transcendent aspect, which is the Origin of everything but stands behind the manifestation.

Because humans and animals are spiritually indistinguishable and come from the same World-Soul , Bruno beholds the theory of *'metempsychosis'*— that is, the belief that souls transmigrate. Metempsychosis differs from the popular reincarnation theory insofar that it allows not only for transmigration from a human (or animal) body to another human (or animal) body, but also from an animal to a human or a human to an animal body. The soul is immortal, but its individuality can change from human to animal or vice versa. To illustrate the process, Bruno used the metaphor of a light source (the World-Soul) that reflects itself in every living being as in a myriad of mirror fragments (the individual souls).[46] After death, each soul returns to and unites with the original Source, rests for a while in it, and emerges with its individuality in a new incarnation cycle.

If one behaves like an animal during one's human life, then in the next life, one will be in an animal's body. Conversely, if a soul in an animal body makes an effort to go beyond its animal nature, it will take on a human body.

Contrary to the rising colonialist mindset, Bruno's metempsychosis and panentheism, which see all beings and all Creation as ensouled, naturally draw him to see, in the conquered American 'Indios', the indigenous peoples of South America, human beings who are equal to all others. Humans are all God's children and are made of the same universal matter as all other living beings.

Another similarity with Meister Eckhart was Bruno's contention that there is no single way to know God; everyone has their own approach to God, and the temporal powers of a church or mediating cleric aren't necessary to realize the Transcendent. His rejection of Christianity went so far as to downgrade Jesus to a 'bad magician'.

Bruno's metaphysical and intuitive philosophy was antipodal to the still anthropocentric and Ptolemaic Christian theology. His theories were so radical that the obscurantist and intolerant religious authorities could hardly ignore them without taking extreme actions. It was a society in the midst of the infamous witch hunts and still struggled to transition from the Middle Ages to the reawakening of the arts, the philosophical thoughts, and the sciences of the Renaissance. But Bruno refused to recant anything and, in February 1600, was burned at the stake by the inquisition for his 'heresies'.

Bruno was a unique figure who could unite high intellectual skills with an intuitive spirituality and natural metaphysics. It is unclear where he got his insights, whether through a rational and intuitive philosophy or due to some mystic experience, but he conveys the impression of a personality with a cosmic awareness coming from another form of cognition far beyond that of an analytic thinker who, of course, brands it 'phantasy'.

j. Baruch Spinoza

We now jump straight to the Renaissance and Age of Enlightenment, skipping the Western Dark Ages and Middle Ages. Let us land with our time machine in the Netherlands of the 17th century and consider the monistic

---

[46] A metaphor strikingly similar to that of the Indian Advaitic school, which compared individual souls to that of the Absolute (Brahman), like the reflection of the Sun's image on the ripples and waves of an ocean.

conceptions of Baruch Spinoza. The Dutch philosopher was exiled at the age of 24 from his Jewish community because of his unorthodox theology. He was initially troubled by the lack of happiness in the worldly human's condition. Motivated by a desire reminiscent of Buddha's quest for liberation from suffering, Spinoza searched for the supreme and unending happiness he could not find. This led him to a conception of God very different from anything that an orthodox and still intolerant religious society of the time could accept.

Fig. 88 Baruch Spinoza (1632-1677)

Spinoza started from a few simple and indubitable principles such as the *'principle of sufficient reason'*, whereby everything must have a reason and each effect a cause. However, because everything can be thought of as the (temporal or logical) effect of a previous cause, we must conclude that in order to avoid infinite regression, there exists a primary and uncaused 'substance', 'res', or 'Nature' which does not require any previous reason to exist but is in itself existent and cause only of itself. This 'substance' or 'Nature', which does not need any prior cause to exist or reason to be explained, it is 'self-caused'– or, as he called it with the Latin term, *'causa sui'*–is, for him, the synonym for God. In this sense, the word 'Nature' must not be confused with the meaning with which the common language associates it. For Spinoza, Nature is not just the material cosmos with its physical laws; it is the primordial uncaused and self-existent as self-explanatory 'substance' which is itself the cause of everything else, be it material, immaterial, or whatever. Nevertheless, this infinite substance, or simply God, is immanent in the universe and can be captured by our sensory perception. God is not something separate or distinct from the insentient matter.

From this assumption, Spinoza proceeds in positing that everything manifest is just a *'mode'* by which this substance presents itself to us. By 'mode', he meant a 'manner' or 'way'—that is, a particular state of being of God. Nature is God self-expressing itself by infinite different modes as a structured pattern of activities. He then distinguishes between an active divine power, Nature 'naturing' *('Natura naturans')* and the things produced by that activity as Nature 'natured' *('Natura naturata')*. Interestingly, this distinction, which was already present in Meister Eckhart and Giordano Bruno, is a recurring theme. This is no coincidence, and we will find it again. It reflects a deeper truth of things that will become one of the central principles of an integral cosmology we will introduce in Pt.III.

Therefrom is the frequent misconception that interprets Spinoza's God as the identity of Nature, with the latter being only the material dynamical spacetime entity of modern science. The identity of Nature with God must be understood as Spinoza's Nature transcending the material universe, though immanent in it. Moreover, these modes are not God, as the waves are not the ocean. Nevertheless, there can't be any modes without God, just as there can't be waves without the ocean. It is how these modes are expressed by God that determines the properties of the World.

According to this point of view, which is a sort of dual-aspect monism, mind and matter are neither two independent things, as a dualist would conceive of, nor the same thing, as the physicalist believes, but the two modes or ways of expression of the very same substance. Spinoza considered the *'res extensa'* (matter) and the *'res cogitans'* (mind) as only two 'modes' of God out of an infinite number of other modes. Mind and matter are the same things appearing with different attributes. They are in God, but God does not exhaust itself only with these two attributes. And the material universe we perceive with its physical laws, its forces acting dynamically in space and time, is only a 'sub-manifestation', not all of Nature. In this cosmology, the human mind is also only a part of the infinite intellect of God. All our minds are minds in the Mind of God.

This concept of mind and matter as two poles of the very same thing is a form of *'panpsychism'*—the view that matter and all its material constituents are a form of mentality or 'proto-consciousness' ('pan' means 'all' or 'everything' and *'psyche'* is a Greek word for soul or consciousness or, as it is somewhat misleadingly translated in modern philosophy, mind). Mind (or consciousness) does not emerge from the aggregation of matter; rather, matter itself, down to elementary particles, such an electron, *is* a form of primitive and involved mind (or consciousness). As we will see later, this view is experiencing a revival in modern philosophy.

Therefore, God is not extraneous to the world but, to the contrary, that which is immanent in all things. All is in God, all lives and moves in God, but this God is itself a deterministic entity without free will. All events in It are the operations of invariable laws. God determines things not because of a free will but from its very nature and infinite power without goal or purpose. It is a non-material but still a 'mechanical-deterministic God', so to speak, which makes it hard to distinguish it from a mechanistic cosmology that has no God at all. This is also the reason why Spinoza was accused of being an atheist in disguise.

Spinoza, however, can be considered an idealist who preceded Kant's transcendental idealism. He recognized how the stimulation of our sense-experience organs is translated into perceptual ideas, which are not only unreliable but don't reveal much about the nature of the things perceived. All the

Knowledge of the first kind	Imaginations
Knowledge of the second kind	Reason
Knowledge of the third kind	Intuition

ideas we derive from this sense-experience are 'inadequate ideas', 'imaginations', which give us an understanding of things he termed *'knowledge of the first kind'*. What we see are only images of things; objective knowledge is impossible and what we call 'falsity' is a deprivation of knowledge. But this falsity of our 'imagination' can be overcome by the second and third kinds of knowledge, namely, reason and intuition, respectively. Interestingly, Spinoza did not place reason but intuition as the highest form of knowledge. By intuition, he meant a form of cognition that can seize immediately and directly the unity between things and the time-less structure of the power of God. Intuition does not require intermediate steps of inference; it reveals things in the light of eternity, the essence of things which reason cannot know.

Spinoza was, to some extent, also a mystic, even though he probably did not ascend the ladder to the ultimate Unity of a Plotinus, a Meister Eckhart, or the Buddhist and Vedantic non-dual realization. His search for permanent happiness led him to a practice of self-control of emotions, passions, and the activity of the mind. He distinguished between *'passive emotions'*, those determined by external conditions over which we have no control if not by overcoming it with the *'active emotions'*, which come from within. Only the latter can give us real joy and, together with reason and intuition, provide a better basis to control our irrational passions, which cloud our understanding and lead to emotional instability. We can gain real happiness only if we look within and understand and know ourselves. This spiritual contempt leads to a state of 'blessedness' characterized by a constant and eternal love for God.

A conception of psychological and spiritual ascesis aimed at the union with the Divine very similar to that of the Neoplatonists and the Eastern mystics based on similar psychophysiological techniques, such as yoga.

Despite the common belief, Spinoza's conception of God can hardly be ascribed to a pantheist worldview. 'Pantheism' is the view that physical nature and God are equivalent (something we would see as a form of monistic materialism, again, as a form of physicalism in disguise as well). It can hardly be emphasized enough how Spinoza saw physical nature only as one of the infinite expressions of the Deity, not the Deity itself. One could eventually characterize him as a panentheist.

Panentheism differs from pantheism because it conceives of the physical nature being part of God, but with God itself being more than that. Some versions of panentheism, however, go further than a mechanistic and deterministic view which characterizes Spinoza's philosophy and which stands, therefore, somehow in between the two forms of monism. Finally, *'deism'* is a philosophical position that rejects revelation as a source of divine knowledge and accepts the belief in the existence of God only based on rational thought. It sees God as completely separate from its creation—that is, the universe or Nature. Deism asserts that God set the universe in motion, then ceased to interact with it. For a review of theories of panentheism as also panpsychism, see [232].

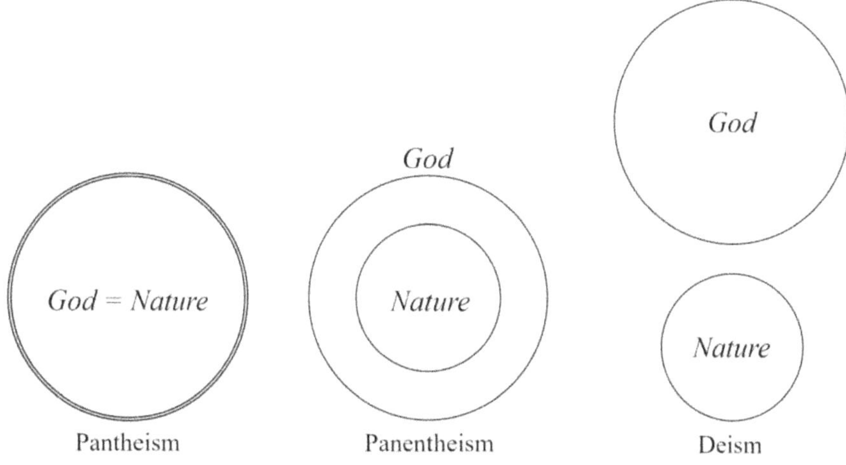

Fig. 89 Three theistic cosmologies compared: Pantheism, panentheism and deism.

At any rate, his Divine was something too foreign and too far from the personal transcendent and not immanent God of the Abrahamic religions. Spinoza was a rationalist but not a materialist. His treatises were written according to a strictly logical and mathematical style with definitions, axioms, propositions, and proofs. His rationalist philosophy is, in hindsight, a quite impressive achievement. It is a nice example of how far the ordinary human mind can go with the help of knowledge of the third kind—that is, intuition—if it is not hindered by external conditionings and dogmas. Also, the intuition that there must be a universal 'substance' whose modes determine the properties of the manifestation has striking similarities with the understanding of modern particle physics: In quantum field theory, every particle is an 'excitation', a 'mode' of a universal quantum field. It is no coincidence that Spinoza's God captured the attention and sympathies of several scientists—among them, notoriously Einstein.

k. Gottfried Wilhelm Leibniz

The German philosopher and mathematician Gottfried Wilhelm Leibniz, best known as the inventor of modern calculus, didn't see Spinoza's ultimate and uncaused 'substance' as the ground of existence; instead, he put forth an atomistic theory—the so-called *'monadology'*—which conceived of fundamental and immaterial units of substance. *'Monads'* are the ultimate, indivisible, immaterial, and true units of Nature. They are the elementary 'souls' without extension or shape and parts, metaphysical points without primary qualities. These elementary 'soul-units' or 'psychic atoms' are causally disconnected; they do not interact with each other. Yet, all

Fig. 90 Gottfried Wilhelm Leibniz (1646-1716)

that exists is composed of these monads, and it is by their combination that objects acquire an extension. Monads can't dissolve and had no beginning, but they can change because of an active internal principle. We experience change because of the internal driving principle of all monads. As elementary souls, these monads have perception and are driven by an *'appetition'*—that is, a sort of Aristotelian entelechy, an internal inborn goal-driven and purpose-directed driving force we might think of striving, a motivation, or a desire. However, monads are not self-conscious or rational beings.

The living body is an aggregation of monads forming a unified 'soul-monad' which, together, give rise to an organism with volition and eventually, as in the human case, to rational reasoning. In the case of humans, Leibniz spoke of a so-composed *'spirit-monad'* as the life-giving and thought-giving organizing principle. Meanwhile, at the top of the hierarchy, the *'supreme monad'* has a perfect perception and perfect appetition—in short, that which qualifies as God.

Leibniz, like Spinoza, could be considered the founder of modern panpsychism, but he was a dualist of a bit of an unusual kind. He saw mind and body as separate entities which have no causal interaction but which are always synchronized. Because, according to him, all these immaterial monads contain all their own past and future, they always know which state of mind is necessary to be aligned with all the external happenings in the body. Mind and body are, therefore, completely separate entities that can't interact with each other, but the state of the former is always a reflection of the latter due to a temporal synchronization. All our perceptions are 'well-ordered dreams'.

Worth mentioning is *'Leibniz's Mill argument'* against mechanical materialism:

"[W]e must confess that perception, and what depends upon it, is inexplicable in terms of mechanical reasons, that is through shapes, size, and motions. If we imagine a machine whose structure makes it think, sense, and have perceptions, we could conceive it enlarged, keeping the same proportions, so that we could enter into it, as one enters a mill. Assuming that, when inspecting its interior, we will find only parts that push one another, and we will never find anything to explain a perception. And so, one should seek perception in the simple substance and not in the composite or in the machine." [233]

Sounds familiar? Leibniz could be considered a forerunner of consciousness studies who was light years ahead. He realized what in modern terms we label as the 'hard problem of consciousness' and followed that sentience might be fundamental, not emergent, leading him to embrace a panpsychist metaphysics. As we will see, a quite similar intellectual path will represent itself in the modern philosophy of mind about three centuries later.

I. George Berkeley

While Spinoza became known for his particular form of monism and Leibniz for his contributions to modern mathematics, the Irish bishop and philosopher George Berkeley became known as the most prominent subjective idealist for a theory he called *'immaterialism'*. Berkeley was also an empiricist, concerned mainly with optics, but most of his philosophical thoughts were devoted to the relationship between sensory experiences and the sensory ideas arising from these experiences. We learn about the world through sensory experience, such as the distance and shape of objects, and then

Fig. 91 George Berkeley (1685-1753)

translate that knowledge into sensory ideas. Following a line of reasoning that we might recognize nowadays in the binding problem (see Pt.I-II.3), he pointed out how all the objects we get to know—that is, these sensory ideas—are the result of a process that somehow collects and binds together, in a unique fashion, all our sensory perceptions. But Berkeley contended that there is no substantial and qualitative difference between the sensory idea and the represented object itself. A sensory idea is an object, and the object is a sensory idea because the nature of objects is to be perceived, and the nature of perceptions is to be objectified. Therefore, Berkeley argued that all exist because they are perceived.

But what if an object, say, a tree in a forest, is not perceived by anyone? It is not clear if Berkeley believed in the persistency of the existence of objects when not perceived. However, he believed that the sensory ideas we associate with our perceptions are ultimately imparted to us by God. God is the ultimate cause of the ideas we form. This somewhat curious claim has led some to interpret Berkeley as saying that even if an object is not perceived by a human mind or other conscious forms of life, still God can—that is, all objects exist as a state of the Mind or thought of God. It is, however, unclear if this was the original understanding of Berkeley.

For sure, he considered all minds as active spirits in God. In fact, we know ourselves very differently than we perceive things external to us. We are aware of our mental contents but not as we are aware of objects. This implies that, while material objects in the ordinary sense do not exist, there is a difference between thoughts and the sensory ideas we perceive as material objects (a sort of distinction we made between mental, sensory, and emotional qualia in Pt.I-I.3). This makes Berkeley also a particular type of dualist: There is no separation between mind and matter; rather, there is the mind experiencing on one side and things that are composed of the sensory qualities we experience. Or, to put it in the modern parlance of the philosophy of mind, the qualia are inherent properties in and of the objects themselves. There is no matter as such, 'out there', but only qualia.

m. David Hume and Immanuel Kant

Berkeley's idealism, like Spinoza's conception of God, didn't receive much favorable acceptance because it seems to imply some form of pantheism or panentheism as well. Like the famous Scottish philosopher David Hume, other intellectuals found Berkeley's arguments confusing because they *"admit of no answer, and produce no conviction"*. [234]

Fig. 92 David Hume (1711-1776)
Immanuel Kant (1724-1804)

However, the skeptic and naturalist Hume had to recognize how empiricism is nothing but a mental representation in form of ideas of the structure of the world and to admit how Locke's distinction between primary and secondary qualities does not hold. Primary qualities are also ideas of the world arising due to our mental associations. Objects are only a 'bundle' of perceptions which we know as mental associations (the binding problem lurking behind the scenes, again). The mind itself is a 'bundle of perceptions'.

He went so far as to say that natural laws do not exist, inasmuch as they are only ideas expressing a likelihood to observe causal phenomena. If a material body falls to the ground 1000 times, how do we know that, in the 1001st trial, it will not 'fall' upwards? In this sense, for Hume, the law of gravity is a probabilistic statement grounded in our habits of thinking. Our understanding of reality is only a sort of phantasy.

The well-known German philosopher Immanuel Kant credited Hume with having awakened him from his own "dogmatic slumbers." In fact, while Kant is known for having erected an impenetrable barrier between our perceptions and the 'thing-in-itself'—that is, the 'phenomenon' and the 'noumenon''—he was inspired by Hume's ideas. As we already pointed out, Kant recognized in his *"Critique of Pure Reason"*, how our objectified ideas cannot be put into a one-to-one relation with the external world. We can't abstract ourselves from our sensorial perceptions. Space and time come into being as formal conditions interpreting our sensory experiences. According to Kant, we cannot tell if space and time are effectively basic structures of the reality 'out there'. With this conception of space and time, Kant was in stark contrast to Newton, who considered them as absolutes and self-evident aspects of reality.

We recognize the particularities (the sensory data) and incorporate them into a category—that is, a semantic whole together with a formal structure (implicit is, again, the binding problem). Without terms and concepts, we would perceive only a meaningless patch of sensorial experiences of the world. We realize the objects of the world only in conjunction with our imaginations, ideas, and concepts which are juxtaposed to those sensory experiences. We categorize things by a sensorial structure with a space-time experiential insight added by terms derived by a mental understanding and a faculty of discernment.

Kant pointed out that also the *'synthetic a priori propositions'* —that is, propositions that are known to be universally true, such as the basic propositions of geometry—are ultimately possible only because of phenomenal appearances and do not account for independent properties of things-in-themselves, which are independent of our modes of experience. For example, for Kant, stating that the sum of the angles of a triangle is 180° is not an a priori truth but a knowledge derived from our experience of geometrical structures.[47]

In a sense, Kant further developed Plato's idealism which divided the world into things and its Ideal, but he distinguished himself (also from Berkeley), insofar as that there are no mental structures in the world to which we gain access. These are only in us, not an immanent property of reality.

But for Kant, Hume's skepticism was too radical. He didn't consider our phenomenal experiences as incoherent dreams. Kant accepted natural laws as being just human intellectual representations of probabilistic statements but considered these laws as existing inherently in the world. There is something real inherent to the external world because it nevertheless has stability and a structure that can't be compared so straightforwardly to a dream. The properties of the world are characterized by an 'objectivity', at least in the sense that they are formal conditions of our experiences. But mind and senses are not enough to grasp reality in itself. There is a frontier that we can't cross between the phenomena deriving from our sensorial experiences and the noumena. In this sense, the noumena are 'transcendent', and therefrom the name *'transcendental idealism'*, not to be confused with a mystic or spiritual form of transcendence.

n. Johann Gottlieb Fichte

On a quite different but complementary line of reasoning came Johann Gottlieb Fichte, who, in contrast to Hume and Kant, abstracted from the issue of whether we can know the things-in-themselves and focused his attention on the nature of consciousness that perceives itself. More on the lines of Plotinus, he asked who or what is the person, that self-conscious 'I' or 'Self', which perceives the world? Fichte rejects Hume's suggestion that it is a mere figment arising from a 'bundle' of ideas created by the stream of impressions. This self-awareness of one's own nature is not

Fig. 93 Johann Gottlieb Fichte (1762-1814)

---

[47] Indeed, as is well known, the 19th-century mathematicians extended geometry to non-Euclidian geometry and this statement turned out to no longer be universally true.

just a byproduct of a mental stream; rather, it is a form of innate self-knowledge that stands prior to ideas and sensory experiences. In his '*Theory of science*' ('*Wissenschaftslehre*'), published in 1794, Fichte argued against Kant, stating that with a philosophy of the Self, we can move beyond the synthetic a priori and have knowledge of the noumenal world. Moreover, he pointed out the binding act of consciousness in an original manner, noting how our body is perceived in our inner consciousness as a whole, not as a bundle of parts. There is an interplay between an act of 'articulation'—that is, that to become aware of a part of the body—and the 'organization' of the whole body. In this sense, articulating is an act of freedom because we are free to decide what to perceive as a part and what as a whole. For example, we can perceive and think of— that is, articulate—an arm separated inside our perceptual field of cognition from that of the whole body. But, also, this articulation that focuses on the arm is itself perceived as a whole again because an arm has a hand with fingers which, in this instantiation, are not accounted for as separate perceptions. It is not a 'bundle' of perceptions but one and a single sensorial and mental experience at once that could not be possible if not for a self-consciousness that allows for this part-whole relationship.

The embodiment of a Self is a necessary condition for self-consciousness, which becomes self-aware by a distinction between the Self and non-Self. We realize this, according to Fichte, by 'listening' to ourselves and our inmost soul. He took a first-person perspective philosophy by going inside instead of looking only at what our mind and senses tell us. It is by this approach that he recognizes how there is a unity of the phenomenological and noumenal and a harmony of the transcendent with the empirical which, at the bottom, leads us to consciousness as the primary Greek '*arché*', the source, beginning, or origin of everything.

o. Georg Wilhelm Friedrich Hegel

Another high-ranking proponent of German idealism was G. W. F. Hegel, also known for his '*dialectic method*'. While for Plato, the dialectic was synonymous with the 'art of speaking', Hegel saw it rather as a progressive process—that is, where a new knowledge points against an older form of knowledge and extends and amends it, eventually replacing it. It is not a rhetoric method but a synthetic cognitive actualization of bringing together polarities. Truth is the result of a combination of extremes into a conceptual unity that cannot be fully captured by words. Therefrom, his famous statement, "thesis, anti-theisis, synthesyis."

Fig. 94 Georg Wilhelm Friedrich Hegel (1770-1831)

The notion of 'progress' was not that evident and taken for granted as we do today, especially in science and technology. In his book of 1806, '*The*

Phenomenology of Spirit', he delineates a theory of knowledge describing the dialectic character of how progress is enacted: a thought process of apparent opposites which, however, finds its synthesis in a new understanding of the world. Hegel did not contradict other idealists; also, for him, the world can't be separated by our mental projections. Rather, it complemented it by pointing out the progressive nature of our understanding. How one sees the world depends on what and how much one knows about it. We may not comprehend the world, but we can nevertheless move meaningfully inside of it. Mind does not portray an extra-mental reality; rather, it is thought that construes the elements of that reality. Forms and contents are related to each other like concepts and reality or subject and object.

Therefrom, Hegel developed the process-philosophical notion according to which the entire historical reality is the process of a *'World-Soul' ('Weltgeist'*, an expression taken from Schelling). In particular, the human arts, religions, philosophies, and sciences are ultimately the expressions of a trans-human divine activity. Natural history also is the self-becoming of this World-Soul, which finds its way back to itself.

The Platonic character in Hegel becomes evident in his equation between Truth and Idea. Reality is the expression of a sort of *'Idea-in-itself'* which self-differentiates as space, time, and matter. This self-differentiation of the Idea follows a threefold method of universality, particularity, and concrete unity. This Idea and its process of distinction are embedded in an absolute Spirit, a divine entity which, by objectification, creates our inter-subjective relationships. There is initially no division between subject and object; this comes into being only with the subjective mental division between the self and the experienced perception. The world is a movement of the Unity that develops itself in particularities. For example, the seed already contains all that determines its development, in the sense that its metamorphosis to a plant can be seen as the expression in several temporal phases of the same universal. It was not a coincidence that Hegel was a close friend of Goethe, as his views resonated with Goethe's principles of metamorphosis; we will consider this in more detail in Pt.II-III.2. The aspect of particular interest for us is that, for Hegel, the course of human history is an evolution in a series of epochs from a relative unconscious state to a progressively greater state of self-awareness. It is a teleological concept of history that we will find again in other authors.

p. Friedrich Wilhelm Joseph Schelling

For this reason, we skip J. W. von Goethe in our semi-chronological order of Western intuitive philosophy and take up the thoughts of Friedrich Schelling, who ventured into a theological worldview that went much further than Hegel. His radical choice was to unify Spirit, Mind, and Nature, considering the different aspects of one unique identity. Nature is the visible aspect of the Spirit, while the Spirit itself hides as the invisible aspect of Nature. Schelling elevated Kant's transcendental idealism to a monistic ontology in which Nature and Reality have an 'original unity' (*'ursprüngliche Einheit'*), the Absolute. Our separation from Nature renders us spiritually empty because everything is a living being and filled with consciousness.

Fig. 95 Friedrich Wilhelm Joseph Schelling (1775-1854)

In his treatise *'On the World-Soul'* (*'Von der Weltseele'*), published in 1798, Schelling recognizes in the human being a somnambulist spirit which, however, is in its essence part of the World-Soul and that identifies itself in the living organism as an object of contemplation. This World-Soul is the universal and all-comprehensive principle in which every soul is integrated. This higher instance is an Absolute that, by seeing itself as embodied, becomes part of Nature and, in it, by successive stages of development, rises in consciousness until it reaches human status.

Nature and Spirit are two complementary parts of a Whole in which the latter struggles to become conscious inside the former. All the properties of things in their form, substance, quantity, and quality are different aspects of an internally differentiated *'primordial totality'* (*'ursprüngliche Ganzheit'*) in its opposites. What we perceive with our senses and translate in our mind as objects, forms, and symbols are only a partial and incomplete one-sided expression of a dynamic Whole. Schelling envisages a *'World-Soul'* as the basis of the whole of reality in which we are, so to speak, embedded and forced to distinguish between a subjective mental world tainted by feelings and perceptions in us, contrasted by what we call an 'objective world' outside us. However, this distinction also arises from the internal differentiation of an 'un-divorced' Unity (*'ungeschiedene Einheit'*) of an absolute-ideal or absolute subjectivity in an 'eternal act of cognition'. For Schelling, though the essential reality is not accessible to our ordinary reason and mental cognition, we can nevertheless conceive of it as a 'state-of-being' without cognition. This Unity may be cognitively inaccessible to us, but it nonetheless exists as 'being-in-itself'—or, in other words, is the sole pure state of existence in which perceptions and mind are non-necessary addenda. The Universe is seen as an organism, as a whole containing the organic principles of mind and spirit. The life-principle is immanent in the whole Cosmos. This 'all-organism' is the manifestation of the Divine, and all things are contained in this divinity.

This latter theological conception is reminiscent of Spinoza's God with its 'modes of existence' but, at least in the case of Schelling, must be identified with a panentheistic (not pantheistic) worldview. The Universe is, in this view, the self-revelation of God and, at the same time, its self-recognition in Nature. In this process, the organic unity enfolds as a dynamic stepwise sequence from the inorganic to the organic—something which might be termed as a sort of post-material understanding of Darwinian evolution.[48] But Schelling extended the image of the material world dominated by mechanical forces—that is, from what we would nowadays call a physicalist conception—to the expression of an infinite spirit and will. Will is the primordial state of being which expresses itself in Nature as a *'World-Will'* behind the appearances (a quasi-spiritual concept that Schopenhauer will make central to his philosophy, as we shall see next).

But besides a theological aspect, Schelling also saw a teleological process in Nature. The primary goal, aim or 'télos' of this 'world-process' is the return of the finite into the Infinite, the recreation of the unity of the origin, which corresponds to the highest level of self-consciousness of the Spirit. The self-organization we observe in the world is only a symbol, a pale reflex of the original Spirit which molds things by its World-Will, a struggle of a Consciousness to reach a higher organization that expresses the pure form of the Spirit. It is a shape-giving Force existing in Nature which is the dynamic principle in all things, and all these things are ultimately the product of this Consciousness-Force. Life itself is the analogue of this Force at a material level which was already contained in it a priori. Everything is ultimately a process of a Soul-Spirit and its World-Will. Schelling is the Western philosopher that, to the best of my knowledge, came closest to the integral cosmology we are going to describe in Pt.III.

q. Arthur Schopenhauer

On not-too-dissimilar tracks followed Arthur Schopenhauer: Everything comes through the filter of the mind, and what we know is in the filter, its representation, and which is not what is 'out there'. The subject imposes categories such as space, time, cause, and effect onto the world-framing with the mind a representation of it. All is purely mental, and that's all there is to it. In his *"On the fourfold root of sufficient reason"*, Schopenhauer describes how the mind makes sense of the world. We construct the world as representation imposing

Fig. 96 Arthur Schopenhauer (1788-1860)

on it a causal necessity—that is, everything must have a cause and effect—by logical laws, by mathematical knowledge, and by the motives that drive us.

He built upon Kant's and Schelling's idealism, stating further that the thing-in-itself is a form of 'Will' independent of knowledge. This Will, which is one

---

[48] Schelling wrote this about six decades earlier than Charles Darwin published his book 'On the Origin of Species' but, contrary to modern popular belief, the notion of 'evolution' had already been introduced by Charles' grandfather, Erasmus Darwin, about four years earlier than Schelling's treatise.

and the same as the thing-in-itself, is the fundamental stuff of which the world is made. His main metaphysical contribution to the Western tradition is his intuition of an *'immanent metaphysics'* which he presented in *"The World as Will and Representation"* in 1818. His advice was that to take the first-person perspective: *"We must learn to understand Nature from ourselves, not ourselves from Nature."* In this vision, everything gets subsumed under the forces known to physics (electromagnetism, gravity, the mechanical forces acting between bodies and particles, etc.) as an expression of a primal Will. The fundamental and ontological basis of the world is Will. We are subjects through which Will expresses itself while all the objects of the world which cause the representations in our minds are 'objectification of the Will'. Every object is the manifestation of that one fundamental and basal stuff that is Will. The whole world stands to this Will as the wave stands to the ocean.

Animals are driven by Will, as we are. Also, a plant or a tree has Will to flourish and grow. Ultimately, the entire Universe has Will. More precisely, the entire Universe *is* Will. But in the animal, the plant, or in the blind physical forces, will is pure instinct not guided by knowledge. This Will is not self-conscious. It is just an impulse, an experience that doesn't have an intellect that can think about its own experiences.

This unifies apparently insentient physical forces with our subjective experiences because everything, at the bottom, is Will—that is, experience, sentience, instinct, urge, motivation—and there is no physical stuff to begin with. In Schopenhauer's words: *"The most universal forces of Nature exhibit themselves as the lowest grade of objectification."* In this sense, Man in Nature is the highest grade of self-consciousness—that is, Will able to look back on itself. Our bodies are the means by which that Will recognizes itself in the world because they are the interface between object and subject. Each individual comes into being by a principle of separation from the One, which manifests in the many. But precisely this involvement in flesh and blood makes humans selfish and greedy beings. This identification of the original Will in matter is what we take for our personality and individual Will and by which the Will recognizes itself in the world. This Will is itself a blind and instinctive longing and desire which becomes aware of itself only in the world.

Schopenhauer was the first Western philosopher to study and explicitly refer to Eastern philosophies such as those of the Indian Upanishads or the Buddhist tradition. In fact, one of his central themes is the unity of the World: The World is, in its essence, a Oneness. The distinction between these spiritual and mystic insights and the mental speculations of a philosophical theory will be explained in further detail in Pt.III. It may only be said that Schopenhauer was not a mystic and never pretended to be one. He was nevertheless greatly inspired by the Eastern tradition and might be considered one of the first authoritative contributors in the West to its propagation.

Not coincidentally, Schopenhauer's main concern was that of the Buddha: the enigma about the origin and meaning of suffering and death in the world. He identified it in the basic structure and inherent features of the world—that which causes suffering and from which we must find liberation. We can find this only by reflecting first on the transcendent Oneness in the apparently polarized and discordant manifestation we live in and by practicing compassion and renunciation. Compassion is the subliminal understanding of the transcendent Unity; the depth of the suffering of other beings is the suffering of our inmost being as well. Contrary to Descartes' theory that animals are just insensible automatons, Schopenhauer considered them worthy of our compassion and respect as well and is therefore often considered one of the philosophical founders of the modern animal rights movements.

Contrary to what most philosophers and scientists did and do, what Schopenhauer pondered more deeply was the limit of reason. In his 1918 work, he included a chapter entitled *"Of the essential imperfections of the intellect"*. Our intellect is like a prison in which we can think only in terms of successive temporal and spatial concepts, whereas Will is not cognizable because it is beyond these categories; it transcends the world of phenomena, the *'veil of Maya'*.

Schopenhauer's intuitive philosophy is a remarkably actual point of view. It is making a comeback in contemporary philosophy, as we will see in the coming chapters.

r. Friedrich Wilhelm Nietzsche

Friedrich Nietzsche was inspired by Schopenhauer but was critical of his metaphysics and theology. And yet, as we shall see, he tapped into the same hidden truth behind the veil of appearances that an intuitional thought can reveal. Here, we recall only his conception of the *'overman'* (from the German *'Übermensch'*), which fell into disgrace when the German Nazis appropriated and perverted it into its well-known anti-Semitic and fanatic ideology.[49] However, there is nothing in Nietzsche's thought that suggests any form of antisemitism or that encourages forms of discrimination against any race, nation, or class.

Fig. 97 Friedrich Wilhelm Nietzsche (1844-1900)

In his famous novel *'Thus Spoke Zarathustra'* (*'Also sprach Zarathustra'*), Nietzsche asserts that *"Man is something that shall be overcome."* – *"All creatures so far created something beyond themselves; yet you want to be the ebb of this great flood and would even rather go back to animals than overcome humans? What is the ape to a human? A laughing stock or a painful*

---

[49] The Nazis coined the term 'Untermensch', meaning 'sub-human' (literally 'under-man'), implying Jews and all 'non-Aryans.'

embarrassment. And that precisely is what the human shall be to the overman: a laughing stock or a painful embarrassment. You have made your way from worm to human, and much in you is still worm. Once, you were apes, and even now, a human is still more ape than any ape." – "Mankind is a rope fastened between animal and overman—a rope over an abyss. A dangerous crossing, a dangerous on-the-way, a dangerous looking-back, a dangerous shuddering and standing still. What is great about human beings is that they are a bridge and not a purpose: what is lovable about human beings is that they are a crossing over and a going under."

This is very in line with our contention that man is a transitional being and reason only a temporary form of cognition. But how can man exceed itself? Nietzsche could not furnish an answer beyond naturalistic means or psychological considerations, which, paradoxically, are much more proper to man than any kind of transitional being, let alone an overman.

s. Karl Lamprecht

Because we will take this up again later, let us make a short mention of the interpretation of the history of the less known German historian Karl Lamprecht. Worth a notice is how he recognized that history cannot be explained away as only the result of the play of forces between the personalities in power. History is more than a clash between kings, emperors, dictators, or presidents; other factors must determine the fate of societies. From this standpoint, cultural and especially economic aspects must be given priority over political factors or the charisma of specific personalities.

Fig. 98 Karl Lamprecht (1856-1915)

Lamprecht forwarded the conjecture that it is possible to recognize different phases characterizing the sequence of historical and social development with particular regularities. Though he focused mainly on the historical, economic, and cultural development of his own country, he contended that these phases or stages could be recognized as a psychogenesis—that is, the explanation of a personal or collective behavior based on its psychological processes, a trait common to all human history. These cultural ages of economic, cultural development were systemized by Lamprecht in five economic stages: the symbolic age (from prehistory until 350, the animistic and occupational economy), the typal age (350–1050, the first collective natural forms of economy), the conventional age (1050–1450, the landowner-based natural forms of economy), the individualistic age (1450–1700, the collective trade and monetary economy), and the subjective age (from 1700 onwards, the individual trade and industrial economy).

Despite his focus on economic aspects, Lamprecht also recognized the psychological factors that determine human history. Perhaps his main contribution was not so much his theory with its speculative subdivision in historical stages but, rather, his intuition that human social development cannot be reduced to a few material factors. It is a much more complex network of many different interdependent aspects shaped by forces acting throughout societies and nations. Interestingly, Lamprecht's ideas found more acceptance in France and the US than in Germany.

t. Rudolf Steiner

Steiner was the founder of a movement called *'anthroposophy'*, which literally means *'the wisdom of man'*, indicating a philosophical and esoteric current aimed at the development of the human toward higher states of consciousness from a spiritual-evolutionary perspective. It is a *'way of knowledge' ('Erkenntnisweg')* supposedly allowing the 'higher Self' to emerge and enlarge man's (*'anthropos'*) consciousness by means of a discipline, philosophy, practices of meditation, and contemplation which allow us to gain contact with 'higher worlds'. In fact, Steiner challenged Kant's transcendental idealism

Fig. 99 Rudolf Steiner (1861- 1925)

that declared the knowledge of the noumenon was something inaccessible. If we rise above our ordinary sensorial and mental perception of the phenomenal world, getting into contact with a 'trans-sensorial world' (*'Übersinnliche Welt'*), we can achieve higher forms of knowledge.

It certainly doesn't go unnoticed how several of Steiner's teachings were derived from Eastern cultures and the theosophical movement, most notoriously represented by the Russian occultist and philosopher Helena Blavatsky. The theory of reincarnation and the existence of 'trans-physical worlds' was a recurring theme in Steiner's writings. The cosmos of Steiner was in line with the theosophical teaching, describing the material universe as the last phase of a succession of 'densifications of the Spirit'. Everything is Spirit *('Geist')* and our material bodies are only an outer layer of more 'subtle' bodies, such as an 'etheric body'' or 'life body', an 'astral body' or 'soul body' and 'mental-body' (*'Ich-Leib'*).

Steiner's views still have large acceptance today, especially his pedagogical conceptions, which supposedly have been implemented worldwide in *'Rudolf Steiner schools'*, also known as *'Waldorf schools'*. It is also noteworthy that he was a fervent supporter of Goethe's 'new science' and 'way of seeing', which we will address in Pt.II-III.2.

But these interesting aspects cannot make us ignore his severe shortcomings. The author must confess to having a personal negative bias toward anthroposophy and the so-called *'Waldorf education'* system. My short but intensive experience for about three years as a Waldorf high school teacher made it clear that this won't be the education of the future.[50] As to anthroposophy, I can't convince myself that we are dealing with something that goes much further than a confused low-level form of occultism. It is one thing to practice a genuine spirituality; it is another to practice occultism, esotericism, or spiritism. Even if Steiner explicitly distanced himself from practices of mediumship, it is nevertheless difficult to escape the impression that his' way of knowledge' doesn't differ much from it. Steiner was more of an occultist in contact with a world of spirits rather than with a spiritual world. In his writings, one finds a questionable mixture of Christian, Buddhist, and Hindu esoterism full of imaginative speculations and mythological phantasies lacking clarity of mind and, most importantly, failing to capture the essential elements of true spirituality. Steiner was not a Meister Eckhart, a Plotinus, a Spinoza, or someone who had attained any enlightened state of consciousness, such as those described by the mystics of the Asian continent. Despite interesting similarities, Steiner's anthroposophy remains far from a synthesis of knowledge that this treatise has in mind. It could be considered an attempt of Nature to bring to the surface a deeper truth. But the vessel and the cultural context weren't ready and lacked discrimination and true spiritual insight.

u. Edmund Husserl

Edmund Husserl, the German philosopher who is known as the founder of *'phenomenology'* (that is, the study of appearances and how they express themselves to our conscious experience), was instead more interested in understanding the limits of the present human creature and took the opposite path of Schopenhauer's or Schelling's theology (let alone Steiner's occultism) by trying to apply reason to its extremes.

Fig. 100 Edmund Husserl (1859-1938)

His aim was to get rid of all mental presuppositions by positing incontestable phenomenal statements in order to ground science and philosophy on a firm, indubitable basis. His approach consisted of describing phenomena as they are experienced. This included all the observed physical phenomena, the sensations of our bodies, and mental events as we actually experience them and how they present to us.

To do this, according to Husserl, we must, resort to what he called the *'phenomenological reduction'* method—that is, we should refrain from trying to explain the phenomena by something which is supposed to be non-phenomenal or beyond-phenomena, as all that we know for sure is only the experience of phenomenal consciousness. For example, we should adopt the *'epoché'* attitude—that is, an attitude that refrains from conjecturing if there is

---

[50] The interested reader can find out how I already dwelled on this at length in [359].

a world existing independently from our consciousness. This epoché helps us to reduce the world to pure phenomena abstract from extra-mental things supposedly explaining things by non-phenomenal entities.

Therefore, in contrast to Kant and others, Husserl does not consider any mind-independent and transcendent object in the sense of the things-in-themselves. It is the act of thinking which is temporal and immanent, while the objects appearing in the conscious experience are the transcendent non-temporal pole of consciousness. These mental acts he called *'noesis'*, in contrast to the object present to consciousness, which he termed *'noema'*. This correlation between the immanent and transcendent—namely, between the noesis and the noema—is consciousness itself. This mental act has the characteristic property of 'intending' or 'being about' an object in its mental dimensionality, something he called *'intentionality'*, borrowing the term from the German philosopher, psychologist, and Catholic priest Franz Brentano whose theories influenced Husserl.

The noema is the object as it presents itself to us, while the noesis emerges through a complex network of *'noematic representations'* from which we extract some invariant feature. The latter is the real essence of the phenomenon. For example, whatever colors we see, we always associate them with some object having a form; there can't be any color without a phenomenal experience of this invariant we call form or shape. By this *'eidetic reduction'* (in Greek, eidos means form), which scans all the particular noemata, we discover the invariances and reduce the perceived phenomena to what is essential to it, to the universal. In this sense, Husserl contends that phenomenology allows us to intuit the essences.

Husserl recognized with Kant that the thing-in-itself is unknowable because it is unthinkable: If there is something beyond the mind, obviously, it can't become an object for the mind. However, this raises the question again: Can a conscious being that has evolved beyond intellectual cognition know the thing-in-itself? Moreover, if there is no realm where we can consider the noesis to exist, where do the noemata come from? Husserl's answer is that it is self-caused; the noesis constitutes the noemata itself, and the subject that remains after the phenomenological reduction is the creative 'transcendental subjectivity' engaged in an ongoing intentional existence: The Absolute or Spirit.

Husserl's dream was not to replace science with phenomenology but to disclose its essential limitations and furnish a platform that could allow us to understand it more deeply and without a return to a non-phenomenological philosophy. It is the last attempt to make sense of the world without resorting to theological or metaphysical reasonings. The strength of phenomenology, as advocated by Husserl, is also its weakness: too much confidence in reason.

v. Henri-Louis Bergson

Perhaps just this extreme application of reason led some to reject intellectualism, as in the case of the French philosopher Henri Bergson, who advocated for more intuition than abstract rationalism in science and philosophy. By an intuitive method, one can come closer to the things-in-themselves and to the Absolute. Bergson took up some classical but still actual philosophical quests, such as free will, creativity in evolution, the nature of consciousness, change, space-time, causality, and spirituality, among several other topics, all being of central interest for our considerations but impossible to exhaust in a short section. Let us pick out just a few of Bergson's ideas.

Fig. 101 Henri-Louis Bergson (1859-1941)

On free will, determinism, and life, he noted that if all our actions were the result of purely mechanical laws determined only by pre-existing conditions, we would behave automatically in all circumstances without effort, hesitations, and resistances. But our life, which is permanently characterized by strife, struggle, and doubts, does not feel like that at all. Yet, we can make choices, be creative, and change our life, even though it often requires sacrifice and hard work. While animals are mostly unconsciously driven by their instincts, habits, and automatic behaviors, we, as humans, can replace these behaviors with new ones, even though this might necessitate some time and a conscious effort against these pre-existing habits and automatisms. In this sense, man is free.

Bergson defends an post-materialistic position rejecting the identity between consciousness and the brain. He goes even so far as to say that all life is conscious, not only those living forms having a brain. In his work "Mind-Energy," he writes: *"It is sometimes said that in ourselves, consciousness is directly connected with a brain and that we must therefore attribute consciousness to living beings which have a brain, and deny it to those which have none. But it is easy to see the fallacy of such an argument. It would be just as though we were saying that because in ourselves, digestion is directly connected with a stomach, therefore only living beings with a stomach can digest. We should be entirely wrong, for it is not necessary to have a stomach, nor even to have special organs, in order to digest. An amoeba digests, although it is an almost undifferentiated protoplasmic mass."*

Bergson might have been delighted by the findings on plant intelligence about a century later. It is only because the sense-mind thinking and imagination are inherently of material and geometrical nature needing strict logical rules based on linear causes and effects that we believe consciousness to be a material epiphenomenon. For this reason, in the footsteps of Spinoza, Bergson considered intuition the form of cognition we must prefer over dry intellectual materialistic thinking. Especially in the domain of consciousness, we must go inward, beyond the pure analytic understanding, by introspection and intuition.

Nevertheless, he recognized how we are, to a large degree, still instinctive animals. There is a vital impulse and principle in life that must transcend the pure material mechanistic processes of matter. In his masterpiece *"Creative Evolution"* of 1907, Bergson noticed how during the evolutionary process, there were creativity, inventiveness, and an everchanging range of new forms that he could hardly believe could be caused only by a blind mechanism. The principles of Darwinian evolution aren't sufficient to explain evolution's creativity. In life, there must be something more, an *'élan vital'*—that is, a *'vital impetus'* or *'vital force'*—responsible for the complexity and morphogenesis of living beings. Bergson did not support the classical vitalistic finalism: There is no "internal finality" nor "absolute distinct individuality" at the origin of the vital force. Rather, he posited it as a compromise between a mechanical and finalistic conception. In Nature as a whole, there is not so much design, a purpose, or an aim acting on life's organisms; rather, there is an entelechy driving it from within. Also, our urges, desires, feelings, and emotions that impel us to action are the same creative impulse. It is this self-impelled force within plants, animals, and humans that determines its effort, vital resistance, and struggles against the inertia of matter. In this sense, reproduction is an invention of Nature to conquer death. Every species as such can survive itself even if the single organism is necessarily destined for dissolution. He even speculates whether life will one day conquer death by realizing immortality.

Bergson was quite famous and honored during his time. He was invited to be a visiting professor throughout the Western world and was even awarded a Nobel Prize for literature for his book on creative evolution. Nonetheless, his philosophical worldview had a scarce impact on an increasingly materialist, industrialized, and commercialized humanity whose minds were much more focused on preparing two world wars than ruminating about its place in the Cosmos.

w. Alfred North Whitehead

Alfred North Whitehead, an English mathematician and philosopher, was less concerned about the nature and essence of evolution but, rather, about that of physical processes. A way to introduce his thought might be to return to pondering: What, ultimately, is the thing we call 'matter'? We already saw how quantum physics or atomic physics, and even chemistry, like any other science, does not answer—and doesn't even try to formulate—this question. In Pt.II-I.2&5 we pointed out that there are no

Fig. 102 Alfred North Whitehead (1861-1947)

particles in the sense of little marbles and that there are only force centers acting upon each other. However, nobody knows what, ultimately, a force is beyond definitions that involve more or less implicitly the very same notion of matter. Physics only describes the process but has nothing to say about its ultimate nature. And it never pretended to do otherwise because it unconsciously knows that it will never be able to do that.

Whitehead realized how everything is about processes. He tried to establish a *'process philosophy'* which posits processes, rather than matter, forces, or particles, as fundamental. In his famous work, *"Process and Reality"* [235], he describes the actual existing world as a network of *'actual entities'*—that is, of events and processes as the ultimate and fundamental reality. What we perceive of this universe of actual entities are its temporally overlapping related *'atomic occasions of experiences'*. Whitehead, who was also well-versed in physics, tried his utmost to develop a philosophy by strict, rational reasoning and a precise set of logical rules and expressions which got to the bottom of physical reality. However, ultimately, he could not get rid of metaphysical categories, as logical positivism couldn't either.

Siding with Hume and Berkeley, Whitehead also recognized the indefensible *'bifurcation of Nature'* of natural sciences that separate reality into primary and secondary qualities, arguing that there is no real distinction. He questioned the tendency of science to exaggerate the abstract which seems to replace the concrete fact. It is science's *'fallacy of misplaced concreteness'*– that is, the error of mistaking the abstract for the concrete.

At the bottom, Whitehead finds how we are always led to the very same ultimate: subjective experience. Also, a process is a meaningless concept if it is not directly or indirectly in relation to some of our conceptualizations that point at phenomenal consciousness. Therefore, the human mind finds itself at a standstill. If it tries to look for the ultimate ontology of things without falling into metaphysics, it will have to admit that it can't go beyond a certain point— never, not even in principle.

His philosophy compelled him to recognize that one can't evade the logical conclusion of the existence of an actual atemporal entity. The ultimate underlying process of unfolding is something creative, and which he couldn't refrain from calling 'God'. From this insight, Whitehead later developed a *'process theology'*, that conceives at the same time that God is permanent and changing, one and many, immanent in the World and the World immanent in God, the creator and the created. Whitehead was clearly tipping into a higher-mind form of cognition, a panentheistic view that transcends the analytic mind, and that would consider this only an incoherent and meaningless self-contradicting wordplay.

x. Pierre Teilhard de Chardin

On another but similar front went the thoughts of Teilhard de Chardin, a French philosopher, paleontologist, and Jesuit Catholic priest who developed a spiritualistic and teleological evolutionary cosmology and cosmogony[51]. In hindsight, we might characterize T. de Chardin's pursuit to unite theology and modern evolutionary biology as an impressive attempt of Nature to impel the mind to look beyond the limiting boundaries of religion and science and search for a synthesis which could harmonize them. Obviously, the times were not

Fig. 103 Pierre Teilhard de Chardin (1881-1955)

ripe: On one hand, the Catholic Church condemned his theology, banned his writings, and ordered him to give up his lectures, while, on the other hand, an atheistic and materialistic science branded his ideas altogether as nonsense.

His main evolutionary concepts were summarized in his most important work, *"The phenomenon of man"*. Evolution is described as a cosmic unfoldment with humanity as the highest expression which will eventually coalesce into a unitary cosmic consciousness, of which Christ symbolized the individual aspect in the earthly context. It predicts the coming of an age when a collective human conscience will appear as a natural consequence of the formation of the *'noosphere'* which, analogous to the biosphere, the layer of life, will be the terrestrial layer of mind (from the Greek' noos', the understanding of mind or knowledge). T. de Chardin did not reject Darwinian evolution, which he knew well as a paleontologist, but, rather, tried to augment it by a vision where the evolutionary impulse is seen as a teleological process that aims at the union with God. This final consummation will be realized at a point in the history of evolution he called the *'Omega-Point'*. The Universe is the material aspect of Christ which, through the evolutionary process, aims at its full realization in the Omega-Point. Evolution is a series of realizations that tends towards an increasingly complex unity. First, there were particles that united to form atoms and molecules, then cells which came together to form organisms in a complex biosphere; the same is now going to happen in the noosphere, which is ascending towards a unity of minds—that is, a human unity with multiplicity and diversity as its fundamental character.

T. de Chardin posits an inner aspect in all matter. He also recognized a 'within', some sort of primitive but real interiority, sentience, consciousness, or basic mentality in plants. The 'within' in the animal and the human is a further development of this interiority that grows and asserts itself over the

---

[51] Cosmology refers to the study of the structure and dynamics of the present universe, while cosmogony is concerned with its origin and evolution in time. However, in the literature, one finds the two terms often conflated as synonymous.

materiality of Nature. It is a constantly increasing complexity that allows for the unfoldment of this interiority. It is a panpsychist view in which each particle has an elementary conscious center that combines with other particles in an increasingly complex structure, beginning with the cell with its own conscious center and progressing to the full-fledged organism in which consciousness, mind, and inner life stand at the highest level of current evolutionary development. Through this process of growth in consciousness of the particle to that of the complex compound, evolution shifts from the exterior material process to the interior experiential life.

This growth from within is what also allows the universal Christ-consciousness to take the lead from above over purely mechanistic processes that dominate the processes below. First, there were only raw mechanical, physical, and biochemical reactions that dominated the scene. The appearance of plants and animals built the biosphere but, after further development, the appearance of reflective thought set a new stage in evolution, and its maturity led to a sphere of thought that began to grow and enclose the Earth's biosphere within the noosphere. Each individual forms a center of thought which extends towards other centers of thought, forming an interconnection of minds which will emerge in this new thinking layer. There are now three layers: The material layer, the life layer, and the mental layer.

Nowadays, we could argue that the development of the Internet is a validation of the hypothesis of an emergent noosphere. However, it is important to point out that in this theory, the biosphere and the noosphere are not just labels standing for the material appearances of living organisms and brains; according to T. de Chardin, they are real layers or spheres enclosing the planet in the form of a more subtle spiritual energy. The life-envelope and the thought-envelope are subtle and non-physical 'conscious energies' which radiate out to other living and mental fields, coalescing together to form a unity that will transcend our instincts of isolation and separation. Humanity will decline, with a consequent dissolution of the noosphere, if it refuses to follow the drive towards a harmonious *'planetisation'*, insisting on its own egoistic privileges of individuals, clans, groups, and nations—a quite actual topic in these times of globalization.

At the apex of this unifying evolutionary movement, the superhuman will appear, emerging from the spiritual unification of Earth's centers of consciousness. A super-consciousness of the collective, which will converge to the ultimate psychic fusion in the Omega-Point. A collective state of super-consciousness in which the now-separate individuality will continue to exist in the whole, not in its separative egocentric life and mind, but rather by feeling and experiencing itself as a part of a larger unity. It will be a soul-to-soul connection in which love will manifest as the primary source of universal energy and the original cause of this drive towards unity. A cosmic love is, at the same time, the ultimate origin and goal of evolution.

y. Maurice Merleau-Ponty

Returning to the phenomenology of perception, the reflections of the French philosopher Maurice Merleau-Ponty are worth a mention. Merleau-Ponty was particularly interested in understanding the naïve element inherent in science, which takes for guaranteed its objectivity without reference to consciousness. He did not really go far beyond the established phenomenological tradition dating back to Kant's and Berkeley's to Husserl's insights. However, he reformulated the connection between the perception of phenomena and our mental representations in more accessible language.

Fig. 104 Maurice Merleau-Ponty (1908-1961)

He reminds us that the things-in-themselves are an a priori that exists before our knowledge and that what science does is an abstract sign-language schematization; it pictures a sort of geographical map that represents a countryside but without any real knowledge of that territory. Merleau-Ponty's critique is similar to that which we expressed at length in Pt.II-I—that scientists are much too often unreflective in their analysis of the perceived world and associate objectivity to the observed phenomena that they do not deserve. Science can't furnish a fully objective account of the world and therefore can't lead us to a fundamental and ultimate understanding.

Phenomenology aims to provide the essence of perception and the conscious apprehension of phenomenal experience as it is presented to us freed from conceptual loads. The world exists before every reflection, and science should be considered secondary compared to the experience of the world. Science will never discover the inherent being of things because it is its abstraction in the first place.

Merleau-Ponty, however, does not entertain himself with a theory of consciousness. Whatever consciousness might be and wherever it originates, it is its content that should interest us beyond the theoretical speculations and ontologies. It is in this sense that he hopes to elaborate a synthesis between science and phenomenology—that is, science as a reflective analysis of contents of experiences—even though we must be aware that such a synthesis will nevertheless remain an artificial construct of perceptions. Because, after all, like it or not, every object emerges from our basal perceptions, sense-data, and mental associations.

However, Merleau-Ponty critiques Husserl's epoché insofar as the phenomenological reduction should not be considered a sort of pure transcendental state of consciousness. Because we inevitably always juxtapose a meaning to the percepts, we can't avoid endowing the objects we perceive immediately with significance. The whole world is always a world of meanings. We can't see it in its factual concreteness in an original act of consciousness on a pre-predicative layer. However, it is also true that the meaningful content that arises in our minds is pre-linguistic. It acquires a meaning before we articulate it linguistically. Analogous to our characterization of 'objectivity' as an inter-subjective agreement, for Merleau-Ponty, rationality is the inter-subjective cross-section of perspectives in which the different individual perceptions and meanings confirm each other and coalesce into a commonly shared unity. The world is not just a dream that can be reduced to an empty void or a purely idealistic construct. To an '*internal subjectivism*', there is also an '*external objectivism*', in the sense that there is also an immanent consistency in our subjective experience of things.

In the footsteps of Fichte's first-person investigation, Merleau-Ponty also reflected on the phenomenology of our bodily perception. We should distinguish between our bodily and corporeal phenomenology. The former starts from an inner perspective and is what we feel directly from our bodily sensations, feelings, or impressions without conceptualizations and projecting an image of it, while the latter arises from an external perspective that objectifies our body or its parts. He noticed how the inner feelings and perceptions we register do not require mental objectivation in order to take any action. For example, by handwriting, we resort to a habit that is more than an automatism because it requires a particular form of knowledge specific to the world and somehow inherent in the hand which writes. It is a sort of 'knowledge of the body' as if the hands have a consciousness of their own. Merleau-Ponty's and Fichte's intuition of a 'body consciousness' might not be just a metaphor but, as we shall see, veils a much deeper truth,

z. Jean Gebser

Again, a very different sort of continental philosopher, but quite relevant to our synthesis of knowledge, was the Swiss philosopher Jean Gebser. His approach took a more intuitive, almost mystic, perspective on the history of humankind. Gebser had a deeper insight than Lamprecht and, as Schelling or P. T. de Chardin, held a teleological conception of history, namely, that of an external manifestation of the evolution of consciousness. Humans underwent different stages of development that he called '*mutations*'—that is, creative

Fig. 105 Jean Gebser (1905-1973)

evolutionary leaps of non-linear progression—corresponding to different forms of consciousness manifested in the different epochs. History is a creative movement of consciousness struggling to come to the surface by manifesting this attempt through arts and language.

Gebser identified the changes in history with the change of the collective '*structure of consciousness*'.

The zero-dimensional '*archaic structure*' was humanity's first and primeval structure of consciousness in which we identified ourselves with the whole of the natural world. This prehistoric stage lived in an imaginal world where we still didn't have a realistic space and time conception—our consciousness was concentrated entirely on the natural environment. It was a tribal life without an idea of a society built by cities, let alone nation-states. There was a 'deficiency' in this state of consciousness that, for Gebser, was exemplified by the primeval painters always depicting the world only as two-dimensional. However, there was an immaterial psychic component in this natural assimilation between Man and Nature, which was felt as spiritual energy.

The one-dimensional *'magic structure'* was the second shift in consciousness where everything is perceived as 'magical', space- and timeless, and where we became aware of Nature as a separate entity. We 'stepped out of the cave', so to speak, and the *'pre-perspectival'* phase began, which is space- and timeless, dominated by an emotional and instinctual consciousness and entirely dependent on the demands of the natural environment.

The two-dimensional *'mythical structure'* came with the beginning of reflection and the awareness of an inner life. Life itself was conceived as a story told in mythologies. Speech and the discovery and experience of the soul become interwoven. It is an *'unperspectival'* phase in the sense that it has latent but still undeveloped tendencies to perspective.

The three-dimensional *'mental structure'* followed when the reasoning logic mind took the lead. Contraries are synthesized by reflective thought and discrimination. The rationalistic age of Western science is born. It is characterized by a more individualistic and materialistic consciousness. Paintings became three-dimensional, Nature lost its spiritual component, and time and space acquired a more concrete and deeper meaning. It is a *'perspectival'* stage in which perspective in painting became rigorously systemized. However, this came at a price. This structure of consciousness has its deficiencies, such as the enslavement from time and the loss of the previous structures. The spiritual aspect of the pre-perspectival archaic structure and the magical and mythological aspect of the perspectival age faded away.[52]

Finally, there is the four-dimensional *'integral structure'* which we are striving towards and which will integrate all the previous phases of collective consciousness. It will be *'aperspectival'* because it will go beyond the spatial distinctions and the division of time in past, present, and future but will perceive it as a unique and undivided whole. It will integrate the inner and outer worlds, with the latter no longer being just an object but a living entity with which we will enter into an intimate relationship. The spatial and temporal will be united with a spaceless and timeless dimension. We return to the original consciousness and will become free of the bonds of time, finding our

---

[52] Gebser identifies the rise of the nationalistic fascist ideals in 20th century Europe as a consequence of the suppression of the former structures from the group consciousness and to the advantage of an exclusive concentration on the rational.

origins and embracing the former perspectives synthesizing them in an integral consciousness. [53]

aa. Martin Heidegger

One of the most renewed (but also most cryptic) existentialists of the 20th century was the German philosopher Martin Heidegger. He took up an old— forgotten for a long time—topic of classical philosophy, namely, the question of the nature of being. Indeed, while that was a central theme of the Hellenic tradition, as we saw, for example, with Parmenides and Plotinus, it went somewhat out of focus. The title of his main work of 1927, *"Being and Time"*, tries to answer questions about the significance and meaning of the

Fig. 106 Martin Heidegger
(1889-1976)

thing which is so familiar to us and yet so difficult to capture rationally: that which we call 'being'. What does it mean to 'exist'? What does it mean to be? What is the fundamental ontology[54] of being? What makes something possible to be at all? It was an investigation he summed up with the famous question, *"Why is there anything rather than nothing?"*

Heidegger was not so much interested in an analytic and logical explanation; rather, he took the phenomenological approach. By taking the first-person perspective, Heidegger noticed first of all how we get to know the existence of things by a sequence of perceptions in time. Rephrasing Parmenides: An apple *exists* at some point in time as an apple but, once eaten, is no more. A wave *is* a wave, but it 'goes out of existence once it hits the shore'. Of course, there are more stable things, like chairs or mountains, but their coming into and going out of existence is only a question of different time lengths, on the order of years and millions of years, respectively; nothing exists in eternity. Thus goes Heidegger's argument, I myself as a sentient being— that is, my center of consciousness—am, so to speak, a 'door' or a 'window' which allows things to come into existence, as we are a spatial region in a temporal interval within which the world arises. Finally, from a purely phenomenological perspective, we can't escape the conclusion that it is us that uncover the world and let things be.

The most fundamental way of being he identified as *'being in the world'* (form the German *'dasein'* which literally means 'being here', with 'da'

---

[53] The analogy with T. de Chardin's thoughts and, as we will see later, with Sri Aurobindo's spiritual emergentism, is not entirely coincidental: Gebser became acquainted with their writings. However, he learned about them only late in life.

[54] It might be useful to recall, at this point, that the word 'ontology' indicates the philosophical study of the nature of being itself.

meaning 'here' and 'sein' meaning 'being' or also 'being present'). We are *'presencing'* beings who make things become present in the here and now, and 'to be' means to become present. To be is to be present in the present time. Or, ultimately, being *is* time.

Heidegger's analysis somewhat recalls Spinoza's different modes of being of the fundamental 'substance'. Yet the objects we observe around us, such as trees, stones, chairs, etc., are 'in being' and are real but come into existence only insofar as we open ourselves to their manifestation. In fact, the word 'existence' comes from the Latin 'existere', which literally means to 'step out' or 'stand out', 'emerge', or 'appear'. It is only by us standing out to the world that we can become the opening within which things become present. It is all about letting things be, allowing them to get into our field of 'beingness' so they can present themselves to us and their meanings come forth and reveal themselves.

Heidegger distinguishes between *authentic* and *inauthentic existence*. We authentically exist when we are allowed and especially allow ourselves to be authentic, meaning we show ourselves and open ourselves to the world, to new possibilities, letting things become alive with their presence, without reservations and prejudices, in a state of pure 'dasein'. If we 'empty' ourselves and become self-forgetful, we are truly ourselves and can *be* authentic. This is reminiscent of the witness consciousness one finds so pervasively in Eastern philosophy and meditation practices. In contrast, we place ourselves into an inauthentic existence when we become absorbed in the everyday stream of thoughts and ordinary points of view of the mass. Most of the time, we are not in that state of 'dasein', but it is only by fully letting things be that we can be authentic.

However, all this tells us what it means to be, not so much what being is. Later, Heidegger 's existential ontology also questions the essence of being. In an article of 1943 entitled *"On the Essence of Truth"*, he first investigates the notion of 'truth'. He comes to the interesting insight that one thing is a propositional truth—that is, the correspondence between an intellectual structure and things—and an essential truth of things. There is something essential that is present before a propositional truth becomes such. Before a propositional truth is recognized, the things it describes are already present to our conscious awareness. He worked on the line of Kant's phenomenon and noumenon but asked for the mechanisms and ways by which our perceptions bring the world into existence. We must conclude, according to Heidegger, that before a propositional truth becomes true, there is a 'wordless' instant during a process that allows things to show themselves as themselves. This process 'unconceals' the concealed; it is a mechanism of revealing and disclosing. The essence of truth is that completely free, un-concealment process that makes the essence of things show itself. But to do that, we have to disclose our own essence, which is the act of total 'dasein', of 'standing open' in pure existence.

bb. Jean-Paul Sartre

One of the main interests of the French philosopher Jean-Paul Sartre was to eliminate any concept of the self and ego. He claimed that there is no such thing as the ego or transcendent self. Sartre, like Heidegger, was an existential phenomenologist but focused mainly on the practical and scientific description of intentionality—that is, the conscious act that references objects in the world and rejects any further notion of phenomenal objects existing in the external world or positing transcendental truths.

Fig. 107 Jean-Paul Sartre (1905-1980)

There is nothing other than intentionality as such, which points to objects and gives them meaning. Intentionality is a meaning-giving act, and we don't need more than that to explain its function and nature. We make objects become meaningful by making them present and focusing on them. In his major work *"Being and Nothingness"*, he claims that our entity, that self, the ego, does not exist apart from this act of consciousness. Or, we might also say that the ego comes into existence only as a figment of a series of intentional processes. What we call 'knowledge' is nothing other than a mode of intentionality. In this sense, having knowledge is an illusion because it does not reveal things as they are; rather, it is the realization of an intentional state. Knowledge is not representational; it is only what is present to me at the instant of intentionality. Every act of conscious experience is a new intentional process that allows us to go beyond what we are now to become something new afterward. The 'I am' is the sum of the past intentional states which make me what I am now. That core that survives as a single subjective entity is the unification of the sum of experiences of the past in every moment. What constitutes that core witness identity arises because of the memory of the moments of intentionality which become united in a single whole. Every intentional act is lived as a unified whole. The 'I' is the ego insofar as it represents a unity of active intentionality formed by states and qualities, and any new reflection leads to a new state of concrete totality, which we call 'me'.

However, at the same time, this intentionality is an attempt to escape from itself towards what is non-itself. The intention projects us from something towards something else. At that moment, the personality disappears. By a pure observation that is absorbed in our phenomenality and does not intervene or evaluate, there is a kind of loosening or even disappearance of the self when the intentionality towards something occurs. In fact, trying to catch our ego, that thing which feels to be 'me' and says 'I' always results in a void. At the precise instant we attempt to bring it into our thought, the ego fades away and disappears with the reflective act.

cc. Gilles Deleuze

The thoughts of the French philosopher Gilles Deleuze are interesting for their attempt to rethink things from scratch. He questioned the role of philosophy as such. Deleuze doubts that philosophy is about a search for objective realities or any transcendental truth. It is not a discovery of reality or revelation of something previously unknown. Philosophy has a much more pragmatic nature; it has always been about developing and creating different conceptual systems describing

Fig. 108 Gilles Deleuze (1925-1995)

different ontologies, which help us live a decently ordered life in a seemingly chaotic world. It is a system of more or less abstract concepts which opens various conceptual windows to a world. In this sense, philosophy is not different from arts; it is a creative practice directed by a *'plane of immanence'*—that is, an immanent logic that precedes the concepts themselves and makes the creation of concepts possible in the first place.

Nevertheless, Deleuze considered himself a pure metaphysician a la Spinoza. There is no transcendence in the sense of a substance and something that transcends the substance; there is only one universal substance that presents itself in different modes. There is no emanation of B from A, which is not A; rather, there are only different expressions of the same and unique thing. Expression is the fundamental process of all existence—an expression that proceeds from inside out. Fundamentally, every process is a substance expressing itself. In this 'ontology of immanence', the Universe is an expression of something within, not a creation from an outside Being. It is a clear attempt to rise to a higher-mind ontology. In his words: *"An ontology that does not seek to reduce being to the knowable but seeks to widen thought to palpate the unknowable"*. A sentence that could have been the subtitle of the present treatise.

Perhaps the most interesting contribution of Deleuze, together with his colleague Felix Guattari, was a theory that envisaged identity as differences through flows.

The first building block is that of the *'machine'* concept. A machine is, broadly speaking, a system of interacting units.[55] A machine can be embedded in another machine. For example, a cell is a machine, but also a collection of cells builds up a multicellular organ and organism. Every individual is a machine interacting with other individuals. At further levels of complexity, a group of individuals can form organizations, companies, social and political

---

[55] In modern parlance, one could call them 'cellular automata' or the mutually interacting 'nodes' of a network as described by system theory.

movements, societies, and nations, all of which can be considered machines seeking other machines by means of flexible interactions.

What moves every machine is ultimately a desire—the desire to impose itself, to interact, conquest, or connect. Therefore, what shapes our society is always desired.

The complexity of this web of machines can be compared to the rhizome, the root structure of trees in a forest, which is an apparently chaotic network of interconnected roots which do not grow uniformly through space and time but, rather, form an extremely complicated but unique and single-living system of which every tree is only a localized outgrowth. This metaphor is applied to our thoughts, seeing it as the movement of ideas forming political systems with their individual ideas, culture, languages, etc.

All the Universe can be considered a rhizome. Ecosystems, ant colonies, termite nests, nervous systems, human culture, and a city and the behavior of its inhabitants are all examples of 'rhizomatic machines'.

In these optics, history is not just a linear unfolding in time of disconnected events; rather, it is the result of a complex rhizomatic interaction of machines. Societies are not simply determined by the people representing isolated units. There is a spreading of ideas, an exchange of material resources, a movement of financial transactions, migration flows, etc. These interactions of all machines at all levels can be understood in a general sense as *'flows'*. It is this continuously changing and redirecting of rhizomatic flows in our society that molds it. This allows us to see world history from a wider perspective, as flows and a network of interacting machines that define it as an identity that changes in time because of its complex multifold internal dynamics. These flows and interactions look chaotic and are unpredictable, but the overall system has a dynamical and historical identity of its own.

It is interesting to see how the question of the meaning and interpretation of history reemerges from time to time. Schelling, Lamprecht, Gebser, and Deleuze, were fascinated by this question, as if there is subliminally lurking an awareness that we are missing something.

However, this is only an apparent paradox in which a universe of differences nevertheless determines a unique identity. Deleuze saw this as proof of the fact that the notion of identity must not be something linked to a uniform and undifferentiated oneness but can also emerge precisely because of difference. Moreover, this sheds doubt on Platon's concept of the universal forms residing in some transcendent world of forms. Because identities change in time, we do not need to imagine it as the expression of a timeless realm. Differences come first, followed by identity, not the other way around. The distinction between being and becoming fades away. There is no static identity because all identities are derived by differences changing in time. Differences come about due to the rhizomatic interactions and flow between machines. The world is in an eternal change of identity, a flux of motion, a process of becoming. Identity exists only because of this constant process of becoming, with the latter being more fundamental than the former. It is an understanding that is reminiscent of Whitehead's process philosophy or the Buddhist teachings of impermanence.

With identity, however, Deleuze not only means human identity but also points at the identity of a group of peoples, such as a city or nation. Any collection of machines seeking interaction with other machines forms an identity. It is only the mixing of identities, with all its diversity, variation, and multiplicity, that can form a healthy society, city, nation, or organization. No city made of the same type of beings, doing all the same things at the same time with no or limited interactions and which refuses diversity, could survive and probably not even come into existence. The world is a perennial motion, a permanent change, a process of eternal becoming that embraces differences (such as diverse peoples, lifestyles, cultures, nationalities, languages, beliefs, ideas, political orientations, etc.). Life embraces difference and change and can't exist by uniformity and conformity.

dd. Arthur Owen Barfield

In Pt.II-I.3, we met O. Barfield, the British philosopher, poet, and writer who could be considered the pioneering father of the Western concept of the 'evolution of consciousness'. Barfield was inspired by Rudolf Steiner's anthroposophy and he also embraced Goethe's way of seeing (which we will describe at length in Pt.II-III.2), calling it *'final participation'*—that is, the fully conscious participative act of perceiving unity with Nature.

Fig. 109 Owen Barfield (1898–1997)

In his work *'Saving the Appearances: A Study in Idolatry'*, Barfield explores the development of human consciousness across history. Unlike Gebser or Lamprecht, who looked upon the evolution of human consciousness from a historical perspective, but still as a process separate from Nature, Barfield embraced a more holistic vision (having some similarities with P. T. de Chardin's ideas) and argued that human evolution is inseparable from that of Nature. He made clear his anti-materialistic stance: Matter and mind are not two inseparable aspects; rather, matter always interacts with the mind and can't exist without it. Noteworthy is his attention to what he called an all-pervasive *'unrepresented'*, the underlying and inherent reality of things which, however, is not directly accessible to our ordinary senses. It is a notion that could be compared to the unknowable Kantian noumenon. However, Barfield contended that the unrepresented could be experienced, provided that we adopt a way of seeing Nature and the world by a participatory movement, which is essentially an extra-sensory form of cognition.

In his book *'Poetic Diction'*, he considered something more than a theory of poetic composition. It was a study of language and a theory of knowledge that revealed the creation of meaning and the evolution of language as the manifestation of the evolution of consciousness itself—language co-evolves with consciousness. He claimed that words once had a 'concrete and undivided meaning' that we have lost because of the separative mind, while poetry could help us, by an act of active imagination, to transcend this limitation. Also, all appearances and natural phenomena contain an inherent unity that our ordinary sensory and cognitive perception has lost the ability to capture, but that was present in the 'primitive' consciousness of our ancestors and therein reflected in language. It is an ability that our analytic and rational mind no longer has but must recover if human consciousness is to continue evolving.

Barfield recognizes history as a process in which mankind started as conscious but not free. History acquires a teleological dimension seen as a divine plan wherein consciousness struggles at becoming increasingly self-aware and free. To do so, we must recover meaning and purpose and open ourselves to new ways of perceiving and seeing with a new power of imagination if we want to overcome the modern collective and individual material and social problems.

Even though not so well known among philosophers of mind, Barfield could be considered one of the pioneers of modern consciousness studies and might get more attention once the still dominating physicalist culture loses its grip on the collective consciousness.

This ends our quick review of some of the Western thinkers who adopted an intuitional standpoint towards post-materialistic ontologies. Let us now move on to more recent metaphysical views that will pave the way to more integral perspectives.

3. The Unexpected Comeback of the Conscious Universe

"The stream of knowledge is heading towards a non-mechanical reality; the Universe begins to look more like a great thought than like a great machine. Mind no longer appears as an accidental intruder into the realm of matter; we are beginning to suspect that we ought rather to hail it as the creator and governor of the realm of matter—not, of course, our individual minds, but the mind in which the atoms out of which our individual minds have grown exist as thought." – Sir James Jeans [236]

A clear symptom that shows how Western metaphysics is rediscovering its own roots is the revival of old metaphysical worldviews like philosophical idealism, pantheism, panentheism, panpsychism, and further developments or modifications of them. This is, among other things, determined first and foremost by the failure of neuroscience and the philosophy of mind to furnish a credible account of the nature of consciousness.

From the 1980s to the turn of the Millennium, the rapid advances of neurosciences made almost all scientists confident that the mind-body problem and the hard problem of consciousness would have soon found a resolution inside a physicalist paradigm. According to this belief, it is only a matter of time, and the progress of the diagnostic and imaging tools, coupled with brain mappings and the exponential capacities of computer simulations, would have led us to the insight into the nature of phenomenal consciousness. The vast majority believed that this was going to happen inside a philosophy adherent to material monism, which would finally have dispelled any dualistic temptation.

As we have elucidated at lengths in Pt.I, the opposite turned out to be true. The lack of any tangible progress and the failure of the modern neuronal approach to the problem of consciousness is now slowly but steadily becoming ever more visible. There is now mounting evidence pointing against a strict naturalistic interpretation of mind and consciousness. Though this new information is still largely ignored because of ideological reasons, this lack of progress is convincing some scientists and philosophers of mind that the purely naturalistic viewpoint can't be exhaustive and needs to be revised. To a lesser degree but, perhaps, as a non-negligible factor, some were also influenced by skepticism towards neo-Darwinism as the ultimate paradigm for a materialistic non-teleological account of evolution.

At any rate, a growing dissatisfaction towards physicalism as the ultimate word in the studies of consciousness and evolution is growing among intellectuals as well as in the academic ranks. While it is still the dominant view, the idea that a strict mechanistic and intellectual understanding of the world and Nature is the only pathway to truth is becoming increasingly unconvincing. What was once an almost insignificant intellectual minority has now grown into a visible splinter group, though still not a majority, let alone a homogeneous representation. Nevertheless, it is to be expected that as time passes, the alternatives to physicalism will become more vocal and will receive more attention and recognition, reaching a critical mass that, at some point, will make ignoring it an impossibility.

a. Panpsychism Strikes Back, but Stumbles

This led to a revival of interest in panpsychism. We have already seen how philosophers like Leibniz, Spinoza, and Whitehead expressed a panpsychist view in one form or another—that everything is fundamentally a form of consciousness or mind. In this view, raw inert matter—a stone, a molecule, an atom, or an elementary particle—has some primitive form of primordial conscious experience. For example, a particular version of panpsychism, called *'micropsychism'*, conjectures that even an elementary particle, such as an electron, has—or rather is—an elementary form of consciousness which presumably, whenever it interacts with other particles, has an inner experience or some form of primitive awareness. Its aggregation with other particles, like nuclei made of protons and neutrons, into atoms and molecules first, then by a combination of cells into living organisms, formed, by successive and cumulative stages of evolution, increasingly complex and conscious lifeforms. Panpsychism conceives, therefore, of the emergence of consciousness as a cumulative aggregation of elementary conscious mental units, sort of 'psychic atoms', leading to an increasingly self-aware entity by a bottom-up process.

This, however, does not mean that the panpsychist believes that an object composed of different parts, like an engine, a car, or an airplane, has a mind of its own or is an entity having a subjective experience. Throwing together a number of stones won't lead to something with a more complex and evolved form of mental and experiential content. Also, connecting and interrelating different objects with specific functional tasks like memory chips and a CPU with other electronic devices does not make a computer more conscious than any of its parts. However, the fact is that we know to be conscious mental beings made of organs, which are made of cells, organelles, macromolecules, molecules, atoms, and, finally, elementary particles which have been assembled in a specific type of hierarchical complexity by means of an evolutionary process of aggregation. Thus, the panpsychist contends that there is some complexity law that, by complicated mutual dynamical interactions and interrelations among the single units, allows for a growth of consciousness that is somehow proportional to the number of units and the complexity of the organism they constitute. This latter viewpoint is called *'constitutive panpsychism'*.

Though panpsychism is still a reductionist understanding of consciousness, it is nonetheless the first step away from the purely materialistic perspective insofar that it reverts the paradigm: Not matter, but mind or consciousness must be posited as fundamental. The hard problem of consciousness is so avoided from the outset.

Over the last decades, this view has been revitalized as a possible alternative to orthodox physicalism. Most notably, panpsychism has been reconsidered in its different forms by modern leading philosophers in the field, like Thomas Nagel, Galen Strawson, David Chalmers, and Philip Goff.[56]

However, panpsychism, does not come without drawbacks. The most notorious issue is illustrated by the so-called *'combination problem'*: How does a combination of a myriad of fundamental experiential entities yield the familiar human conscious experience? Why is this happening at all?

There is no apparent reason to believe that combining low-level forms or elementary microscopic experiences (think, for example, of Leibniz's monads) should result in a unified high-level form of experience and cognition. Even if we contend that only some specific types of combinations lead to high-level forms of consciousness (combining microchips does not make a PC conscious, but, somehow, combining billions of neurons does), this still does not explain why a particular functional combination, however complicated, special, and unique, results in a mind and a conscious subject that is, in all respects, very different from the sum of the elementary experiences from which it is supposed to arise. Or, to put it in terms of the so-called *'subject combination problem'*: How do several micro-subjects combine to yield a single macro-subject?

A famous formulation of the combination problem was given by William James who made the following observations:

"Take a sentence of a dozen words, and take twelve men and tell to each one word. Then stand the men in a row or jam them in a bunch and let each think of his word as intently as he will; nowhere will there be a consciousness of the whole sentence. We talk of the 'spirit of the age' and the 'sentiment of the people,' and in various ways, we hypostatize 'public opinion.' But we know this to be symbolic speech and never dream that the spirit, opinion, sentiment, etc., constitute a consciousness other than, and additional to, that of the several individuals whom the words 'age,' 'people,' or 'public' denote. The private minds do not agglomerate into a higher compound mind." [237]

b. The Cellular Basis of Consciousness Model

Whether unicellular forms of life can be considered conscious in the sense of having some form of a sentient experience is something hard to establish given the inherently subjective nature of consciousness. Nevertheless, the evidence of cellular intelligence prompted hypotheses regarding the cellular origin of consciousness as well. Noteworthy is the *'Cellular Basis of Consciousness'* (CBC) model, championed by American cognitive

---

[56] For a modern review, see [372].

psychologist, Arthur S. Reber, and which challenges the assumption that subjective awareness emerged late in evolution [238]. Conscious experience may already appear in cellular form, co-terminus with life. Processes taking place in excitable cellular membranes may lead to a basal form of proto-consciousness with intentional and cognitive capacities [239]. In this theoretical framework, the plasma membrane, together with the nuclear envelope and the internal cellular complex, are seen as the two different '*nanobrains*' generating a 'cellular self' with a purposive agency. In this view, individual cells maintain their identity by measuring information of the outside world to sustain a self-referential homeostatic equipoise in reaction to the external environment [240]. A self-referential awareness that *is* self-organization, and in which every cell embodies an elemental cognition. From this perspective, one could conceive of a '*Cognition Based Evolution*' (CBE), complementing Darwinian natural selection processes, in which receptive intelligent cells measure the environment–that is, assess information–and consequently learn and react to uphold self-identity. In a sense, life could be seen as a 'continuous measurement of information'. For a thought-provoking summary of the CBC theory, see also [241], [242].

The CBC model and the CBE theory place biological development and morphogenesis in terms of an information architecture. The multicellular organism is considered beyond the material aggregation of material units: It is a whole complexity of a communicative assembly of cells seen as an integrated cellular information field. Individual cellular information fields aggregate into an architectural matrix that enables organism-wide information management ([243], [240], [244]). From this perspective, multicellularity is collaborative cellular management directed towards the optimization of information quality through its collective (internal and external) measured assessments. The biological organization represents a dual heritable system: It constitutes a biological materiality and also a conjoining information architectural matrix, with morphogenesis deriving from the reciprocations between these two inter-related facets and thereby yielding coordinated multicellular growth and development. Hence, an information architecture could serve as a morphogenetic template. This could be deemed a natural bridge between vitalist concepts and physical/energetic realities. The logic is that in the cognitive frame, information is physical, not ethereal.

c. Biopsychism

Evan Thompson, a philosopher at the University of British Columbia, Vancouver, followed a similar line of argument introducing the idea of '*biopsychism*' (a term borrowed from the German zoologist and philosopher Ernst Haeckel) [245]. It is the idea that all biological life, even a single-cell organism, is always sentient or minded—that is, conscious and capable of feeling. Biopsychism distinguishes itself from panpsychism insofar as the first form of phenomenal experience does not arise in elementary particles, atoms, or molecules; rather, it is an exclusive property of life. Thompson was inspired by the theory of *'autopoiesis'* (from Greek: 'auto' = 'self', 'poiesis' = 'production') of Chilean biologists Humberto Maturana and Francisco Varela. An autopoietic system is a metabolic system capable of self-reproducing itself and self-maintaining its parts and is adaptively regulating its interaction with the environment. Every autopoietic system is already incipient in mind, and mind belongs inseparably to life. For biopsychism, feeling is a vital ability of all organisms. Every form of individuality and agency so instantiated becomes sentient. That is, life regulation processes determine the feeling of being alive with an experience of bodily existence. It is the self-individuating dynamics that put every organism into an interactive engagement with the environment and of which the bacterial cell is the most elementary example.

The question is how and why such a cybernetic system—that is, a unit defined by its internal recursively interacting networks (think of the complex molecular signaling pathways and biochemical activities inside a cell)—however complex it might be, can interact and modulate its engagement with the environment to enhance its vitality or flourishing or to gain some advantage, without having some motivation. There is a missing link between a purely mechanical system that strives for its survival and behaves by seemingly goal-driven volition, and that is nevertheless supposed to have no volition, feeling, sentience, or affection of some sort.

This motivates the biopsychist to bridge this gap by admitting that sentience is a necessary ingredient that must arise once such a self-regulating, self-reproducing, and adaptive system comes into being. Once something becomes a self-individuating, self-reproducing, and self-delimiting system, it automatically must acquire some form of consciousness, feeling, or mentality. The difference between the functionalist approach (recall Pt.I-II.4&5) and the biopsychist one is that the latter does not consider functionality to determine a conscious experience; rather, it is the organizational and structural dynamics of the system.

The biopsychist contends that this could also suggest a way out of the combination problem. Sentience emerges only in bound and self-individuating subjects—something that a particle, an atom, or a molecule is not, but living cells are. In general, every boundary-forming system that comes into being also becomes a unified sentient being. This is what multicellular organisms do: By association with several bounded and small self-individuating subjects— that is, the cells—they end up forming another unique subject on a larger scale, having a private experience of phenomenal consciousness distinct from that of its constituents. There is something about the interdependent aggregation proper to the multicellular system that supersedes the constituent cellular consciousness, but that does not necessarily erase it. There are still a huge number of cells, each with its own basal cognition and a microscopic conscious subjectivity, but also a body or a brain composed of all these micro-conscious micro-entities, instantiating another conscious subjective experience because, also, this large-scale aggregation is a bounded and self-individuating entity.

d. Dual-Aspect Monism

A form of ontology that transcends the reductionist worldview proper to the previous models towards a wholistic outlook is *'dual-aspect monism.'* It isn't a new worldview since, as we have already seen, dates back to Spinoza. As you might recall, Spinoza didn't conceive of mind and matter as two separate entities but rather as the two 'modes' or 'attributes' of the very same substance. In his view the one and only existing substance could present itself in an infinite number of modes, two of them we call mind (res cogitans) and matter (res extensa). In Spinoza's view, this substance, Nature and God are one and the same thing. In a sense one should call this a 'multi-modal' monism because it allows for many modes, not just mind and matter.

The modern version of dual-aspect monism, revived by Wolfgang Pauli and Carl Jung in the mid-20th century, is, nowadays, championed by German physicists Harald Atmanspacher. It follows the steps of Spinoza, but restricts itself to a dualistic ontology–that is, to a psychophysical neutral substance or reality ('base-reality'). It also starts from the whole of the universe but focusses mainly on a mind-matter dichotomy. The key idea is, again, that mind and matter are not two distinct substrates but the two (dual) aspects of the same underlying neutral substance. Neutral in the sense that it is neither material nor mental. It is a wholistic theory of universal consciousness inasmuch as this neutral psychophysical substance is not made of distinct parts but is one and the same all over the universe. It can't be divided or separated into parts since it is made of only one indivisible component, that is everything. Therefrom also the name *'decompositional dual-aspect monism'*.

In this view, Atmanspacher likes to conceive of the notion of the archetype as an entity that is neither mental nor physical, yet part of this base psychophysical reality. However, when active, it generates a mental and physical counterpart. By doing so it creates a correlation between the two because they share the same origin which is one whole.

e. Reflexive Monism

A step forward toward an integral cosmology that can integrate the many 'isms' of consciousness studies can be found in the *'reflexive monism'* of British psychologist Max Velmans ([246], [247].) It rejects reductive physicalism and functionalism and reframes the problem of consciousness similar to the approach of Pt.II I.3&4. In the simplest terms, reflexive monism is a form of neo-Kantian idealism reformulated with the lens of a modern neuroscientific reasoning (yet maintaining that the noumenon can still be apprehended, against Kant's claim) and that expands to a cosmology of universal consciousness. It is a map of how the mind relates to the physical world by a reflexive act of perception, emphasizing our instinctive and unaware act of 'projection' of our conscious content into the world. Materialist reductionists separate physical objects as perceived from experiences of those objects which are 'in the brain'. Velmans makes the example of an external observer observing the brain of a subject that sees a light bulb 'out in the world.' The reflexive model highlights that the contents of consciousness appear to be 'out there' only because it 'projects' the experience into a construction we call the '3D-space', and that seems to be beyond the subject's body surface. What makes the process reflexive is that once the stimuli are processed by the subject's visual system, the physical processes originating within a light bulb are perceived as light out in the world–that is are 'reflexed' outside of the brain. But they are not, because all what we perceive is inside our skull. Similar reflexive loops occur in the body. For example, a pin prick to the finger activates complex pain circuitry inside of the brain but, nevertheless, results in a feeling of pain located where the prick occurs. This is a sort of reflexive illusion, because the truth of the matter is that all experiences are always situated *in us*. Reflexive monism recognizes that, at the bottom, all observations can be "objective" only in the sense of being "intersubjective." In line with our monitor metaphor, Velmans realizes how this has an absurd consequence for reductive physicalism. If an external observer looks at the brain states of someone else, these brain states are only the other's own brain states. Thus, they cannot provide any information on how and why the subject's experiences arise, undermining the case for reducing first-person experienced phenomena to third-person observed phenomena. The paradox becomes even clearer once we conceive of the case where one would observe one's own brain states. There is a lack of conceptual closure between the first- and third-person observation that we cannot bridge, even not in principle. We always have to reflect a sensation or a thought into the world with figures images and figments that are always subjective or inter-subjective.

But one of the main conclusions of reflexive monism, and of primary interest from a cosmological perspective, is that the reflexive aspect of perception also exemplifies a deeper reflexivity in the universe.

Consciousness is compared to an iceberg where our conscious appearances relate to an *'unconscious ground of being.'* The three-dimensional world that we think of as the "physical world" is only a partial approximation of the greater universe. Human consciousness is embedded in and supported by this greater universe of which we see only a 'tip.' The contents of human consciousness are themselves a natural manifestation of the embedding universe. The thing-in-itself is the one universe with relatively differentiated parts in the form of conscious beings like ourselves, each with a unique, conscious view of the larger universe of which it is a part. Human knowledge is one manifestation of a wider reflexive process by which the universe comes to know itself. There is ultimately no separation between knower and known, and knowledge becomes a form of self-knowledge. A form of ontological monism close to the Eastern tradition, or to Huxley's Mind at large and that anticipated Kastrup's view of the dissociated alters (see next). An approach that tries to integrate the 'isms', by viewing different aspects of the universe as sentient manifestations of itself, from within itself.

f. Cosmopsychism

Some philosophers of mind, moved by the combination problem of panpsychism, but still intentioned to not fall back to the physicalist standpoint, considered the opposite viewpoint of micropsychism, that of *'cosmopsychism'*, which replaces a bottom-up with a top-down approach and which conceives of the whole cosmos as the ultimate conscious being (see, for example, Itay Shani [248]).

From this perspective, the Universe is a sort of *'cosmic Mind'*, a unitary conscious entity (something not really new, as it is a concept appearing also in several Eastern traditions). Assuming a universal consciousness as the ontological primitive solves the combination problem because it is the high-level consciousness, rather than the low-level subject, which is the starting point. For example, the human mind, its mentation, and our subjective experiences are a fragment, a localized spot of a much larger cosmic consciousness and cosmic Mind.

If one extends subjectivity not only to matter but also to all space and time and sees the Universe as a 'super-Subject', then we don't need to explain why the bottom-up combination of material objects or particles yields a subject endowed with a unified phenomenal experience—because it doesn't. It is a top-down decombination that does the job, namely, a process of self-individualization of this universal consciousness in space and time by a fragmentation in itself.

This, however, immediately raises the opposite question, that of the *'decombination problem'* (or *'decomposition problem'*): How do these segments of universal consciousness, what we experience as an individuality, an 'I', having private experiences and individual mental contents, come into existence? After all, we cannot (at least not consciously) read each other's thoughts. If my and your minds are two different specks of a larger Mind, one would expect mind-reading to be part of our everyday experience. Because this seems to not be the case, the cosmopsychist must tackle the opposite problem of the panpsychist and explain how this process of decombination or separation from a universal consciousness yields seemingly isolated individuals with private experiences.

g. Analytic Idealism

An interesting proposal on how the decomposition problem can be solved comes from the Dutch idealist philosopher Bernardo Kastrup [249]. In his theory of *'analytic idealism'*, also termed *'Cosmoidealism'* [57], everything is mental, the inanimate world is an 'extrinsic appearance of the thought of the cosmic Mind', what he called *'Mind at Large'*, borrowing the term from Aldous Huxley, and all living organisms are *'dissociated alters'* of this cosmic consciousness. He was also largely inspired by Schopenhauer and his vision of the world as Will [250]. Let us look at this in more detail.

Kastrup identifies every experience, be it cosmic or individual, as a 'pattern of self-excitation of cosmic consciousness' analogous to modern quantum field theory, which describes every elementary particle as a harmonic oscillator localized in the space and time of a fundamental quantum field. One might even go so far as to question whether it is more than an analogy—that is, one might conjecture whether a quantum particle is indeed the self-excitation of a cosmic entity in itself. In this view, every particular experience corresponds to the particular pattern of self-excitation of this cosmic consciousness.

The most curious aspect of Kastrup's theory is that our 'discrete centers of self-awareness', the *'alters'*, arise due to some dissociative process that psychiatrists know as *'dissociative identity disorder'*. A person suffering from dissociative identity disorder exhibits multiple personalities as if the same body contains (or is possessed by) several different subjects with different minds and personalities. Persons affected by this rare state of consciousness, officially recognized as being a real psychological 'disorder', exhibit multiple discrete centers of self-awareness. Physically, from a third-person perspective, one sees the same physical body but the individual acts as though driven alternatively by completely different subjects with different personalities.

Kastrup submits that these different subjects—that is, different alters—are not fictional personalities created by a brain disorder but real dissociated centers of awareness caused by the same dissociation process in cosmic consciousness, which leads to the formation of our subjective individuality, the 'I-ness' we experience as sentient beings. Our identity as a seemingly separate individual having the illusion of separation arises due to a cosmic dissociation at the mesoscopic level of living beings, between the microscopic realm of elementary particle physics and the macroscopic size of cosmic structures.

Furthermore, Kastrup identifies metabolic systems—that is, our physical bodily boundaries—as the *'dissociative boundary'* inside which we experience our separate phenomenal consciousness. Metabolic processes in closed biological systems, such as a living cell or a multicellular organism like a plant, animal, or human being, delimit the dissociative boundary between our

---

[57] The term 'cosmoidealism' we borrowed from D. Broderick who cites D. Chalmer in [379].

individualized phenomenal experience and the cosmos from which it has dissociated. This is something on the line of the bounded and self-individuating system of biopsychism, but with the decisive difference that, here, consciousness arises not due to a complicated autopoietic process but rather from a dissociative process in an a priori existing Mind at Large.

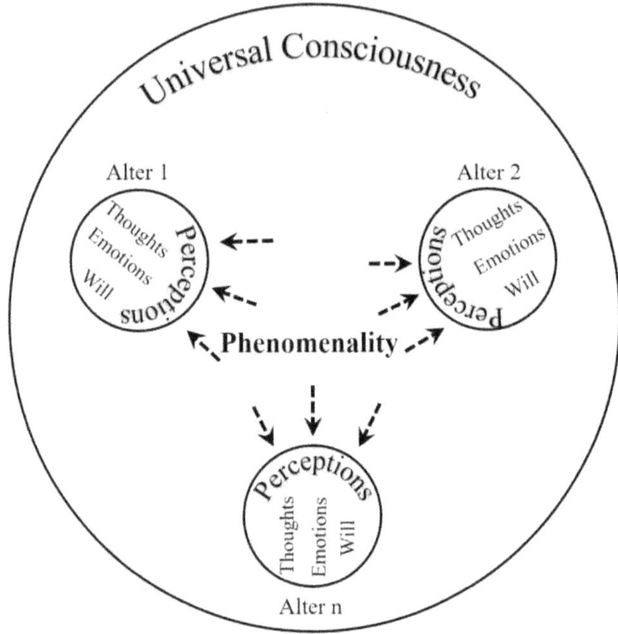

Fig. 110 B. Kastrup's analytic idealism in a nutshell.

In this view, all the matter and the complex structures we observe in the universe are thoughts of Mind at Large. Also, all the physical phenomena we perceive, such as light, electromagnetic, and nuclear forces acting in space and time, are thoughts of this cosmic Mind impinging on our sensorial organs and bodies, the alters' dissociative boundary. In fact, all our physical senses are mediated by sense organs at the dissociative boundary, which is our body. Analogous to the monitor metaphor that we illustrated in Pt.II-I.3, Kastrup frames the hypothesis that all the qualitative experiences that we become aware of on our 'screen of perceptions', such as colors, sounds, flavors, textures, etc., are 'desktop-representations' of the qualities experienced by a segment of cosmic consciousness. This is to say that our experiences and all the qualia we perceive are not the same qualities that Mind at Large perceives; rather, they are a reduction and figment of it. For example, what we apprehend as a tree with a green leaf is a thought of this cosmic consciousness which it experiences as mental, not qualitative, content. Our experiential qualities are coded representations of thoughts of a cosmic Mind but from the perspective of a dissociated boundary. They do not feel like thoughts at all because of this informational process of reduction and compression that occurs due to us observing through the dissociative boundary.

Finally, in this cosmoidealist view, the cosmic consciousness is not really a self-aware superconscious being that is capable of a higher cognitive process; rather, it is a reactive and instinctive being that does not really have a goal or preconceive a cosmic evolutionary project. It is a sort of borderline between a pantheistic and panentheistic conception of a Spinozian God which, as in Schopenhauer's world-conception, simply is and acts only by a blind Will that is not self-reflective but grows in its self-awareness by being and becoming in an undirected evolution process. This also answers the question about the existence of suffering in this seemingly unconscious cosmic play. Mind at Large has no moral or ethical values; it is just an instinctive Being that slowly becomes self-aware through painful material processes but can't do anything other than express and follow its own automatic instinctually.

h. Panspiritism

Another variant of cosmopsychism, which also envisages all of reality as the expression of a *'fundamental consciousness'* is *'panspiritism'*, an approach of American psychologist Steve Taylor [251]. Panspiritism means "all is spirit" or "spirit is everywhere". In a nutshell, it is a panentheistic conception of the universe with Spirit pervading and flowing through all things.

Fundamental consciousness is still the ultimate essence of all that there is, but there is a distinction between spirit and matter. Matter has its own ontological status and is distinct from spirit. It is not just a 'projection' or a 'mirage' as the Advaita school teaches, or an 'excitation' of a *'universal Mind'* of the cosmopsychist or analytical idealist; rather, it is a 'product' and 'emanation' of—and pervaded by—spirit, being at the same time of the same nature of it (sort of another state of the Spinozian fundamental substance). For the panspiritualist, matter is not a purely mental or spiritual entity; it has its own reality, just as the wave is distinct from the ocean or different degrees of solidification and 'crystallization' of a fundamental substance are distinct forms of the fundamental substance itself.

However, matter is also pervaded by fundamental consciousness, from which it originated and which flows through all space as well. Therefore, consciousness is in all things, but not all things; molecules, atoms, or particles have their own consciousness or mind, as the panpsychist contends. Non-living objects are not individuated conscious entities but nevertheless are immersed in the workings of the universal Spirit. Only structures having a certain degree of complexity can receive and 'canalize' fundamental consciousness into themselves. There is a difference between a universal all-pervading consciousness and things through which this fundamental consciousness is 'transmitted' by means of its material structure having an appropriate degree of organizational complexity to receive it. For example, a cell is a sufficiently complex structure that can receive and canalize from its inside fundamental consciousness.

In this sense, a cell, a multicellular organism, or a brain each possesses an interior nature which, however, is not 'produced' by the physical processes of the cell, organism, or brain; rather, they are the result of the 'canalization' of the all-pervading Spirit in—and through—it. It is this localized inflow of the Spirit that generates an individual consciousness. The Spirit flows into our inner beings via our brains. This is what provides us with subjective sentience by an 'internal animation' of an organism—a conception of the flow of spirit reminiscent of the animistic beliefs of the native American culture or the flow of the Holy Spirit in the Christian religion.

Also, mind emerged once a physical structure became complex enough. However, contrary to matter, the mind arises due to the interaction of fundamental consciousness with matter in the brain and is not another state of fundamental consciousness as matter itself is. Only in this sense does panspiritism align with the 'radio metaphor' or James' prism analogy we considered in Pt.I-IV.1: There is a 'transmission' of a cosmic subject via the brain, which, in the material manifestation, is expressed as mind. This means that, for panspiritism, the physical preceded the mental and can exist without it while the reverse doesn't hold: There can't be the mental without the physical.

Also, cells, organisms, and all metabolic structures 'receive' and 'canalize' fundamental consciousness. But there is no evidence that non-living material objects, molecules, atoms, or particles' receive' or 'canalize' such a fundamental consciousness; therefore, contrary to panpsychism, panspiritism considers them to be non-conscious, inanimate, and without any form of sentient or mental content. In fact, death can be considered the inability of an organism to further prolong its reception and canalization of that consciousness.

i. Interface Theory of Perception

A different approach to idealism that starts from evolutionary biology comes from the American cognitive scientist Donald Hoffman.

Hoffman's Interface Theory of Perception (ITP) proposes that our perceptions of the world are not a direct reflection of objective reality, but rather a 'user interface' designed to guide adaptive behavior evolved for survival rather than convey truth. He argues that what we perceive—shapes, colors, objects—are akin to icons on a computer desktop; just simplified tools designed for efficient interaction. Just as a blue folder icon doesn't reveal the underlying binary code or hardware, our sensory experiences don't reveal the true nature of the world; they merely present simplified, useful representations that help us navigate our environment effectively. Evolution, according to

Hoffman, favors perceptions that enhance fitness, not necessarily those representations that are truthful.

At the core of Hoffman's theory is the idea that natural selection doesn't reward veridical perception (seeing reality as it is), but rather adaptive perception. Organisms that see a 'useful fiction' are more likely to survive and reproduce than those that perceive the world accurately but inefficiently. For example, perceiving a ripe fruit as bright red doesn't necessarily mean the fruit is red in an objective sense—it means that color signals nutritional value in a way that's beneficial to the observer. This challenges the conventional view that human senses evolved to approximate reality, suggesting instead that space-time and physical objects are constructs of our interface, akin to a desktop's icons hiding the complexity of a computer's circuitry.

Hoffman extends this idea to consciousness itself, proposing 'conscious realism'—the notion that reality is fundamentally composed of underlying interacting conscious agents, with physical phenomena emerging as secondary to these interactions. Rather than emerging from complex arrangements of physical matter (as in traditional neuroscience), Hoffman argues that what we call the 'physical world' arises from the interactions of 'conscious agents.' These agents are formal systems that perceive, decide, and act—not in spacetime, but in a deeper, non-physical reality. In this model, spacetime and physical objects are not fundamental; they are merely projections or 'user interface icons' created by the activity of these conscious agents. Hoffman and collaborators have explored connections between the structure of conscious agents and modern mathematical physics, particularly the geometric structure introduced by Nima Arkani-Hamed that enables simplified calculation of particle interactions in quantum field theory known as 'amplituhedron' and 'decorated permutations'. The amplituhedron is a geometric object arising in quantum field theory that encodes particle interactions without referring to spacetime, suggesting that spacetime itself might not be fundamental. Hoffman suggests that the mathematical elegance and timeless nature of the amplituhedron might reflect the deeper realm in which conscious agents operate and interact. These agents combine and evolve, forming more complex networks that give rise to what we experience as physical reality. This radical reimagining of the universe posits that what we see as particles, forces, and even spacetime geometry are emergent phenomena—shadows cast by an intricate web of conscious interactions in a deeper, more abstract domain governed by the mathematics of geometry and symmetry.

The ITP bridges cognitive science, philosophy, and physics, suggesting that reality might lack space-time or causality as we perceive them, with implications for understanding consciousness and the universe.

Hoffman supports his theory through computational models and evolutionary game theory, showing that agents who perceive reality accurately are consistently outcompeted by those with tuned, distorted perceptions. He extends this concept into metaphysics, suggesting that space, time, causality, and physical objects themselves may be part of this evolved interface, not fundamental elements of the universe.

j. My Critical Assessment

If we indulged particularly in these modern forms of idealistic and post-material theories, blending out other proposals coming from various fields of the modern philosophy of mind[58], it was because, in some respects, they came closest to the integral cosmology we will outline later.

Thompson's idea of biopsychism, or the CBC model and the CBE theory—that single cells are endowed with a mental or conscious experience, engaged with an informational exchange with the environment, and which admit, at the same time, that their aggregation leads to a sentient subject—is an interesting hypothesis that we will take up in Pt.III, although from a very different spiritual perspective.

However, these do not even begin to tackle the hard problem of consciousness. While panpsychism, dual-aspect monism, reflexive monism, cosmopsychism, analytic idealism and panspiritism posit consciousness as fundamental a priori and, therefore, don't have to explain phenomenal consciousness in the first place, biopsychism and the CBC/CBE approaches take a step back into the direction of physicalism. The question is why the dynamics of a self-regulatory, self-individuating system, however complex, networked, informational, interacting with—and adapting to—the environment, should become conscious. Sentience is posited as an almost magical property that emerges once certain structural or dynamical conditions of a systemic organization are met. These theories posit consciousness as an emerging property from material processes. There is no real distinction between these speculative views and the physicalist approach other than the size and scale where sentience is supposed to emerge first. While the physicalist postulates consciousness as emerging from the dynamics of a neural network, they assume it to emerge from the autopoietic processes. Matter remains fundamental, and consciousness a derived epiphenomenon as well.

Moreover, its solution to the combination problem doesn't sound satisfactory. It only shifts it from the elementary particle to the cell, adding boundness and self-individuation as necessary but sufficient conditions to create a sense of subjectivity. This raises again not only the question of why split-brains maintain their subjectivity but also why the combination of cells constituting a brain instantiates the totality of our phenomenal experience and sense of subjectivity, while the liver, the heart, or a limb does not?

---

[58] Such as consciousness research inspired by the Asian tradition, especially the Advaita-Vedanta; see, for example, Miri Albahari's take on 'perennial idealism'. [370]

These discrepancies do not emerge from a cosmoidealist perspective. For example, Velmans' self-knowing universe or Kastrup's intuition that a cosmic dissociative process is at the base of the subjective individuation and that our subjective experiential sense-mind contents are a 'redux' of something having a cosmic dimension is in line with the mystic tradition of Eastern philosophies.

A positive aspect worth noticing is that, together with Spinoza's dual-aspect monism of the substance modes theory, analytic idealism and panspiritism are not affected by the '*mind-body interaction problem*' of mental causation. The interaction problem only arises if we conceive of mind or consciousness as purely unphysical and matter as an unrelated purely physical substance. If one posits a priori this assumption, one is automatically forced to face a paradox, namely, the question of how a completely unphysical entity can interact with a material substance and maintain a causal relationship with it, such as that of determining our brain states. An immaterial consciousness and mind could not, even in principle, affect a material brain or whatever material body. And even if they could do so, that would violate the principles of energy conservation. This is because, by such an interaction, a material object could change its energy content or status of motion without a physical energy interaction.

Most physicalists see this as the ultimate and conclusive argument that definitely refutes any form of alternative thinking. But if we take into account that matter is simply another form of consciousness, the different 'aggregation state' of the very same 'substance', rather than thinking of matter and consciousness as essentially two different things, this dual-aspect monism offers itself as a solution to the interaction problem. To use an analogy from physics, consider the various degrees of solidity or physical aggregation states of the same substance—say, vapor, water, and ice. They all can interact with each other because they are all made of the same substance—H_2O molecules—only in different states, and there is no energy loss if there is an energy exchange between the three phases. The energy conservation paradox only arises if we regard one phase as having nothing in common with the other phases. Then, obviously, one sees the energy 'disappear' from the only phase considered real. In this analogy, heat exchange could be taken as an image of the flow of Spirt from the panspiritist perspective. Of course, these are only analogies, but they offer a possible logical solution to the mind-body interaction problem.

It might be useful to recall that events apparently violating energy conservation are well known in particle physics. For example, one can observe in particle accelerators how, on rare occasions, an electron can suddenly change its state of motion apparently without any force interacting with it. This behavior might at first suggest a violation of energy conservation but, with the discovery of the neutrino particle, which is almost impossible to detect with normal detectors, it turned out that the electron was kicked from its path because of the action of a particle that interacts only via the weak nuclear forces. Yet, electrons and neutrinos, like any other particle, are considered two different excitations of the same universal quantum field.

The interaction problem is an example that shows how some conceptual issues appear as seemingly unsolvable paradoxes only because one sees reality from an exclusivist standpoint, a 'tertium non datur' mindset leading to a false dichotomy. The understanding of reality based on a universal consciousness perspective can lead us a step further.

Nonetheless, we believe that these are still too-simplistic models of reality and that we will extend and amend them in a more complete and integral cosmology.

For example, Velmans' approach does not go much beyond a modern reformulation of Kant's idealism, in a more optimistic form, and the idea that we are the differentiated parts of a One-being. The latter is a point of view reminiscent of Eastern traditions. Something to welcome in the Western philosophy of mind but that does not go beyond the already known either.

Moreover, I find it a pity that modern dualism falls back from a Spinozian 'multi-modal' monism to a dual-aspect monism made of two modes only. More than ever, modern science shows us that the universe is not black or white but is made of a variety of shades in between. As we shall see with the integral cosmology later, it is much more in line with modern conceptions of the world and a less superficial psychology, to consider mind and matter only two aspects of reality inside a much vaster (infinite?) number of modes. Besides mind and matter there might be also a subconscious, and several superconscious domains that escape a simplistic dualistic notion of reality.

While, I contend, Kastrup's monistic idealism (all is mental), which sticks at misplaced principles of parsimony, imprisons itself in a one-dimensional theoretical framework that tries to capture a multidimensional reality. The most evident stumbling block is the semantic, or at least linguistic conflation between mind and consciousness, which we consider a serious fallacy hampering deeper insights, as we amply discussed at the beginning of this treatise. This is a fatal flaw that prevents one from progressing and looking further. It is like Alice in Talbot's 'Flatland' novel, who tried to determine all the properties of a 3D-spherical surface in terms of the circles she observes appearing in her intersecting 2D surface on which she lives. In these regards, the exclusively idealist approach is questionable. The fact that we perceive everything physical in terms of mental phenomena on our 'screen of perceptions' raises the question of whether this authorizes us to conclude that everything must be mental? The fact that we perceive, at our evolutionary stage, which is the stage of mind, everything through the mind does not imply that everything is mental.

Moreover, while Kastrup's approach solves the combination and decombination problem, it raises the question of how Mind at Large could form the first metabolic system from raw insentient matter in the first place?

According to this model, before the formation of the first metabolic unit, namely, a biological cell, there was no mind or consciousness at work on a spatially and temporally localized scale. There was only an instinctive Mind at Large without dissociated discrete centers of self-awareness. How can mere instinctive impulses create such an amazingly complex, efficient, and self-reproducing thing like a living cell? A cell is a much more complex object than the most sophisticated supercomputer, airplane, rocket, or spaceship. Supercomputers, airplanes, rockets, or spaceships don't self-assemble themselves by mere instincts, let alone reproduce. And it is implausible that this could happen even after 13.8 billion years of instincts. This issue is reminiscent of the combination problem: One can't see why and how putting together gazillions of mechanistic instincts, no matter how much time passes, should supposedly lead to anything creative and cognitively more developed other than yet another blind super-instinct.

If we don't embrace a panpsychist view which could, in principle, make a bottom-up '*abiogenesis*' —the theory of the origin of primordial life from non-living matter—plausible, and where we conceive of some consciousness locally at work meddling with macromolecules in a primordial soup and leading to the appearance of the first biological cell, what then led to life's appearance?

But the point in question does not regard only the origin of life; rather, it regards its whole evolution, from the first eukaryotic cell to the full-fledged human. It is hard to believe that a purely instinctive and blind will without any form of intelligence could master an evolutionary process with such a fantastic complexity, leading to such an incredible variety of lifeforms. This is a huge explanatory gap in Kastrup's theory.

The only alternative the author can think of is to fall back to the purely naturalistic perspective of material microscopic and macroscopic processes driven by genetic and natural random selection principles. That would be a rather disappointing outcome for a theory that was supposed to get rid of a physicalist account of phenomena and processes. Because then, one wonders, why should we posit a universal or cosmic mind in the first place?

Instead, the shortcoming we see in panspiritism is that it distinguishes between fundamental consciousness, mind, and matter as three ontologically distinct entities. But with these ontological distinctions, it steps out of a strictly monistic view without explaining how and why they came into being. Moreover, panspiritism does not furnish an understanding of the mechanism by which individuality, the subject, or the soul emerges. If it is generated by an inflow of the spirit in a complex material structure—that is, a living organism that canalizes it—should we then suppose that by the body's dissolution, no subject, or soul, or whatever psychic individuality, will survive? Taylor also posits matter as the necessary ingredient for mind to come into existence. Is fundamental consciousness just a state of 'mindless' pure being and pure passive existence without any form of intelligence? Then, again, as in the case of analytic idealism, here, the evolutionary aspect of life would remain an unexplained mystery unless one falls back to a purely mechanistic neo-Darwinian conception of evolution.

This is what vitiates analytic idealism and panspiritism. These are theories that fixed their attention too much on resolving the hard problem of consciousness but lost sight of the evolutionary emergentist dimension of life and consciousness—something which is naturally harmonized in an integral cosmology, as we shall see later.

While, Hoffman's ITP merits attention for bringing to a broader audience the principles of an idealistic worldview. The metaphor of the 'user interface', designed to guide adaptive behavior rather than convey objective truth, serves as an insightful introduction to the realization that our perception of the world is not a reflection of reality. However, framing the argument through the lens of evolutionary biology is misleading. Because, even if our sensory perceptions were a 'veridical' representation of reality, we still would not see the world as it truly is. Idealism embodies the profound, almost 'transrational' understanding that a 'truthful' depiction of reality is unattainable, even in principle. Our inability to perceive the world as it exists in itself doesn't arise from evolutionary necessities, even though perceptual distortions introduced by natural selection may play a role on the top of that. As we have seen, we cannot grasp the true structure of the world because no phenomenon can, even in principle, represent the noumenon.

Moreover, Hoffman's portrayal of space-time and reality as a network of conscious agents (something reminiscent of Leibniz 's monadology,) is intuitively appealing. However, I would exercise caution in attributing a Platonic ontology to abstract mathematical entities such as amplithuedrons and decorated permutations. These constructs are simply tools used by theoretical physicists to calculate scattering processes between particles, and I would argue that they are no more or less real than any other mathematical construct or the existence of the square root of five.

Nevertheless, these new metaphysical trends in the modern Western philosophy of mind are to be welcomed because they undoubtedly constitute a leap forward compared to a narrow-minded physicalism. Probably they constitute only a first step, a beginning from which we can expect new developments. Though the author of this book feels they still need to be detailed and amended better, they come nevertheless as fresh air and as a new hope that might help humanity go beyond the void conceptions that modern materialism still imposes on our minds and hearts.

This ends our pinpointed tour of some of the intuitionally inclined philosophies of Western civilization. It reflects more a personal and subjective selection and preference of the author and is by no means a representative or exhaustive overview. It certainly had no pretension of doing justice to such a complex and deep subject. Nevertheless, hopefully, it was a useful selection. It aimed to pick out, here and there, some of the higher-mind aspects of the ancient and modern philosophies that reveal its more intuitional or phenomenological, and eventually also spiritual character that, in the modern, strongly materialistic-oriented worldview, are at best ignored or, worse, have been long forgotten.

The next chapter will further focus on another Western intellectual: Johann Wolfgang von Goethe. Of particular interest to us is his 'way of seeing' as a possible alternative approach to science rather than philosophy. It will be another ingredient that we will employ in our synthesis of knowledge in Pt.III, because it complements the philosophical approach with a more pragmatic natural philosophical understanding of reality. Then we will widen our scope beyond the mind to an evolutionary perspective of consciousness and its related cognitive layers within an expanded paradigm in the frame of Eastern philosophy, especially in light of Sri Aurobindo's cosmology.

This will finally allow us to acquire the grand vision and synthesis of knowledge we are seeking. This chapter and the following are the pieces of the Western puzzle that are usually missing when people talk about the connections between science and spirituality. The Western philosophy that 'palpates the unknown', as Deleuze used to call it, and Goethe's scientific phenomenology are the two indispensable fragments that will allow us to connect the wholistic and mystic experience to the purely physicalist and reductionist perspective.

III. Towards a New Way of Seeing

During our journey through the mystery of consciousness and our inquiry into the nature of reality, we encountered a common theme: the inadequacy of present science in tackling these questions. We found that science, especially that science that clings to a naturalist reductionist paradigm based on a naïve local realism, when it comes to questions of consciousness, existence, and meaning, fails to deliver a credible answer and is unavoidably affected by humans' limited cognition. A purely sense-mind approach is revealing its limits.

The natural question, then, is whether a new science or, at least, a somewhat different scientific approach can lead us a step further?

We have no definite answers, and you won't find anything about an ultimate new science supposedly replacing the conventional scientific method. However, we would like to try looking further, taking a step in a novel direction, which might help us go beyond the present paradigm. We contend that many problems and questions with which the present scientific conceptions struggle—be they from philosophical quests such as the origin and nature of consciousness to more pragmatic issues such as globalism, the clash of civilizations, or the environmental impact caused by economic and technological progress—can be tackled only if we begin to raise our state of consciousness and, first of all, question the background assumptions we are working with—that is, generally, how we think. Only by becoming aware of and learning to better know ourselves can we understand the enigma of consciousness and reality as a whole and develop a wider form of scientific inquiry that transcends a narrow-minded empiricism. A sense-mind-based perception supported by rationalism, reason, and analytic thinking are necessary but not sufficient. Our perception of the world and ourselves must include something else.

1. Perceptions, Assumptions and Fallacies in Science

A very common belief is that the exact sciences do not make assumptions and hypotheses and that cultural or psychological tendencies don't influence them; they simply ascertain facts. We still tend to give credence to the idea that the way science 'sees' the world comes only from strict inductive procedures of evaluation and observation of brute facts and data without any outer conditioning and influence. This apparently ideal epistemology is taken as the cornerstone of Galilean-Newtonian empiric science and is thought of as the fundamental guarantee in the search for objective truth.

However, after some practice and experience, every scientist who is honest with themself realizes that things are not as simple as that. Also, exact sciences based on raw empiric data, exact mathematical models, and rigorous theories have always been influenced by different interpretations, opinions, and disputes. This has been the norm throughout all sciences and all times.

A simpler historical analysis shows how a basic background belief system that sees science as an objective and perfect expression of truth disengaged from human preferences and cultural backgrounds is misplaced. Historians and philosophers of science already pointed this out many times. The history of science and its developments, theories, discoveries, and past and present worldviews is not a simple linear outgrowth of empirical and technical facts. Science was and still remains strongly affected by the cultural beliefs of the time. Empiric data alone aren't sufficient to discover truths. Whether or not we are aware of it, interpretations that depend on our continuously present unaware assumptions, hypotheses, and metaphysical beliefs always creep in, in one form or another. As an ideal observer of brute facts collected through empiric experiments, the scientist, unaffected by prior assumptions and hypotheses, may never, and perhaps can't even in principle, exist.

In 2015, an article in the prestigious journal Nature reminded the scientific community that humans are remarkably good at self-deception and that scientists are no exception in fooling themselves. Even if we are the most honest scientists, we can nevertheless be masters of self-deception: *"In today's environment, our talent for jumping to conclusions makes it all too easy to find false patterns in randomness, to ignore alternative explanations for a result or to accept 'reasonable' outcomes without question — that is, to ceaselessly lead ourselves astray without realizing it."* [252]

Whether or not we admit it, it is a well-established fact that we are all, scientists included, affected by cognitive fallacies and biased worldviews.

This does not imply that science is only a social construct or a simple belief. Of course, the events we observe are in some sense 'true' and somehow describe a reality we cannot simply deny. Many 'truths' that science established are and will forever, at least in a limited dominion, remain undeniable facts. The predictive power of exact sciences such as physics is not in question, the effectiveness of mathematics is as mysterious as it is overwhelming, its enormous practical scientific and technological success is undeniable, and the models it works with, without a doubt, fit truth much better than a world-bearing turtle cosmology. If science were only a social construct, it could not observe several phenomena our theories predict or build functioning technological devices based on its insights.

However, that said, what we would like to underline here instead does not concern the metaphysical idealistic questions of the things-in-themselves or the level of 'truthfulness' of science (whatever that might mean) but, rather, its supposed independence from methodological approaches and cultural conditionings.

Let us clarify this distinction with a little historical anecdote that could serve as an example. What does the trajectory of a projectile, say, a cannonball which has been launched from the earth's surface (with a given velocity and non-vertical angle and neglecting air friction) look like? Which of the two cases in Fig. 111 would you opt for?

If you are not a physicist or engineer, you may not know the precise equation of motion and hesitate for a moment; however, also without being a ballistics expert and without any prior knowledge of physics, almost everyone would probably have no difficulty intuitively determining which of the two depicted possibilities must be the correct one.

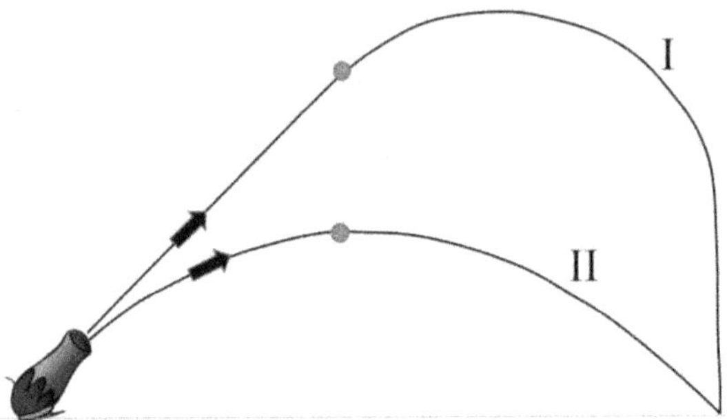

Fig. 111 Projectile trajectories: Which is the right one?

Trajectory II is the correct one—that is, the curved path of a parabola. Nowadays, almost everyone, even if they lack a background in physics, recognizes the parabolic motion to be the 'true' one. However, before the exact measurements of Galileo, this was not at all regarded as an obvious fact. In 1537, the Italian mathematician and engineer Niccolò Fontana Tartaglia was the first to investigate the paths of cannonballs in his treatise *'Nova Scientia'* (*'A New Science'*). His observations led him to prefer path I: First, the cannonball flies along a straight line, then follows a circular path, and finally falls again along a straight line to the ground. Soon, everyone adopted Tartaglia's model as the foundation for ballistics (because he correctly found that the projectile reaches farthest with the cannon's inclination of 45°) until Galileo recognized, with his strictly empirical measurements, that the projectile paths are always parabolic with whatever initial angle the cannon launches it. Tartaglia argued, based on Aristotelian physics, which was the dominant view of the time, and not based on empiric data and mathematical reasoning, which has been the scientific approach until today and which Galileo first adopted systematically.

Strangely, however, no one questioned Tartaglia's conclusions for a century. What makes one wonder isn't so much the fact that people before the 17th century didn't use the abstract mathematical notion of a parabola for a falling body[59], even though the use of objects of analytic geometry such as a parabola was possible, at least in principle, as it had been introduced earlier by the ancient Greece philosophers of the Platonic tradition and only later found its rigorous application in natural sciences beginning from Galileo, Descartes, and Newton. The curious fact is that, at the time of Tartaglia, nobody seemed to doubt that path I might be the wrong one. Today, almost all of us instead see path II, also without any mathematical and physical knowledge. Why do we see something completely different today from what well-educated people saw five centuries ago? Our perception of the world is—from the physiological and biological point of view—exactly the same as that of the human beings of the Middle Ages. The reason for this can't be methodological or scientific but must have been of a cultural and sociological nature.

May modern science be subjected to similar social biases? One could argue that modern empiric science has been introduced as a working tool precisely to avoid such naive mistakes and that the example above does not hold in the frame of any modern research approach. This is correct to some degree. Today, such an illusion would have been exposed by a precise quantitative measurement as, indeed, Galileo illustrated eloquently. But this does not at all get to the point of the question. The question is: Why, today, would most of us doubt path I and agree on path II? We nowadays see something different even without resorting to empiric measurements.

This shows that in the past, people had *not* been deceived by an optical illusion that betrayed their senses. It is *not* an observation that comes from our imperfect and misleading sensory organs of perception. Otherwise, we would continue to see path I today, despite our knowledge that it is false. What is happening here is completely different.

Today, we see something else simply because our prior knowledge of things has changed. It depends on what we know and what we believe. In other words, it depends on the education we got and not just because of our sensorial imperfection. Aristotelians believed in something very different from what we do today. Therefore, they convinced themselves that they indeed saw just that.

What we see does not depend exclusively on what is happening 'out there' in the world. What we see is also not solely due to our more or less imperfect and deceiving sensory apparatus. To a good degree, our seeing also depends on our mind and our belief system, as well as our psychological, cultural, and historical contexts. We may switch from one way of seeing to another, even though the senses and sensory data are the same. It is an even more subtle cognitive process than that which we experience with the Gestalt figures.

---

[59] By the way, it might come as a surprise to some that, to be picky, this is false as well: A falling body follows an extremely eccentric elliptic orbit. The parabola is an exceptionally good approximation, but not the mathematically exact one.

Therefore, before coming to any conclusion about our observations, we must ascertain whether our own beliefs and cultural influences may have somehow not crept in already. Because if 'seeing is believing', we must become aware of the fact that sometimes also 'believing is seeing'.

In what sense does this have to do with our modern epistemological approach towards the world and reality? Are we free and immune from our beliefs and personal and subjective convictions? Modern science got rid of Aristotelian science, which wasn't entirely aware of the power of perceptual illusions on our sensory organs, but did it also get rid of its mental illusions? Modern exact sciences could certainly eliminate the former uncertainty but seem not to be very aware of the latter. The power of the psychological and cultural influences on our most enlightened erudition and disciplines is still strong and healthy.

Unaware assumptions and a priori working hypotheses resting on shaky grounds remain unacknowledged today, as in the times of the Aristotelian sciences. This becomes apparent beginning from Newton. Indeed, in the two final paragraphs of his *"Mathematical Principles of Natural Philosophy"* (*"Philosophiæ Naturalis Principia Mathematica"*), published in 1687, and commonly also abbreviated as the *"Principia"*, Newton says:

"Hitherto, we have explained the phenomena of the heavens and our sea by the power of gravity...But I have not been able to discover the cause of those properties of gravity from phenomena, and I frame no hypotheses; for whatever is not deduced from the phenomena is to be called a hypothesis; and hypotheses, whether metaphysical or physical, whether of occult qualities or mechanical, have no place in experimental philosophy. In this philosophy, particular propositions are inferred from the phenomena, and afterwards rendered general by induction."

Newton knew that any attempt to reflect on the real and ultimate causes as he was asked to do regarding gravity—that is, to frame hypotheses as to why and how a massive body generates a gravity field—is a philosophically dangerous task that almost certainly would end in debates and controversies. Therefore, in his official statements, he retired to a pragmatic position declaring that any speculation about the true nature of things, beyond a crude observation of facts and a pure induction from these, must be considered fruitless. Such an epistemological proceeding was not entirely new at the time of Newton. It was advocated, more or less implicitly, among others, for instance, by Francis Bacon and Galileo, but modern scientists remember this historical declaration very well and cite it openly and repeatedly as the foundation of today's epistemology, placing Newton as the central example of the perfect inductive empiricist.

Since the 18th century, scientists have taken Newton's famous passage *"I frame no hypotheses"* as the foundation of a presumably rigorous epistemology on which every scientific research should be based and indicating Newton as the methodological paradigm to follow. This passage tremendously influenced modern science and our way of perceiving reality as a cognitive activity that aims to separate the objective from the subjective and the primary qualities from the secondary qualities. Nowadays, it is a common conviction that science should not 'frame hypotheses' and that it simply evaluates facts. We tend to believe that doing otherwise is, per definition, 'unscientific'.

But did Newton himself follow his advice? From many of his unpublished papers (most of which were discovered only after World War II), it appears that framing hypotheses and working with a priori assumptions was one of his preferred intellectual exercises. For example, before any evidence, he already had in mind the hypotheses of the corpuscular nature of light, which he intensely applied in his conceptual, experimental constructions. It is a well-known fact that Newton was deeply involved in alchemic and occult studies, and it might be instructive to also read what he wrote in *"The Principia"* only two paragraphs earlier: *"This most beautiful system of the sun, planets, and comets, could only proceed from the counsel and dominion of an intelligent and powerful Being... As a blind man has no idea of colors, so have we no idea of the manner by which the all-wise God perceives and understands all things... We know him only by his most wise and excellent contrivances of things... And thus much concerning God; to the discourse of whom, from the appearances of things, does certainly belong to Natural Philosophy."*

This doesn't sound very 'Newtonian' in the sense of which we think of him nowadays. Newton seemed to not embrace the worldview of a physicalist who conceives the cosmos as mechanical clockwork. A closer historical look into Newton's life reveals that he was not really a modern Newtonian. Perhaps he was so successful precisely because of this worldview, not despite it. Contrary to his own claims, historical documents show that he resorted to intuitive and inspirational thoughts, not limiting himself to a purely analytic and mechanistic intellect. This does not diminish the genius of Newton but makes it clear that the ideal of a science based simply and only on experimental facts is misplaced. This was certainly not the case with Newton and has never been elsewhere. If Newton had really taken himself seriously, probably today, his name wouldn't shine among history's greatest scientists. And if he really would never have framed hypotheses, working without any metaphysical assumptions, many of his discoveries wouldn't have been made.

We can also try to examine this from the opposite point of view: We can question whether, in those cases in which it is possible to abstain from framing hypotheses, in front of a huge amount of data and empirical facts, we would become free to see phenomena independently from our beliefs?

This seems not always to be the case. It is clear from the history of science that sometimes the contrary happens: Looking at the data without an underlying hypothesis, assumption, or conjecture can lead us into overlooking reality as it is, and we might then be unable to reveal phenomena and facts occurring in plain sight. A first curious anecdotal example of this is how Galileo observed Uranus and, assuming it to be a star, failed to become the discoverer of the seventh planet of the solar system.

The idea of abstaining from framing any hypothesis, like that of not multiplying pluralities using Occam's razor, is not necessarily an epistemological positive and fruitful approach. This belief, which holds that science must not frame any hypothesis and must concentrate only on empirical facts, betrays not only and not so much a profound historical ignorance but also especially reveals a lack of awareness of how one's own psychological and intellectual self-delusions are at work.

Let us consider two more examples, which, as is now well known, signed the last century with a decisive change in our scientific paradigm: the birth of relativity and quantum mechanics.

In principle, the theory of relativity could be brought to light before Einstein's great intellectual effort. The clear signs that classical mechanics wasn't complete and couldn't be a definitive theoretical framework for the physical world were evident. It was already known that Maxwell's equations of the electromagnetic field remain invariant under a change of coordinates only if we assume that the speed of light is a Nature's constant—that is, independent from observers. Another anomaly was the precession of the perihelion of Mercury's orbit, which could not find an explanation in the frame of Newtonian celestial mechanics. Moreover, the experiments performed in 1887 by the American physicists Albert Michelson and Edward Morley suggested that a *'luminiferous ether'*—that is, a medium for the propagation of light (as water might be for water waves)—does not exist. Great physicists and mathematicians, such as the British physicist H. A. Lorentz and the French mathematician J. Henri Poincaré, had the necessary mathematical, physical, and intellectual backgrounds to potentially formulate the theory of relativity before Einstein did. For several years, they searched for a physical theory that could make sense of the absence of ether. But despite their undeniable genius and intellectual excellence, they could not find the root of the problem.

Moved by Michelson's and Morley's experiment, Lorentz pondered, correctly as we know today, about the possibility of length contraction—that is, the fact that for two moving observers, lengths (say, the length of a ruler) appear differently. But he insisted on preserving the notion of ether and was profoundly influenced by his own electronic theory of matter, with which he tried to justify such a length contraction. Ultimately, he could not frame a convincing theory. Meanwhile, Poincaré came close to the right conclusion. He went a step further, contemplating whether something like a *'principle of relativity'*, which states that all laws of physics are invariant for different frames of reference (that is, for observers moving at different speeds), might be plausible. Unfortunately, among other things, this also implies that the speed of light is always the same for any observer, independent of the frame of reference we use to measure it. Poincaré could not get rid of a wrong assumption that we all regard as perfectly obvious due to our everyday experience—the assumption that, as for any physical object we observe, the speed of light must be relative to our own velocity. But the speed of light is always the same with whatever speed with which you try to chase it.

This is certainly tremendously counterintuitive for our view of the world. A genius like Einstein was necessary to overcome this belief. Therefore, he 'framed the hypothesis' that the speed of light might, indeed, be a constant of Nature that is always the same for all observers, independent from their state of motion. Interestingly, Einstein's conviction that the speed of light is always the same and that light does not need an ether didn't come from empirical facts, as from Michelson's and Morley's experiment. This historical version, which is sometimes also told in physics courses, is false. Einstein, unlike Poincaré, was always convinced of the principle of relativity, independently of Michelson's and Morely's experiments, whose existence he acknowledged only afterward. The principle of relativity was instead the hypotheses and background assumption he always had in mind. By accepting these working hypotheses, he could successfully create his well-known theory of special relativity.

What makes all this of interest for our discussion is the role of our beliefs and the assumptions we are working with. The only empiric facts were not sufficient for giants like Lorentz and Poincaré to come to the right conclusion, as they were working with wrong background assumptions. Instead, Einstein could reach the goal even without such empiric information because he was working with the right assumption. Moreover, the anomaly of Mercury's orbit could be explained ten years later with general relativity, which is a much more complex theory than classical mechanics and one that Einstein would never have been able to formulate by sticking to principles of parsimony.

So, the common belief that relativity was simply and only discovered from the recognition of pure facts and the observation of anomalies does not hold. Things are a bit more complex: The history of science is guided by psychological and social factors.

Did the other revolution in physics, that of quantum mechanics, have a different history in this regard? Quantum mechanics also came to life due to some strange anomalies. In the frame of classical mechanics, physicists found that a body with a finite temperature should emit infinite energy of electromagnetic radiation if a wavelength beyond the visible light frequency is considered. This was known as the problem of the *ultraviolet catastrophe* and obviously could not be correct because, as every one of us can observe, the universe did not explode in a blast of infinite energy.

We do not discuss here the technical details of the problem; it may only be said that initially, the puzzle seemed to be of an exclusive mathematical nature.

The formulas of classical thermodynamics and statistical mechanics describing the energy emission of a hot body showed inevitably and always to diverge in some way towards infinity. Many attempts were made to avoid this in order to find a correspondence between our calculations and physical reality. But they always failed.

It was Max Planck who discovered the underlying assumption that we all are naturally working with, and that has been introduced in the mathematical formulation: energy absorbed or emitted as a continuous quantity. If we let such an assumption fall and accept energy as a discrete entity—a physical quantity that can only have specific values and manifest only 'quantum-wise'—the infinities in the calculations disappear like magic, and the mathematical predictions fit very well with the observations. Today, we know that Planck's intuition was correct and that Nature works with the emission and absorption of discrete quanta of energy throughout the universe. But it might be of historical interest that Planck himself had to struggle to accept his own insight, which he continued to consider only a provisional formal solution and not an attribute of the physical world.

What all this shows is what the famous philosopher of science Thomas Kuhn called *'theory laden observations'*. What we see is laden with our own theories, unaware assumptions, and our ideological background.

Thus, the way we believe the world is made and how it functions profoundly influences our intellectual progress or, otherwise, the temporary suspension of it. History clearly shows how empiric data alone are not sufficient and how any great revolution was more than a discovery of facts, as Newton liked to make us believe. If we carefully control and reconsider our assumptions, hypotheses, and belief systems and remain open to changing them without taking them as an absolute, progress is possible. Otherwise, insisting on the idea that science must progress only and always through a crude evaluation of observations exempt from any hypotheses, conjectures, or philosophical backgrounds and abstracting from humans' natural tendency, which asks for 'how' and 'why' things come to happen, leads to a halt in normal and healthy progress and, in the worst case, to stagnation. We must become aware of the fact that if our ideal is to 'believe in what we see', we cannot escape the fact that sometimes we might not only 'see in what we believe' but also 'neglect what we see'.

There is nothing wrong with having a belief system, nurturing some unverified ideas, framing hypotheses, advancing conjectures, or even cautiously multiplying pluralities. It becomes a bad practice only when we pretend not to have any, becoming unaware of working with an ideological background or that we are sticking dogmatically to some principle. The question is not whether we can abstain from our own preferences. We can't. The question is whether we are aware of having personal preferences. It is doubtful that a science that does not proceed from previous assumptions may ever exist, and even if it does, it would probably be severely handicapped and end up in apathetic stagnation. The supposed rigorous social and historic independence of today's science does not hold at a closer analysis—especially when we try to find an explanation for phenomena like consciousness, evolution, the emergence of life, and the universe itself, which are issues highly dependent on our personal and subjective philosophical, religious, and cultural convictions. It is psychologically almost impossible to avoid subjective and personal interpretations that 'color' the facts. The hope to disclose reality as it is independent of our assumptions and human ideas, dispassionately observing the facts, is a vain chimera.

This doesn't mean that science is just a collective delusion. It means, however, that the progress of science depends strongly on our belief system. The only way out is, first of all, to become aware of these psychological mechanisms and, secondly, to begin reflecting on whether modern science shouldn't find a new basis that goes beyond a naive naturalist physicalist rationalist reductionist positivism on one side, as, of course, also beyond an even worse naive pseudo-science based on scriptures or new-age mumbo-jumbo, which stands at the other extreme.

2. Rediscovering Goethe's Phenomenology

"Man discovers the law tablet of the universe by a
'flash of memory' falling into the dark."
Wolfgang von Goethe

a. Seeing Differently

The question at this point is: How can humankind go further than where we are? If the ordinary materialist analytic and reductionist science can lead us only up to a certain point but not further than that, then what can? As a first step, two things must be kept in mind if we want to proceed from where we are.

First, we must recognize the limits of an exclusively physicalist and reductionist science and become aware that the reality we perceive with our senses and the reality that science describes are illusions. The material appearance of the physicality of life should not be ignored or neglected. Nurturing the material aspects of our lives and trying to explain it in terms of mechanical and physical processes isn't wrong. However, it becomes incorrect once it is embraced as the unique and only possible worldview. Otherwise, we fall into a one-sidedness that sees only an outward appearance, takes it as the whole of reality, and misinterprets a superficial momentary wave for the ocean as a whole. Even if we aren't, at this stage, able to directly see the whole and can only intuitively receive intimations of it or feel its vastness, the fact that we become intellectually aware of the limitations of the materialistic science, as it is actually conceived, leads us in the right direction. Being aware of not being aware is the first step towards a fuller awareness.

Secondly, and equally important, is to realize the evolutionary context in which science manifested in human history. Science is a cognitive activity of a specific species, the Homo sapiens, and because there is no reason to believe that humans are the last and ultimate lifeform appearing in the earthly evolutionary process, there is, therefore, also no reason to believe that science, as it is actually practiced, is the ultimate tool to unveil the truth of things. The human being, with its cognitive faculties, is a transitional being. Reason, rationality, and all the powerful cognitive skills that distinguish humans from animals are also only transitional, limited, and incomplete faculties for investigating the world and reality. Materialism isn't wrong as such but is only one possible way of seeing—a way that developed from an evolutionary process. It would be quite an anthropocentric attitude (and a negation of evolution itself) to believe that science is the only and final tool of knowledge leading us to the ultimate truth of things. Science is a 'species-specific' form of cognition.

Nowadays, many speak of the necessity of inventing a new science that should unite conventional science with spirituality and look for a post-materialistic change in paradigm. The word *'paradigm'* comes from the Greek word *parádeigma* (παράδειγμα), which stands for 'model' or 'example'. The philosopher of science, Thomas Kuhn, introduced the notion of the *'scientific paradigm'* to designate an intellectual breakthrough that bases our old worldview on a new and different model. The typical example is the shift from a Ptolemaic Universe, with the Earth in its center, towards a heliocentric Copernican conception.

In this sense, the theory of relativity and quantum mechanics also determined a new scientific paradigm. But the revolution occurring about four centuries ago that brought us from an Aristotelian science towards the empiric one of Galileo, Descartes, and Newton was more than a change of paradigm. It was also a methodological change: the process and the procedure through which knowledge is acquired had changed. The underlying idea of how Nature should be investigated experienced a complete overthrow. Philosophic or religious reasoning was considered no longer an essential ingredient or point of reference, and the practice of exact quantitative measurements of the phenomena has taken its place as the ultimate arbiter of the 'truth'. This determined much more than a change of paradigm; it not only gave birth to a different cosmology or confirmed a theory, it also determined the way we look at the physical realm, switching from a 'way of seeing' of the natural philosopher to the quantitative method of the modern empiricist.

Therefore, we could ask ourselves if there could also be other 'ways of seeing' that do not contradict and not even replace science but eventually complement and enrich it. Present science is intrinsically limited by the material sense-mind perception that instinctively reduces, particularizes, atomizes, and separates. Again, there is nothing wrong with an intellectual and rational activity that analyzes the whole by a bottom-up process beginning from reduction, polarization, differentiation, and separation. However, the question is whether we can equally develop a 'complementary sense' that sees things differently, namely, by a top-down vision of the whole first where particulars are not the parts of this Oneness but, rather, manifestations of it at a different level. Can we develop another 'sense of seeing' leading to another 'understanding' of the world, which transcends the mind and reason without abolishing it?

Who writes firmly believes that the answer is affirmative. Even though the ordinary human consciousness still has a long way to go before being capable of fully realizing such an alternative way of seeing and a new science based on it, we can already find intimations of this approach.

As is so often the case when we forget about our roots and the achievements of the past, it might sound paradoxical that the most interesting example hinting towards a new science comes not from a vanguard research or sci-fi vision but, on the contrary, from a distant past—that is, from the phenomenological approach of the German poet, writer, and natural philosopher of the 18th century, Johann Wolfgang von Goethe, who is well known for his literature but somewhat less for his

Fig. 112 Johann Wolfgang von Goethe (1749-1832)

attempts to found a new science that was supposed to, in his view, supplant the Newtonian conception of the world. Though he considered himself a staunch adversary of the already predominant Galilean, Newtonian, and Cartesian science, in hindsight, one can interpret Goethe's science not as something in opposition but, rather, as a complementary cognitive activity to the ordinary reductionist worldview.

In fact, readers unfamiliar with Goethe's phenomenology will probably wonder why someone should reconsider and eventually try to resurrect an over two-centuries-old approach to natural phenomena that most modern scientists consider as no more than a historical curiosity, if not a fancy imagination of a talented writer and poet who, however, is regarded as a failed scientist. Goethe's science may appear as an appendix to the history of natural philosophy or as something we should set aside in scientific considerations, as one does with alchemy, astrology, and mysticism, or that is belittled, comparing it to a geocentric or flat Earth doctrine. But a less superficial and more informed knowledge reveals how these sorts of comparisons completely miss the point. For example, nowadays, we can prove, with observations, experiments, and crisp, clear reasoning, that the Earth isn't flat and isn't even the center of the universe. Geocentrism and flat-Earth theories are simply wrong. Meanwhile, Goethe's science is not a theory about the world; rather, it is a different way of seeing it, and his description of the plant kingdom or his color theory of light has never been proven to be wrong. It remains, until today, contradiction-free.

It is just another way to observe the same phenomenon of Nature, which could go hand in hand with Newtonian science. The present way of doing science is based on dissecting, atomizing, and reducing Nature to elementary components through which, with a bottom-up approach, it explains reality as a whole. Goethe's phenomenological approach, instead, first 'senses' the whole and regards the elementary components becoming visible by a top-down approach, not as summing up into the observed reality but, rather, still as the manifestation of that very same whole. While Newton's science looks for the fundamental laws of Nature that govern the manifestation from a microscopic to the macroscopic level of existence, Goethe looked for the *primal phenomenon* (the *'Urphänomen'*) from which all other phenomena can be derived. Newtonian science looks at the quantitative aspect of the world with the sense-mind, which by mental, intellectual, rational, and analytic cognitive acts, reduces everything to elements and laws that can be described by the 'mother-science' of quantity: mathematics. Goethe looked at the qualitative aspect of the world with a more intuitive and comprehensive sight of things representing the whole by describing it in terms of primal phenomena. Therefore, the two approaches are not competing for the representation of truth; rather, they resort to two different compenetrating forms of cognition.

One might also question why, then, Goethe's approach received so little attention? There are several reasons, which we will discuss at the end of this section, once we have, hopefully, conveyed a better picture of it, though we can't do justice to it and can only furnish a superficial overview inviting the reader to resort to the specialized literature (such as [20] and [253]). Nevertheless, to furnish the reader with at least an intuitive understanding of what this alternative way of seeing is about, let us briefly outline the main principles of Goethe's approach, especially how he applied it in biology to plants and in physics to light and color.

b. Life From the Perspective of the Unity of Knowledge

As humans, we simultaneously possess sensory and imaginative seeing. Seeing imaginatively does not act as a substitute for seeing sensorially; rather, it exists beside it, meaning that we can alternate between the two ways of seeing or that we are even able to see simultaneously in a 'dual manner'. Contrasted with a regular dual vision that switches between the sensorial and imaginative seeing of the exact same object, this particular 'dual vision', which incorporates sensorial and imaginative seeing at once, allows us to view things in a complementary manner, in which our sensory vision enables us to view the separateness of distinct parts while our imaginative way of seeing allows us to view the relation and wholeness of these distinct parts. This twin way of seeing is an irreducible one. It is polar and yet integral.

The advancement of such dual vision is Goethe's manner of viewing, which enables us to establish direct relationships of a different type than that of the mechanical, material connections which the ordinary scientific single-minded vision introduces at an exclusively sensory level. Goethean science enables us to experience the wholeness of the phenomenon that adds a novel and additional element to the appearance.

As humans, we have always been aware of the twofold to some degree; it is something applied daily, yet usually, we are not aware of it. We use this dual way of seeing in our everyday communication as we speak, listen, read, or write. In fact, just like the abstract concepts we regularly deal with, we already pointed out how the objects of our conscious awareness of the world are bound together into a single and indivisible whole. This cognitive binding skill is what led to the binding problem and the emergence of meaning in the philosophy of mind we discussed in Pt.I-II.3. Our cognitive system has already a natural and innate tendency to see in universals instead of particulars.

For example, we pointed out how numerous letters come together to form a word, and in the same way, numerous words, all of which are unique, come together to form a sentence, from which emerges a unified and unique meaning. However, this hierarchical process of knowledge continues: These sentences form paragraphs; paragraphs then form sections and chapters, and ultimately, chapters form books. At each stage, we reach a new level and a 'unity of knowledge' that sees comprehensively. At each stage, the unique aspects are all viewed at once, leading to a new and broader emergent meaning that differs from these distinct aspects when viewed singularly and can't be explained as the sum of it. Even if we are mostly unaware of this process, we always view things collectively, using a 'dual vision' that relates the parts into an object of cognition that can't be reduced to the parts themselves. Recall how, also, Fichte recognized a similar cognitive instantiation when we perceive our body as a whole, not as a bundle of parts.

Goethe's scientific method is based on this *'twofoldness'* and applied to natural sciences, which enables us to view it by *'reading the phenomena in the book of Nature'*, as Galileo used to say. But, while Galileo saw this book written in mathematical terms, Goethe opposed the conventional numerical science that analyzes its constitutive components. The word 'reading' refers to addressing Nature's sense-perceptual elements in a manner similar to that in which we read letters that make up words and the words that build up sentences constituting texts. Once we have read, we have a meaningful representation in our minds of the message that was supposed to be conveyed but then forget about the single words and letters. Similarly, we should look at Nature like an open book, not by analyzing the single letters it contains but by silencing our minds and allowing it to capture the significance of the phenomena it conveyed. In contrast, modern science employs an approach that tends to examine the particularities individually and reads the book of Nature by examining the ink or the thickness of the paper while missing its content. The meaning of the sentences and words is ignored, and thereby, the chance is missed to gain a more comprehensive understanding.

An example of particular interest is Goethe's way of seeing living organisms from a perspective whose central theme is the multiplicity in unity. The way of understanding unity isn't an aspect of sense-perception. Unity arises by how things are associated with our cognitive act. For instance, we may display an image of a specific plant, yet we are unable to display the plant's unity. While the conventional manner of viewing unity eradicates discrepancies and encourages similarities, Goethe's biological unity identifies discrepancies and incorporates them, preventing multiplicity from being flattened into a dry uniformity. It avoids a disintegration of reality into mere multiplicity and enables the individuality of the specific subject to be visible inside the light of the unity of the whole. We face something intrinsic in the creature when this happens. Multiplicity is viewed in the light of unity, as opposed to attempting to obtain unity from multiplicity. It is from the multiplicity of disparities that unity is seized (*not* obtained), which is the opposite of uniformity.

Goethe extensively studied the plant kingdom and specifically showed how his way of seeing could be applied. For example, adopting the standard way of seeing characteristic of natural sciences, when we look at the leaf of a plant, the awareness of it captures the abstract notion of leaf in an instant, eventually analyzing the particulars of the sensory experience, such as focusing on the differences between the parts of the leaf or what distinguishes it from other leaves. Then an inner, almost automatic perception arises that leads to the concept of a universal that we call 'leaf'. This cognitive process becomes an active one and looks for the commonalities in the

Fig. 113 *The leaf of a Tilia tomentosa.*

particularities (something reminiscent of Husserl's eidetic reduction.) We look not just at one leaf but many and recognize some underlying commonality or invariance that is, more or less implicitly, contained in all of them. One proceeds from the many to the one. Once we have realized this 'one over many', we can go into deeper levels of abstraction with this active process of seeing, which looks for commonalities and the one feature (physical or abstract) binding them all together. An example is recognizing several leaves as all being parts of a particular tree, such as the leaf of the *'silver linden'* or *'silver lime' (Tilia tomentosa).* Then, in turn, the trees are part of a particular living organism, and all the organisms taken together are the expression of something we call 'life'. By looking at Nature in this way, we reveal the multiplicity in unity. It is about seeing the phenomena and the properties they express, looking for the multiplicity in unity not only by embracing commonalities but, to the contrary, by seeing all the differences as an expression of the diversity of the very same underlying unity.

An analogy that could clarify this is the example of the hologram. A hologram displays a 3D image from a 2D photographic plate (illumined by laser light), which can retain the 3D features of an object even if we look at it from different viewpoints (see Fig. 114). If we cut the hologram into two pieces, we won't be left with half the image, as we would expect with conventional photography; we will still be able to see the whole image, though with less resolution. Cutting the hologram again into pieces, we will always see, in each piece, the whole image but with a decreasing resolution until it becomes unrecognizable because only a few pixels composing the whole image are left.

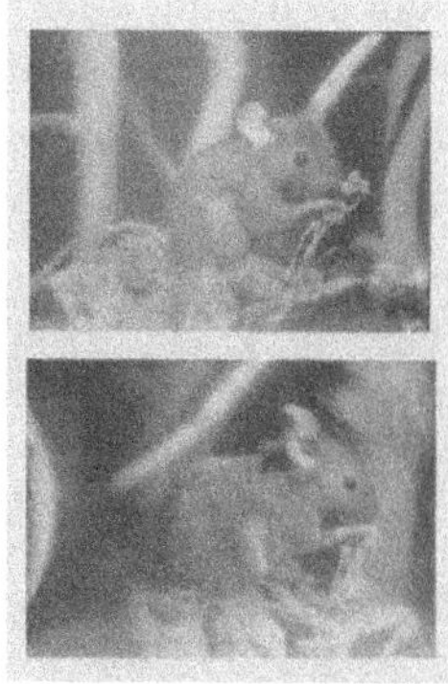

Fig. 114 The same hologram seen from different viewpoints.

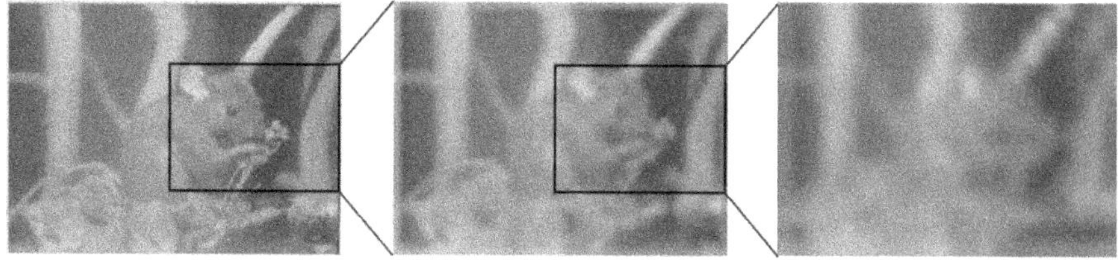

Fig. 115 The hologram and how it appears by cutting out its parts.

Fig. 115 shows how, by cutting out a piece of the hologram (two times, from left to right), the picture that remains still illustrates the whole visual field but with decreasing resolution and, therefore, decreasing sharpness until it becomes almost unintelligible. The whole is still present in its own parts but with a lesser degree of information content.

The hologram can be taken as an example that clarifies the difference between the multiplicity in unity vs. the unity in multiplicity. The former way of seeing always realizes an inherent unity *within* the multiplicity that does not break the unity, and that can express itself in the multiplicity but isn't dependent on it—that is, the whole is not the sum of its parts (each part of the surface of a hologram already contains the whole). The latter comes into being by the analytic and separative way of seeing, which binds the multiplicities and particularities into a unity—that is, the sum of the parts makes the whole (for example, the image arising due to the combination of the pixels from the surface of a conventional photographic plate).

The common scientific understanding of Nature by a bottom-up approach proceeds by an '*extensive way of seeing*' things—that is, it is a process of summation of the parts. We could, however, adopt a top-down approach that proceeds by an '*intensive way of seeing*' —that is, it recognizes the inherent unity in things that is independent of the level of distinction and fragmentation. It corresponds to a quite different logic with which the reductionist mind works. It can even lead to non-sensorial suspension of the divisive and fragmenting sense-mind activity altogether. At the bottom, this is the psychological reason why so many scientists are suspicious, if not even afraid, of adopting this cognitive act of seeing, as it apparently invites us to quiet the rational activity. But it is not about shutting down our intellectual skills. On the contrary, the silent way of seeing the wholeness at once without indulging in its parts is a cognitive state that complements and even empowers the rational and analytic mind. Recognizing the intensive dimension of the One is a powerful form of cognition that goes beyond the mind but does not abolish it.

Goethe applied this to the plant world. While he didn't know anything about holograms, he nevertheless recognized the same principle being already present in Nature. Indeed, it is a well-known fact that if we cut from a plant, say, a branch or a stem or a twig, we can grow an entirely new plant out of it. As in the hologram, the whole is already contained in each part. It is only a matter of time before the plant will regrow itself from one of its parts. There is something that contains a wholeness, and that manifests in our conscious experience, participating in it in the form of a unity of knowledge. There is something that reproduces an 'original plant', though it never reproduces exactly the same. All plants belong to the same One in the many at the same time—that is, to the multiplicity in unity.

In doing so, we don't use the empirical and quantitative divisive sense-mind but, rather, some other form of cognition that knows by seizing the indivisible unity behind the form and number. It is an effortless, intuitive way of seeing that waits for the intimate substance and meaning of the object or organism to be recognized. It is not the One made of the many that we realized; rather, it is the many as the display of the very same One.

Also, the process of reproduction can be seen in this light. When a plant flowers from another plant, it is the same One plant that forms as many plants.

Seen from this perspective, even humans could be considered as the One being reproducing itself in the multitude of humans. However, we feel as though we are separate beings, individual subjects, sometimes also very different and, not seldomly, at war with each other. Where is the unity? Again, we will argue that it is the mind that brings into the play separation, division, and numerical differentiation. Hidden in our subliminal depth, we know to be one family and, at the bottom, the One without a second. We will take up this important aspect later.

At this stage, it is critical to become aware of the fact that there are also other ways of seeing and looking upon reality. The ordinary scientific sense-mind consciousness begins with the spatio-temporal material multiplicity of life and its processes and, eventually, builds some unity by a bottom-up approach, while Goethe's way of seeing begins from the One and, eventually, recognizes its identities in the form of bodies inside a spatio-temporal conceptual structure.

For example, we already pointed out in Pt.I-IV.3, how trees use a network of soil fungi, the mycorrhizal fungi, to communicate with each other. We can't see the underground network of the tree's roots, but we can imagine it. Can we consider it as the unity of one species, where each tree is the expression of the same unity in its manifoldness? Or, to rephrase the question, is all this just allegorical, a metaphor, just poetry, or does this way of seeing perceive a more fundamental truth that we didn't notice with the ordinary sense-mind way of seeing? Conceiving, or even literally 'seeing' the whole present within its parts, may sound like a play on words. After all, we are told how all this is supposed to come into being: The sameness of the organism and its variations can be explained away by genetic inheritance. There is this belief that our DNA contains all the information encoded for assembling all the proteins necessary to build a fully-fledged organism.[60] But shifting our seeing from a macroscopical to microscopical observation of the process does not abolish the phenomenological way of seeing; it only presents it under a different perspective. The molecular processes themselves could be seen as the expression of an idea, the Idea, which is at work behind the superficial appearances of a purely material process. Looking at microscopic processes doesn't mean that we have to adopt a reductionist perspective; we are still allowed to see the dimension of the One expressing itself in a microscopic multiplicity of the many. If the reductionist mind perceives the holistic approach as a philosophical abstraction of no scientific value, the holistic mind perceives the reductionist approach (as, unfortunately, Goethe did) as a meaningless abstraction because it fails to seize the One. Instead, the correct attitude should be that of embracing both ways of seeing, not that of taking one approach against the other. Both ways of seeing are mutually complementary powers of cognition that, eventually, might lead us to a third yet unknown form of knowledge. Seeing comprehensively is not opposed to seeing selectively, but they are two possible forms of cognition that are not mutually exclusive.

Goethe went further with this form of seeing comprehensively by also investigating the wholeness of the mammals—that is, rodents, bats, ungulates, cetaceans, primates, and carnivores. In this scheme, diversity appears to be the 'self-difference' of the single organism. One can, again, resort to the duck/rabbit metaphor but thought in its many-sidedness instead of only two appearances. There exists only one rodent, bat, ungulate, cetacean, primate,

---

[60] We will see, in section Pt.III-II-5g, that, contrary to common belief, modern science showed that this isn't correct; it is a misconception that survives in the collective imagination.

and carnivore and, finally, only one mammal of which these organic orders are all a one-sided manifestation.

An intuitive sight of life reminiscent of Plato's universals: this identification is, however, contested. Plato conceived of two separate worlds—the world of sensible phenomena and a mental world of forms and universals beyond the sensory experience—while, from the Goethean perspective, there is no such distinction. On the other side, we could also combine both and see a third possibility in which the One is transcendent and also immanent, a dynamic Essence in latency in the manifestation and in all that there is. This would unite the Goethean with the Platonic way of seeing.

This new form of seeing 'sees' differences and relations in a wholistic and instant manner, it is non-spatial, and it realizes the One in a single cognitive act that does not arise due to a logical inference or a combination of elements but because of a simultaneous all-comprehensive view. It is in this wholistic manner that we should read the book of Nature. We may start from the details and seize the One but should not consider the details limiting it. Again, it is like reading a text. We see the single letters, bind them into words, and read several words forming a sentence. Its combination leads to one meaning and only a single perception of the significance that the sentence contains. But the letters, words, and sentences do not limit the semantic object. In fact, the same significance can be expressed by different words in different languages and characters. In contrast, the exact same sentence can convey different meanings, as we have shown, in Pt.I-II.3, discussing binding and the emergence of meaning.

On the other side, this does not imply that the single elements (here, for example, the letters and words) do not exist. There is no contradiction or antagonism between the two cognitive phases, that of realizing the letters and words separately first and the awareness of the meaning they convey simultaneously. The problem arises if we get stuck in the first phase: Analyzing the single letters and their relationship would cut us off from the possibility of understanding the meaning that the whole is trying to convey by it, and we will forever wonder how and why all those symbols are arranged in the sequences as they are. The comprehensive reading is not a bottom-up act of cognition; it arises due to a top-down apprehension of the meaning of the sentence, which is not contained in the combination of the letters and the words they build. The perception of the meaning of a sentence is a form of 'higher' cognition than counting the single letters or making a statistical analysis of the frequency with which the words appear or that of finding links and connections between the single elements.[61] It is 'higher' in the sense that it is a

---

[61] An aspect of our own cognition that modern AI research still struggles to become aware of, leading to the insurmountable complications with machine language recognition and translation that we described in Pt.I II-4.

comprehensive top-down insight that the bottom-up analysis will never be able to convey and yet allows for the bottom-up cognition as well—the same fits for Goethe's imaginative non-sensorial way of seeing Nature and its phenomena. The reductionist approach isn't wrong or inadequate, but it is only one possible way of seeing.

However, seeing comprehensively is not just a sensorial act. The comprehensive seeing is realized by sensorially perceiving the multiplicity in unity that stands behind and is immanently present in the unfoldment of the phenomenon itself in a subtle non-sensorial manner. This realization can't rest on the materiality of things. It arises due to something 'pre-material' contained in the phenomenon and of which the phenomenon itself is only a symbol, a figure presenting itself in our awareness due to a mental transcription. It is not even an imagination, at least not in the ordinary sense of the word. Imagination still requires an effort that takes the elements of an object of cognition and links them into a concept, image, or meaning. If we use the word 'imagination' in the context of Goethe's science, we mean an intuitive act of seeing that grasps the multiplicity in the unity of the observed phenomena or object in a simultaneous act at once. We don't construct the whole by an analytic and intellectual operation that links things; rather, we comprehensively see the Idea that stands behind the phenomenon or the symbol that the material objects represent.

Seeing comprehensively offers a complementary way of understanding evolution. Evolution is not only a process that proceeds from one species to another, building up onto the past by natural selection, random mutations, and other mechanism such as recombination, hybridization, lateral gene transfer, etc., but is also the emergence of the One organism in time and space. Speciation can be seen in Goethe's way—that is, metamorphically from the perspective of the multiplicity in unity. In Henri Bortoft's words: *"This means that the sequence is One organism and not a sequence of different organisms connected in an external way. They are different manifestations or actualizations of the same organism, not different organisms that have evolved from a common ancestor as in the standard theory of evolution. Once again, we have to turn our way of seeing inside-out. As one leaf does not transform into another one in the growth of the plant, so one kind of animal does not turn into another kind. They are not descended from one another, either directly or from a common ancestor, by procreative connection. As with the plant, what we see here is the development of One organism out of itself, which has the dynamic unity of self-difference. So the sequence is really the progressive expression of the whole itself and not one stage turning into another one. This is an evolution in the perspective of the intensive dimension of One."* [20]

The diverse species are not a fragmentation from a previously isolated entity; they are the expression of a single unity that is dynamic. This dynamism manifests in the expression of the diversity that is still a manifestation of the very same unity. The bodily entities of each organism exist; they might even be profoundly diverse and separated in space and time, and yet all represent the expression of the One that ultimately defines its dynamism. As in the hologram, the whole is in the parts, and at the same time, each part is one expression of the whole. We might say that the different plants can be seen as the modifications of a unique plant, not as separate bodily entities that were related in the past. This connection still exists; our fragmentary mental way of seeing cannot see behind the external appearances and realize *the* plant that lives in all plants. It is like our inability to see how a forest might well be only one huge single organism that, however, we don't realize because our superficial sight isn't able to capture the complex network of the interconnected roots below the surface and that, if we could imagine comprehensively, would reveal to us the single and undivided organism of which the single tree is a local manifestation.

The difference, however, is that by adopting the phenomenological approach, we don't just see material connections, like those of the underground network of roots of a forest. Rather, we also realize an indwelling unity that transcends materiality. We 'see' the Oneness in things even though the external appearances might suggest a separation and material fragmentation. The one plant is an Idea that we can apprehend intuitively and that has a concrete reality in itself. This *'plant-type'* or *'archetypal plant'* —or, as Goethe called it, the *'primal plant' ('Urpflanze')* —manifests and reveals itself to us in the many plants. Just as an object reflected in multiple ways by a system of mirrors remains unaffected and the same, so are the different plants a phenomenal expression of the same plant type that remains unaffected and the same. Nevertheless, if this immanent and inherent archetypal plant is immutable, its expression on the material plane is dynamic and observable in a variety arising from the multiplicity in unity in the manifestation.

This way of seeing is not limited to the plant world. The reality of which it makes us aware is inherent in each living and also non-living manifestation, phenomenon, and process. Ultimately, it is Nature as a whole that reveals us by a dynamic unfolding of the One. If we enlarge our way of seeing even further, we realize how the archetypal plant is itself, again, the expression of an 'archetypal life' inherent and dynamic in Nature, of which, however, we see only a cross-section in space and time due to the limitedness of the intellectual and analytic sense-mind.

This way of seeing should not be confused with a conjecture, speculation, or mental abstraction. It isn't something like, for instance, a mathematical abstraction or complicated conceptual structure. Goethe's 'seeing' is more affine to our notion of intuition, sight, or foreknowledge that captures, in a single and almost instantaneous act of cognition, a qualitative aspect of the phenomenon and that can't be expressed by a quantitative formula, a set of symbols, or whatever complicated description. It requires the contemplation of the phenomena and things, not an analytic dissection that formalizes, compares substructures, and links them together by concepts, logical inferences, or definitions. It is a sudden, effortless, and unmistakable awareness that is quite the opposite of the intense thinking of the ordinary mind.

But it isn't even the intuition we associate with the word 'instinct'. Fear, anxiety, fright, and anger are psychological instincts, and hunger, thirst, sexual drive, or uncontrollable nervous reactions of pain and pleasure are bodily instincts. The intuitive way of seeing about which we are speaking here has nothing to do with our inherited animal instincts. It is a much subtler higher-mental awareness of an inherent truth in things and life that does not arise due to an automatic and mechanical bodily or psychological instinct. Rather, it transcends these and the ordinary mechanisms of the intellectual reason. What we realize by an inspiration, vision, and insight coming from a contemplative seeing is the opposite of what we can reach by an instinctual nervous reaction. For this reason, we will never associate the word 'intuition' with the word 'instinct'.

The archetypal plant is not a poetic or artistic conception, either (although poets, artists, and even scientists may well resort to poetry or arts to express the 'impalpable'). It is the realization of a reality standing behind the superficial material appearances that manifest it. It is an ontological fact immanent and inherent in reality that manifests in its materiality to our sense-mind but presides beyond it.

We could speak of the archetypal plant or any archetypal life form and object as the intrinsic 'mode of being' that visibly manifests in the domain of our five senses. The evolutionary process seen from this perspective becomes a dynamic metamorphosis of the One that is expressing something of itself in a diversity of forms. The One, however, is in itself indeterminate and cannot be fully understood in determinates like color, shape, or other sensorial features but can, nevertheless, express itself by these determinates (sort of like Kastrup's 'qualia redux', which we discussed in Pt.II-II.3g). The indeterminate One is a self-determining entity that determines itself in a determinate plane of existence out of itself. From this, Goethe concluded that the living cannot emerge from the non-living; rather, the living determines its state of being *in and through* the non-living.

In other words, we may say that there is more to seeing than meets the eye. There is a non-physical dimension that works behind the scenes and goes beyond the mere sense perception and mental experience. After all, subliminally, we know that all too well. For example, we know that it's one thing to see a natural landscape through an artificial medium, however perfected and technologically advanced it might be, and another thing entirely to have a direct experience of the landscape on site. A video conference—however broadband and high-resolution it might be, with hi-fi sound—still lacks something that makes us aware of how meeting and knowing each other physically brings us to a psychological dimension that even the most advanced medium can't deliver. The materialist might brush aside this as an irrational and emotional aspect of our human nature that still clings to socialization (because, you know, socializing has no meaning and purpose other than being a Darwinian process with an evolutionary advantage supposed to reproduce and spread the selfish gene, right?). We contend that there is more than that and that one can become directly aware of this cognitive extra-dimension if one becomes conscious of what is seen and how one is seeing.

Investigating the physical world from a phenomenological approach is not restricted to the spatial wholeness of the phenomena; it also includes its manifestation in the temporal dimension. In fact, Goethe extended his new way of seeing to the *'movement of metamorphosis'* of the phenomena unfolding in time. It is about not only what the form represents but also what its dynamism manifests. Also, a movement, a modification in time, a change in form eventually occurring beyond the human lifetime reveal something about the nature of things. It is this process of metamorphosis that, in its temporal development, expresses a multiplicity that underlies a single unity. The archetype emerges in our awareness when we see the unity that the movement expresses through the whole series of appearances.

To understand the plant world from the phenomenological standpoint, we must shift from our reductionist and material way of seeing forms to its perception based on an intuitive temporal non-sensorial, such as a spatial holistic dynamic way of seeing. The temporal process reveals what Goethe used to call the shift from the '*Gestalt*' (the static form, shape, or organization of the object or living body) to the '*Bildung*' (the development, forming, or reorganization of the perceived object or living body). This process of transformation, change, and movement shouldn't be treated as secondary. It is part of the temporal expression of the archetype, no more and no less than its spatial manifestation in form and structure. The never-ending spatial movement or temporal modification of forms is inherently the expression of a change in itself, which transcends the material domain that expresses it only in a special and temporal universe. It is a concept of metamorphosis reminiscent of the Heraclitan statement 'everything flows' or the Buddhist notion of impermanence. For example, in the case of the plant, we should not just look at its morphological details but also participate in its metamorphosis in time (something that nowadays is nicely revealed to our senses by time-lapse videos of plant growth) *and* as a whole (for example, including the observation of the growth of its roots, as depicted in Fig. 116).

But Goethe's subtle perception of change isn't something that can be reduced to its change of forms. It is the realization of a change that transcends the physical transformation of which it is only a subsequent and not even necessary expression and of which we can become aware by a conscious participatory way of seeing the metamorphic process.

Fig. 116 The metamorphosis of the plant in time and as a whole. Credit: [254]

It is a comprehension that participates with the dynamical change of the plant, but that doesn't arise by looking only at its material dimension or by thought alone; rather, it arises by participating in the plant's action in form and time from within. We can say that, in a certain sense, we try to identify and become the plant. It is neither a sensory experience nor an intellectual inspection but a subtler cognitive perception that works subliminally and must be brought to the conscious surface awareness. It is a different type of knowledge that goes beyond the observation that our senses and organizing scientific mind can give us. It is the science that Goethe called the *'Wholeness of Nature'* and that H. Bortoft has marvelously reframed in modern terms [20].

What distinguishes this phenomenological and trans-phenomenological approach from the analytical and reductionist one is not so much conceptual but, rather, characterized by the meeting point of our consciousness with the object of investigation. While the reasoning and divisive mind analyzes the particulars of the determination of the archetypal, the phenomenological sight directly catches the indeterminate archetype itself. It is a method of cognition that doesn't stop at the materiality of things but identifies the primal phenomenon standing behind the outer material phenomenon. It isn't even an exclusively first-person perspective; rather, it is an integration of the third-person with the first-person perspective. Insisting on the third-person perspective, which only apparently abstracts from our consciousness, may make inter-subjective communication easier but automatically loses the deeper meaning and ontology of the phenomenon and the archetypal nature it represents. This is precisely the reason why the material universe appears, to the scientific mind, to be so meaningless and devoid of purpose, not because it is, but because we chose a priori to stick to a form of cognition that expunges the archetypal—that is, any meaning and dynamic significance—from the outset.

However, having said that, a living organism like a plant or an animal is not only a spatio-temporal expression of something but is also determined by the surrounding environment. This is the difference and commonality between

Goethe's and Darwin's understanding of Nature's workings. Both share the view that life is determined by the environment (the climate, natural resources, the soil's fertility, Earth's geological activity, etc.), but until recently, Darwinian evolution dispensed with any hypothesis admitting that the organism plays a role in the evolutionary process as well. For Darwin, the organism was a passive subject molded by the environment, natural selection, and genetic determinations. Goethe, instead, saw this interaction occurring in space and time in both directions: from the outside to the inside and also from the inside to the outside of every living being. Modern scientific evolutionary theories still maintain only the former interaction as the primary force standing behind the evolutionary process, with the latter having no role or, at best, playing a very marginal role. It is, however, becoming increasingly clear on a completely different scale how the human species is a vivid example contradicting this belief: Humankind is destroying the environment, decreasing its biodiversity, and even changing the climate. This is a clear (tendentially self-destructive) action of the species on the environment. Moreover, as we will discuss in Pt.III-II.5g, modern findings are now beginning to confirm Goethe's insight.

At any rate, there is no doubt that the environment plays an important role. It is here where the infamous random mutations, the laws of statistics, and coincidental environmental events could profoundly shape the environment itself and, with it, all the living organisms it contains. For example, sudden catastrophic events determined by natural climate changes, volcanic eruptions, or cosmic 'coincidences' like the impact of asteroids or comets, in some cases, wiped out almost all living forms on Earth. These facts were still unknown at the time of Goethe, but he was well aware, already before Darwin, of how his insights also imply that the archetype cannot manifest itself fully and must come to terms with the physical environment through which it wants to manifest. The archetype, immanent in the plant or animal or even in an inanimate material object, 'struggles' to appear in its purest form, which the outer conditions prevent it from fully manifesting. It is only through a different form of cognition that we can 'see' or sometimes only 'glimpse' its phenomenal expression in the manifestation. The external material appearance of this process will later be called 'evolution'.

This 'delicate empiricism' resorts to a series of experiments and experiences of phenomena that sees by a 'binding act' of particularities, parts, and relations how the organism no longer appears as something conditioned only from the outside environment but realizes how it is also driven by an internal creative soul-power. This internal 'primal-force' (*'Urkraft'*) determines its form as well. This primal force of Nature is a reflection of the wisdom of a thinking Being, of the power of a Creator. Every organism exists not only to be but also to become.

Also, what already is must come to expression through a never-ending metamorphosis that tries to express the unity of the whole and that we can become aware of if we 'ask' Nature to adopt a participatory cognitive move inside of us. By this move, we become able to recall and see in front of us the 'ideal unity' that Nature is trying to express. This metamorphosis must be seen as the development of an organization by a progression, transformation, development, a series of transitory steps of Nature. What is seen, then, is not the phenomenon or the structure but the phenomenon of the organic structure as a whole—an epigenesis of forms in an evolutionary process that reveals a wholeness both visible and invisible at the same time.

The decisive factor that allows for this seeing is, guess what, again, consciousness. Only through this subtler form of conscious cognition can we apprehend and become aware of something that the reasoning and differentiating mind alone cannot. It is a higher form of cognition that does not abolish reasoning but, on the contrary, empowers it further. Goethe openly spoke of this form of knowing as something that "belongs to a highly evolved age".

c. Light and Darkness as Primal Phenomena

Goethe also applied his way of seeing Nature to light and the color phenomenon, giving birth to what is known as *'Goethe's color theory.'* He considered his theory and his overall approach to optics in stark contrast to Newton's theory of light.

As most of us have learned in school, Newton was the first to show that, by means of an optical glass prism, one can obtain from a white light source the typical rainbow color spectrum, ranging from dark blue, light blue, green, yellow, red, and dark red.[62] Famous is his dark room experiment, also remembered as Newton's *'experimentum crucis'* (crucial experiment), in which he let sunlight shine through a pinhole in a wall, through an optical prism positioned behind the little aperture, displaying a vividly colored light spectrum—that is, the images of the

Fig. 117 Isaac Newton shining the sunlight through a prism.

pinhole itself displaced according to their colors. Therefore, Newton concluded from this experiment that white light must be composed of several different colors. White must be considered a color mixture—that is, a derivative phenomenon emerging due to the combination of the more fundamental phenomena that colors are—but it is not a color in itself.

---

[62] He did this not really with white light because it is quite complicated to produce a perfectly white source but, rather, with sunlight. These technical details, however, aren't essential here.

According to this point of view, the prism separates, distinguishes, and differentiates the colors, which, however, should already be considered contained in the white light. The color dependency of the refracting index of the prism causes the colors to be dispersed differently, thereby making them visible to the eye. This idea that white is composed of all the colors of the visible spectrum has been, until today, the commonly accepted wisdom. We get to know that white light can be reduced to the sum of the spectrum colors because the whole (white light) is conceived as the sum of its parts (the colors). How could these empiric facts be interpreted otherwise?

Goethe, however, did not agree and considered Newton's conclusion fallacious. First of all because, from a holistic perspective, one refrains from imagining the whole as the sum of its parts. From this standpoint, the multiplicity of phenomena (the colors) can be thought of as belonging together and as the manifestation of an underlying unity elicited by a primal phenomenon rather than resulting from a process of fragmentation and separation. Secondly, observing the phenomena comprehensively, in order to find the primal phenomenon that reveals it as the multiplicity in unity, means that one should look at it from all its possible perspectives, not limiting oneself to one experiment and viewpoint. We can intuit the primal phenomenon only from the set of overall experiences seen together.

In fact, his counter-argument was that Newton's dark room experiment could also be turned upside down, suggesting the opposite conclusion. Instead of considering a bright spot on a dark background as Newton did (see Fig. 118 left), if one observes through the prism a dark spot surrounded by a white background (see Fig. 118 right), then one observes a color spectrum arising from the vicinity of that dark spot as well (even though the colors appear in a different order). Specifically, if we are looking through a prism from a pinhole on a dark surface or, on the other hand, a dark spot on a white surface, we will see in both cases the same phenomenon, namely, the circles filled out with colors of the visible spectrum. [63] Therefore, so argued Goethe, using Newton's same logic, we could claim that also darkness is composed of the multitude of colors of the spectrum. Since this is something we could hardly believe to be plausible, Goethe considered this to be proof of Newton being wrong and led him to advance a completely different theory that was supposed to explain the observed phenomena.

---

[63] Readers are invited to find a simple optical prism and perform this and the following experiments by themselves. Experiencing these phenomena from a first-person perspective can be quite impressive and conveys a much more authentic understanding of them than a written description can do.

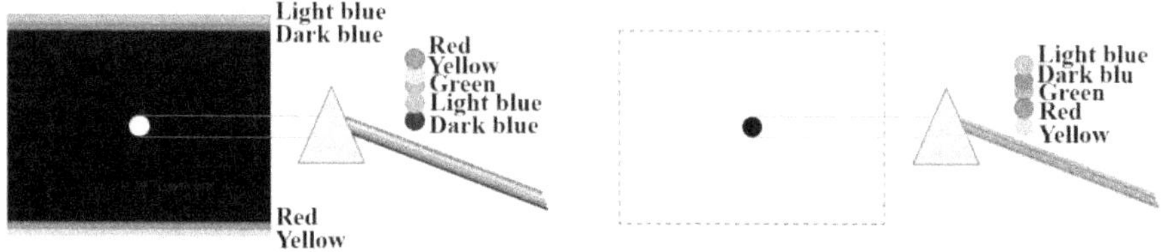

Fig. 118 Left: White spot against a dark background as seen through a prism.
Right: Dark spot on a white background as seen through a prism.

Of course, modern optics is capable of explaining the effect of Fig. 118 right (as a complicated superposition of white light rays coming from the entire white surface surrounding the black dot and undergoing a wavelength dependence diffraction at its boundaries) but, for the Goethian scientist, it is much easier to think of a phenomenality that comprises both effects—that is, thinking of colors as being the derivative phenomena of a more fundamental one.

Note that the foundation of this different theory relies on an empiric approach that isn't fundamentally different from that of modern science: a repeated set of experiments which, by precise but comprehensive observations, looks for the most fundamental aspects of it. While Newton looked for the constituents of the observed phenomenon and the fundamental laws governing it, Goethe searched for the primal phenomenon that is capable of explaining all the others—the primordial, primary, or basic phenomenon within which the manifold is to be seen and where all other phenomena can be traced back to its origin.

Thus, for the Goethean scientist, there is not much left in these experiments with the prism other than opting for light and darkness itself as being the primal phenomenon. Goethe's optical 'Urphänomen' is the tension between the polarity of light and darkness. Darkness is no longer considered simply the absence of light but a distinct and concrete entity and universal essence. Light and darkness are not two opposite and separated phenomena but dual aspects of the same entity. This duality is also suggested by the simple fact that we can't even think of darkness without light, and vice versa. One cannot come without the other, and both must be thought of as a multiplicity in unity.

Unlike Newton's standpoint from which white light emerges as the interaction of all the fundamental light colors, for Goethe, the opposite is true: White light and darkness are, at rock bottom, the only fundamental phenomena, while color emerges due to an interaction of it. In this view, darkness is as real as light, with both representing a natural dichotomy from which all the other optical phenomena can be explained.

What confirms this further is the fact that, generally, colors appear only at the boundaries or edges between light and darkness. See, for example, how the upper and lower horizontal boundaries of Fig. 118 left appear when seen through a prism and compare it with the upper and lower colored spots of Fig.

118 right. Only where light and darkness come together can colors arise through the prism: Colors appear when light is extended over darkness or darkness over light. With this way of seeing, one conceives the lightest yellow containing more darkness than white and the darkest blue (or violet) containing more light than darkness. The yellow 'radiates' into the white, and the red 'encroaches' on the black, while the light blue is 'drawn' into the white and the violet is 'drawn' into the black. These *edge spectra* display the primary phenomenon in its fullness: Colors arise only at the meeting zones of light and darkness. The same happens for a grey surface over a black or white background, even though the colors will be less intense.

Therefore, colors are not developed out of or driven from light; rather, they can be considered a phenomenon called forth by darkness from whiteness. They are a manifestation or the excitation of light (something reminiscent of the excitation of the quantum fields that represent elementary particles in QFT). Colors are *qualities of being of Nature* or, as he couldn't fail to express as a poet, the *deed and distress of light*.

By this method, one can explain all optical phenomena without resorting to the conventional concept of a refraction index of a transparent medium (the prism) associated with an electromagnetic wave (the colors). The prismatic and color phenomena can be derived even without inventing abstract light rays having diverse refrangibility. Goethe considered the modern idea of the light ray as a useful symbolic tool, a 'hieroglyph' that replaces the true phenomenon, which, however, obfuscates its true cognition. From this perspective, it is the modern, scientific, mechanical, and materialistic conception of reality that ends up having a much more transcendent character than this phenomenology, because ideal lines and calculations replace the real phenomena with abstract (almost Platonic) mathematical forms–that is, mistaking the map for the territory–what Whitehead called the science's 'fallacy of misplaced concreteness'.

Moreover, searching for the underlying principles by repeated experimentations with different optical semi-transparent media and observations, Goethe noticed how, for example, looking at a white (dark) background through a darkening (light-flooded) turbid medium leads to a yellow to red (violet to blue) coloring. The interplay between light and darkness as the primal phenomenon that originates colors can be reduced to the following:

Light through darkness → yellow – red
Darkness through light → violet – blue
Light and darkness in balance → green

Armed with these principles of his color theory, Goethe went further in applying them concretely to explain the sky's coloring. The blue color of the sky at noon and the yellow/red/orange sky at sunrise or sunset can be explained as the tension between the darkness of the sky behind the atmosphere in space (the sky in space appears completely dark; ask astronauts) and the sunlight-flooded medium (the Earth's atmosphere). When the Sun is at its highest point over the horizon, and the atmosphere layer flooded by its light is thinner, the darkness dominates light, leading to the blue (or indigo or violet) color becoming visible. Meanwhile, when the Sun is low above the horizon, and the atmosphere layer flooded by its light is thicker, the light dominates darkness, resulting in a yellow (or orange or red) coloring. The same phenomenon is also explained by modern physics, depending on the traversed thickness of the atmosphere but, of course, with an entirely different and much more complicated process that involves the scattering of light (so-called *'Rayleigh scattering'*) and the atomic absorption and emission lines of the substances contained in the atmosphere.

d. Phenomenology: A Relic or a Legacy for the Future?

"If a science seems to falter and, despite the efforts of many active people, does not seem to move from the spot; then it can be noted that the fault often lies in a certain type of imagination as to how objects are conventionally seen, in the once adopted terminology, to which the great crowd unconditionally submits and follows and to that which even thinking people escape only in individual cases."

J.W.v.Goethe [255]

However, Goethe's color theory isn't that simple. These were only a few examples, a brief sketch of a more elaborate phenomenology.

Summarizing Goethe's approach, we could say that it is about a vision in which we recover an intimate relation with Nature, between its phenomena and our sensory perceptions, and between things, plants, animals, and humans. But it would be a mistake to misinterpret the intuitive Goethean approach as non-empiric and non-rational. He was not alien to forms of precise and meticulous observation, description, ordering, and categorization. It would also be a misconception to see Goethe as averse to other forms of investigation. He accepted the other sciences like anatomy, physiology, chemistry, orthodox zoology and morphology, natural history, etc. What did come under his severe criticism was the fact that empiricism stopped there, at a purely quantitative, anatomical, and descriptive level, without an attempt to find the 'Ariadne thread' through the maze of natural shapes and figures ('Gestalten'). By doing so, we dissect life into particularities and miss the wholeness of the living beings.

He also cautioned about jumping to conclusions by performing a single experiment. One experiment, however crucial, can't be enough. Phenomena must be understood and explained only after a long series of experiments and observations that have to be seen in their entirety as an expression of a state of things that cannot be revealed by a single and limited fact. Every experiment must be connected and united with others. Only by seeing all the similarities and gentle divergences taken together can we frame a hypothesis describing the true nature of the so-observed phenomena. Nothing in the living Nature happens without the existence of a connection with the Whole, and we would incur an extremely limited if not entirely false conception by focusing on a single isolated fact without realizing the connection of all the phenomena. In his words: *"One cannot be too careful, therefore, not to conclude too quickly from experiments. For in the transition from experience to judgment, from knowledge to application, is the passing border where the inner enemies lie in wait for man: Self-suggestion, impatience, rashness, complacency, stiffness, thought-form, preconceived opinion, comfort, carelessness, changeableness, and whatever the whole host with their retinue may be called, all lie in ambush here and suddenly overpower both the acting man of the world and the quiet observer who seems secure from all passions."* [255]

It is only with this spirit, where we have to perform a large number of observations and experiments that connect with slight variations, and by adopting different points of reference that we can gain a 'higher kind of experience' of what is seen and that thereby emerges as a single cognitive phenomenon departing from a variety of natural phenomena. Otherwise, we will be prone to picking out the single observation, fact, or empiric datum that confirms our biases and false premises while leaving out all the other findings—something we believe is a quite actual state of affairs in modern science, as we amply documented with the mind-brain fallacy of modern neuroscience.[64]

With this intention, Goethe developed his theory of metamorphosis. His aim was to find the 'general typus' in complete and matured animals. We should take a higher standpoint of understanding that contemplates Nature's formative diversity and expresses itself in a creative impulse of a multiplicity of forms. It is precisely this huge variety of forms and its parts that expresses the 'One-Form' and the whole power of nature, triggering in us the non-sensorial sense of bewilderment and awe. What our 'inner sense' captures through our outer senses is the formulation of only one 'typus'. Though Goethe embraced an evolutionary perspective, he didn't see the animal as the original form of man; rather, both man and animal were the creation of the same original form invisible to the purely analytical sense-mind. It is about developing a general picture in which we take a higher standpoint where man naturally appears to be subordinated, not the peak of the natural processes. Then we can apprehend, in a sensorial and non-sensorial single vision, a general character inside an organized world of connected elements and ruled

---

[64] In fact, in Pt. I, we presented not just one fact but a whole series of facts and invited the reader to see them in their entirety. It is by this wholeness of all 'cerebral phenomena', seeing them comprehensively, that the physicalist mind-brain doctrine becomes untenable.

by a universal harmony of all the parts of a living being. This puts us into the right cognitive status through which we should observe the developing forms and phenomena of Nature.

This approach could potentially be applied not only in the macroscopic realm of life but also within the microscopic world of biological cells. After more than a century and a half of observing cells under a microscope, biologists still struggle to coherently categorize living cells. [256] Interestingly, Nature resists being confined to our anthropomorphic classifications, which may come as a surprise to materialists who seek to explain everything in naturalistic terms. However, adopting an archetypal perspective can provide clarity and coherence to these complexities.

Let us turn back to the question: why was Goethe's approach not as successful as Newton's? One of the possible answers is that, because a purely phenomenological explanation relies only on qualitative observation and description that excludes any quantitative mathematical formalization, it could not develop into a practical science, while Newton's physics lends itself naturally to a practical and technological utilization via mathematical tools. Building skyscrapers requires precise calculations to guarantee static and structural stability. Building electronic circuits would be impossible without quantification of its electric current-flows while launching a satellite into orbit can't succeed without knowing Newton's law of universal gravitation.

Moreover, Goethe's phenomenology is a fascinating science that tells us, from another perspective, how the phenomenon we observe works, but it might not look like a particularly useful methodology to predict new phenomena. Galilei's and Newton's scientific approach could predict the existence of a planet (Neptune) that nobody suspected to exist before, and led (much later) to Maxwell's equations and the modern theory of light, which predicted that visible light is an electromagnetic wave and that it even extends beyond the visible spectrum—that is, into the infrared or ultraviolet domain. Modern optics can equally account for all the optic phenomena that Goethe described, though complicated calculations are sometimes necessary—a fact that, yet again, shows that science was never driven by principles of parsimony but, rather, by the predictive and practical effectiveness of its way of seeing and describing reality. If scientists would really adhere systematically to principles of parsimony, Goethe's approach would have had to be preferred by far over the conventional, rather complicated physical models. A science that looks for archetypes, primal phenomena, or a unitary vision of Nature and reality making testable predictions has yet to be born.

And yet, we will see next that it is a science that has indeed a still undiscovered predictive power that several conventional scientists even use subliminally, resorting to a higher-mind sight without being aware of it. It is something on its way to being discovered that will show us how Goethe found the tale of the lion and was still struggling for completeness.

But the excitement for the dawn of a new philosophy of Nature, what we nowadays call 'science', was far too great to allow the subtle call of a finer cognition to be noticed. Nevertheless, Goethe's way of doing science remains a contradiction-free alternative to the present paradigm. Branding it as a mere historic remnant of a nostalgic humanist would be a superficial and blind understanding of its merits and the nature of his investigations. This complementary way of conceiving and practicing science may have to be developed further to reach the status of a convincing and viable alternative to Newtonian science. We tend to believe that it has met the same fate as the natural philosophy of the ancient Greek philosopher: Hellenic culture came very close to the modern scientific approach. Aristotle, Democritus, Aristarchus, Ptolemy, and several other ancient Greek philosophers were clearly making their way towards a quantitative and empiric science, though they were not fully there yet. Unfortunately, their tradition was discontinued and forgotten for about fifteen centuries. We dare to maintain that Goethe's approach may have suffered the same fate. It was still not a full-fledged science and had to cope with the invasion of an industrial revolution and the increasingly materialistic and sense-mind-based age of the Enlightenment, which quickly swept it aside. We don't have a crystal ball and can't predict the future but contend that the last word has still not been said.

However it is, Goethe's theory of colors is only an example of how a highly mathematized, exact science like physics also has its non-mathematical, qualitative side. Modern physics, especially the theoretical physics of the last century, is a manifestation of Nature's quantitative aspect reflected by the intellectual mind of our species. It is not surprising that mathematics is so effective in describing natural phenomena: The materialistic and analytic sense-mind, which we call reason, is in its essence a cognitive tool that deals with differences and distinctions and that discriminates between things. By adopting a way of seeing that, instead of seeing the whole, distinguishes in space and time by separating things numerically, it obviously can see nothing other than a mathematical universe. But the mind is also able to differentiate between qualities. Adopting the phenomenological way of seeing the very same reality will then present another face, that of a phenomenal qualitative universe that, if we transcend the divisive tendency of the sense-mind and see comprehensively, will reveal its unitary nature as a display of primal phenomena. The emerging unity appearing from the bottom-up cognition of the reductionist sciences, which bind and connect the pieces only by means of external relations, fails to recognize the 'relationless' Unity of which they are only an expression.

Natural phenomenology does not negate, let alone replace the former approach, but adds a new dimension to it. It is the rational reductionist scientific materialism which assumes that its own way of seeing is the only right and possible one and, more or less subconsciously, holds even the analytic mode of consciousness as that which knows the world as it is in itself as if a human's scientific way of seeing is the last word of the cognitive evolution of the species.

This new way of seeing could play the same role in the transition between natural philosophy and the empiric Galilean and Newtonian science. The latter didn't emerge due to new facts but because the human species switched to another way of seeing, to the mathematized reductionist empiricism. Similarly, a new phenomenological science may well emerge, not because of some groundbreaking discoveries of new facts but, rather, because the human species may again be bold enough to take a step further, namely, that of switching to a holistic mode of consciousness. Once we realize how the change in the way of seeing also implies a change in what is seen, this becomes an unavoidable and logical conclusion. As H. Bortoft succinctly wrote:

"Now, it is usually supposed that this is the only kind of science which is possible, and hence that any alternative to the analytical perspective can only come from outside science. But another possibility is that there could be a transformation of science itself which would be the development of a different kind of science that is not only analytical. We have seen how Goethe showed the way toward such a science and that this is not an alternative, in the sense of seeking to replace analytical science, but a way of being complementary to it. The point is that analytical science is really a one-sided development, and it is this one-sidedness which needs to be removed. So the aim is not to replace one science with another but to overcome a one-sided development that is historically founded and not intrinsic to Nature in the way that it is imagined to be. This new kind of science, which is holistic instead of analytical, is the science of the wholeness of Nature." [20]

e. Phenomenology: From Theory to Practice

Some conclusive observations about Goethe's philosophical background are necessary and useful for further considerations later.

To his own admission, Goethe was not particularly interested and skilled in philosophy. Nevertheless, he was in personal contact with philosophers, especially the German idealists like Schelling, Schopenhauer, and Hegel, who pressed him to become acquainted with Kant's ideas.

In particular, Hegel's ideas overlapped to a good degree with his conceptions. In the philosophical work of 1806, *'The Phenomenology of Spirit'*, Hegel, who embraced Kant's transcendentalism, came to similar conclusions that were already fully present in Goethe's way of seeing the world. Hegel went a step beyond the conventional notion of the mental and philosophical concept, the abstract term as an object of cognition that can be captured by a word. Beyond words, there is an original Idea. Hegel's 'Idea' is no longer a human concept but something transcending it, something that has always existed even before the world and that is *in* the world, independent from the human concepts itself. It is THE Idea, which could be that of divinity or deity and which already preceded the creation. With this intuitive way of looking at the world, Hegel constructed his dialectic method, which proceeds from apparently contradicting differences and particulars which, however, should finally be transcended and recognized at once as the expression of the Idea that represents the Whole.

Given these analogies with Goethe's way of seeing, it was no coincidence that Goethe and Hegel became close friends. However, Goethe was not at all a Kantian. He truly believed that his way of seeing could unveil the 'thing-in-itself', as if Nature unveils itself to our senses by fully speaking to us with its innumerable manifestations. And we can't understand it simply by describing it with words, abstract symbols, and mathematical quantifications but only through direct observation and contemplation.[65] Instead, Kant is well known for his claim that, whatever we do, there remains a hiatus between things-in-themselves, our perceptions, and mental representations that can't be bridged. We know the phenomenon as representative of a noumenon, but we can't know it directly. Schopenhauer tried to convince Goethe that any seeing mediated by human senses and the mind can't reveal the inherent and intrinsic nature of things, not even in principle. Nevertheless, Goethe continued to firmly maintain his conviction until his last days. That's why, in present-day philosophy, Goethian approaches to natural philosophy are considered antipodal to Kant's transcendental idealism. However, if we expand our horizon beyond the anthropocentric assumption that mind and reason are the ultimate forms of cognition, we are going to show that Kant's idealism and Goethe's phenomenology reveal themselves as two sides of the same coin.

Another important aspect to keep in mind is that Goethe was a pantheist—someone inspired by Spinoza's conception of the identity between Nature and God. The Divine *is* the eternal metamorphosis of Nature. All the permanent becoming, formation, and disaggregation of forms and organisms in Nature is the metamorphosis we can capture with Goethe's way of seeing. In this sense, Goethe's conception of Nature was similar to Hegel's view that all natural processes are a development that unveils, in time, the absolute Idea—a 'divine thought' manifested in the process of development of the forms and organisms. There is, however, a crucial distinction: The leading force and dynamic factor in Nature is an instinct, an inherent primal impulse, not a Mind or higher almighty and omniscient cosmic consciousness. There is no transcendent God—that is, a 'trans-sensorial God'. Nonetheless, his science was supposed

---

[65] The author could not forget the experience of observing, for the first time, the planet Saturn through a telescope. The blurry image that barely revealed its rings had a much deeper impact on my conscience than all the high-resolution images I had seen before from space probes. There is something that first-person observation of an object conveys to our subliminal awareness that not even the best representation can disclose.

to capture and unveil the divine process. The 'Goethian pantheist' looks for the divine Principle and Idea inherent in things, which we must learn to see, first with the senses and then with an intuitive form of refined non-sensorial 'seeing'—that is, beyond the sense-mind. It is a non-rational and sudden cognitive experience that discloses to us the creative process of Nature in a completely different light from that which our sense-mind can give us.

This insight is not of a mental nature. Also, Hegel's absolute Idea is not something emerging only through the ordinary sense-mind that apprehends and analyzes the material universe in its space-time continuum by process of reduction and unification. Rather, it is already the intimation, the first step towards a more intuitive, evolved way of seeing that is still not fully developed in human consciousness—that which we term the 'higher-mind seeing'. An intimation, a first step, is precisely only that, not the last and finished coronation of the evolution of consciousness. The failure to see this was probably one of the main reasons why Goethe couldn't find the closure with Kant's transcendentalism. And, his difficult character, together with a stubborn unwillingness to also consider Newton's way of seeing as a viable option, which led him to brand Newton as a 'charlatan', certainly did not help him win sympathies among an already established scientific and secularized academic power structure.

Moreover, his teleological worldviews weren't much in line with the scientific materialism or the religious anthropocentrism of the times and, while Europe's academia of the 18th century was largely free from the state and the church, it could still be dangerous for one's career and reputation to openly express these philosophical inclinations. After all, what is the practical purpose of this way of seeing? How can Goethe's way of seeing produce something tangible beyond a philosophical theory?

In this regard, it is worth noting the striking similarities of this understanding of Nature's workings with Darwin's theory of evolution, which arrived only about two decades after Goethe's departure. In many respects, Goethe could be considered the philosophical precursor to Darwin. Perhaps his way of seeing has, after all, some predictive power that wasn't given an opportunity and enough time to flourish.

For Goethe, the human being must be related to animals because every creature is a 'tone', a 'shading' of a larger harmony that can't be recognized by looking only at the particular. This 'shade' reveals itself as a self-evident fact if we apprehend the whole. Otherwise, the particular will appear as only a dead letter of a random stream of other symbols with no meaning. This relation between human and animal is worked out inside a (Spinozian) godlike molding process of a natural metamorphosis, with humans being the sole species able to willingly and consciously foster this process. Mankind is, in its mental nature, able to improve itself. Through this unitary sensory and intuitive apprehension, we can 'see' how all the species are related in a natural metamorphosis of increasing complexity.

Indeed, applying his dialectical process, Hegel came to similar conclusions. Hegel recognized the process of life as driven from the inside-out of the animal and, at the same time, with the interaction of the environment, with the sexual reproduction process symbolizing the concrete unity of two poles.

This anecdote alone shows how a different way of seeing, while it may not deliver certitudes and facts, as the empiric sciences can do, nevertheless could be a powerful, intuitive tool, pointing us in the right direction where we should look for further progress. Goethe and Hegel offered a hypothesis to test that humans are a descendant species, rather than standing apart from the rest of creation.

In fact, Goethe thought he had discovered what he then called the *"intermaxillary bone"*, now called the *"premaxilla"*—a pair of incisive cranial bones at the front of the upper jaw part of the maxilla that bears the incisor teeth of animals. He considered this as being the discovery of the missing link that supposedly linked humans to apes. This, however, wasn't correct. The intermaxillary bone was already known to anatomists, and it could hardly be considered a decisive disproof that could shake the still-strong anthropocentric beliefs of the times. However, the interesting aspect of this story[66], was not so much the discovery of a bone and its significance but, rather, Goethe's way of seeing that naturally drew him to look at Nature from an evolutionary perspective. It is the higher-mind way of seeing that determines what we will conjecture and then look for. This way of seeing Nature and phenomena predicts something that can be tested and explored further.

It is not at all implausible that modern science is already largely driven by an unaware non-reductionist thinking and holistic vision of life, including among those who would hardly admit it. An anecdote along this line could be that of Peter Mitchell, the British biochemist, who did not limit himself to a peeking down into the molecular details of the workings of a cell, but who also wondered about the relationship between the cell and the environment as a whole—or, to put it into Goethean parlance, decided to go the other way around and look at the wholeness of the phenomenon from a top-down perspective. He once summarized his way of seeing in scientific terms as follows: *"I cannot consider the organism without its environment... From a formal point of view, the two may be regarded as equivalent phases between which dynamic contact is maintained by the membranes that separate and link them."* [257] This holistic way of seeing led him to formulate the *'chemiosmosis'* hypothesis—that is, the movement of ions across membranes for cellular respiration or photosynthesis. The hypothesis was confirmed in the 1960s by the discovery of the *'ATP synthase'* enzyme (more on this

---

[66] For a modern and more detailed historical account, see [378].

mindboggling object in Pt.III-III.2). In 1978, Mitchell was awarded the Nobel Prize in Chemistry.

Most scientists have no idea of Goethe's science and may even react with a disinterested shoulder shrug when someone talks about 'seeing things holistically'. But approaches like Goethe's are becoming more prevalent. Scientists tend to resort to it without being aware of this.[67] Only those few who begin to perceive that there must be more than the lower-level physicalism begin to consciously apply different forms of seeing.

Perhaps one of the most dramatic and recent cases of a scientist who consciously employs a new way of seeing Nature is the already mentioned plant and marine biologist Monica Gagliano, whose research on the associative learning on pea seedlings we encountered in Pt.I-IV.3. Gagliano adamantly admits that her research approach and methodology rely on a direct inner and spiritual engagement with plants [258]. By calming the mind, creating a focused, intentional space, a meditative state which gives full attention, presence, and devotion to the plants, meeting them in their own space and dimension, Dr. Gagliano claims to receive their 'messages' and wisdom. By opening the mind to the 'voice of the heart', she perceives the plants' talking' to her and receives information that is not hers or of her mind. By doing so, she gets at what she contends to be the 'essence that is in everything'. It is an inner soul-interaction between us and the 'other' being with its environment, where mind is not abolished but, rather, has the function of following instructions from the heart. Finally, she realized that what is really talking to us in that state of consciousness is neither the plant nor something in us but the Soul that inhabits everything and is everywhere, and that materializes itself in different forms and expressions of itself through the plant and ourselves. It is a way of gaining contact with the natural world that she describes as being 'shamanic' and 'animist' (we guess, to the dismay of her skeptical and scientific-minded colleagues). It is a science without objects where knowledge is allowed to emerge in the space between the different creatures and the environment—a different approach to knowledge that is closer to wisdom.

Though it sounds like more of a 'soul-based' way of seeing Nature, rather than a higher-mind science, in Pt.III-I.4e&f we will see how this confirms an evolutionary cosmology in which the soul and higher states of consciousness beyond the mind are connected and appear as the two poles through which knowledge of the world can come. Seeing with the higher mind and soul-based wisdom are the two aspects of a 'masculine-feminine way of seeing', so to speak, which are not exclusive but, to the contrary, are complementary and even necessary to each other.

---

[67] An example of this can be seen in this YouTube video: [380] . One could call this "Goethian microscopy".

As far as we know, Dr. Gagliano was not inspired by Goethe. However, her approach is certainly very much in line with his way of seeing and does not lure her into abstract imaginary realms where she might 'lose her mind'; on the contrary, it led to several scientific experiments with peer-reviewed publications. Hers is a way of seeing that isn't really new, but that has been forgotten, perhaps for too long. These instances of scientists who more or less subliminally resort to these new forms of knowing are signs of reemerging in the collective consciousness.

We might see a comeback of a way of seeing Nature not too dissimilar from that which was already present in indigenous cultures and has been branded 'primitive' by a purely technological culture of 'progress'. It is a conception of Nature from which the environmental movements, which are too fixed on a scientific and technological resolution of the ecological crisis, may have much to learn.

For example, in the native American culture, there is no disconnection between humans and the land or between natural phenomena and life. Humans are seen as part of Nature, not something separate from it. In their philosophical system, there persists a relationship with the places, a deep regard for Nature's cycles, and natural phenomena are seen not solely from a mechanistic standpoint, but as the expression of consciousness. Mother Nature is not an object but a sacred and unifying spiritual Entity—an Earth Goddess of creation—whose power runs through all living beings and non-living places. Several indigenous tribes still nurture a sense of responsibility for the safeguard of the environmental resources, which, in their view, should not be owned or privatized by the few to the detriment of the many. Nature's balance must be kept because she is a Spirit that we need and can teach us. It is a sustainable culture and a post-materialistic ecology that values biodiversity and inclusion in a system of reciprocity in which we are not seen as isolated beings that can take from the land without allowing for regeneration. It is a cultural and spiritual connectedness with the environment meant to protect future generations.

Despite the various efforts to eradicate their culture, to say nothing of the genocidal attempt to eliminate them physically, they still hold a connection to the land and a sense of unity with Earth and Nature.

Their animistic worldview—which sees in landscapes, trees, plants, and all natural phenomena an indwelling spirit, with unique powers, characters, and personalities that speak to us—may look like an obsolete and superstitious belief system. But how much does this perspective differ from the modern biologist speaking of an indwelling intelligence in plants and cells? How far is this animistic faith from the contemporary philosophy of science that returns to a panpsychist or cosmopsychist view? How different is the indigenous idea of a symbiotic interconnectedness between humans and the natural environment where everything is related to everything else from a '*Gaia theory*' that conceives of Earth as a synergistic, self-regulating, and complex system?

An example of the wisdom of native American peoples that unites the analytic perspective with a spiritual connection of humans with Nature could be that of the figure of the native American Professor of Environmental and Forest Biology Robin Wall Kimmerer. Her leitmotivs are ethical, moral, and spiritual principles of sustainability and reciprocity as inescapable ingredients necessary to regain respect, harmony and balance with Nature [259]. People and the environment are not seen as two separate entities but, rather, as a whole in which both live

Fig. 119 Robin Wall Kimmerer © John D. and Catherine

a mutually beneficial relationship. But humans still regard themselves as separate from Nature and are busy primarily in taking without giving in return. It is this disparity that an indigenous sustainability practices can heal–a sustainability through reciprocity that will determine the shift in consciousness beyond a purely technocratic understanding of the environment and towards a more comprehensive view of our mutual relationship with the planet. It is an inner contact and heart-centered worldview of indigenous agricultural practices as an act of kindness in which one not only takes but also gives back resources to the land and acknowledges the mutually beneficial interconnectedness of all life. A system of values and a wisdom that should be passed to the next generation –something that our contemporary parenthood, immersed in an exploitative culture and economy, doesn't care much about.

At any rate, without falling into the temptation of romanticizing indigenous cultures, we contend that it may well turn out that history will flip destiny upside-down. The scientific mind, which is based on a reductionist and anthropocentric analytic way of seeing Nature, will turn out to be the primitive and barbaric ones along with a rapacious capitalist consumerist society that sees the environment and its natural resources as an object to exploit. Meanwhile, the millennia-old traditional indigenous spiritual ecology and its local economics with a non-anthropocentric environmental culture are based on principles of sacredness as part of a more integral philosophy and spirituality, revealing itself as the path towards a future of real progress.

Why, then, did this ancient wisdom fall apart and become almost forgotten? This wisdom wasn't about romantic ideals of Nature. Everything indicates that the Goethean way of seeing, the heart-based perception of life, and the indigenous connectedness to the land are just a few remnants of a lost state of consciousness. It was a completely different state of being and perceiving Nature and reality, which went beyond an intellectual understanding or a refined emotional feeling.

There are good reasons to conjecture that it was once a common state of awareness that already went beyond the present human cognitive abilities but was lost and left behind along the path of the evolutionary process and of which the indigenous culture retains only a fragmentary but vivid remembrance. There may have been a prehistoric age when humanity effectively lived in a perfect state of harmony with Nature and itself, having acquired a higher and non-mental spiritual consciousness. From anthropological studies, we know of rare but still existent indigenous groups with an egalitarian, non-patriarchal, and decentralized social structure without competition, leaders, hierarchies, or wealth disparity. That's how we humans may have been for 95% of our time on the planet before agriculture came along.

Perhaps precisely because this spiritual awareness lacked a full development of the mental pole, a relapse was necessary. Humanity had to learn other lessons before it would be able to synthesize spirit with mind and go beyond both—a relapse that may have been triggered 100,000 years ago when the last glacial period began and subjected a vast part of the prehistoric human community to the need for blank survival issues which couldn't be dealt with other than by resorting to the skills of the mind. The species had to fall back to what might appear to us to be an inferior evolutionary step but one that was necessary to recover what it left behind. The biology and the brain of the species were fully developed, but the mental consciousness still didn't shine through. Mind was not yet established and had to grow by defocusing from an inner spiritual life favoring an exteriorized and material consciousness to allow the mental component to emerge.

We are still in this evolutionary phase but have come a long way and are almost ready for a further leap and achieving the great synthesis. This understanding of the human adventure of consciousness on this planet would plausibly explain that strange and particular human dichotomy where its most beautiful achievements coexist with its abysmal perversions.

3. Speculations on a Future Integral Science

However it is, Goethe's approach was an attempt to build an integral view of reality that, more or less subliminally, was driven by an impulse to transcend the sense-mind inside a unitarian perception. It puts the dominant Newtonian approach upside down: Not the derived phenomena (for example, colors) should come first and all the others construed from it (such as white light); to the contrary, the primal phenomena (for example, light and darkness) should be the base from which everything else comes (such as colors being a 'tension' between the two). Misapprehending and replacing the primal phenomenon with the derived one can be fatal, as he wrote in his color theory:

"The worst thing that can happen to physics, as well many other kinds of knowledge, is treating a derived phenomenon as the original one and, since

one can't deduce the original from the derived phenomenon, one seeks to explain the original from the derived. Hence, an endless confusion, verbiage, and a constant struggle to seek and find subterfuges anytime the truth emerges and wants to impose itself arise" [260]—a prophetic description of present-day materialistic naturalism.

But where do we go from where we are?

A new science that takes into account the idealistic worldview and, eventually, goes even beyond that by integrating the spiritual component of the universe could be called an *'integral science'*. It will consider not only the external but also the inner world. It presupposes the knowledge of the material as also our inner being, resorting to both the third and first-person perspectives.

However, it can't be just the knowledge and union of the present materialistic conception of science together with some idealistic or spiritualistic worldview. Putting together the sense-mind science with a spiritual investigation won't work. That alone would not change the dominant modalities of the physical sciences, which are based on an almost exclusive mathematical and reductionist empiricism and bottom-up theoretical models.

Accepting and uniting the physicalist with the spiritualist points of view might be the wrong path to go and dangerous. Many spiritual-minded people have not yet developed a full-fledged analytic mind, while those scientific-minded usually have not yet established contact with their own inner spiritual natures. The attempts to cross their respective boundaries, by the former, usually results in a sequel of pseudo-sciences or, at best, eccentric spiritualistic-philosophical constructs, while the latter did not perform much better with naïve and bizarre attempts to drag down the spiritual into a materialistic and rationalistic conception. We must come to terms with the fact that before attempting a false synthesis, mind and spirit must both be conquered and establish themselves in the being. The easy shortcut to accept and unite both standpoints has been attempted several times but didn't lead us much further. Despite all the crises and facts that should have led us to a change of mind, materialism is still alive and well in its full glory and remains the dominant paradigm.

Eventually, most people who talk about a 'spiritual science' envisage a conceptual but not real methodological paradigm shift. This sometimes goes so far that a 'spiritual science' is considered to be just the ordinary scientific approach that is supposed to confirm the spiritual experience. Examples could be neuroscientific data of meditative states, medical assessments of the benefits of spiritual practices, new physical theories supposed to prove the existence of so-called 'paranormal phenomena' or 'extrasensory perceptions' or subtle matter and transcendental worlds, etc. Though we believe serious research on the paranormal exists and is worth attention (for a summary see [261]), we don't consider this as being sufficient. The path towards a post-material or integral science or anything going beyond the present established conventional scientific practice needs more. These attempts to connect with something that doesn't fit into the orthodox materialistic framework are a first step but, still, aren't essentially different from the traditional approach, being only an extension of the empiric domain of investigation. It is the object of study that is enlarged, but the method and way of seeing remain substantially the same.[68]

A modern trend among some spiritually-minded people who speak of 'uniting' science with spirituality tries to explain the mystic experience in terms of brain states. Recent research investigating how a specific spiritual practice correlates with brain areas lightening up or that shows how neural synchronizations correlate with spiritual wellbeing could be used to develop some therapeutic methods to cure stress or psychological and mental disorders, but can hardly lead us beyond the present state scientific way of seeing and doing. In Pt.I, we amply elucidated why any form of third-person empiric 'neuro-philosophy' is doomed to failure. But what is worse is that the modern tendency to explain spirituality in terms of brain states, and even purporting it as a new 'spiritual science', subconsciously nurtures and continues to keep alive the physicalist narrative. It strengthens in the collective a neuro-centric image of our spiritual nature. A 'spiritual science' of this kind only reduces, again, everything to what 'spiritual' is not. Deep down, it still believes spirituality is only a 'neural phantasy'—that is, the opposite of what it claims to be inspired by.[69]

Most of the so-called 'spiritual science' approaches are an attempt to rationalize the spirit. This is not what we are trying to put forth here. Eventually, it should go the other way around: One should instead spiritualize rationality.

A post-material and integral science must have an aim to also change the methodology—that is, the way we investigate and see things. It is something else, a third position that is not the result of simply combining mind with spirit. That's the reason why we dedicated an entire section to Goethe's phenomenology as a possible (though undeveloped and yet to be established) paradigmatic example.

The distinctive sign of such an integral science will be to include the subjective approach and not limit itself to a superficial quantitative knowledge

---

[68] That is the reason why we didn't consider studies on the paranormal as, perhaps, some readers may have expected.

[69] Research that tries to reduce spirituality and religion to a cerebral secretion, or a mere genetic trait supposedly having an evolutionary advantage, is abundant throughout the scientific literature. Curiously, however, no one ever felt it necessary to consider also the opposite hypothesis: If materialism and atheism could be traced back to similar neuronal or genetic mechanisms as well. This speaks volumes about the unaware premises and biases of modern neuroscience.

but, instead, widen its awareness of the qualitative and the ideal aspect and dimension of reality via a higher cognitive function together with the subjective inner domains of our being—a giant leap from physicalism towards a much more comprehensive knowledge that could be compared to the leap that quantitative empiricism of the Newtonian sciences made with respect to the Aristotelian natural philosophy, and beyond.

In the last decades, we assisted in an outburst of countless writings, conferences, and discussions about the relationship between science and spirituality. Why all this sudden interest in such matters? Is it only a momentary cultural fashion, or is there more behind it? Some argue that this is due only to uneasiness and a lack of certitude that science could not deliver us. It is only a psychological, irrational reaction that manifests in the form of an unreasonable return to surpassed superstitions and beliefs that help us overcome our fear of the future and death.

Instead, we contend that such a superficial analysis doesn't capture the essence of the real reasons standing behind this state of affairs. It is time to interpret the roots of this growing uneasiness also as an inner evolutionary psychological and collective social process manifesting itself in an awareness that slowly comes to the surface. We are becoming, directly or indirectly, aware that things are not as they appear to be, as mind and reason, with its rational sense-mind consciousness, so convincingly suggests to us. At the same time, an inner source indicates to many that, somehow, something is wrong with our usual scientific naïve realism—a pseudo-realism that has fallen prey to the illusion that 'explaining' the processes of Nature means knowing what they are. This is a delusion typical of the mathematical-physicalist mind.

For example, knowing everything about the human anatomy and the biochemical processes of the body and its organs abstracting from the inner life and the psychological and spiritual dimension of an individual, and eventually even denying that there could be any, separates us not only from ourselves but from the universe as a whole. This is the self-deception that lured us into the delusion that knowing the 'how' also means knowing the 'what' and 'why'.

But this fallacy is increasingly becoming evident under the pressure of the growth of the collective consciousness. We must look beyond and learn to see how physical processes unfold themselves beyond their superficial physical appearances–that is, not only by knowing by which mechanism but also for what and why these mechanisms stand. It is about developing an 'intuitive sight' that looks behind, below, above, and especially beyond external physical facts. It is about two spheres of comprehension and apprehension. One is mediated by our senses and the intellect, the domain of physical sciences, and the other pertains to the non-sensorial sphere of intuition, inner knowledge, subjective perception, and spiritual sight of the mystic. Are these two spheres forever condemned to remain separated, or might there be some way to bring them into a 'third dimension' that encompasses them both but, again, without being a simple summation of two polarities?

Science is born as a natural outgrowth of the rational skills of the human being who consequently discovered the existence of a rational order in Nature. There is no reason to think that when we can ascend towards a higher trans-rational sight, it won't become possible to discover the existence of a trans-rational order in Nature as well. The physicalist sticks to the external physical aspects of Nature, not for a rational and scientific reason but because it sees only those aspects and denies other possible perspectives from the outset due to a lack of inner contact with their own inner spiritual nature. Science, as it is actually conceived, is like a one-eyed vision without depth. It looks at the world with a two-dimensional vision and doesn't see the third dimension; it even denies that this dimension could exist. Looking at the world from multiple perspectives is considered an unnecessary complication that calls for Occam's razor.

Once contact with our inner spirit is established, the appearances of the physical world will progressively disclose themselves as an undeniable aspect of that very same spiritual cosmos. Once we have entered a conscious relationship with ourselves, the spiritual dimension of things will present itself as a self-evident and obvious fact, no less and even more than the existence of the physical world. The sharp division between spirit and matter will diminish until it appears as having no fundamental reality. Both are two aspects of the very same thing.

This would unite the original rigor of the mental sciences, avoiding the limits of reductionist and positivistic instrumentalism, with the thirst for a more profound knowledge of our place in the universe and the meaning of our existence. On the one hand, it would allow us to avoid falling back to untenable religious beliefs or superstitions and, on the other hand, prevent us from sticking to a concept of a giant mechanical, meaningless, and dead universe that popped into existence just by chance.

However, it wouldn't be mere philosophy or another abstract conceptualization. Rising to a more comprehensive vision and perception of reality would automatically imply a completely different relationship with the world and our practical doing and behavior. An example could be how our deeper understanding of Nature would lead us to a much more respectful relationship with it. We wouldn't see the solution to the ecological crisis only in terms of material and technological measures (reduction of CO_2 in the atmosphere, incentivizing solar and wind energy, electric cars, new economic models, etc.). These measures alone, as necessary as they are, won't solve the problem if we continue to think of our relationship with Nature only in materialistic and technocratic terms. That won't stop our egoistic and rapacious animal lower nature from continuing to cause harm to the Earth's environment and, ultimately, to ourselves. The ecological problem also reflects our inner spiritual disconnect from Nature but also, first and foremost, from ourselves.

The environmental issue is only one example among the many that one could make that exemplifies the divide between the sense-mind and spirit. The same could be said about the economy, international relationships, education, health, and medicine, etc. (some of these aspects we will take up again later in Pt.III-III). We might even go so far as to claim that what is at stake is the survival of humanity as a whole.

One of the fundamental differences between us and the animal kingdom is our mental and intellectual ability. The difference between animal and vegetable life is not only the different states of consciousness but also the development of new natural sensory means in the animal that the plant does not have. What would happen if we acquired new senses? The creation of artificial sensorial devices such as telescopes and microscopes changed our worldview: It opened new worlds of knowledge to humankind. However, new sensory means and the change in the morphology of a species alone do not exhaust the evolutionary process. Also, a new awareness, a different form of cognition, and a new level of consciousness appeared.

Therefore, if we accept that evolution did not come to an end, it is logical to conclude that this ascension will also be set forth with respect to both our present state of consciousness and sensory perceptions. This new consciousness and sensory tools may put us in contact with realms that we cannot imagine or understand or that we even refuse could exist. The first intimations of this came from the mystical experience. But transcendence is not the whole thing either. We are looking for something that synthesizes materialism with spiritualism, the sense-mind with a higher-mind, matter with spirit.

Therefore, as the appearance of mind gave us reason and the intellectual knowledge that expresses itself in the different sciences, so the appearance of a more comprehensive and evolved consciousness may express itself in an integral science. The first steps necessary at present can lead us to the awareness of the limits of science and see where we can go from there. But before a new scientific paradigm and methodology may become a reality, a new education in science, in the philosophical practice of mind, in the way we contemplate Nature, and in the actualization of psycho-physical techniques will be necessary. The future scientist will no longer be someone having only an intellectual preparation in a scientific discipline but will also be someone who has gained—or at least aims to gain—a higher state of consciousness and more developed means of intuitive sight, subtle perception, non-sensorial inner participation with the phenomena.

We do not need profound mystic realizations to do this. Possibly only an initial inner development, through the emergence of a subliminal awareness, will automatically manifest outwardly. If a sufficient number of scientists would realize the narrowness and limitation of the materialistic approach and would open themselves to only slightly higher levels of consciousness and the limited reasoning mind, this alone would presumably be sufficient to cause a revolutionary change not only in science but in any other fields and our entire material life as well.

The vast majority of the scientific community simply avoids these sorts of discussions, saying that science's duty is not to discover the profound nature of things but to simply limit itself to providing a set of rules that predict an experimental outcome. In other words, they retire to some form of pragmatic *'instrumentalism'*. From this perspective, Nature should be regarded as a 'black box': We know what will come out of it when we feed it with a specific input. To question the real nature of this 'box' is useless and unscientific, per definition.

That would be fine if they would stick to this limited desire for knowledge. However, when asked specifically but indirectly, most of these scientists are more or less unconsciously convinced that they know what reality is, as it is. The unaware assumption that what science represents is the Kantian 'thing-in-itself' always lurks behind (though most scientists have no idea what Kant's transcendental idealism is about). The instinct to get to the roots of an ontology in terms of particles and spatial-temporal primitives resurfaces regularly.

Also, the quests about the foundations of science and the natural philosophy from a trans-physical or even spiritual perspective have reemerged, and with new, intense vigor as time goes by. An example of this could be the debate between a mechanistic neo-Darwinian conception of evolution and one that includes teleological considerations. This debate never ceased; to the contrary, it is gaining momentum. Many still tend to interpret this cultural renaissance, which seems to no longer be satisfied with a superficial description of reality and searches for the deeper meaning of the material world, as an emotional outburst or, worse, a return to religious beliefs or as an unconfessed desire and hope to discover that there is more than matter—that is, life after death, etc. But these sorts of objections are often used too easily and are based mostly on logical claims and subtle metaphysical assumptions whose foundations are at least as doubtful as their opposite claims. Upon closer scrutiny, it always turns out that the physicalist arguments are based on rational prejudices and subjective modes of a perception whose burden of proof isn't less necessary than that of any metaphysical claims.

However it is, the grounds on which classical science walks are becoming unstable and may become even more so if the holistic nature of the universe continues to be denied, eventually even under the pressure of more solid experimental evidence, which indirectly already suggests, with some experiments in psychology and biology, non-local phenomena in quantum mechanics, as in chaos theories and synchronicity in biological ecosystems. The discovery of the 'Oneness' can presumably cause the most disrupting crisis that the reductionist mindset might have to confront within the next decades.

Whatever the attitude will be, an inner change with which we are looking at the world is clearly visible on the horizon and is producing an increasing number of personalities that are no longer satisfied with this dry understanding and representation of reality. Some physicists, as well as other scientists, are beginning to understand that asking the subtler and deeper 'why' and 'how' questions about material phenomena is not a waste of time but, instead, can lead to interesting and sometimes definite breakthroughs in our understandings of the universe we live in. Something seems to drive the human race inwardly; there is in us a force that compels us towards a new perception, a new consciousness. Many perceive the real necessity of a new science that goes beyond the controversies of monism vs. dualism, reductionism vs. holism, materialism vs. spiritualism, scientific realism vs. illusionism, etc. A 'third thing' is needed where any form of duality vs. monism disappears into a position in which the materialistic and non-materialistic views merge into a single entity that contains them both as the approximation and manifestation of a central and more fundamental integral truth.

What an integral science will really look like may be too far from our intellectual understanding, but we can see today that science is moving with more insistence towards a direction that may reveal itself to have no return. Many scientists show clearly the beginnings of receiving, more or less consciously, inspiration and insight from a higher consciousness, a Knowledge that 'shines' through the mind but does not have the mind as its source. Some of them are searching for new approaches and ideas, or look at the world differently, trying to capture the undefinable that they can't even explain themselves but clearly feel a drive towards something else. We are beginning to feel and perceive the first intimations of higher states of consciousness, though without being directly aware of it and without understanding what it really means. It is a sign that signals to us how it is time to proceed towards a more evolved and comprehensive way of seeing the world. It is this new thing that is slowly emerging onto our evolutionary path, and that creates both discomfort and excitement.

An integral science will become a primary force that can help, serve, and eventually boost the evolutionary process of life towards an expanded awareness. The ultimate goal is to harmonize the essential truth-seeking nature of mysticism with the materialistic scientific attitude and go beyond both. An integral science is no longer satisfied with a symbolic description and the predictability of dry empiricism but will long for a wholistic knowledge becoming intimate with a dynamic reality of things and the world.

We believe that, in the long run, it will develop into a search for the truths within the world-phenomena and which can become aware of its creative significances—something that can become one with a spiritual essence pervading all things and their processes and that will reveal the purpose and meaning of ideal forms of which we and the cosmos are an expression. This is a new seeing beyond a mentalized reality that also sees the subtler forces and principles dwelling in creatures and objects through an 'inner eye'. In short, it is a discovery of the workings of a divine consciousness in everything.

Whatever future approach reveals itself as the new conceptual and methodological paradigm, it must be a new science that can overcome the apparent irreconcilability of matter and spirit through a transcendence and immanence that does not deny materialism but, to the contrary, recognizes it as an integrating part of its own essence. It is no longer a blind physicalist and mechanical conception of the universe and ourselves, but neither is it a mystic escape into an insensible transcendence good for only a few. We had to live and experience these extremes, going through both of them and even with both negating each other before we could reconcile them. Extreme and opposite experiences helped us stretch our faculties in both directions. In hindsight, it becomes clear that it was an evolutionary necessity to develop the species' consciousness. The time has come to recognize this trick of Nature, step out from it, and synthesize these two poles of existence in an advanced form of life and knowledge.

Part III
A Synthesis of Knowledge

"All now seems Nature's massed machinery;
An endless servitude to material rule
And long determination's rigid chain,
Her firm and changeless habits aping Law,
Her empire of unconscious deft device
Annul the claim of man's free human will.
He too is a machine amid machines;
A piston brain pumps out the shapes of thought,
A beating heart cuts out emotion's modes;
An insentient energy fabricates a soul.
Or the figure of the world reveals the signs
Of a tied Chance repeating her old steps
In circles around Matter's binding-posts.
A random series of inept events
To which reason lends illusive sense, is here,
Or the empiric Life's instinctive search,
Or a vast ignorant mind's colossal work.

But wisdom comes, and vision grows within:
Then Nature's instrument crowns himself her king;
He feels his witnessing self and conscious power;
His soul steps back and sees the Light supreme.
A Godhead stands behind the brute machine.
This truth broke in in a triumph of fire;
A victory was won for God in man,
The deity revealed its hidden face.
The great World-Mother now in her arose:
A living choice reversed fate's cold dead turn,
Affirmed the spirit's tread on Circumstance,
Pressed back the senseless dire revolving Wheel
And stopped the mute march of Necessity.
A flaming warrior from the eternal peaks
Empowered to force the door denied and closed
Smote from Death's visage its dumb absolute
And burst the bounds of consciousness and Time."

Sri Aurobindo – Savitri, Book I, Canto II

I.   Towards a New Consciousness[70]

A short preamble to this third part might be necessary.

Readers acquainted with other 'integral', 'holistic', or 'universal' theoretical frameworks may recognize some similarities between the integral cosmology we are going to present here and the *integral theory* of American philosopher Ken Wilber. There are, indeed, similarities and overlaps, due mainly to the common source of the integral yoga of Sri Aurobindo. But the approach, viewpoint, and purpose of the present work are quite different. Wilber's integral theory is more concerned with developing a comprehensive representation and conceptual organization of all human knowledge in a *'four-quadrant grid'* with its 'levels' and 'lines' of development, its different forms of intelligence and states of consciousness organized in a structural framework of developmental psychology (the *'pre-personal'*, *'personal'*, *'trans-personal'*). Here, instead, we will not attempt to establish yet another model, conceptual organization, or 'structure' other than adopting Aurobindo's cosmology as it is and interpret modern scientific knowledge, the past and present spiritual philosophy, and the philosophy of mind from its perspective.

Our task is to develop a post-material synthesis of knowledge that already works with a given 'grid', namely, the integral vision of Aurobindo, which allows us to 'see' the universal phenomena, evolution, consciousness, physics, cosmology, life and matter, the human's social development, etc. from that spiritual perspective. In our opinion, it is 'integral' and 'trans-rational' enough and doesn't need further modeling, mental reorganization, or intellectual extension. We will emphasize much more the 'way of seeing' things, bringing into our awareness the forces standing behind the phenomena, rather than trying to build another conceptual organization that is supposed to contain and systematize them all. Though a theoretical framework with its conceptual background will also be necessary in the present context, our aim isn't to build another 'integral theory' rather an 'integral seeing'. It is in this spirit that the following must be read—not an elaborated intellectual systematization of knowledge leading to a grand scheme, but a synthesis of knowledge that resorts to a higher-mind way of seeing that reveals itself spontaneously as a grand vision.

1. The Question of Meaning and Purpose

Many are growing towards an inner awakening and beginning to perceive the world with awareness from a spiritual perspective that is slowly changing

---

[70] If you are looking for an overview of this third part I published a short version (in a long essay) here [393].

their lives. In most of these people, this inner awakening is a source not only of new discoveries but also of dissatisfaction, disillusionment, and even suffering. To live an externalized life, which our modern society has developed to its extremes, perceiving it from an inner standpoint, can cause tension between what is and what we feel, between what could be and what is not, and ultimately between what we see and what we know.

In particular, the quest for meaning, purpose, and significance becomes more than a philosophical musing. It becomes something very concrete and tangible that necessitates an answer with an urgency we have never felt before. It is something that science, religion, or philosophy can no longer give us. The physical account of the reality that modern materialism furnishes is no longer satisfying. Physicalism proudly presents us with a meaningless and purposeless universe as a given a priori. It is proud of its purely materialistic and soulless ontology because, in its blindness, it misapprehends this attitude as proof of intellectual and analytic maturity without being aware of its anthropomorphic belief in reason as the ultimate arbiter for truth. Asking for meaning and purpose is considered an old form of superstition that the analytic mind should overcome. Any temptation for a teleological conception of reality must be ignored.

This temptation, however, strangely arises in us continuously, almost with an annoying constancy. Even the hardcore materialists must frequently control themselves by not falling into a language or use of concepts that might inadvertently indicate a purpose or a goal in Nature. According to the naturalistic narrative, whatever facts and discoveries are presented to us, we must dismiss this tendency, as there is no direction in evolution, there is no meaning in our existence, everything is just a giant cosmic machine without meaning and purpose that must be explained away by a *'blind watchmaker'* mechanistic worldview. We must permanently 'mind the gap' and avoid falling back into more or less implicit formulations that might hint at anything that is not in line with this conception.

Especially in biology, this sense of 'agency'—that is, the ability of microscopic living entities to act as if seemingly driven by what appears to us as a purpose that suits an agenda—or the evolutionary theory itself based on concepts like 'advantage', 'adaptation', 'fitness', or 'survival' are already words relying on an implicitly goal-oriented semantics. But any finalistic interpretation is usually dismissed and brushed aside, declared as simply an annoying human tendency to anthropomorphize or a linguistic misconception we should avoid at all costs. Because this narrative tells us that cells aren't really 'trying' or having an 'intention' to do anything, organisms don't evolve 'in order to' achieve something. We should never forget that everything *must* boil down to genes and molecules, chemistry and physics, and any description of events must rely on a mechanism that unfolds with no aim or design, though appearances seem to suggest persistently otherwise.

However, if this idea and premise that this universe, with all life and all its creation, is just a void and senseless machine would be correct, then this raises the question of why it is so hard to get rid of the opposite feeling. If there is no meaning and 'télos', why is it so hard to disprove it once and for all? The most materialistic minds must sometimes struggle to not resort to this language they consider so misleading and must reprimand themselves when they do inadvertently use it. Sometimes it looks like they are at war with themselves. Is this continuous struggle of a naturalistic conception against the temptation to fall into teleological interpretations a sign of our subconscious mind that clings to old and outdated religious beliefs? Is this just an automatic irrational secretion of our brain or an inborn habit arising due to thousands of years of cultural, religious, social, or philosophical conditioning that still molds our collective mind?

And yet, we can't avoid this annoying and more or less subliminal feeling or intuition that a universe that self-organized itself with increasing degrees of complexity pretty much looks like a goal-directed entity. Could this be an intuition of a deeper truth that the sense-mind can't seize? Where does this deeply engrained necessity of the materialist's conception to stay permanently on guard against itself in raising questions about meaning and purpose really come from? Might it not be an unconfessed fear to look beyond its own narrow limits? Or the fear of the worst-case scenario: that of losing one's own mind and rational abilities? Arch-materialist Francis Crick, the discoverer of the DNA molecule, once declared: *"You, your joys and your sorrows, your memories and your ambitions, your sense of personal identity and free will, are in fact no more than the behavior of a vast assembly of nerve cells and their associated molecules."- "You are nothing but a pack of neurons."* [262] Nobel physicists Steven Weinberg famously noted that *"The more the universe seems comprehensible, the more it also seems pointless."* [263] Biologist and well know bestseller author Richard Dawkins thinks that *"The universe that we observe has precisely the properties we should expect if there is, at bottom, no design, no purpose, no evil, and no good, nothing but blind, pitiless indifference."* [264]

If so, the unanswered question remains: If the universe is purely deterministic, with no consciousness, no agency, no freedom, no volition, no aim and no purpose, why did it produce beings such as us, having the sense (or at least the illusion) of being conscious, with agency, freedom, volition and aim, and acting with purpose? Talking about evolutionary advantages or the instinct of survival only restates the fact without explaining it. Pointing at selfish genes only kicks the can down the road; it explains nothing.

The feeling that behind all the events of life, of Nature, and of the cosmos, there is a Force, a Consciousness that doesn't admit coincidences, if not only apparently, is becoming more intense and diffuse. The materialist interprets this as a self-delusion resulting from one's own frustrated desires in a life after death or the hope in the existence of an almighty God that compensates for our pain and sufferings. This might be the partial psychoanalytic explanation that fits for some, and perhaps most, of the conventional religiously motivated minds. But there is also a knowledge, an inner feeling, a deep intimation coming from within that is much more concrete, powerful, and precise than our intellectual strength allows for, and that makes us understand that this cannot be the whole story. Once one gets to that point, the question of the meaning of life is no longer perceived as a waste of time of self-improvised 'weekend philosophers'. Quite the contrary, it becomes a regular quest one can't get rid of. The awakening to a new but thin 'vibration' becomes tangible. It manifests first in the form of a distant and dull call but sooner or later emerges into knowledge and certitude that nothing will be able to dissolve. Life begins to appear dull, the persons who surround us boring, and some lifestyles that, for others, are absolutely normal and even delightful appear as something that becomes suffocating, empty, and slowly but steadily unbearable. At that point, some previously barely visible and subtle truths become an obvious and self-evident fact.

Those who are in such an inner condition that feels the senselessness of their lives are looking for a 'new air', a 'new perception', a new life that is inwardly more than externally or intellectually satisfying. This leads the spiritually dissatisfied personality in search of something radically different from any previous philosophy, religion, or spirituality. One looks for points of reference that go deeper and tell us not just how things are and how they work, but also something more about why they are, and finally tells us something about the roots of the enigma of our existence.

Yet, it is not necessary to achieve a particular consciousness or spiritual height to understand that something is beginning to impose itself in many of us, and that forces us to see and perceive things in a different way from what most do. Indeed, there are always more and more persons who feel that, despite everything, there is a unity that ties human beings, animals, plants, rocks, and all the universe into a big movement of consciousness. All this lack of unity is felt like a falsehood. Many are searching for something without exactly knowing what it is. The number of persons who feel the need for a new harmony is constantly growing. They feel that a new ability of intuition is beginning to manifest in them and that they often have sudden revelations. Some go in search of peace, silence, and calmness but find none.

This struggle and stress is part of an evolutionary process and change of humankind. The sooner we realize that it is not coming from the outside but from the inside of us, the better we will be able to go through this process with less friction and resistance. The fundamental message we should capture from the turmoil through which the world and ourselves are going, as individuals and as a collective, is that we won't be able to qualitatively change our human condition only by means of external reforms, scientific achievements, and intellectual constructs that abstract from meaning and purpose. If there won't also be a change in our inner spiritual condition—that is, a change of consciousness—there can't be real progress of humankind.

The still quite strongly prevailing belief that only reason can solve every problem, that science can furnish the solution to all our defectiveness, or that only the rational mind can explain the mystery of the existence of life, evolution, matter, and consciousness is holding us back from growing, expanding and evolving in consciousness. It is necessary to go beyond the present positivist, instrumental, and reductive sciences, but it is equally important to avoid a relapse into pseudo-sciences or sub-rational forms of knowledge or post-truth temptations, which, unfortunately, became quite popular in the last decades, including among those who have developed this new inner intuitional seeing of things, life, and the world. We need to know better ourselves, ask the right questions, and attain a deeper knowledge of ourselves. This should lead us to a real improvement, first and foremost of ourselves, which is based on a new understanding of the world as well.

This uneasiness will also lead us to search for deeper and more profound reasons in the way we conceive of scientific knowledge, even if, probably, by going through intermediate chaotic stages. Science will not understand matter without the soul. Evolution will forever remain affected by the mystery of life if it doesn't admit to some form of teleology. Psychology won't be able to understand the human being by analytic knowledge alone. There are other layers of reality, much subtler and plastic, that impose laws and forces of their own from within.

2. Spirituality and the Notion of 'Progress'

"Mysticism is a state of consciousness within which the conventional scheme of space, time, causality, matter and self, including subject-object separation, is transcended."
Jochen Kirchhoff

Before going on in an outline of the relationship between science and spirituality, we need to clarify the real and deeper meaning of the mystic experience. The term 'spirituality', as used here, should not be confused with a religious doctrine or a cult. While we would like to go beyond a blind physicalist understanding of the world, likewise, we would like to go beyond a blind belief system. While it is true that spirituality and mysticism might be colored by our religious convictions, education, intellectual erudition, and cultural context, by 'spirituality', we don't mean a mental or emotional construct of moral or ethical teachings or a set of rules whose truth should not be questioned. Spirituality is neither a religion nor a philosophy; it is a technique that searches for truth by the first-person approach in an inner realm, as science does in the outer material realm. For example, the investigation we pursued previously into our conscious experience through a first-person approach can also be called 'spiritual' in contrast to a third-person analysis. Other, more sophisticated, inward-oriented psychological practices like contemplation, concentration, meditation, and yoga allowing us to get into contact with other states of consciousness can disclose an entirely new universe to us, leading to knowledge that we can't fathom if we are limited to a sense-mind perception.

In this sense, a genuine mystic experience acquires the same value and status as the empiric scientific approach. One is not asked to believe in anything but, rather, is invited by a spiritual practice and ascesis to have the same experience and gain the same inner realization. We are allowed, as in a scientific experiment, to control, check, and accept or reject the wisdom someone else reached through the same ascesis and experience. What we are going to examine is not about philosophy, let alone about scripture or dogma that we must accept as a religious belief system. It is about a lived experience and a realization of higher states of consciousness, through which someone can apprehend and comprehend the world from an entirely new perspective. The true spiritual experience, realization, and its mystic insight must not be a matter of belief, no more and no less than the account of the discovery of a new continent by a navigator. We might not be able to build a caravel big enough to cross the ocean and verify the navigator's claims, but we are at least given an account such that we know how to confirm or falsify when we will be able to do so. The same fits for the spiritual non-religious experience: Even though we might not be able to immediately transcend our present psycho-physiological condition to higher states of consciousness, we at least have an account that might furnish us with an indication and lead us a step further. There is nothing that we have to take at face value; everything is a matter of experience and realization of a reality other than our limited ordinary mind and sensorial perceptions.

Moreover, the spiritual knowledge goes beyond philosophical speculation. Philosophy, especially Western philosophy, works primarily from a mental state of consciousness—its analysis, hypothesis, theories, and speculations about the world, Nature, ethics, or theology is based primarily on the power of reason, its intellectual skills, and deductive reasoning. In some instances, philosophy was able to rise beyond a mere analytical conception, which we called the 'higher-mind philosophy', as we illustrated in Pt.II-II. It is a sort of insight and intuition that recognizes the value of the inner introspection and, more or less consciously, realizes or gets intimations of something that transcends reason. We laid some emphasis on the 'higher-mind philosophers' because they help us take a step towards a recognition of even higher forms of cognition and states of consciousness.

However, most of what we term 'philosophy' was and still prevalently remains an intellectual activity or, at best, a higher-mind way of seeing a la Goethe. The philosopher might receive, from time to time, the sparks of light from supra-intellectual domains, but these are usually only rare and discontinuous experiences or realizations. Philosophy is, therefore, mainly a mental activity, whereas the spiritual gnosis aims at much higher forms of cognition and 'seeing' that transcend by far the sense-mind physicalist conception and its separative and divisive way of perceiving things.

Fig. 120 Sri Aurobindo (1872-1950)

In this sense, we found particularly interesting the accounts of the Indian mystic, yogi, poet, and nationalist Aurobindo Ghose, more commonly known as Sri Aurobindo. Born in 1872 in Kolkata (former Calcutta), at the age of seven, he was sent to Cambridge, England, to study English literature and to protect him from what his father considered a backward mystic Indian culture. Ironically, Aurobindo would become one of the most influential yogi—that is, a practitioner of yoga[71]—of the 20th century. After perfectly absorbing the British language and culture and returning to India 14 years later, he worked as a teacher and, in about 1906, became directly involved in nationalist politics by joining the Indian revolutionary movement against British rule.

However, it soon turned out that something much more interesting and urgent would determine the rest of his life. In 1908, Aurobindo was arrested for his political engagement against the British and spent about a year in jail. During that time, he had powerful spiritual experiences, which became the basis of a vision of human progress and spiritual evolution. In 1910, Aurobindo left politics and moved to Pondicherry, in the south Indian state of Tamil Nadu. Here, he developed the principles of an *'integral yoga'* (also called *'purna yoga'*), a form of ascesis that incorporated the former Indian spiritual practices and, most importantly, expanded them to a discipline also aimed at physical transformation. He would no longer leave Pondicherry, where he founded an ashram (a spiritual hermitage for yoga practitioners), the 'Sri Aurobindo Ashram'.

Fig. 121 Mirra Alfassa, called 'The Mother'. (1878-1973)

In 1926, Mirra Alfassa, a French lady who became his collaborator and spiritual partner, also called by

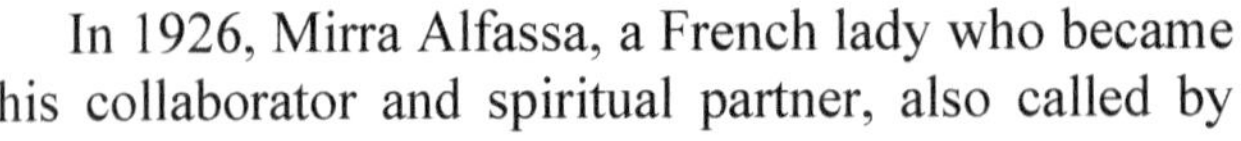

---

[71] Contrary to common belief, especially in Western culture, yoga is not only and also not necessarily a physical discipline consisting of postures ('asanas') or breathing exercises ('pranayama'), but is first and foremost a mental and spiritual practice aimed at uplifting the state of consciousness.

their disciples "The Mother", would take over management of the ashram. This allowed Sri Aurobindo to dedicate himself more intensely to his '*sadhana*' (the yogic spiritual discipline) and to what would later be called the '*cell yoga*'—that is, an attempt at spiritual transmutation of the physical body. This determined his seclusion for the rest of his earthly existence. However, from his room in Pondicherry, he maintained an intense written correspondence with his disciples and the rest of the world.

Aurobindo was an extremely prolific writer and was the author of an enormous amount of texts, mainly about spiritual topics, which highlighted his spiritual emergentist evolutionary vision that envisaged the destiny of the human race and the unity of spirit and matter. His three main magnum opuses are the "*The Synthesis of Yoga*" [265] (the practical guidance to Integral Yoga), "*The Life Divine*" [266] (the description of his spiritual vision based on his inner experiences framed in a philosophical language), and "*Savitri*" [267] (a spiritual epic poem of about 24,000 verses).[72] Sri Aurobindo might be described as a 'mystic empiricist'. He noted his inner spiritual experiments in a diary, the 'Record of Yoga', a difficult text to read because of its extensive use of Sanskrit terminology but which vividly resembles a scientific procedure (see [268] or, for an easier read, see also [269]). A less known, but to some extent relevant for the type of integral cosmology we are going to discuss, are also his commentaries on the Isha Upanishad [270].

In a nutshell: These works most eloquently reiterate a vision of the evolution of human life into a life divine with the destiny of humankind of a collective spiritual realization that not only liberates the spirit from the bonds of matter but can transform human nature and the material body itself into the image of the spirit, which will enable us to live a divine life on Earth. The significance of the experience of higher states of consciousness and how these affect our perception of the inner and outer physical world, as the following evolutionary consequences of a cellular yoga and transformation of the physical body, will especially be considered here given their scientific interest in the view of the problem of consciousness, evolution, and spirituality.

What differentiates Sri Aurobindo's evolutionary spiritual vision from other paths is the concept of the existence of spiritual evolution and progress in Nature. It is an inner progress that will not be a privilege for just a few but will become generalized to humanity as a whole.

The contemporary notion of progress is usually associated with a material, scientific and technological progress. Nowadays, we speak of 'scientific progress', 'technological progress', or, more generally, 'material progress'. What we mean by that is a going forward, a growth, an improvement, a development and an advancement to higher stages of material wealth and of more sophisticated technological and functional devices or machines. For

---

[72] For a more 'soft' introduction, the reader might prefer to resort first to [368] or [382].

example, there has been a progress in building airplanes beginning from the first models created by the Wright brothers to, nowadays, jet fighters. This concept of progress has become a central idea starting from the scientific and industrial revolution.

Historically, however, before the industrial revolution, this idea and ideal of a material and technological progress was absent. The word 'progress' itself originally meant only 'moving forward' or 'making the next step' or just a 'change', but had no qualitative connotation of improvement. It acquired the modern significance when the human race, impelled mostly by economical motives, began to build increasingly efficient machines and realized how a technical and material improvement is possible. Before this collective insight, the idea of an advancement, refinement or, as we would term it nowadays with an even more recent nomenclature, an 'upgrade', was not conceived of as a possibility, except, perhaps, by some occasional idealistic and philosophically inclined mind. During the millennia preceding the industrial revolution and until the end of the so-called 'dark ages', humanity did not believe in a material, scientific and technological progress and even had no concept of it. There was no concept of a possibility of a gradual betterment of one's physical conditions by a progressive science. Living in circumstances of material misery and poor medical and appalling hygienic conditions was a normal state of affairs that perpetuated itself throughout the generations and was considered intrinsic to the human condition and nature. This mindset has curbed a material advance and progress for centuries. Only when the human race began to refuse the inevitability and unchangeability of this state of affairs did things begin to change materially.

But it is highly questionable whether this has also led to the spiritual progress and betterment of human nature. Despite the enormous material advancements of humankind, we can hardly say that this has made us morally or ethically a better species (whatever 'morality' and 'ethics' or the word 'better' might mean). The human has gained better control of Nature, of the environment, and, thanks to technology and science, lives a materially more comfortable life (something that, by the way, is true only for a limited part of present humanity, and that achieves these comforts not infrequently to the detriment of others or the environment). However, it is now clear that this external material progress has always been abstracted from the internal spiritual progress of mankind. Our thoughts on how our lifestyle, institutions, and economy might improve are almost exclusively concentrated on a notion of material and financial progress. The idea that the human being might be able to progress also inwardly, mentally, emotionally, and ultimately at the soul level is, at best, relegated to being a secondary matter. And, more importantly, the idea that once we have gained sufficient material wealth, all the rest will follow automatically turned out to be an illusion.

Nevertheless, we nowadays continue to nurture the same kind of disbelief towards progress which is by no means different, other than being related to a spiritual progress. We believe that a material, scientific and technological progress is possible but we aren't willing to admit that a similar idea and ideal with regards to our psychological, moral, ethical and spiritual nature makes sense. Nor do we have a concept of a possibility of a gradual betterment of one's psychical conditions by a progressive spiritual advance, and living in circumstances of psychological misery and poor spiritual conditions is considered a normal state of affairs that has perpetuated itself throughout the generations and continues to be regarded as a natural human condition. Because, so goes the narrative, we are essentially selfish beings (or, worse, genes), there can't be any spiritual progress, as the human being is intrinsically what it is and no technology or machinery will ever change that. That's how life always was: a dangerous competition based on the survival of the fittest. There can't be any spiritual progress in humankind, only change, modification or even transformation but no advance for the better. This narrative is further supported by a widespread relativism. Because concepts like 'ethics', 'morality' and 'spirituality' are subjective and their meaning is in the eyes of the beholder, or, worse, has been perverted by religious or secular authority for means of subjugation of the individual and society, we therefore conclude that these are only humanmade concepts that ultimately have no meaning other than to impose some sort of juridical or societal order. There is no good or bad, progress or regress, improvement or decline, let alone anything ethical, moral or spiritual advancement. We should stop thinking in these terms and consider only change, difference, eventually transformation, but without assuming there could be a psychological and spiritual advancement and progress of humankind with a forward process in human life beyond materialistic conceptions.[73]

Aurobindo contends that precisely this mindset continues to curb a spiritual advance and progress of the race as the disbelief in a material progress curbed a material progress. Only when we will get out of this self-fulfilling prophecy and refuse the inevitability and unchangeability of this state of affairs can things begin to change and make a difference also spiritually.

The conception that Spirit and Matter are two separate domains of existence and that they must be considered separately or, at best, as one functional to the other is still a (more or less subconscious) Cartesian assumption. The naturalist's refusal to allow for any purpose or aim in Nature is only one of the many modern cultural reflections of this conceptual separation. Also, the idea that the body can be elevated and transformed to spiritual heights and can be

---

[73] By the way, this is also the ideological background that impels biologists to define evolution as a "change in the heritable characteristics" or a "descent with modification" carefully avoiding any terminology suggesting functional or qualitative improvements.

infused by a spiritual force is rarely a matter of consideration. With few exceptions, most of the spiritual teachings are focused on an inner realization, where the improvement of the mental, emotional, and bodily personality is only accessorial and functional to that aim. These are welcome as a positive effect of the spiritual realization or are posited as a necessary condition for that realization, but the evolutionary concept of progress and a transformation of our mental, emotional, and physical existence as a process occurring due to an emerging Spirit in Matter is rarely addressed, let alone made the central aim and purpose of the ascesis for a collectivity.

While religions may have had a role in avoiding catastrophic relapses into previous stages of evolution, they did not make us more evolved beings. Also, the more refined forms of spiritual realizations that, for example, the different meditation techniques pursue, aim at an individual inner realization, not so much at an outer transformation, which should ultimately reflect the inner divinity by a collective spiritual and material ascension. We continue to distinguish between material and spiritual progress, keeping it separate—as having nothing to do with each other. Whereas, for Sri Aurobindo, we should synthesize the two aspects of life into a third position, a sort of *divine materialism*. It is this synthesis of a notion of progress that acts on both planes of existence and is aimed at a long-term spiritual and physical transformation of the human race on a planetary scale that distinguishes the integral yoga from other spiritual paths, which, however, are not denied but, rather, embraced and enlarged.

What follows is a brief introduction to Sri Aurobindo's vision of humankind, with a special emphasis on its evolutionary aspect and its physical and scientific implications, his conception of man as being transitional and mankind evolving towards a 'gnostic humanity'.

We chose this particularly interesting case of the integral yoga of Sri Aurobindo because it appears to us to be the most comprehensive and insightful approach that encompasses and generalizes all the previous scientific as spiritual approaches and, without denying its value in the social evolutionary context of humankind, expands its vision beyond it, revealing the meaning and purpose of the whole of existence in an emergentist and evolutionary perspective of spirit in matter. It is in line with modern science, makes sense of the hard problem of consciousness, and gives meaning to evolution in a synthesis of knowledge between East and West. This results in a cosmology that leads us far beyond the current scientific theories, philosophical thoughts, or spiritual accounts.

3. Is Man a Transitional Being?

The world in which we live is undergoing a tremendous transformation. Nothing is as it was before. On one side, incredible discoveries have been made, new rights conquered, new hopes appeared on the horizon. At least from the material point of view, this species— homo sapiens—did make some progress since the time of the stone age. On the other side, reforms, revolutions, new political and economic orders, new philosophies, technological and scientific progress itself, and whatever attempts to make things work better did not really improve humans' psychological condition, still plagued by anxiety, problems, complications, disappointments, and suffering. War, potential self-annihilation with weapons of mass destruction, economic systems that have created new forms of exploitation and slavery, hunger, violence, overpopulation, environmental destruction, crime, folly, social divisions with the rich getting richer and the poor getting poorer, and (self-inflicted?) pandemics remain a daily reality or dangerous possibility on this little blue planet. This world continuously goes through a long series of crises and contradictions, and the destiny of humankind remains uncertain. It is felt that the Age of Reason and the cherished values of the Enlightenment triumphed and yet failed us. Life seems to have lost its original balance and harmony.

If we look at the Universe, with all its billions of galaxies, each one containing billions of stars, one of which (our Sun) certainly has a planet with life on it, we see a reality in which everything has an order and a law. Looking further to animals' lives, we can see how, despite the Earth's natural catastrophes and a natural selection that, to our eyes, may appear to be a harsh and cruel method of Nature, it is nevertheless founded upon an existence ruled by natural laws that, at least before human's interference, were based upon a balance and harmony that we apparently have lost and forgotten. Now, instead, this world is in turmoil. Externally, everything has changed but, somehow, we feel that something continues to remain unchanged.

So, we wonder: Why? Why does this little homunculus that we are put the evolutionary process on Earth so upside down? No other species has been so self-harming and, in a certain sense, so stupid. Are we a mistake of Nature? A coincidental error in a meaningless and random creation? The 'wrong' random genetic mutation? After all, what are we doing in such a world? Who are we? From where do we come? For what reason are we here? To do what? And, above all, where are we going?

Science, philosophy, and religion did not furnish us with a satisfying answer. Science tells us how things work but remains silent on what and why things are. Philosophy has turned out to be a nice intellectual exercise, at best a luminous higher-mind stream of thoughts that can inspire and indicate a direction and furnish an intuitive intimation but that did not deliver a real answer. Religions may have maintained some social order, have sometimes given birth to great souls, and were and remain a source of comfort for many, but failed as well in telling us who and why we are, let alone in really improving human nature.

We will argue that this chaotic situation can be fully understood neither in terms of purely social and historic factors nor in mere scientific and materialistic explanations of the natural processes of evolution. The present circumstances can be understood only by realizing that humankind is actually going through a spiritual evolutionary crisis. Such an evolutionary crisis is not new in natural evolution. There is plenty of scientific evidence that the transition from one species to new forms of life went through sudden and sometimes catastrophic changes. But everything becomes clearer and much more intelligible if we include the 'soul-factor' in the evolutionary process. Trying to understand and explain the confusion and trouble of humanity, working out some sci-fi technocratic wonders or new political revolutions, won't improve our condition if we aren't able to look beyond the material and flimsy intellectual surface and go to the root of the problem.

Sri Aurobindo's starting point was that to recognize that all the human knowledge, all the wonderful scientific progress, all its philosophies, and even the spiritual systems he himself embraced and further developed still didn't grasp the real reason for human misery. If there isn't a collective change of the human consciousness, we will continue to see how everything will change, yet without anything changing. Governments after governments, new political orders will replace older systems, and material progress will make us live even longer and allow us to reach for Mars and beyond, but the inner dissatisfaction, that sort of 'divine discontent' and outer chaos won't substantially change.

The human condition is the result of its inner state of consciousness. The exterior material perfection never was and never will be a sufficient element to lead us towards a happier and more harmonious life. The idea of achieving a fair distribution of material resources and wealth, of reducing humans' instinctual egoism and greed, stopping the environmental destruction, and annulling the risk of violence and war by creating new political and financial world orders but without considering the inner, spiritual, and 'soul-dimension' of humankind is an empty chimera. If we don't change ourselves, our outer conditions won't change much either. It is an illusion of the modern political and materialistic mind to believe that betterment of the outer material conditions will automatically lead to the improvement of everything else. This has been proven to be wrong over and over again. How long do we want to continue this story?

The Buddha's spiritual search for a way out of suffering, the Indian non-dual teachings that lead us to a personal realization, or the Western logical inquiry a la Spinoza for the certainty of being happy, went in the right direction but fall short because they stopped too soon with too little. That can only change the individual and never will lessen the collective suffering. If we don't look further, the world won't change much. It will continue in cycles that, like in an eternal merry-go-round, might deliver a few momentary satisfaction that, however, will not last long. Something is missing, something we refuse to recognize and take into account, and this is consciousness.

Sri Aurobindo used to say, *"Man is a transitional being"*, meaning that our present human crisis is an evolutionary crisis that won't be settled by non-evolutionary means. The evolutionary crisis is about an old consciousness breaking down and a new consciousness emerging. We will have to acknowledge that the whole idea of transformation must come from within. A material or intellectual 'upgrade' of humankind or personal salvation for few that will leave behind all the others won't work if there isn't also a collective inner growth and an effort to raise all the other's state of being by some spiritual practice, by some form of 'yoga'. In fact, he used to say that, at the bottom, *'all life is yoga'*.

The question is: Do we want to participate in this yoga of Nature consciously—that is, by a common effort of transformation—or unconsciously by being subjected to the severe and harsh methods of natural selection?

"The generalisation of Yoga in humanity must be the last victory of Nature over her own delays and concealments. Even as now by the progressive mind in Science she seeks to make all mankind fit for the full development of the mental life, so by Yoga must she inevitably seek to make all mankind fit for the higher evolution, the second birth, the spiritual existence. And as the mental life uses and perfects the material, so will the spiritual use and perfect the material and the mental existence as the instruments of a divine self-expression." [265]*(Intro., Ch. III)*

The beginning of an inner cleansing may be a key through which we can change things outside of us. The term 'yoga' [74] doesn't imply gymnastics, an exercise of breathing, or any technique that scratches the surface of our being but does not go in-depth. Sri Aurobindo's integral yoga takes up the traditional yoga but aims to go beyond exercises of contemplation, let alone a personal salvation and escape into an undefinable ecstatic heaven. It is about a personal and collective transformation of the whole being, up to the last physical cell of our body, going through the most remote and dark zones of our personalities, which have not, or have only scarcely, been explored. The peculiarity of this spiritual transformation that distinguishes it from other spiritual traditions is its collective character and its attempt to fully embrace the materiality of existence with the aim of a physical transmutation as well.

But what precisely is it that must change in us?

Firstly, we must become aware of our true inner and outer condition, as it is really, without rosy glasses. If we look inside ourselves honestly, we see how we are an admixture of habits, attached to a surface personality, with some dispositions and tendencies we are mostly unaware of and, even if we become

---

[74] The word 'yoga' comes from the Sanskrit 'yuj' and means 'to unite', 'join' or 'connect', implying a practice and method that aims at the union with the Divine.

aware of them, have not much control over them. The recognition that we are conscious only of a sort of 'bubbling surface' mostly dominated by irrational impulses, desires, and automatic thoughts and behaviors would already be a giant leap towards self-consciousness. When we become able to look inside of ourselves only with a little bit of introspection under this surface of bubbling appearances and don't tell ourselves nice fairytales or invoke some excuse that tries to justify, with seemingly rational arguments, what we really are emotionally and mentally, one soon realizes how we are conditioned by something that does not go much further than the nervous animal nature. Then we realize how we are a complex amalgam of nervous and automatic habits that may also be embellished by some rational ideas and rules, but that doesn't make us self-aware beings. Sri Aurobindo's sharp and, at times, caustic language clarifies this as follows:

"Man twitters intellectually about the surface results and attributes them all to his 'noble self', ignoring the fact that his noble self is hidden far away from his own vision behind the veil of his dimly sparkling intellect and the reeking fog of his vital feelings, emotions, impulses, sensations and impressions." [271]*(Vol. II, pg.184)*

Does it make sense to believe that we will ever find the solution to our worldly problems, ignoring our own state of consciousness?

Therefore, we should recognize that humans are neither special nor a final species. We are just transitional beings who have not terminated their evolutionary journey and who are presently moving towards a new state of consciousness that transcends our mental one.

It is reasonable, almost inevitable, to suppose that humanity is, most subliminally, conquering this higher status step by step and is raising its existence slowly but definitively to a new form of knowledge that will change our way of looking at the world. It is perfectly in line with the logic of the present scientific conception of Darwinian evolution. Unless humanity terminates its existence by self-destruction or some natural catastrophic event such as a giant asteroid hitting the Earth and annihilating all life on it, inevitably, the homo sapiens and its way of cognizing the world will be exceeded. It is not a question of whether this is the case but, rather, a question of when and especially how we are supposed to do that.

We are going through an evolutionary process that must lead us to a refined form of knowledge that transcends reason and the analytic sense-mind to a form of (non-religious) 'gnosticism'[75], leading toward the appearance of what we might call the 'homo gnosticus' or a 'gnostic being'. It is an evolutionary process toward a 'gnostic humanity' and that, among other things, will inevitably have to develop an integral science.

---

[75] It could be useful to recall that the Greek word *'gnosis'* (γνῶσις) means 'knowledge'.

One might object that, even if so, this would still take thousands if not hundreds of thousands of years, making these ruminations irrelevant for our everyday life. We might talk about something that is so far in the future that it doesn't make any difference in the here and now.

This argument, however, works with two unaware assumptions. One is that evolution progresses linearly, only step by step, without perceptible changes in a lifetime. This is unscientific in the first place, as modern evolutionary theories have now recognized. Evolution can and does frequently proceed by abrupt and sudden changes, and speciation occurs with sudden big leaps, even though after long times of apparent stasis and inactivity, not rarely preceded by turmoil and eventually catastrophic events. Everything indicates that the phase we are actually in consists of times of transitional evolutionary labor pains.

The second assumption is that we are only passive subjects inside a much wider evolutionary context and that we can't change this process. This is also a misplaced belief because, if the mind is limited and not the final term of cognition, it is also true that, for the first time in evolutionary history, a form of cognition appeared that can actively participate in the very same evolutionary process it came from. The human mind has a powerful property that distinguishes it from other animal species: that of recognizing its own limitation and participating willingly in the evolutionary effort. We are not just passive automatons but beings that can tap into a Will-Force that can accelerate evolution by orders of magnitude.

Once the above two (more or less unaware) assumptions are set aside, things appear in a new perspective. We realize how near we are at this transitional age and how actual this could be beyond some philosophical musings. The evolutionary process that is leading us from the rational age to a gnostic one is long and will certainly not be without incidents, oversimplifications, and misunderstandings, and, particularly, the confusion of a mixture of mental, intuitive, and emotional preferences is actually the biggest stumbling block and most evident sign of the times. But in an evolutionary breakthrough, it is typical to have a period of disorder that precedes the subsequent new order.

Seen from this perspective, what once looked like meaningless chaos becomes intelligible. Something is lurking beneath the surface and waiting to arise more clearly in our actual consciousness and knowledge.

But words cannot express the idea of what transcends our observable world. We can ascertain that we are only the cognitive apex, if at all, of the terrestrial evolution, not the final point. It is about overcoming our present evolutionary state. If we will be able to ascend to higher states of consciousness, a new horizon may disclose us inwardly and outwardly and will change our way of doing and conceiving science, arts, music, etc. We would like to show here that this is not a far-off utopia but something that, at least as an initial movement of a long-term journey, can be put into practice here and now.

Man is a transitional being and not the last step of the evolutionary process. There is a meaning and purpose not only in the Creation or in a cell, but in our life and existence as tiny beings walking on the surface of this beautiful planet. We are not only on the verge of the next evolutionary leap but also the first species that can actively participate in it. A new symbiosis between gnostic scientific reasoning that begins to accept that mind, intellect, and our sensory means are only transitory tools of a lower state of consciousness and an inner experience, detached from religious beliefs, superstitions, and the cultural environment may project us toward unexpected and unexplored new frontiers.

4. Planes and Parts of Being, the Psychic and Spiritual Transformation

"People say I have created things. I have never created anything. I get impressions from the Universe at large and work them out, but I am only a plate on a record or a receiving apparatus — what you will. Thoughts are really impressions that we get from outside."

Thomas A. Edison [272]

a. Transforming the Ordinary Consciousness

Every mystical tradition tells us that behind the visible and material world that our senses perceive and that we call the 'real world' exists an immense reality of which we are not aware. Occasionally, we may even have a very transitory and superficial experience of it, but most of us tend to confuse it with an illusory, indefinite, and empty realm that our mind typically takes for its own creation. Instead, the mystic who practices a spiritual, psychological, and physical discipline that harmonizes the mind, the emotions, and the physical, detaching from the natural and their innate movements, instincts, and appetites, slowly learns to look behind the appearances and realizes that not only is our reality completely different from what our sense-mind perceives, but also this reality is only a tiny wave on a much vaster ocean. There is something more behind that 'noble self' and the world that we actually experience.

We can already understand why this must be so without going beyond the material boundaries. Just consider how the history of science developed. It was once believed that the electromagnetic spectrum of visible light is all there is. We now know that the light our eyes can detect is confined to a tiny electromagnetic spectral region. Radio waves, Infrared, X-rays, γ-rays, and many other wavelengths we previously didn't suspect to exist are invisible to our visual perception. The same can be said for soundwaves: Several species hear the ultrasounds we cannot detect. We have discovered that even the material stuff we see in the universe must be only the tip of the iceberg: Astronomers now know, through indirect evidence, that the visible matter (stars and galaxies) represents only 5% of the Universe, with the remaining 95% being a still mysterious *'dark matter'* and *'dark energy'* about which we know almost nothing. Therefore, if we do not resort to an anthropocentric concept of ourselves and refrain from placing the human at the center of all there is, it is quite natural and almost obvious to expect that what our human-centered sensorial and cognitive system discloses to us must be only a thin surface of a much wider reality. A large part of our inner and outer existence escapes our sense-mind as well.

Nonetheless, as ignorant as we might be, our real distinctiveness from the animal kingdom is not the bunch of neurons we have in our skulls or our abstract thinking. It is our mental ability to know what we don't know, realize our own limits, and precisely for that reason become able to transcend it.

But most are happy with what they are and have no urge or feeling to transform, evolve, or improve something in themselves. This is not because many haven't a secured material existence. Even if all our basic needs are satisfied, rare is the impulse to look inside of us. The vast majority of humanity has no concept of going beyond what it is and doesn't feel the need to change anything other than its surroundings or accumulate and consume material goods that assure its survival and physical appetites, the satisfaction of its ego, and economic and financial security, and eventually pursues some external goals that are supposed to fill the gaps of an inner feeling of emptiness. But sooner or later, an inner dissatisfaction that translates into an aspiration to go beyond will certainly arise. The future generalization of a self-enacted evolutionary process in all of humanity might well be the destiny of the race.

But what does it mean to 'evolve'? What is it that is supposed to be 'transformed' in us?

The way through which any concentration, meditation, or mystic technique works begins with calming and even silencing the mind and our negative emotions. The mind produces an interminable chain of uncontrolled thoughts that most of us are completely unable to stop and control consciously. The simple concentration on a single mental object for a period longer than a minute appears to be an impossible task for the vast majority of humans. Calming our restless minds and emotions is the first fundamental step toward a higher and clearer consciousness.

The second step is to proceed with a similar work of self-transformation on what Sri Aurobindo called the *'vital'*—that is, that part in us that expresses itself with positive or negative emotions and feelings, such as love and hate, inner peace or anger, fearlessness or fear, happiness or depression, emotional excitement, etc. To start becoming acquainted with the nomenclature of integral yoga, let us label the vital manifesting positive or negative emotions as *'higher vital'* or *'lower vital'* respectively.

As a side note, it may be shortly pointed out that whenever we speak of 'calming', 'silencing' or 'controlling' the mind and vital, we don't mean a practice of suppression and repression. It is about a transformation, a transmutation, a metamorphosis of our mental and still animal psychological traits, not about any practice of violent suppression or inhibition. And when we adopt labels such as 'higher' or 'lower', we, again, do not articulate a moral judgment but simply realize and acknowledge the existence of polarities such as pleasure and pain, love and hate, bliss and agony, etc.

This, or one of these transmutations, is the original goal of any practice of meditation, concentration, and spiritual discipline that can open new doors to our awareness of the world and beyond. 'Beyond' means that once this widened awareness is achieved, new states of consciousness beyond the mental disclose themselves, allowing for greater seeing and knowledge. New realms can be perceived and sensed that were previously not in the reach of our ordinary senses because they were subjected to the 'noise' of our ordinary mental and vital being. When our consciousness 'awakens', we discover that there are subtler realms than the material one, but these are interdependent with and in it, as in a fractally structured universe. The transcendent ladder that goes from matter to soul through different, subtler, and plastic layers of existence is a sort of universal 'matryoshka embedding' in and of the cosmos.

But many will think that, after all, the ephemeral and evanescent pleasures of life are necessary for our happiness. Why should we give them up? The set of vital instincts from which we have moved is a natural thing that, at least apparently, makes our life fuller. Why should the progression toward a greater consciousness be desirable? Those who were able to go deep enough through these layers of our mental and vital restless movements reported:

"If mankind only caught a glimpse of what infinite enjoyments, what perfect forces, what luminous reaches of spontaneous knowledge, what wide calms of our being lie waiting for us in the tracts which our animal evolution has not yet conquered, they would leave all and never rest till they had gained these treasures. But the way is narrow, the doors hard to force, and fear, distrust and skepticism are there, sentinels of Nature to forbid the turning away of our feet from less ordinary pastures." [273]*(Jnana, aph.5)*

b. Satchitananda and the Divine

What, then, is beyond our superficial sensory appearances and mental conceptualizations? What is the base of the whole material and non-material universe? At the level of mind, these are questions that remained without answers, while for mystics, these are not mysteries hidden forever from us but, rather, find a resolution in a transcendental experience. It is from the report of their insights, which is neither religious nor philosophical, but an account based on a direct and lived experience of a multidimensional reality that arises due to free inquiry and spiritual practice, that we can reconstruct a cosmology going beyond religion, philosophy, and science.

According to Eastern spiritual philosophies, everything has its foundation in the One, the ultimate principle, and support of all that exists, the *'Self'*, the *'Absolute'*, or in Sanskrit, the *'Brahman'*. It has a transcendent unmanifest aspect, that of a pure existence in itself that is above the manifest cosmos: a spaceless and timeless and unchanging impersonal Absolute. Its essence is the triune Existence ('Sat'), Consciousness ('Chit'), and Bliss ('Ananda'), commonly known as *'Satchitananda'*. This trinity must not be misunderstood as a summation of three qualities but, rather, as three aspects of the same undifferentiated Brahman. There is no difference between 'existence', 'consciousness', and 'bliss'. These are only three words that describe the very same Thing and ultimate undifferentiated, featureless, and relationless Quality. It is only in the manifestation that these appear as separate qualities. That's also the reason why it is said to be the *'Nirguna-Saguna Brahman'*–that is, the impersonal Brahman without qualities and, at the same time, the personal Brahman with qualities–being equal co-existent aspects of the Eternal.

Something that didn't remain unnoticed among Western mystics and philosophers as well, as we have amply illustrated in Pt.II. It is no coincidence if this spiritual experience has several points of contact with some Western philosophies, those we termed as higher-mind philosophy or that also appeared in other cultures throughout the ages. Though a mentalized consciousness does not have direct access to these domains, it can nevertheless intuit their presence by rising beyond the sense-mind and tapping into a spiritualized mental state. This is something that can be done independently from culture, nation, religion, or historical age.

Of course, one might also call this Absolute simply 'God'. However, especially in Western culture, 'God' is associated with a concept of a personal God of the Abrahamic religions or a judgmental Father figure. For this reason, we will refrain from using the term 'God' in this context and alternatively speak of the Transcendent as the Satchitananda, or the Self (with a capital 'S'), while including its immanent aspect in the temporal and universal manifestation as the *'Divine'*.

According to advanced mystics, or 'seers', and practitioners of the spiritual psycho-physiological techniques of self-control and self-transcendence, especially those of Eastern descent (such as the non-dual Vedantic or Buddhist tradition), one realizes that at the origin of all things is a supreme Self that manifests itself in its cosmic aspect—that is, in the form of a multitude of dynamic and infinite extensions. One of these cosmic extensions is manifest in the form of space and time, matter, forces, and energy, or the physical universe of science. But there is the individual aspect of the Divine, the *'central being'* and the individual soul of every living being, the subject with its individualized consciousness, and that by a downward 'projection' or 'emanation' from the Satchitananda projects itself throughout the planes of consciousness.[76]

All these three aspects—the transcendent, the cosmic, and the individual—are the one and the same Self, manifesting itself in different forms on different planes of existence.

c. Planes and Parts of the Lower Hemisphere

The traditional Indian Upanishads and yoga philosophy conceives of the being made of 'vehicles', 'bodies', or 'sheets of consciousness', called *'koshas'* covering our soul like the rings of an onion. Matter is not the only substance that exists. The coarse-grained metaphysical dualistic model that reduces life to just a soul and a body is expanded. Once the consciousness of the mystic ascends beyond the material realm, subtler forms of matter, substances, and energies, which exist beyond the physical, become part of experiential reality. Our physical body appears as the outer layer of other, subtler physical and subtle energetic 'bodies'. In descending order (with the correspondent 'aurobindonian' plane, see later for more details):

Anandamaya kosha,	Satchitananda
Vijñānamaya kosha,	supramental
Manomaya kosha,	mental
Pranamaya kosha,	vital
Annamaya kosha,	physical

Later we will address the Vijñānamaya kosha in more detail—that is, what Sri Aurobindo called the *'supermind'*. For the time being, it should be noted that, according to the yoga philosophy, these sheets are made of 'substances' of different grossness. The annamaya kosha is our familiar physical body made of matter. Meanwhile, the pranamaya kosha, or the *'vital'*, which, as already mentioned, is the life-sheet and plane responsible for our emotional consciousness, is not made of matter. Rather, its substance is of the nature of a subtler 'vital energy'[77] also commonly known as *'prana'* or as the Chinese *'ki'*, which exists independently from the permanence of the physical body. The mystics describe death as a 'dismantling' of the physical body with the other sheets remaining intact. Also, the manomaya kosha—that is, mind—is a subtle 'body' of its own, not something that will cease to exist with death or

---

[76] The reader might find it useful to resort in advance to *Diagram 1* on page 404.

[77] The word 'energy' in the spiritual context should not be confused with that of physics, meaning something capable of producing physical work. Nonetheless, the word 'energy' is not so misplaced because we know all too well from our everyday experience how our 'e-motions' can set us pretty much 'in-motion', as the etymology betrays.

the cessation of brain functions. Mind and brain are two sheets to be distinguished.

In integral yoga, the five kosha-structure of classical yoga was further refined. Sri Aurobindo's integrality can also be recognized in extending this Vedantic structure of 'planes' (not to be confused with spiritual 'centers', which we will describe later) to the *'subconscient'* and *'inconscient'*. Interestingly, while Western psychology for a long time abhorred anything that went beyond a positivist and materialist understanding, it also recognized the importance of the subconscient and inconscient aspect of human nature (though it maintains a naturalistic understanding of these being nothing else than a physical brain-epiphenomenon).

More specifically, in integral yoga, the subconscient is the submerged part of our being dominated by incoherent thoughts, irrational fears, impressions of traumatic experiences, habitual movements, and disorganized emotions that we are mostly not aware of in our surface waking state but, as is well known, can influence us (sometimes even dominate us) in our mental, emotional, and physical everyday lives and in our decision making. The subconscient records and stores experiences, impressions, and emotions throughout our life, especially those of particular intensity or traumatic character. Our surface being might have completely forgotten an experience, but the subconscient might well toss up impressions and memories that have not been retrieved for years or even decades, in the form of sudden remembrances or incoherent thoughts or emotions.

This subconscient is analogous with the corresponding Western as Eastern notions. An example is the Buddhist or Hindu notion of *'samskaras'*—that is, the individual dispositions, tendencies, character, and behavioral traits—which are related to the theory of *'karma'*. However, in Aurobindo, it takes a much larger dimension and significance down into cellular processes and is seen in a wider evolutionary perspective.

The inconscient is a plane that goes even deeper than the subconscient insofar that its movements become even more mechanical and reactive and almost impossible to change.

At the lowest rung of the ladder of consciousness, Sri Aurobindo placed what he called the *'nescient'*. It is the most inert, passive, obscure, mechanical, and obnubilate involved consciousness, the bottom of creation, the root of matter's inertia itself. It is a substance almost completely forgetful of its original source, and that is the characteristic trait of matter.

These are the planes and parts of being of the *'lower hemisphere'* of consciousness. However, each of these parts—the mental, the vital,and the physical—is not just one homogenous undifferentiated block. Only a bit of introspection reveals that there must be more.

d. The Concentric System of Being, Soul and Nature

By proceeding with the spiritual psycho-physiological ascesis, one begins to discover other inner realms of being that are concealed from the surface awareness of the ordinary human being. As testified to by all spiritual traditions, saints, and mystics, one comes into closer contact with what Sri Aurobindo called the *'inner being'*, distinguishing it from the *'outer being'* of which we are aware in our surface state of consciousness. This inner spiritual consciousness is what we termed the subliminal in Pt.I. This *'subliminal being'* is a much vaster and luminous *"place of deep peace, light, happiness, love, closeness to the Divine or the presence of the Divine."*

Our outer or surface personality is dominated by half-controlled mental and emotional impulses, desires, fears, and more or less reactive mental patterns, restless thoughts, intellectual sense-mind reasoning, and all those psychological traits we call our 'personality' or ego or self (lower case 's').

The *'outer mental'* is what we know well as our sense-mind, which is responsible for mentalizing and organizing all the perceptions it receives in forms and shapes, [78] presenting it to the *'physical mind'* (the mind that isn't able to conceive of anything other than what the sense-mind suggests to it)— that is, it takes the physical world and its mechanical workings for all that there is. There is also the half-controlled emotional mind, which Sri Aurobindo termed the *'vital mind'*. The vital mind is dominated by the life-impulses; it is that irrational part of the mind that thinks instinctively and imagines emotionally in an unreflective manner and that can possess us by moving us to actions we may later regret. While the *'outer vital'* or *'lower vital'* is the source of what we term, in our everyday parlance, 'negative emotions' such as anger, fear, selfishness, greed, etc., but it can also have a semi-positive side insofar that it can express somewhat nobler emotions of human love, albeit tainted by possessive tendencies or driven by subconscious egoistic motives.

The *'outer physical'* is simply the gross physical body.

In contrast, the *'inner mental'* is a silent, calm, unbound, unattached mind free from the influences and error of the ignorant thought and will of the outer mind and is directly open to the knowledge and guidance from above. It is capable of thoughts detached from the more animal instinctive and emotional impulses. It is the analytic thinking of the intellectual. That's why the Enlightenment enthroned it as the ultimate and final form of cognition. However, it must also be said that almost all intellectuals are far from having developed skills of self-control and fall prey to the impulses of the vital mind as well. Only rarely do they receive guidance from above cleared from all the lower impulses.

---

[78] The author conjectures that the brain is only a sort of 'preprocessor' of the sense-mind, not much more than that.

The *'inner vital'* or *'higher vital'* is that emotional part in us that expresses itself with finer emotions of love, beauty, peace, calm, etc. once it comes to the front. Meanwhile, the *'inner physical'* or *'subtle physical'* is a body made by a subtler substance that is much more plastic and has freedom and power that the gross physical body does not have. It is a physical existence parallel to the material one that, however, is accessible only by developing subtle senses and rising to the highest levels of consciousness.

The distinction between 'subtle' and 'gross' matter and the different degrees of 'subtleness' of substance is not so unknown to physics. Physical sciences conceive of–besides the usual different states or *'phases of aggregation'* of matter, such as the solid, liquid and gaseous state–other, more exotic conditions that are realized with extreme heat (*'plasma'*) and/or pressure. Also, physics introduced, in its conceptual repertoire, the 'field'–that is, a physical entity that serves as a substrate for the description of dynamical forces, such as the gravitational field or the electromagnetic field in space and time. The field and the force field lines describing it are, however, considered more of a mathematical abstraction than a real object. Nevertheless, something being immaterial in the sense of having no mass exists. For example, the photon–commonly known as the light particle–has no mass, unlike electrons or protons. It is just an electromagnetic massless oscillating field that travels throughout space at the speed of light. The same can be said for the gravitational field, and in Einstein's general relativity, gravity is considered an immaterial but deformable space-time manifold. Then, it is now well known that particles such as neutrinos exist and interact with ordinary matter (electrons, protons and neutrons) only extremely weakly. Neutrinos are almost massless (their mass is six orders of magnitude smaller than that of an electron) and are so evanescent that almost all are perfectly able to penetrate the entire Earth and Sun without being perturbed. Moreover, as already mentioned, the existence of a universal quantum field that fills the empty space of the entire universe, sort of an *'ether'* or *'akasha'*, as the Indian cosmology termed it[79], and that is ultimately responsible for all the interactions between particles, is now an accepted fact among particle physicists.

Thus, modern physics has already become accustomed to considering the existence of more immaterial objects than that kind of naïve materialism conceived and that envisaged only marble-like massive particles swirling

---

[79] A word of caution: the 'ether' or 'akasha' of the Indian cosmology should not be confused with the *'luminiferous ether'*–that is, a supposed medium of propagation for electromagnetic waves–that physicists of the 20th century searched for. The former is more analogous to the field of quantum field theories–if any direct comparison could be made at all–while the existence of the latter has been ruled out by experiment.

around and kicking each other.[80] It is also to be expected that it will find even more 'subtle' forms of matter and/or fields and forces or 'substances' than those actually known.

Nonetheless, the more subtle or 'elemental' forms of substance making up the different koshas and planes of being considered by mystic traditions should not be confused with any aggregate or material condition that physical sciences speak of. Here, we are talking of conditions of matter, or just 'substances', and forces that are not perceptible (and, eventually, even denied) by science. The typical example is the 'life-force' or 'prana' that many would swear to feel acting in their bodies but that allopathic medicine resolutely denies the existence of and brands as a popular phantasy. Whatever the truth may be, the conceptual conflation between physical matter and forces that science acknowledges and these much more subtle substances of which spiritualism talks about is to be avoided.[81]

While the outer being is dominated by mostly nervous habits and instinctual reactions, such as the habitual obligation to suffer, the inner being—that is, the subliminal—can suspend these mechanical habits of the waking state. This is what one does by means of hypnosis. What is ordinarily called a 'hypnotic state' is the suspension of the dominion of the outer being reasserting the power of the subliminal, which holds off the body's habitual nervous reactions to pain or the ordinary instinctive emotional and mental responses. As is well known, under hypnosis, the hypnotized subject can be ordered to stop feeling pain by suspending the habitual waking consciousness reactions of suffering properly to the nervous outer being. If the subject wills it, hypnosis appeals to the subliminal mental being, which can become the real master of the nerves and the body. But this freedom, effectuated by an external hypnotizer without true possession of the subject, may be attained much better by one's own will. This is what, among other things, the practice of a contemplative method like yoga does.

However, our inmost and true being and individuality is not the inner or outer mind, vital, or body. Also, the noblest thoughts and emotions of the inner

---

[80] That's also the reason why the modern philosophy of science prefers to speak of 'physicalism' instead of 'materialism'. Modern physics acknowledges the fact that immaterial objects and phenomena exist.

[81] As a side note, it might be worth recalling that the Indian mystical tradition conceives of different stages of an original matter that are beyond the sensible activity of our sense-mind and scientific enquiry. These are the five '*tanmatras*' of the Samkhya philosophy. The first stage is a primordial state of matter as a 'shadow of spirit' and the second stage is of a more ethereal nature called 'causal matter' ('*Vayu*'-metaphorically termed as 'air') from which three other states of progressively gross matter emerged (named '*Agni*' or 'fire', '*Apah*' or water and '*Prithivi*' or 'earth'.) All but earth are sensible to the first-person inquiry but still invisible to the present scientific investigation.

being are only the reflection of the qualities of the '*inmost being*', our true evolutionary soul, which Sri Aurobindo called the '*psychic being*' (the '*chaitya purusha*', as the rishis, the ancient seers of the Indian Upanishads, termed it). The psychic being needs special attention and will be described in more detail in the next section.

The outer and inner being both compose part of the manifestation, of the dynamical universe, and belong to what the Vedic yoga system called '*Prakriti*', the universal cosmic Nature, or just Nature. One's true being is the '*Purusha*', which supports all the actions of Nature. Purusha is the Soul or Spirit side as opposed to the Nature side. Purusha is the noumenal while Nature the phenomenal. Prakriti translates Will into phenomena in terms of space, time and causality. The psychic being is the Purusha in the heart center, but besides it, the Purusha can be mental or vital, standing behind these planes as an unaffected 'soul-witness' to the activities on these planes of being. By stepping back and going inwards, one learns to distinguish the phenomenal universe (Prakriti, Nature) on its physical, vital, or mental plane with a mental, vital, and physical 'witness awareness' that observes but is not affected by whatever clash of forces it experiences. In the ordinary human consciousness, the Purusha is the witness-consciousness, also called the '*witness-Purusha*', the part in us that seems to be a passive and inactive consciousness that does not intervene in the manifestation and stands back as an unperturbed observer. However, it can become active once the inner will has been developed and we ascend the steps of the ladder of consciousness.

This triple system of the outer, inner, and inmost beings is called the '*concentric system of being*' and is summarized in Table 1.

Prakriti		*Purusha*
Outer being	**Inner (subliminal) being**	**Inmost Being**
Outer mental (sense- and physical mind + emotional or vital mind)	Inner mental (reason, intellect)	Psychic being (Chaitya purusha)
Outer vital (lower vital)	Inner vital (higher vital)	
Outer physical (gross body)	Inner physical (subtle physical)	

Table 1 The concentric system of being.

Let us take a closer look at the nature and functions of these parts.

While for the ordinary human consciousness in the waking state, the outer being is all that we believe to be and is indicated as 'me' and 'I', during the sleeping state, we retire from the outer into the inner or inmost being or into the subconscious or supraconscious planes (the higher hemisphere we will describe later).

As it also appears in the writings of Sri Aurobindo, dreams are (in most cases) a distorted symbolic transcription of the experiences of our consciousness moving throughout the different planes of being. In this respect, he aligns with traditional Indian psychology. While the number of possible states of consciousness is virtually infinite, each of these states can be related to four main categories.

1) The common waking state (*'jagrat'*) where the sense-mind and our physical mentality are most active and almost completely identified with our bodily existence.

2) The dream state (*'svapna'*) where the psychic being enters into a series of life-planes and mind-planes, detaching from its corporal identification that was predominant during the waking state, but still with a physical mind active and which translates all the experiences on these planes in the form of an incoherent jumble, wondering phantasies, disordered associations from brain memories, etc. We most easily recall these experiences on vital and mental planes as vague reflections in form of confused dreams because they are filtered by the same physical mind through which we interpret the physical plane.

3) The (deep and dreamless) sleep-state (*'susupti'*) where the soul is liberated from the outward-going senses and enters the supramental plane, also more commonly called the *'casual body'*. Here the self becomes what Aurobindo describes as the 'Master of Wisdom and Knowledge' (*'Prajna'*).

4) Finally, the supreme or absolute self of being, the 'fourth' state of consciousness (*'turiya'*). It is a state of pure and absolute self-existence where the soul rests only for a relatively short period of (earthly) time before climbing back from this divine state to the ordinary waking state.

The gulf between our waking state and these other progressively deeper states of consciousness arises due to an untrained psychic being that experiences this state of 'trance' as a blank to the waking mind. The aim of classical yoga is precisely to bring these deeper dimensions to the surface and even rest in the fourth state during the waking state.

Thus, dreams are experiences on the inner planes filtered through our mind. If the consciousness dwells in the inner being or higher planes, the mind translates it into vivid and refreshing symbolic dreams, while an experience on the subconscious plane is reported as a series of incoherent and heavy dreams and eventually nightmares. The short time interval during which we really find ourselves in a state of sleep without dreams is that of 'psychic rest'—that is, when the psychic being retires into itself and momentarily detaches from all the outer and inner parts—and is the most refreshing part of sleep. Therefore, generally, dreams are the mind's interpretation of experiences on the subconscious, vital , mental, psychic, or higher spiritual planes. If we find our waking-state daily experiences reflected in a dream, it means that they have been registered in the subconscious, which tosses them up once again when the waking mind is quiescent and does not filter them out.

In this view, spiritually speaking, sleep is a necessary retreat of the soul from the burden of the material plane. Not only is it a physiological necessity, but it also has a psychological function ensuring that the soul finds back to itself from time to time. Sleep prevents our inmost essence from losing contact with its source and with other subtle non-physical realms. In fact, as is well known, sleep deprivation leads to severe psycho-physiological disorders and, in the extreme case, to physical death.

Another spiritual function of sleep is that of 'recharging' the life-sheet, or 'vital body' with its vital energy. When we don't sleep enough, we increasingly feel a sense of lack of vitality, energy deprivation, and inertia that no substance, food, or physical means can regenerate other than the sleep cycles, which restore a balance in the vital sheet with a vital force. This is, obviously, something modern science resolutely rejects. Nonetheless, intuitively we feel and know this from our daily waking-sleep experience but misinterpret that feeling and sense of weakness as purely physical because that's what we are told to believe.

Modern science is, in some sense, right in describing dreams as having the functional role of processing our emotional waking-life experiences to avoid an informational and experiential overload that we could otherwise barely handle. While scientists agree that sleep has the purpose of repairing and reorganizing neural pathways, consolidating memories acquired during the waking state, filtering out redundant information, restoring the body and mind, and serving other physiological and psychological functions, there is nevertheless a shared consensus that this can't be the whole story and that the real function and purpose of sleep remains elusive (for a review, see [274], [275]).

In fact, we don't know why all the organic functions of sleep mentioned above couldn't be performed in a waking state as well. But most puzzling is its evolutionary function. From a naturalistic evolutionary perspective, sleep looks like a bit of an outsider. According to Neo-Darwinism, everything— including sleep—must have evolved out of natural selection and random mutations to allow for the best survival and reproduction chances. But sleep is a risky habit in a prey-predator environment, especially if you are at the bottom of the food chain. Yet, there is no living organism, from cyanobacteria (photosynthetic bacteria), all the way to humans, that isn't subjected to a circadian clock that, in brains, expresses itself as what we call 'sleep'. There is no evidence to suggest that natural selection favored those able to stay awake longer. But there is evidence that even jumping spiders display REM-Like sleep behavior, suggesting spiders dream [276].

Furthermore, if sleep is a physiological necessity of the brain, having the function of reorganizing and repairing neural networks, one would expect to observe some direct correlation between brain size and sleep duration. Humans need approximately eight hours of sleep a night, but an elephant or a sperm whale that, as we have seen, has a three-to-six-times-larger brain, gets away with only three to four hours of sleep. And mammals such as mice and shrews, whose brains are much smaller than the human brain, sleep much longer than the length of the night. This is yet another fact that does not fit into a mind-brain identity theory. The naturalistic theory that identifies sleep as a brain cycle and a cerebral necessity is as questionable as the physicalists' mind-brain identification itself.

In fact, waking-sleep activity is not limited to organisms with a brain; those without a brain also show waking-sleep cycles. A sleep-like state has been observed in the cnidarian Hydra vulgaris, a small freshwater polyp with only a primitive nervous organization [277]. It is now known that sleep is also present in animals, such as the jellyfish Cassiopea, that possess neurons organized into a non-centralized nerve system but that have no brain. Their pulsing behavior, alternated by periods of quiescence at night, is consistent with waking-sleep cycles. When deprived of these quiescence periods, their activity and responsiveness decrease, indicative of a sleep-like state and supporting the hypothesis that sleep arose prior to the emergence of a centralized nervous system [278].

Of course, by speculating about any sort of evolutionary advantage and adaptive benefit, one can always imagine something that could explain away sleep's function in the orthodox evolutionary Darwinian paradigm. For example, we might think of sleep as something that keeps us fit by 'charging our physiological batteries' so that we can escape the predator during the day or avoid dangers that hide in the darkness. But what is it that 'recharges' during sleep? Why can't we recharge in a waking state? And the day/light cycles alone can't be the only reason for the wake/sleep cycles. There are also nocturnal animals. Then there are animals living on the poles, in the deep sea, in caves, or under the earth, where no such cycling exists and all, without exceptions, are nevertheless subjected to some circadian cycle. Fill this in with dozens of other hypotheses and conjectures. Imagining so-called 'evolutionary advantages' is easy and is like finding the culprit to blame for everything. You will always find something to cling to, but proving it to be the real cause is another matter.

The fact is that the evolutionary function of sleep remains unclear to science. When confronted with the hard fact that non-neuronal forms of life also sleep, at the time of this writing, some opined that the most plausible hypothesis is that sleep serves metabolic functions [279]. That might well be the case, but we predict that there will always remain an explanatory gap, a deep mystery that refuses to disappear. The true nature of the sleeping consciousness and its dreamlike character will be precluded forever to science if it doesn't accept and open itself to a first-person inquiry and research into other planes of being.

Perhaps, the first step in this direction could be that of thinking the other way around: Sleep isn't an adaptation; rather, it is evolution that adapted to sleep because it is a non-physical necessity of the inner being to temporarily disengage from the outer gross plane and reconnect with subtle domains. But, we believe, as long as science continues to ban the soul and vital forces as anathema, positing it a priori as inexistent, not because of a scientific rationale but because of an axiomatic and almost unquestioned assumption, it will hardly be able to go beyond a superficial understanding of the function of sleep and will continue to circle a mystery it can't make much sense of.

According to Aurobindo's integral system, another important aspect of our inner beings is that they are open to cosmic influences. The subliminal is in direct contact with the mental, vital, and physical environment surrounding it. Every individual receives subliminally subtle energies, forces, thoughts, and emotions via this *'environmental consciousness'* or *'circumconscience'*. This environmental consciousness is in relation to others beyond the physical body and is in contact with the universal forces. It is part of the inner consciousness surrounding the body, also popularly known as *'aura'*. It can catch emotions, thoughts, passions, suggestions, or forces of illness, affecting our subconscient and physical beings, but eventually crossing to the outer being into our outer conscious awareness.

For example, not all the emotions we are aware of are our emotions. The emotions, feelings, and more or less subtle vibrations of excitement we are normally aware of are those of the outer vital, but these impulses can be initiated by something coming from the outside or by 'vital waves' in the collective. The so-called *'empaths'*—that is, those who claim to be highly sensitive to the emotional state of people around them (eventually, also far away) and who sense the high or low vibrational vital energy they emanate— have become aware of an aspect of this environmental consciousness. However, this doesn't necessarily imply that the empath is empathic in the conventional psychological sense. One might be open to subtle energies in the environment without having particular empathy for others. Real empathy is not a quality of the vital but of the soul—the psychic being—that feels and realizes, inwardly, an intimate identity and connectedness with other humans and all living creatures beyond the separative surface appearances and personalities.[82]

Similarly, not all thoughts are our own; many come from outside of us as well. Several thoughts, ideas, and insights we take for our own come instead

---

[82] That's also why some show a high degree of empathy for members of their own ethnic group but none for others or feel an intimate connection with animals and none for humans, or vice versa. It is a 'vital empathy', not a soul or 'psychic empathy'.

from a collective thought or even from the cosmic environment. Besides the surface causes such as genetic or physical and social factors, this is what shapes our environmental intelligence. It is also the more occult effect that stands behind the multiple independent scientific discoveries made independently and almost simultaneously by different scientists who didn't know each other and had no apparent external contact and access to each other's information. This is something the ancient Greeks already knew and called the *'egregore'*, a non-physical collective thoughtform or psychological entity shared by a group or nation at a subtle non-visible level that, nevertheless, can be felt inwardly or somehow perceived intuitively.

Our soul, the inmost being, the psychic being, is also not a separate and isolated entity. Rather, it is in contact with the rest of the collective consciousness that forms families, clans, ethnic groups, or nations. It is no coincidence that one speaks of the 'soul of a nation', such as the 'American soul' or the 'Russian soul'. Of course, one can take these only as metaphorical images and not literally. However, if we ascribe an inherent reality to it, the interpretation of historical facts and the happenings of the human collective acquire a much deeper and more transparent significance, as we will see in Pt.III-III.8.

This environmental character of the inner being—that is, its openness to subtle forces and energies that come from beyond the individual boundaries—is, after all, not surprising. It has always been reported, more or less implicitly, by people throughout all cultures and times. Indeed, recent research tends to support how this is reflected in our brain activity. The nature of human cognition during social interaction triggers inter-brain neural phase synchronization across the brains of those involved in these interactions [280]. This inter-brain synchronization is described by the participant subjects in the form of social connectedness, engagement, cooperativeness, and experiences of social cohesion and 'self-other merging'. These findings challenge the idea that we are isolated and independent brains whose inner first-person experience is exclusively personal. We could speak of an 'extended consciousness' that crosses, subliminally, the boundaries of our discrete and personal bodily existence by a functional integration across individuals. Building a bridge with the above-mentioned modern spiritual cosmologies, we could also say that this environmental consciousness, which Aurobindo describes as having an individual as well as collective dimension, can be interpreted as being the missing link connecting the dissociative boundary of Kastrup's alter to Mind at Large.

At any rate, it is only by transforming, opening, and widening our consciousness that we can begin to more concretely distinguish between our personal and surrounding mental, vital, and subtle physical impulses. It is by becoming aware of this inner consciousness that we also extend our ordinary sense-perception to inner senses and get into contact with all those supra-physical and subtle domains that are popularly labeled 'paranormal'.

Therefore, according to this hierarchy of non-physical structures of integral yoga, the psychological existence of the ordinary human is a mixture, an interdependent cross-interaction of all these beings. We have many different intermixed personalities at different levels but what bubbles up on our outer surface nature is what we call our 'personality'. It is a 'concentric' system of being in the sense that all these parts are wrapped around the central soul, which is our true essence, identity, and individualized truth. However, as humans, we identify and define ourselves mostly by the outer mind, outer vital, and outer physical body misinterpreting it as the whole of our being. Our waking consciousness is only a small selection from several levels of consciousness, with the true conscious being the 'true person', far removed from that surface awareness. An inner subliminal consciousness exists behind our waking consciousness, with the subconscient and inconscient below and the superconscient above.

This fact is now accepted by Western psychology, though the recognition is based only on a vague concept derived from a third-person approach. For example, we misinterpret our surface emotions, such as love and hate, joy and sadness, euphoria and depression, for the true emotions and as something defining us, as that ego, that 'personality' that we take for ourselves, our true essence and nature. But once one starts becoming aware of how we are embedded in several layers of different forms of consciousnesses, we realize the illusion and the falsehood of such a simplistic conception of ourselves.

"This ego or "I" is not a lasting truth, much less our essential part; it is only a formation of Nature, a mental form of thought-centralisation in the perceiving and discriminating mind, a vital form of the centralisation of feeling and sensation in our parts of life, a form of physical conscious reception centralising substance and function of substance in our bodies. All that we internally are is not ego, but consciousness, soul or spirit. All that we externally and superficially are and do is not ego but Nature. An executive cosmic force shapes us and dictates through our temperament and environment and mentality so shaped, through our individualised formulation of the cosmic energies, our actions and their results. Truly, we do not think, will or act but thought occurs in us, will occurs in us, impulse and act occur in us; our ego-sense gathers around itself, refers to itself all this flow of natural activities. It is cosmic Force, it is Nature that forms the thought, imposes the will, imparts the impulse. Our body, mind and ego are a wave of that sea of force in action and do not govern it, but by it are governed and directed. The sadhaka in his progress towards truth and self-knowledge must come to a point where the soul opens its eyes of vision and recognises this truth of ego and this truth of works. He gives up the idea of a mental, vital, physical "I" that acts or governs action; he recognises that Prakriti, Force of cosmic nature

following her fixed modes, is the one and only worker in him and in all things and creatures.

But what has fixed the modes of Nature? Or who has originated and governs the movements of Force? There is a Consciousness—or a Conscient— behind that is the lord, witness, knower, enjoyer, upholder and source of sanction for her works; this consciousness is Soul or Purusha. Prakriti shapes the action in us; Purusha in her or behind her witnesses, assents, bears and upholds it. Prakriti forms the thought in our minds; Purusha in her or behind her knows the thought and the truth in it. Prakriti determines the result of the action; Purusha in her or behind her enjoys or suffers the consequence. Prakriti forms mind and body, labours over them, develops them; Purusha upholds the formation and evolution and sanctions each step of her works. Prakriti applies the Will-force which works in things and men; Purusha sets that Will-force to work by his vision of that which should be done. This Purusha is not the surface ego, but a silent Self, a source of Power, an originator and receiver of Knowledge behind the ego. Our mental "I" is only a false reflection of this Self, this Power, this Knowledge. This Purusha or supporting Consciousness is therefore the cause, recipient and support of all Nature's works, but he is not himself the doer. Prakriti, Nature- Force, in front and Shakti, Conscious-Force, Soul-Force behind her,—for these two are the inner and outer faces of the universal Mother,—account for all that is done in the universe. The universal Mother, Prakriti-Shakti, is the one and only worker." [265](Pt.I-Ch.VIII)

As a side consideration, it should be briefly mentioned that integral yoga naturally integrates the *'seven chakra system'*, the spiritual centers of the ancient yoga meditation practices. Just for completeness, as a reminder, let us recall them briefly.

At the base of the spine is the *'muladhara chakra'* or *'root chakra'*, where the dormant *'Kundalini'*—that is, the spiritual cosmic energy, the universal *'Shakti'*—resides and that, once awakened by the yogic practice, runs along the spine opening the other centers above. The Muladhara chakra is the center of the physical, the gross material body. In the Aurobindonian classification, the 'Svadhisthana chakra' at the root of the sexual organs is the center of the lower vital (you may have noticed how it tingles after intense emotions of anger or rage). The navel 'Manipura chakra' is the center of the higher vital (emotions are felt in the belly, aren't they?). The fourth center, the *'Anahata chakra'* or *'heart chakra'*, is another emotional center that is of utmost importance in integral yoga because it is the center *behind* which the psychic being resides (the pure emotions of love are felt in the heart, right?). The *'Vishuddhi chakra'* is the center in the throat of the externalizing physical and mechanical mind, while the *'Ajna chakra'*, the famous *'third eye'*, is the center of the dynamic thought and analytic reason, will, and vision (when we think we feel the thinking in the head). Notice how the yoga philosophy identifies the mind with a trans-physical center, not with the brain. Finally, above the head, the '*Sahasrara chakra*', also called the *'thousand-petaled lotus'*, opens to the higher hemisphere of consciousness (higher thinking mind, illumined mind, and intuition). When the Kundalini rises to this point, the practitioner goes into *'samadhi'* —that is, establishes their consciousness in the Absolute, where our subjective identity coalesces with Satchitananda. In other words, the object and subject lose their distinction.

However, in Sri Aurobindo's yoga, the awakening of Kundalini and its gradual ascension upwards opening these centers is neither a necessary nor an encouraged practice. Integral yoga relies on the purification and transformation of the being performed by a spiritual Shakti that acts from above downwards. There is no forced attempt to awaken an ascending Kundalini; rather, one prefers to surrender to the descending spiritual Force and Wisdom that, if allowed, will deal with the purification and transformation acting from the psychic center and the centers above the head. One lets the Divine force descend into us from above instead of awakening an energy from below. This is a safer path because, after all, the Divine is supposed to know better than us what must be done, as opposed to a universal force, like Kundalini, which is still part of a cosmic vital and mechanical energy. Integral yoga aims to transform the outer surface consciousness to the image of the inner Divinity and live in it permanently. This can be done only by a psychic and spiritual transformation, as we will describe in the next sections.

e. The Psychic Being and Psychic Transformation

Yoga is a series of psychophysiological transformations based on spiritual practices aimed at obtaining a self-perfection towards higher states of consciousness that can lead us beyond our present mentalized nature. Sri Aurobindo spoke of his own process of ascesis as a *'triple transformation'*: the psychic, the spiritual, and the supramental transformation.

The first transformation, *'the psychic transformation'*, brings the soul, our inmost spiritual individuality, our 'mystical heart'—that is, the psychic being—to the front. It will then take the lead and determine the transformation of all the other parts. The qualities of the psychic being are primarily of a higher emotional nature, such as love, beauty, a perception of the truth that does not need mental inferences, a sense of good, harmony, emotional equanimity, sweetness, peace, and an intimate relationship with the Divine. It is the psychic being that gives the saint or mystic the feeling of being a 'child of God' in intimate communion and in love with the Divine, the *'bhakta'*, as it is called in Sanskrit terminology. These qualities are always present in our intimate nature but hidden behind the veil of our untransformed mental, emotional, and physical restless activities.

Of course, the soul is, per definition, not a physical entity. Nevertheless, it can act from a physical place and from there transmute our whole mental, vital, and physical individuality. In fact, the psychic being is centered behind the heart region but is not the physical heart. The ordinary human being is only vaguely aware of it (the popular 'inner voice' in our heart whispering to us is sometimes the dim and barely intelligible communication from the psychic through the inner being to our surface mind). This heart center is also described in different forms throughout many traditions, particularly in Sufism, which speaks of the *'five levels of the heart'*. The Trappist monk, mystic, and author Thomas Merton called it the *'point vierge'* —the virgin point— describing it as a *"center of our being, a point of nothingness which is untouched by sin and by illusion, a point of pure truth, a point or spark which belongs entirely to God... which is inaccessible to the fantasies of our own mind or the brutalities of our own will"* [281].

But, in Aurobindo's evolutionary cosmology, this pure 'inner flame' is not just a static and eternal entity; rather, it is the very evolutionary essence of each individual that evolves life after life in the process of transmigration. The psychic being is the evolutionary outgrowth of the *'psychic entity'* or 'divine spark', our true inmost individuality, an individual Self that links the impersonal Absolute with a personal identity in a material, vital, and physical individuality and that knows us better than we know ourselves through our outer personality. During its life experiences and its evolutionary journey of transmigrations, it is this psychic entity that forms the psychic being. It grows and gains *"strength every time there is a higher movement in us, and, finally, by the accumulation of these deeper and higher movements there is developed a psychic individuality–that which we call usually the psychic being."* [271](Vol. I, Pt.II, Sec.II)

However, the psychic being should not be confused with what we call our 'character', 'personality', or 'ego-sense'. While its distinctive trait is of a divine emotional nature, our surface emotions are mostly determined by our outer vital impulses, mechanical habits, physical and emotional preferences, mental tendencies, unaware prejudices, convictions, etc. These are only aspects of our surface untransformed personality that are influenced by the socio-cultural context, our education, the environment, genetic factors, and, especially, the maturity of the soul and its experiences accumulated in previous lives, which become the blueprint for our present life (popularly known as 'karma'). What we call the 'ego' is a shadow cast on our superficial being, not something that defines our true inmost nature. But our surface personality also has precisely those psychological traits necessary for this evolutionary soul to grow and evolve during a lifetime. Only once this inmost true personality has sufficiently developed itself can it come to the surface and impose itself over our outer superficial nature. At that point, it can begin to govern and transform our whole being according to higher directives without the lower-Nature as the tyrant external ruler, but by a direct action from the inside-out, which aims to turn every plane and part of being towards the Divine. It is this soul that knows what is good for us even though our mind, most of the time, thinks otherwise. In other words, according to Aurobindo, the psychic being is our evolutionary soul that progressively develops out of the nescience of the mineral, first in the plant, then the animal, now in the human, and finally in the future, overman or '*superman*'.

The interesting aspect that distinguishes this spiritual evolutionary perspective from a Darwinian conception that sees life as a struggle for survival and a ruthless selection process is that, at least for humans, evolution does not necessarily have to be a painful process of battles and sorrows. We can consciously and willingly participate in this process by taking up our own evolution and allowing this true soul to come to the front to take control of our lives and their further evolution. Only by a transformation of our internal and external nature can our true spiritual individuality emerge and convert the Prakriti of ignorance into a Prakriti of knowledge.

However, as long as the psychic being remains hidden behind the veil of activities of the material existence made of bodily sensations and appetites, the gross emotional impulses, and the mental activities, it is subjected mainly to a passive evolution, and all that we are (or believe to be) is subjected mostly to the clash of the outer mental, vital, and physical forces and the outer events that escape our control in a domain of ignorance made of alternating pains and pleasures, rest and stress, ups and downs determined by the mechanical aspect of Nature or, in short, by what we call 'destiny'. The psychic being always supports, behind the veil, the body, mind, and vital but, as long it does not become active on the surface, it will have to be subjected to the play of blind forces. This is the ordinary natural evolutionary process from life to life by a series of transmigrations, as has been the case so far in human history, and that is the normal state of affairs in the animal kingdom.[83]

The centrality of this evolutionary soul-dimension in Sri Aurobindo can't be emphasized enough. In the traditional Indian philosophy, such as Vedanta, one speaks of the '*Self*' or the '*Atman*', which is the reflex of the eternal and immutable Brahman itself. In this non-dual philosophy, one rises above the ordinary state of consciousness and awareness by psychological, mental, and eventually also physical practices, such as yoga and meditation. This culminates in the realization of the non-existence of the ego—that is, the awareness that this also is a figment and an unreal entity that is itself part of phenomenality. Non-dual masters refrain from speaking of a 'soul' and, if they do, they don't mean the personality identifying itself with an "I" having a feeling of selfhood. Rather, they mean something that transcends it and is

---

[83] One could point out, however, that because the animal mind is not as active as the mind in the human being, a spontaneous and harmonious adaptation to the natural environment is much easier for the animal than for the human. The animal kingdom not held in captivity by the human may be less subjected to suffering than the human, with its history of self-inflicted pain.

always equal to itself, that is an 'atemporal beingness' or something that words can't express other than saying that it is 'nothing' or 'emptiness', 'vacuity', the Buddhist *'sunya'*, and once realized, leads to 'liberation', *'moksha'*, or *'nirvana'*, the release from suffering due to the absence of mind's activities. Here, the concept of the soul, if accepted at all, is usually considered a static, passive, and immutable witness representing an immaterial and spiritual individuality in us and that preserves its existence after death. At least in the Eastern tradition, it then continues its journey in a cycle of reincarnations but is never conceived of as a spiritual individuality able to change and evolve through this process of transmigration. It is a witness that essentially does not influence and determine our life. The notion of an evolutionary soul is absent.[84]

The metaphysical assumption of a changeless soul also conditioned modern forms of Western psychology. Due to the secular approach of most psychological schools, the hypothesis of an immaterial soul-entity is ignored. However, also among those forms of psychological theories and practices that are open to spiritual perspectives (for example, the *'transpersonal psychology'* of Ken Wilber or the *'psychosynthesis'* of Roberto Assagioli), the tendency is to attribute little or no importance to the soul as having any potentiality to influence our psyche and life, let alone ascribe to it an evolutionary nature and purpose. This a priori assumption, like so many rationalistic and naturalistic gospels, may have severely hampered further progress of psychological sciences (for an interesting review of this aspect, see the research of E. M. Teklinski [282], [283]).

In contrast, the psychic being is quite different from the soul concept of the ancient Vedantic schools or modern forms of Western metaphysics, philosophy, or psychologies. The particularity that distinguishes integral yoga from other paths is that it specifically introduces a spiritual evolutionary dimension in which other traditions stop at the spiritual realization. But the integrality of Sri Aurobindo's spirituality does not abolish or contradict the past knowledge. Rather, it extends, complements, and integrates it.

Moreover, it distinguishes between the psychic being evolving in a temporal manifestation and a non-evolutionary self-existent and timeless spiritual individuality beyond space and time, the *'Jivatman'* or just *'Jiva'*. The Jiva is the individual self that does not change and is the same throughout all the outer changes. It is the spiritual individuality that consents to the play of Nature but is not affected by it. It is the inmost individual witness-consciousness that does not evolve but presides over all the personality. Both the Jiva and the psychic being are a 'spark of the Divine' and are two aspects of the same individuality.

---

[84] The so-called 'higher self' that is nowadays in fashion among modern 'channelers' might have analogies with the psychic being. But we prefer to refrain from too straightforward comparisons.

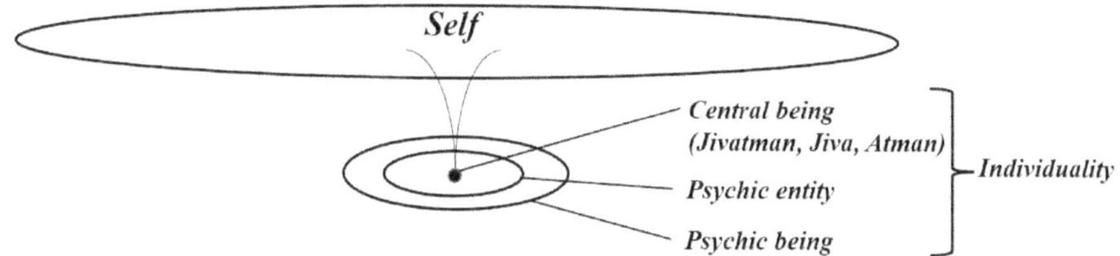

Fig. 122 The evolutionary soul as a 'spark of the Divine'.

But the Jiva is the *'central being'* above evolution and projects the evolutionary psychic entity into Nature (*not* the other way around). It is the Jiva that merges itself in identity with the Divine but knows itself as a center of the multiple Divine. As an analogy, we might use the holographic principle (recall Pt.II-III.2b): We can metaphorically see the Divine as the entire hologram, while a portion or reflection of it is the Jiva. This portion or reflection, however, still contains the image of the whole (just at a 'lower resolution', so to speak). It is still the image of the same divinity but in its individualized form in Nature reflects only part of its potentiality. This is an aspect we will take up again later.

Thus, from the perspective of this integral evolutionary cosmology, death is not a malady or an inevitable accident. Death composes part of the evolutionary process that allows an evolutionary soul to create different experiences in different forms and lives. Physically, it stems from the inability of the material sheet—that is, the physical, biological cells—to keep up with the inner pace of transformations. The untransformed body that still has not been subjected to a higher spiritual transformation isn't able to keep up with the pace and plasticity of the psychic being in the development it could achieve due to its experiential life journey and all of its transmigrations. Death is a transition that allows for a further transformation, not an annihilation or a reset that throws us back to the point of departure.

f. Planes and Parts of the Higher Hemisphere and the Spiritual Transformation

While the first transformation—the psychic transformation—is primarily concerned with bringing our soul qualities such as love, beauty, equanimity, good, and peace to the surface, the second transformation, which Sri Aurobindo called the *'spiritual transformation'*, has as its main objective an ascension to higher states of cognition beyond the ordinary mental one. [85]

---

[85] This is not to say that the psychic transformation isn't 'spiritual' in the ordinary sense of the term. Aurobindo, however, distinguishes the two types of mystic transformations as 'psychic' and 'spiritual' because of their distinct source and action: One coming from the heart center, the other from the centers above the head.

But what does it mean to 'ascend' towards higher states of consciousness beyond the mind? Is it about simply switching from a rational to some trans-rational wisdom or from ordinary virtue to sainthood?

As usual, words and our preconceived ideas can be misleading. Moreover, reality is not black and white. Several gradations exist between truth and falsehood, beauty and ugliness, or, as is the case here, between unconsciousness and supra-consciousness. It is only the ordinary intellect or the sense-mind in its limitation that sees simply a body, soul, and mind but nothing else. The coarse-grained cognition of the physicalist, and most of the Western philosophers as well, have in common the fact that they both despise any further gradation as an unnecessary hypothesis. However, by doing so, they are blinding themselves again with the trap of the law of parsimony. But once we ascend to higher states of consciousness, a vast universe discloses itself to us which couldn't care less about Occam razors or any anthropomorphic self-imposed rules. Once someone experiences reality in its multiplicity, gradations, shades, and diversity, and especially its complexity, any binary or parsimonious thought pattern becomes insufficient to embrace it.

Sri Aurobindo described these gradations that surpass the mental consciousness as a series of sublimations of The Consciousness. There is not just mind and a trans-rational mind but several steps on the ladder of consciousness. He described it according to an ascending scheme that, at the bottom, places the mind proper or the conventional reason and analytic mind. This is followed by the *'higher-mind'*, which we have amply referred to throughout this treatise, but also continues beyond it, towards even higher cognitive states, namely, an *'illumined mind'*, an *'intuitional mind'*, an *'overmind'*, and, at the summit, the *'Supermind'* or *'divine Gnosis'*.

Let us unpack this. First, each stage of this ascent must not be confused with a magnification of our present intelligence. It is not about being able to parallel the calculating power of a supercomputer or skyrocket our IQ. All these (suffixed) 'minds' are not the aggrandizement of rational analytic skills. Rather, they represent a deeper 'seeing' and a further intuitive as spiritual knowledge of things that the mind cannot conceive, let alone achieve.

Moreover, this ascension is not limited to the 'seeing' of the world but, rather, aims at a transmutation of our entire nature, converting our being by a new light and power. Aurobindo contends that, as the appearance of mind in the first humans led to the species of homo sapiens, so will the appearance of these higher cognitive faculties in the ladder of the evolutionary process of consciousness lead to a species beyond the human. After all, it is in perfect agreement with modern Darwinian evolutionary theories, though it adds the inner cognitive and spiritual dimension that conventional science ignores.

Without going into too many details,[86] we can characterize these different levels of consciousness and cognition as follows:

The first layer above mind is the *'higher-mind'* which we have already become a bit familiar with. The peculiar aspect of the higher-mind (which must not be confused with intuition, which we will deal with later on) is a large *"clarity of the spirit"* capable of a spontaneous inherent knowledge by a *"luminous thought-mind"*, a mind of *"spirit-born conceptual knowledge"*. This kind of cognition is the first that meets us when we rise from a conceptive and rational mind and is considered the first step out of ignorance. It is a knowledge-power that tips into the first layers of a spiritual realm. It no longer proceeds by a stepwise logical process of inference and deduction but, rather, by a greater thought which does not need a seeking ratiocination because *"this limping action of our reason is a movement of ignorance searching for knowledge"*. Instead, the higher-mind proceeds by a *"harmony of significances put into thought-form"*. It apprehends things in a single view establishing relations not by logic but by the vision of an integral whole, not an acquired knowledge. As Plato and many others after him more or less subliminally figured out, form appears in this higher mind as an initiation of something beyond the form itself. However, this higher-mind is not just a passive tool of cognition. It also possesses an inherent power of action and creation that expresses itself in feelings, in life, and even in the body through the power and force of the idea it presents (an example of the latter could be that of a more effective self-healing enacted by thought power). Therefore, it also has an active and creative aspect which, once realized and established, would inevitably manifest in our emotional and physical life with ideas and ideals becoming concrete and tangible.

But the higher mind is only a rung of the ascending ladder towards a full gnostic knowledge. The *'illumined mind'* surpasses the vision of the higher-mind and goes beyond, from the higher thought into a cognition guided by a 'spiritual light'. It is characterized by a *"clarity of the spiritual intelligence ... and illumination of the spirit: a play of lightnings of spiritual truth"*. It is a spiritual-cognitive calm and wide enlightenment that operates not by thought but by vision. In this spiritual light, there are no verbal representations; it is a consciousness that proceeds by an inner and more direct sight, way beyond any intellectual thought-conception. It is a *"spiritual sense that seizes something of the substance of Truth"*. Like the higher mind, the illumined mind is not just a passive container but brings with its spiritual vision a light and energy into the heart, the emotions down to the physical a truth and inspiration that expresses itself as a spiritual life-force. It infuses itself in the whole being and transmutes even the physical mind—that is, the mind of the physicalist that

---

[86] A detailed description is given, for example, in [266], Ch. XXVI, "The ascend towards Supermind". The short citations in this section are derived therefrom, if not indicated otherwise.

believes in the physical materiality enlarging it only by *"replacing its narrow thought-power and its doubts by sight and pours luminosity and consciousness into the very cells of the body"*.

At a higher summit, we find, then, the 'intuitional being'. A word of caution is needed here with regards to the word 'intuition'. It is not about an instinct or the proverbial 'gut feeling' that, most of the time, is an impulse of opposite quality and nature. Sri Aurobindo, by intuition, does not even mean the same thing that we can perceive from time to time even in our ordinary state of consciousness and that, sooner or later, everyone has experienced as the little 'voice' of our conscience vaguely whispering to us. Although, in rare instances, this might also have its origin in the higher domains of cognition or come from the psychic in the heart. But most often, this 'whispering' is subjected to an invading mixture or a mental coating or interception and substitution with subconscious or nervous instinctual movements. What we commonly call 'intuition' is usually a seeming or camouflaged intuition—a communication rather than intuition—or, at best, an inner intimation coming from the psychic being or related to the higher mind.

Whereas, in this hierarchy of consciousness, the term 'intuition' is of a quite different quality and power of consciousness, which stands opposite the 'instinct' proper to an animal consciousness or our subconscious mechanical nervous reactions. Intuition, in the terminology of integral yoga, refers to something that comes closer to an original knowledge by identity. The *"consciousness of the subject meets with the consciousness in the object, penetrates it and sees, feels or vibrates with the truth of what it contacts"*. In other words, intuition is the first step towards a direct contact and 'seeing' of Kant's 'thing-in-itself'. It is a consciousness that begins to be able to do what reason (especially Kant's reason) believed to be impossible, namely, to intuit the noumenon standing behind the phenomenon. This intuition, this 'lightning-flash' that emerges from the contact between the subject's consciousness and the object it contacts, is also a cognition of the hidden forces behind the appearances it meets—a union between subject and object, the first spark of intimate 'truth-perception', a penetrating and revealing touch beyond sight and conception. It is a light of truth, a 'truth-remembrance', so-called because it does not know by learning things but, rather, puts us in contact with something we always already knew to be inherently true. It is an automatic certitude that does not proceed from any reasoned conclusion or verified conjecture. It is a supra-rational (not infra-rational) source of pure and native intuition in which the lower intelligence of reason can be only an observer or registrar. Aurobindo describes it as having four powers: a power of revelatory truth seeing, a power of inspiration or truth-hearing, a power of immediate seizing of significance, and a power of true and automatic discrimination. It can, therefore, perform all the actions of reason and much more than that. Also, intuition is not limited to a passive cognition of things but takes up and transforms the other parts and planes of the being, namely, the mind, the vital and the physical senses imparting to it a deeper perception, leading us towards a divination of the body and a greater integrality and perfection. It can thus change and recast the whole consciousness from a state of ignorance into the stuff of intuition transmuting our will, feelings and emotions, actions, and the very workings of the body consciousness.

However, intuition is still not the true knowledge by identity. Our subconscient and inconscient basis is too vast, deep, and solid to be altogether penetrated and transformed by it. The next step in the ascent brings us to the overmind, a power of '*cosmic consciousness*', a principle of global knowledge. When the overmind descends, the centralizing ego-sense is finally abolished and replaced by a *"wide cosmic perception and feeling of a boundless universal self"*. *"Many motions that were formerly ego-centric may still continue, but they occur as currents or ripples in the cosmic wideness"*. Thoughts are no longer perceived as an individual act of cognition of a separate self but are realized as 'cosmic Mind-waves'. *"Feelings, emotions, sensations are similarly felt as waves from the same cosmic immensity"* with the body being a mere point for the action of vast cosmic instrumentation. The sense of individuality, even that of our body and mind individuality, entirely disappears and reveals itself as the delightful play of cosmic forces alone. The overmind consciousness is an unlimited consciousness of unity with the sense of the universe in oneself, or as oneself, as an extension constituting a cosmic being, a universal individual.

The overmind is governed by the directions of the cosmic Self, where the body is a physical instrument recognized as *"something instrumental to the action of a Transcendent and Universal Being"*, a cosmic center of the action of the Infinite. It is experienced as a consciousness of Light and Truth with a sensation of beauty and delight. Here also, *"all essential experiences belonging to the mind, life, body are taken up and spiritualised, transmuted and felt as forms of the consciousness, delight, power of the infinite existence"*. The other forms of cognition and active powers below, those of Intuition, illumined sight, and thought, are themselves enlarged by the action of the overmind. The nature of the being with all its thoughts, feelings, and bodily activities becomes more universal, all-understanding, all-embracing, cosmic, and infinite. It is a principle of separate dynamism, a source of creative power that *"though luminous itself, keeps from us the full indivisible supramental Light, depends on it indeed, but in receiving it, divides, distributes, breaks it up into separated aspects, powers, multiplicities of all kinds."* [271](Vol. I, pg. 138)

This overmental transformation completes the spiritual transformation.

g. The Supermind

But the overmind still contains the seeds of a separative and divisive character in the cosmic play. Despite its basis in a cosmic unity, its action is one of division and interaction in the play of multiplicity. It is only the Supermind, the *"supreme self-determining truth-action and the direct power of manifestation of that Transcendence"*, that can accomplish the final transmutation–that is, the supramental transformation. While the overmind allows the individual consciousness to be universalized and, to some degree, transforms the lower parts of the being, a basis of untransformed nescience in matter remains, which only the Supermind can transform completely. The Supermind is *"the supreme power of the principle of unity taking all diversities into itself and controlling them as parts of the unity, which must be the law of the new evolutionary consciousness"*. [266](Ch.XXVI) Otherwise, the pull of the inconscience would prevent a divine or 'gnostic evolution'. The transformation of the inert depths of the subconscient and inconscient is possible only by the descent of the supramental light into matter, life, and mind, penetrating it down to the material basis. Even overmind would begin to be transformed because everything is a power of the Supermind in its origin. The spiritual significance of our whole existence would be revealed by the ascent of the individual consciousness to Supermind and the descent of the Supermind in life, mind, and matter with the 'gnostic Light' effectuating a complete transformation of our ignorant base.

The teleological character of Sri Aurobindo's vision is that matter will reveal itself as an instrument of the manifestation of Spirit. The Supermind is the highest level of consciousness, a *'Truth-Consciousness'*, *'Truth-Knowledge'*, or *'Knowledge by identity'*, which would be proper to the new species, the *'gnostic being'*. It is with the Knowledge-by-identity that one knows the things-in-themselves, with all due respect to Kant, who claimed that to be impossible. The supramental vision brings us back to our original unity in the creation— a holistic unity in which nothing is separated and where all is One. *"It sees everything from the stand-point of oneness and regards all things, even the greatest multiplicity and diversity, in the light of that oneness"*. [266](Ch.XXVIII) The supramental sense is that of an ultimate Unity, Oneness, and Wholeness of the Infinite, where everything we perceive on our ordinary mental level is like a hologram, a *"drop that is yet a concentration of the whole ocean and inseparable from the ocean"*. [265](Ch.XXIV) Yet it is not the unity of the mind but a greater sense of unity inherent in all things as a manifestation of the only one thing.

This gnosis goes, therefore, much further than a practice of philosophical idealism. In the supramental consciousness, one becomes aware of all phenomenal events in Nature and the cosmos as the result of a creative Idea, the *'Real-Idea'*, or *'Seed-Idea'*, a power of Consciousness expressive of the real Being. The Supermind is the Truth-Consciousness that knows itself and sees the universe as a subordinate reality resulting from an overmental creative process of differentiation of its own Real-Idea. The Real-Idea is the 'ultimate archetype' that the higher-mind philosopher captures as the hidden and yet clearly perceptible 'idea-form' in things and Nature. The mind interprets it in some name and form. By 'names' Aurobindo meant not just the words describing objects, but in its deeper sense, the powers, qualities, and characters (features and traits that distinguish one thing from another) of the form, and being caught up by our consciousness. In this sense, all 'names' are already latent and inherent in the nameless and timeless Absolute but are expressed by the Supermind in the temporal manifestation as qualities, powers, and characters of its forms. Aurobindo's 'names and forms' are something reminiscent of the Aristotelian hylomorphism extending it to inherent powers and qualities in all things.

The Real-Idea is an expression of a potentiality enacted in the manifestation by a supreme *'Will-Force'*, as an aspect of a *'Consciousness-Force'* (the *'Chit-Tapas'* or *'Chit-Shakti'* of the Vedanta), in the form of cosmic energy which, with its creative formative processes, manifest that Archetype. According to Aurobindo, once we rise at supramental heights, we recognize how it is ultimately this immanent Real-Idea with its expressive, active Will-Force that stands behind all the workings and phenomena of an apparently unconscious and mechanical universe. In the Unmanifest, consciousness and will and force are the very same and undifferentiated thing. There is no distinction between these three aspects of the Absolute. It is when Satchitananda plunges into the manifestation of polarities and division that consciousness, will, and force appear as three distinct qualities seemingly independent from each other.

One can't fail to notice the similarity, even in the terminology, of Schelling's, Schopenhauer's, or Hegel's 'World-Soul' and 'World-Will'. However, the decisive difference is that a supramental transformation is no longer a philosophical higher-mind speculation but, rather, a lived experience that knows a truth by direct sight, vision, and knowledge by identity. It is one thing to think, ponder, and philosophize with the limited power of reason, eventually also backed by an inner perception of deeper truths and subtle spiritual or psychic intimations. It is another to ascend the ladder of consciousness and become fully aware of the inherent reality that human cognition cannot fathom.

This also subverts our understanding of how mind works. Mental processes are not only determined by a bottom-up mechanism that makes the thought, the mental construct, and the mental idea emerge as a combination and interaction of elementary neural units. In Aurobindo's accounts, the brain is not much more than the physical mind, with its subconscious thoughts or hardwired mental and physical reactions. The most illumined and deepest inspirations upon which our mind seizes are caused by a projection from above–that is, by the mind that captures a fragment, a distant and dull representation or coarse-grained image of the Real-Idea of the Supermind or other spiritual planes. It is a top-down process in which a divine intuition percolates through the layers from the Supermind down to the mind and, thereby, is reduced, filtered, and eventually twisted into a mental reduced knowledge and meaning that has lost most of its original insight and knowledge by identity, retaining only a distorted and relative truth of things. This is what happens when we speak about 'intuitions' and generally wonder how 'meaning', the semantic content of things, concepts, and ideas, comes into being. Mental ideas and meanings are not 'created' or 'generated' by either the brain or the mind itself. Mind captures them only during their transit downwards from the most refined to the coarser forms of consciousness. What we seize as meaning is a redux and pale diminished shadow of the original supermental Real-Meaning of the Real-Idea. The semantic content of a word, symbol, or something we perceive through our senses and mind is something that was originally a 'Truth-Seeing', which, in a fraction of a second, becomes a mental half-ignorant partial image that the mind translates in its own categories and takes for the whole truth. This is, deep down, the true reason why AI seems so utterly incapable of 'understanding' the world. AI research trying to create a machine that has human semantic abilities is a naïve attempt to reproduce something by a bottom-up approach that is, instead, a cognitive process that, in its origin, works the other way around. But AI scientists don't know that, and even if they did, in their physicalist and neuronal cage, they would continue to deny that and, therefore, would have to go through many other disillusionments. My thesis (that most disagree with) is that a machine, however complicated and advanced, will never understand anything, not even in principle.

Whatever the case, according to the integral seer, the main gnostic trait of the Supermind, a 'delegate' of an extra-cosmic divine power, the immanent Godhead in the manifestation, is a vision far from any divisive mental cognition that works by polarities. The supramental vision is always guided by a principle of unity, oneness, wholeness, and, yet, knowing and acting in diversity and multiplicity.

"The fundamental nature of this Supermind is that all its knowledge is originally a knowledge by identity and oneness and even when it makes numberless apparent divisions and discriminating modifications in itself, still all the knowledge that operates in its workings, even in these divisions, is founded upon and sustained and lit and guided by this perfect knowledge by identity and oneness. The Spirit is one everywhere and it knows all things as itself and in itself, so sees them always and therefore knows them intimately, completely, in their reality as well as their appearance, in their truth, their law, the entire spirit and sense and figure of their nature and their workings. When it sees anything as an object of knowledge, it yet sees it as itself and in itself, and not as a thing other than or divided from it about which therefore it

would at first be ignorant of the nature, constitution and workings and have to learn about them, as the mind is at first ignorant of its object and has to learn about it because the mind is separated from its object and regards and senses and meets it as something other than itself and external to its own being." [265](Ch. XIX)

Not a knowledge that proceeds by logical steps, inferences, or hypothesis to be tested; these are the inferior methods of reason, of an *'inflexible machinery of the rational intelligence'* —that is, a form of cognition proper to the domain of an obscure ignorance and that is compelled to search for truth by an inductive process or piecewise analytic procedure revealing the truth of things only partially and bit by bit. The supramental gnosis instead directly reveals the truth of things by an immediate, direct, and inherent knowledge. All that is mental appears as a deformed, suppressed, pale, and partial imperfect figure of that original divine cognition. Nonetheless, the Supermind in the lower nature is felt most strongly as intuition, and it is for this reason that by exercising the intuitive mind, we can foster a process of ascension towards higher domains than the ordinary mental, vital, and physical.

Moreover, the supramental consciousness has a completely different time perception in which past, present, and future are seen unitarily in one and a single continuous map of knowledge. This *'three-fold seeing'* of time (*'trikaladrishti'*) links past, present, and future in an indivisible whole (more on this later).

This is the ascending ladder of the higher hemisphere of consciousness from mind to Supermind according to Aurobindo's scheme.

h. The Vertical System of Being and Cosmic Consciousness

To summarize, Diagram 1 complements Table 1, displaying the spectrum of the vertical system of consciousness according to Sri Aurobindo's cosmology.

The enormous hiatus that separates a divine superconscience down to the physical (and eventually even below that) should be considered as a continuum, not a fixed and well-defined number of steps having sharp boundaries.[87] In a certain sense, the number of planes and where these lines of demarcation must be drawn is a matter of subjective evaluation and may vary and depend on the cultural context or the mystic experiencing it. This, however, doesn't make it less real or is proof of the illusionary nature of what is experienced, no more and no less than a conventional demarcation between day and night makes the sunlight unreal or an illusion.

---

[87] This, by the way, could shed even further light and resolve the famous and longstanding mind-body interaction problem we shortly addressed in Pt.II-II-3j. But we will not dwell further on this aspect here.

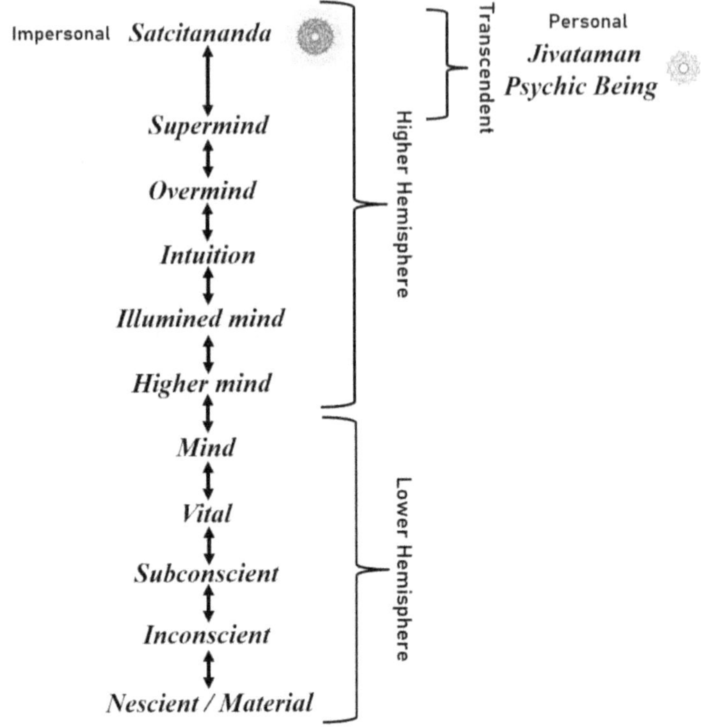

Diagram 1 The vertical system of the planes and parts of being.

Note that this is not just a structure of consciousness related to an individual being. It is a cosmology of consciousness applying to the universe as a whole. Like the psychic being that connects the transcendent with the individual, the Supermind links the unmanifest Divine, the Satchitananda, the Vedic Brahman, the Absolute beyond space and time with its immanent and dynamic cosmic universal manifestation. The Supermind is the 'delegate' of this Absolute in the temporal and spatial manifestation. The physical universe we are aware of in our ordinary mental consciousness is only an aspect of this manifestation.

As we will see later in more detail, in this integral cosmology, the individual mind is a localization in a universal plane of existence of a cosmic Mind or universal Mind. This is something we already found in cosmopsychism, analytic idealism, and panspiritism in Pt.II-II.3. But, there is a decisive difference: An integral cosmology detaches from the coarse-grained dualism between matter and mind (or consciousness) or body and spirit and embraces reality in its full spectrum with all its universal (or cosmic) gradations, which we roughly illustrated in Diagram 1 together with the concentric system of Table 1. This universality fits all the planes. We can generally speak of a cosmic consciousness incorporating not only a cosmic mental, cosmic vital, and cosmic physical, but also a cosmic subconscient and inconscient.[88]

Perhaps the most difficult aspect to accept for some, especially the scientifically minded Neo-Darwinist, is that all these planes of existence are *not* the creation of an evolutionary process. They preceded the creation of earthly life and even the physical universe itself. This is to say, for example, that the mind is *not* a form of cognition resulting from an evolutionary process. Mentality and life emerge in the material world—that is, the cosmic physical— but are not created by or do not originate from it. As paradoxical as that might sound, the terrestrial physical life is also the outer manifestation of a life-force that pre-existed, in a subtler form, the material universe itself. Goethe 's inner perception that there is a wholeness of Nature standing behind the phenomenal universe was not coincidental. Even the subconscient and inconscient, being universalized and of which our individual subconscient and inconscient are only a particularization, a segment, are states of consciousness that pre-existed the material creation. This amounts to saying that life and mind were already immanent and existing before the Big Bang.

The planes and parts of being according to integral yoga are more complex than the structure we presented here. However, this was intended as a first overview, which will suffice for our purposes. For a more detailed description of Sri Aurobindo's 'topography of consciousness', the reader will have to resort to his writings (especially [266] and [271], or for a specialized redux, see also [284]).

The central message of the integral yogi is that we are made of many parts, most of which we are completely unaware of, but each of which influences our thinking, feelings, and actions. Almost every one of our conscious thoughts and impulses is the result of a confused pell-mell that bubbles on a surface beneath which we consciously know almost nothing about. We mistake the part for the whole and know only a little part of that part. There are several parts of our being we don't know of and eventually deny could exist. Once one gets into conscious contact with it, one realizes that the mental, vital, and physical are not just one independent block but, rather, a complex amalgam of other interconnected layers, such as an 'outer', 'inner', and 'true' being for each of the mental, vital, and physical with the inmost center, the psychic being, which stands behind them all. These planes and parts are intermingled and interdependent and have no clear-cut boundaries. To add a further layer of complexity, they are also subjected to universal forces that we are not aware of or that we mistake for our own. We are a very complex combination of inextricable layers of consciousness that are all non-physical (except, of course, the cosmic physical and the outer physical—that is, the physical

---

[88] Something reminiscent of a Jungian 'collective unconscious' but we prefer to refrain from too simplistic equivalences.

universe and our body, respectively). We should never mistake the surface personality we know from our ordinary waking state for the whole and not even for our true personality (something we do much too easily when we talk about 'consciousness', meaning the surface waking consciousness). In other words, we as humans, with all our marvelous sciences and achievements, are still beings immersed in deep ignorance and self-forgetfulness.

5. The Gnostic Being

a. Involution and Evolution

> *"He sleeps in the atom and the burning star,*
> *He sleeps in man and god and beast and stone:*
> *Because he is there the Inconscient does its work,*
> *Because he is there the world forgets to die.*
> *He is the centre of the circle of God,*
> *He the circumference of Nature's run."*
>
> Sri Aurobindo – Savitri, Bk. XI, Canto I

The word *'evolution'* comes from the Latin' evolvere', which means to 'unroll' or 'open out' or 'uncover' (the prefix 'e' means 'out' and 'volvere' stands for 'rolling'). Interestingly, contrary to the current understanding of 'descent with modification', which is the modern Darwinian accepted notion of biological evolution, the term 'evolution' betrays lost wisdom. It is an emergent process that unfolds something inherently already contained in the thing that unrolls. Also, the word 'development' implies that something develops by coming out of an envelope. It seems to suggest that there is something already fully developed but concealed or hidden and that evolution unveils—something that was previously veiled by an opposite process of evolution, an 'involution'. But it is not something material, molecular or genetic.

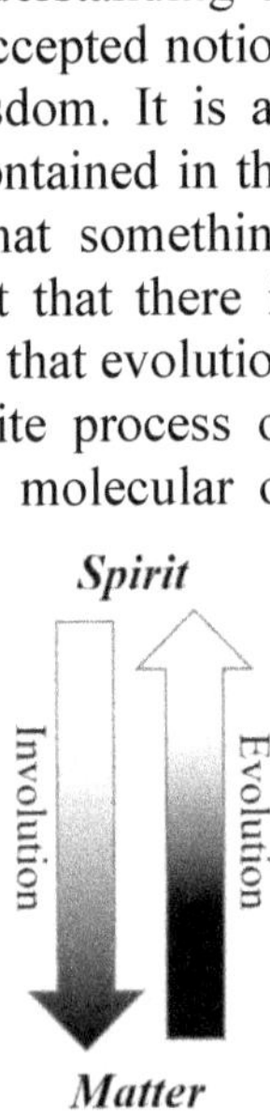

In fact, the peculiarity of Sri Aurobindo's cosmology is that all the physical manifestation is an involution of the Godhead in matter—a divine super-consciousness that subjected itself to oblivion, to what he called a 'sacrifice', by plunging itself into a sort of 'self-concealed' and 'masked' still divine consciousness but latent and 'asleep' in the inconscient matter. The real truth and significance of this process are beyond mental knowledge, but we might gain an intuitive understanding of its meaning by analogies, metaphors, or images. Such as saying that the One without a second wanted to become the many and acquire a self-

Fig. 123 Involution and evolution: from Spirit to matter and back.

awareness in multiplicity. Satchitananda fell into its opposite casting out of itself the many, and which led to the creation of a space where the original Consciousness is no longer aware of itself, but only individually in the many. By doing so, it lost its original bliss of unity which it is now struggling to get back by the opposite evolutionary movement.

Thereby, matter, which is a projection, a formulation, a 'shadow' of the same Consciousness in the manifestation, is, however, inherently still an expression of an original divinity. Nevertheless it behaves as a mechanical and inconscient substance, in a gross material universe—that is, in and as 'cosmic ignorance'—and from which it is re-emerging ('unrolling') by an evolutionary principle.

This resulted in a gradation of several planes and parts of being, of the individuality and the cosmos. Matter is a form of veiled life, and life is a form of veiled consciousness. It is the One that has self-involved itself in the depth of a descending ladder of consciousness.

Though this is not what science regards evolution to be, note, however, that this spiritual perspective of 'spiritual emergentism' does not contradict any scientific finding. The integral vision does not deny the orthodox Darwinian conception but complements it with a polar process of emergence. On one side, after an involution of a super-conscient spirit descending from the higher planes into the planes below, it now grows by a bottom-up process as an inconscient spirit in matter, which unfolds itself from a nescient state to progressively higher states of consciousness. On the other side, it is also immanently guided by the action of a top-down process of the same superconscient spirit from above.

What differentiates the integral view from Darwinian evolution (and echoes Schopenhauer's 'World as Will') is, first of all, that there is a will and urge in the inconscient, initially dumb, slow, and obscure but that, sooner or later, must inevitably reveal itself. This original primitive non-physical impulse comes not from a mere biophysical blind and mechanical force but, rather, acts from below, behind the veil of these material forces. As long as the concealment of the Spirit is almost complete, the consciousness dwelling in matter is forgetful of itself and subjected almost completely to mechanical forces. It emerges by a bottom-up process, slowly and painfully, by an eonic process of Darwinian evolution based on the laws of natural selection, mutations, and habitual automatisms. However, once consciousness begins to manifest ('unroll') itself out of matter, it progressively takes the lead. Despite everything, there is also pressure from above. The Supermind, which does not intervene directly, nevertheless orders, controls, and acts in the evolutionary drama indirectly, behind a veil by a top-down influence.

Once one becomes aware of this trans-physical-spiritual polar mechanism in Creation, contradictions dissolve. There is no irreconcilable duality, or Maya, separating mechanical matter from the spirit or the mechanistic conception from teleological ones. Both are the two aspects of the very same process that we can reconcile in this integral vision. For instance, the blind instinctive God of Spinoza or Schopenhauer's undirected 'World as Will' and that also inspired Kastrup's Mind at Large is Sri Aurobindo's cosmic vital. They captured a particular aspect of an omniscient and omnipotent Divinity that, however, on that plane, appears to be an involved instinctive universal will. This cosmic vital is only a small interval of a large cosmic spectrum of the planes of being—that is, layers of involved consciousness.

Thus, the mind is also not a creation resulting as an epiphenomenon of an aggregate of matter. Rather, it is already involved in it and re-emerges—that is, it evolves—out of matter during the evolutionary process once the physical aggregate is ready to hold it. Mind is a universal plane already existent and pervasive in all of the cosmos before any evolutionary process started. However, the mind is a derivative and subordinate power of the levels of consciousness above. This means the universe resulted from a process of creation in which matter was the last appearance, not the first basis of existence. Evolution is a process and progress of the emergence of the Spirit in matter, with matter being a 'densification', 'condensation', or 'precipitation' of the very same self-involved and self-limited Spirit. From this perspective, a universal Consciousness shrinks and confines itself in a seed that we call 'matter'. Matter is a 'self-focusing' and 'self-demarcation' by differentiation of the One. Just as a seed, a cell, or an egg already contain in potentiality the full organism they will manifest in time by development and growth, matter also already contains all the potentialities of the Spirit. But, at this stage of evolution, matter's conscious aspect is still involved, inert and inactive, except for its almost entirely mechanical and deterministic character.

The forces that drive this evolutionary process come from below, in the form of a material process that conditions it from the outside by the laws of physics and by the known Darwinian evolutionary principles of natural selection, genetic drifts by (apparently) 'random' mutations, environmental factors, adaptation, survival, and reproduction of the fittest, etc. But, at the same time, there are other types of forces that come from above and guide the process from within by an inherent consciousness—that is, more or less directly or indirectly according to the plane of existence, in contact with an original superconscience.

From the integral perspective, this is the bottom-up emergence of the Spirit in the manifestation from its involved condition—that is, from the inconscient. Whatever the physicalist might say and believe, we are not robots subjected to mere external physical stimuli. There are also forces acting and conditioning us from higher planes of consciousness and from inside of us: From our evolutionary soul, the psychic being, mediated by the subliminal sheet (who doesn't know the 'inner voice of wisdom' whispering to us in our heart?) Our soul in the heart center is a source of higher emotions and feelings but also presides the transformation that drives the evolutionary process. Most of what we think and do are indeed conditioned by the subconscious, primitive and mechanical impulses below, but there are also intimations in us coming from the higher-mind and above (who didn't experience the sudden intuitive flashes of wisdom?) and that, subliminally, always conditions us. At our evolutionary stage, these influences are mostly very partial, indirect, and distorted, or even resisted and refused by our subconscious and unconscious nature. In particular, the Supermind has to hide itself acting only indirectly, behind thick layers of gross substance, but, nevertheless, works out our earthly conditions.

Thus, evolution is a bottom-up process and, at the same time, a top-down causation. Once we rise to higher states of consciousness, this teleological aspect of Nature becomes an obvious and self-evident fact.

In a certain sense, the Christian concept of the fall of mankind from a state of perfection under the grace of God and obedience to Him to a state of disobedience and corruption, as described in the biblical Genesis, is not coincidental. It is a sort of distant subliminal remembrance in the collective of a truth. Unfortunately, as is so often the case, the human mind twisted a fundamental truth into its opposite and made a doctrine of sin and guilt of it which, instead of leading to salvation (that is, to the evolution of consciousness), on the contrary, prevents it.

While the concept of involution might sound as though it is in greater contrast with the modern understanding of evolution and cosmology, it doesn't contradict it. There is no inconsistency.

For example, the dichotomy between the growth of consciousness and the organic change in evolution—that is, speciation—is already contained in this integral view. One is the expression of the other, bidirectionally. It is now an accepted fact in biology that sudden large mutations in one or few generations can occur, which can lead to a new species. These discontinuous jumps in evolution are known as *'saltations'* (from the Latin, saltus, "leap"). This view of the evolutionary temporal dynamics that contrasts with a slow and gradual evolutionary progress is already inherent in the integral evolution. When the species rests for a long time without externally visible phenotypic changes (*'phenotype'*=the external observable characteristic traits of an organism), this does not necessarily imply an evolutionary stagnation; it could also mean that it is developing new powers of consciousness internally. The physical surface-form remains substantially almost unaltered, but the being may grow inwardly in consciousness and in its cognitive abilities. Once this internal process has come, by a sort of 'interior psycic accumulation', to a threshold, it manifests the inner change in a sudden external leap by creating a new physical form. Humans may well be at this threshold between one and the other evolutionary stages.

This is only one of the many examples one can make to show how the integral view only turns the perspective of the very same evolutionary and cosmic creative process upside down. The validity of the Darwinian evolution and the Big Bang theory is not questioned. What is seen differently is their significance. Evolution is true; however, it is not a blind and automatic clockwork without meaning. Mind needed millions of years to emerge on the Earth-scene as the scientific theories of evolution tell us, but it was not a process of creation but of emergence—the emergence of the reality of a mental consciousness that was already inherent but concealed in matter. In the same manner, other planes of existence are going to impose themselves by their emergence in the physical realm until, finally, the supramental consciousness will take full control and emerge and transmute all terrestrial life.

In this sense, strictly speaking, it would be incorrect to speak of an 'evolution of consciousness', as all the levels of consciousness were already immanent in the creation. However, what evolves is the individual soul, the psychic being, the individualized aspect of the Divine in the manifestation and which, by process of rebirth, grows in consciousness from life to life. It is this individualized 'soul-evolution' that expresses itself in the physical plane as a metamorphosis of the species from inert matter to the plant, from the plant to the animal, from the animal to the human, and now from the human to the gnostic being. But mind, overmind, Supermind, let alone the Absolute do not evolve; they are ever-present realities on their own and that are not subjected to any evolution (Aurobindo called the cosmic mental, vital and overmental planes of being *'typal worlds'*). If, with some abuse of language, we use the term 'evolution of consciousness', we mean the emergence of the Spirit in matter by the metamorphosis and ever-increasing complexity of the organic form, which becomes progressively self-aware and capable of containing and manifesting these powers.

Once we see things from this perspective, the perennial question of whether we have free will appears in a new light—that is, it reveals itself as an ill-posed question from the outset. In the 1920s, Aurobindo pointed out the wrong assumptions and fallacious premises which modern science and philosophy still work with hundred years later when it comes to the question of free will.[89] The question does not specify who is supposed to be free from what? And, after all, what is 'will'? Are we talking about the freedom of our true soul, the inmost being, the psychic being, the true Purusha? Or are we talking about the privilege to indulge in whatever lower instincts of the outer self, the 'desire-soul', the ego? Is it about the Will of the Supreme, and of which our real soul is a spark, or do we mean, by 'will', the mind's volition that clings to the appetites of the egoistic lower vital? These are all questions that are ignored in the orthodox scientific and philosophical debate that conflates all planes and parts of being into a misleading monistic or dualistic theoretical framework

---

[89] For example, see [265] Pt. I, ch. III or [362], first series ch. XXI+XXII.

and that, instead, in the integral paradigm, gains a new meaning and clear interpretation.

Here, the answer is that the true soul in us is free from the subjection to Nature but, as long as our mind identifies itself with the outer personality and the ego, which are caught in the web of Nature, we will continue to be subjected to it. In particular, the identification with the mechanical determinism of Nature dictates a mechanical and deterministic subjection of our being to its rules and laws. We can free ourselves by learning to detach from this identification—that is, to learn by means of spiritual practice and discipline like yoga—to stay back and observe first, and then sanction and govern the workings of Nature from above.

In between these two poles—the total subjection to Nature's instinct on one side and, on the other side, the complete release from it as the result of the ascension to progressively higher planes of consciousness—life finds itself in many intermediary stages of development and degrees of detachment from a binding and mechanistic to a free and luminous existence. Humans are creatures largely under a spell of the material nature and far from a real freedom of willed action, thought, and feeling. We are free only inasmuch we can free ourselves from the deterministic aspect of the universe. Though we still live in the illusion of freedom, most of our thoughts and choices are the result of subconscious impulses we are not aware of. If we live, as most humans still do, in the instinctual and animal lower vital dominated prevalently by a subconscious and nervous individuality or, at best, a half-conscious mentality, we are still chained to a mechanistic and impulsive life with a lesser degree of freedom. However, it is also true that we are no longer the nervous animal that a worm or an insect is. We are partially free and partially not. But, ascending toward higher states of consciousness, transforming and purifying our being in the light of a higher spirit, the psychic freedom begins to dominate the determinism of matter and the blind instinct of the lower life.

So, we see that what appears to the mind as an exclusive binary and inescapable dichotomy because it does not know itself and is oblivious to the complexity of the whole being, in the integral view finds a resolution that can reconcile these polarities naturally. The materialist was partially right in contending that we are determined by physical laws, while the spiritualist was also partially right in affirming the contrary. Or, in other words, both were partially wrong. Once we do not limit our ideas to exclusively monistic or dualistic ontologies, but realize the complexity of the universe in all its planes and parts and understand who is supposed to be free from what, by raising our seeing to a comprehensive view and knowledge which realizes the richness of our being in all its facets and on all planes of consciousness, only then can we formulate a contradiction-free and consistent answer to the question of free will.

The upshot is that what we call our 'mind' is only a little speck of an emanation of a superconscient Spirit in the manifestation. Therefore, the reader may now understand the author's emphasis on the distinction between mind and consciousness. The mental is only one plane of the manifestation, just as matter or life. Mixing these together, or mistaking one for the other, or conflating them as synonymous isn't just a semantic error or a matter of definitions; it is a categorial mistake that leads to conceptual confusions and abstractions that end up having no correspondence with reality and, thereby, leads to endless speculations preventing people from progressing toward a clear understanding and the framing of better theories and models of the truth of things.

We can therefore see our nature—that is, the dynamics of our thoughts, feelings, and physical—as being determined by a dual and complementary action and influences from below as above. As modern psychology is also beginning to realize, our thoughts and feelings are determined, to a large degree, by the involved planes beneath. These planes are an obscure and untransformed semi-subconscious vital and an instinctive subconscience which is dominated by an automatic inconscience down to nescience and immersed in complete darkness where consciousness is reduced to complete inertia—that which we call 'matter'. Also, these sub-, in-, and nescient planes are not only proper to an individual but are cosmic—that is, they are something inherent in all the cosmos.

Something which, however, psychology still struggles to accept is that, on the other side of the ascending ladder of consciousness, it is also true that we can and do receive intimations from the more enlightened domains above, namely, from the higher-mind upwards with the psychic being in the background, and which can become the base of one of the more powerful expressions of mind (and beyond it) or beautiful creations of artistic nature. That's the origin of the popular belief that 90% of the brain's potential is latent but not developed. Indeed, we have such potentiality in ourselves. The fact that there is no neuroscientific evidence that substantiates such a claim is because such a potentiality isn't in the brain. Nevertheless, the popular belief reflects an inner wisdom of something that has its truth.

But what is this divine Consciousness trying to achieve in the material existence? What is the ultimate goal of realization? Why did an omniscient Being impose a plunge into ignorance, self-forgetfulness, and duality on itself?

According to Sri Aurobindo, it is the 'delight of being' that seeks to realize itself as the 'delight of becoming'. Delight of being is the intimate nature of Satchitananda in a timeless Absolute but still unmanifest in the universal play of manifestation. Delight of being is the background of all backgrounds, from which pleasure or pain, bliss or sufferings, anguish or joy are different forms and gradations emergent in our temporal earthly experience. When delight of being is no longer superconscient but subconscient in matter, it seeks in mind and life to realize itself by emerging in the becoming. Also, at the bottom, a protozoan seeks with an intimate, concealed, and subconscious aim and urge for its original delight of being in the flux and cycle of life experiences, and which expresses itself in what we call the 'instinct of survival'. The finding of an absolute, eternal, and infinite Self-existence, Self-awareness, but timeless and unmanifest Self-delight of being in its self-involved temporal manifestation is what secretly supports the universe and is its hidden truth, force, motivation, aim, and goal.

It is by seeing the evolutionary process out of an 'involution' as a bottom-up evolution guided by a top-down Consciousness-Force influence that life, mind, matter, and consciousness, seeking the delight of being, acquire a completely new significance. This spiritual framework furnishes us with a vision of things that has more explanatory power than a naturalistic or limited idealistic conception.

b. The Supermental Transformation

> *"The Spirit shall look out through Matter's gaze.*
> *And Matter shall reveal the Spirit's face"*
> Sri Aurobindo [267](Bk. XI, Ct. I)

In Sri Aurobindo's yoga, the psychic and spiritual transformations are still not the final consummation of the evolution of spirit out of matter. Realizing the supramental consciousness is still an inner realization of a spiritual vision, not a physical transformation of the species. The ascent of the mental being who climbs up to the Supermind might realize a universalized awareness and lives in a transcendent bliss, but this does not lead to a descent of the Supermind in matter and which is the only that effects a complete transmutation. In this vision, Nature is the unfoldment of an indwelling spirit in matter, which gradually emerges by an instinctual force from below but is also determined by an action of a supreme Consciousness from above. Matter is the bottom of this descending involution and, therefore, also the substance which is most resistant and inert to an awakening call from its inner source. Only the Supermind can bring the full dynamism of the spirit into matter. Even the overmind alone cannot lead to a complete transmutation of these inconscient and nescient planes of existence involved.

Cognitively, we might speak of a new being, an intermediate being between man and the envisaged gnostic being, but if there isn't also a physical transformation, we can't speak of a new species. One thing is rising to the supramental consciousness; another is to bring it down into the material realm and transform the physical body as the crowning of this adventure of consciousness. The transformation can go from the bottom up, as most spiritual traditions have done, but still lacks the top-down action of the Supermind, which will finally manifest and unfold its full powers on Earth. It is this integral transformation, where the Spirit calls matter, and that does not leave aside the material existence, that can be called a 'divine materialism'. It acknowledges the fact that a spiritual mystic realization is not the true meaning and purpose of our existence but that matter is the basis and instrument for the full manifestation of the Spirit.

Aurobindo and his spiritual partner, Mirra Alfassa, tried this ultimate and most difficult transformation after having had the highest spiritual realizations; to bring down the Supermind in their bodies—that is, a transmutation process in and of the cells—by what they called the *'cell yoga'*. *"A difficulty which amounts almost to an impossibility,"* as he later reported, but the inevitable evolutionary destiny of humankind. It is a colossal enterprise because matter is rooted in the deepest of the unconscious, the nescient. Its inertia, negation, and opposition to change are the most inalterable, blind, mechanical, and stable ones. Death is rooted in this inconscient; it is a nescient refusal to progress, change, evolve. It is this refusal that makes the physical universe appear to be so mechanical and apparently ruled by inalterable and fixed laws that we call the 'laws of Nature'. A Spirit forgetful of itself, growing and emerging in matter by a subconscious bottom-up evolutionary principle, has to subject itself to the mechanical laws of matter first or, one might even say, has become the laws of physics. The mechanistic aspect of the physical universe is not something from which consciousness emerged. It is itself the outer aspect of an involved consciousness and which, thereby, has become mechanical, automatic, and impersonal.

However, the secret of matter relies precisely on this concreteness and stability: It is only this solidity that can hold the full power of the divine Supermind that would transmute the physical body to its image and even convey physical immortality. The most extreme contraries—matter and spirit, life and death, or omniscience and complete ignorance—can be reconciled only by the unitary power of the Supermind.

As is well known, so far, nobody could realize this ultimate state of immortality. Sri Aurobindo and Madame Alfassa reported that they were able to realize the inner supramentalization and began working upon the material change. They left us huge documentation that described not only the inner states of consciousness they attained but also the first intermediate steps of a physical transformation. According to their accounts, a communion of the Spirit with matter is the very secret of Creation. We are becoming an instrument of a spiritual manifestation if we consent and participate, finally surrendering to a Power that knows much better than we know. The aim of the yoga of the cells is to let the body become an instrument responsive to this supreme Consciousness. The gnostic change will also lead to the supramental transformation in the cellular structure, which, in the previous evolutionary stages of Nature, was effectuated almost exclusively by the environment and the blind clash of material forces. This is the difference between humans and animals. The appearance of mind is a necessary intermediate step in evolution because it endows the human species with the ability to consciously and actively participate in this process.

But almost all that our body is and does is ruled by the laws of a subconscient or inconscient plane. The difficulty resides in making these obscure lower planes of existence conscious and subject to the supramental control, light, and action. The meaning of such a transformation has to be found in a supramental emergence. It is about subjecting the body—that is, the outer physical stuff we are made of, down to the single biological cell—to the governance of the Idea and Will-Force of a divine Power. The Idea and Will of the Spirit are in direct dynamic action on all levels—that is, in our mind, soul, emotions, and down to the still subconscient, which is the cause of an ignorance dominating our bodily existence. This will lead to the emergence of a new species, the gnostic being.

Therefore, what distinguishes Sri Aurobindo's yoga from other spiritual traditions is that it not only effectuates a mystic awakening or a spiritual transformation to realize some extra-cosmic Reality but also aims at transforming our subconscient and inconscient parts down to the very physical. A transformation enacted by the power of the spiritual consciousness can also turn the body into a responsive instrument of the Spirit. It is about the awakening of a 'body-consciousness' which, at this evolutionary stage, is still dominated by the inconscient in the cells. The aim and purpose of evolution turns out to be that of manifesting the power and delight of the Spirit in matter and realizing a divine life on Earth.

But the manifestation of a Consciousness-Force in the mental, vital, and physical being is actually resisted by the inconscient and even nescient abyss that are unable to receive it. This is what, at the bottom, causes all of the pain and suffering. This is also the reason why new world order, policies, technologies, sciences, philosophies, religions, or whatever other mental-based spiritualities will never be able to cancel suffering and pain. Only by a change in and of ourselves and, finally, by a spiritual ascent that will allow this Will and Power to emerge can things truly change. The ascent toward a spiritual mind is a necessary but insufficient step because if it does not change our exterior nature, the material will continue to drag us downward into the subconscient and inconscient. Only if the supramental vibration acts and transmutes the body-consciousness—what Sri Aurobindo called the 'mind of the cells'—can a spiritual Ananda (bliss) flow into the body, inundate cell and tissue and finally lead to the divine transformation. It is a vibration that comes not only from above, but also from the heart center by the active participation of the psychic. Also, other spiritual practitioners or mystics must have had similar experiences, such as, for example, Meister Eckhart, who speaks of *"a likeness of the light in the ground of the soul* [that] *flows over into the body, which is then filled with radiance."*

Only then will we be able to ascend to a state of consciousness where a divine bliss, love, delight, unity, harmony, and deathless physical perfection become a natural state of being in a stable and permanent eternal ecstasy of a divine life upon Earth.

"It is such a change and such a reshaping of life for which humanity is blindly beginning to seek, now more and more with a sense that its very existence depends upon finding the way. The evolution of mind working upon life has developed an organization of the activity of mind and use of matter which can no longer be supported by human capacity without an inner change." [266](Ch. XXVIII)

In fact, our organized mental and technological society is becoming increasingly complex and escaping our control. The external sciences and technologies supported by reason alone won't be able to self-sustain themselves for long if they are not also guided by wisdom from within as a consequence of the ascension toward higher planes of consciousness of the race.

What Sri Aurobindo tried to do was to put in direct contact the Spirit of Creation with the raw matter of which our body is made and awaken a hidden and 'crystallized' consciousness, a mind of the cells. This is, however, not supposed to be either a purely material or purely spiritual transformation. It is both and something else entirely. It also reforms and enlarges the view of the classical spiritual teachings that consider the physical transformation of secondary importance or even an illusion, the cosmic illusion of Maya, the veil of Ignorance. It goes against our scientific as spiritual understanding that matter and spirit will never find a point of contact. Aurobindo rejects the idea of the polarity of matter and spirit as being irreconcilable opposites. The spiritual transformation is seen as something not limited to an inner realm but ultimately aims also to an exterior and material metamorphosis, with this being the real reason for the manifestation: the fusion of Spirit with Matter, the transformation of our body to the image of the Divine. This evolutionary concept of an emergent consciousness in matter, which will infuse and transmute the species into a new being, is essentially what makes the integral vision of Sri Aurobindo stand apart from the other forms of yoga or mystical traditions. It integrates the spiritualism of the East with the materialism of the West and completes, widens, and goes beyond both toward a synthesis where both appear to have their final significance.

c. The Mind of the Cells – Part II

That's where Sri Aurobindo's account ends. What is known is that he went into seclusion for 24 years until his death to try this difficult spiritual and material transformation in his body. What this yoga of the cells was really about, he never spelled out explicitly, other than in a few short remarks throughout his writings or in his epic poem Savitri.

However, his cellular work was taken up by Mirra Alfassa, the "Mother."
She also was largely hesitant to speak openly about something that still had to
be accomplished and whose success or failure was far from clear.
Nevertheless, one of her disciples, Bernard Enginger, also named Satprem,
who was frequently in correspondence with her, collected from 1962 to 1973,
in a 13-volume series entitled "The Mother's Agenda" [285], several accounts
which outlined her exploration of the body consciousness. It was an
exploration in the consciousness of the cells through a myriad of experiences—
a documentation of the discovery of a mind of the cells, capable of reforming
the conditions of matter and the law of the species. It is a document of
experimental evolution and a question about our future, which attests to how
we are at the end of an evolutionary cycle. It is the report from the heights of
a supramental consciousness of an attempt at a cellular transformation. It is not
an easy read or a message that we, as humans, can really understand. After all,
what would our ancestors—say, an Australopithecus—'understand' if they
could get into a conversation with modern humans? Nevertheless, as far as our
limited reasoning mind can go, we can at least try to get a glimpse of what this
transformation might be about.

It seems that this still 'sleeping' consciousness in each biological cell is at
work through a kind of 'mind in matter' or 'mind of the cells'. This mind of
the cells bestows a mental ability to each cell. But, in its still untransformed
inconscient state, it is mostly dependent on a reactive, unconscious, and
subconscious instinct that can sicken us, cure us miraculously, or let us die
suddenly.

The reason for this, despite the millenniums of materialistic realizations on
one part and spiritual realizations on the other part, all that had not been
recognized, can be found in the fact that a physical transformation is revealed
to be an enterprise of extraordinary difficulty. The biggest obstacle to such a
realization was shown to be the subjugation of the lowest layer of
consciousness to the supramental Light. Without our being aware of it, this
'dark side' is ingrained in the most recondite zones of our being. In her words,
the Mother expressed this as follows:

*"I went down into a place ... a place simply in the human consciousness,
thus necessarily in my body ... I have never seen anything more timorous,
fearful, feeble and mean! It's ... it must be a part of the cells, part of the
consciousness, something that lives in apprehension, fear, dread, anxiety ... It
was truly, truly dreadful. And we carry that within us! We aren't aware of it,
it's almost subconscious – for you see, the consciousness is there to prevent us
from yielding to that – it's cowardly, and it can make you fall sick in a minute."*
[285](Nov. 5, 1960)

Aurobindo's cryptic poetical verses suggest he already had the same
experience:

The mind of the cells is still unawakened and subconscient, moved only by shocks, pain, and catastrophes. It has an intelligence of its own, but only a negative and catastrophic imagination that always imagines the worst. It is this subconscient, inconscient down to the nescient in matter that is revealed to be the strongest resistance to change. Billions of years can pass without it changing at all. And nothing can move it, except the supreme Power of the Supermind. The physical universe is the Divine nescient of itself, the lowest stage of a downward plunge from which it now recovers a self-awareness in matter by evolutionary leaps.

The modern scientific findings of the basal cognition in cells discussed in Pt.I-IV.4 came as no surprise to those who were already acquainted with Sri Aurobindo's cell yoga. Even though it is still scratching the surface, science is beginning to discover this cellular mind.

It turns out that even death is not inevitable but a sort of subconscious cellular habit, an accident that must not remain forever part of the eternal law of life. This is confirmed by modern cell biology, which discovered that, indeed, at least in principle, cells are immortal. Curiously, it seems more difficult to explain death than immortality, as modern science still struggles to satisfactorily explain the aging process.

But conquering death is not the purpose of integral yoga. This is only a consequence, an intermediary step toward a realization that should be more far-reaching and decisive. It is not that kind of immortality that, maybe, one day will give us genetic engineering or medicine, nanotechnologies, or transhumanistic dreams, as some hope, and that would perpetuate a race which, inwardly, would not be much different from actual humans. Instead, it will be the direct consequence of a state of consciousness and of an inside and outside perfection.

This physical supramental metamorphosis will not happen at once; it will have to transform and break through different layers. There is a substrate of the inconscient which has been conditioned by the physical mind for millions of years and won't easily give up its habits, automatic reactions, and primitive and reactive instincts. The physical supramental transformation is about the transfiguration of matter which, however, is the most obstinate part of the being that opposes the most recalcitrant resistance to any form of change and progress. The physical body is the part most firmly entrenched in the unconscious and nescient plane, and its change and evolution take a slow, invisible and hard underground labor–like a battle that might need centuries to change the present physical consciousness. But, sooner or later, someone had to begin working on it.

These habits, mechanistic aspects, and repetitive inclinations of the physical mind of the cells are what determine our health, nutrition, physical reactions, bodily instincts, etc. These have such a solid and concrete reality for us that they appear absolutely irrefutable and unmodifiable. It is the mind of the cells which is full of fears and anxiety, receiving permanently the suggestions of our mental understanding of the world that permanently whispers to us: 'be careful', 'the doctor says…', 'the professor says…', 'the scientist says…', 'physical laws tell us', 'you can't do that', 'if you do that, it will end in a catastrophe', 'be realistic!' and, finally, 'don't change anything'. It is a sort of mechanical and automatic materialistic 'guardian consciousness' that anxiously and meticulously prevents anything from changing beyond the limits of strict physical determinism. This deterministic and mechanical mind which foresees every possible disaster, disease, accident—and, above all, death—is not our mind or in our mind. It is in the cells, and deep down, it is the mind of matter. It is in every molecule, atom, particle. The quiet inertia of the stone—that is, materiality itself, what we call 'matter'—is a Consciousness that involved itself by descending into the night of the inconscient and has, thereby, become automatic, deterministic, and what appears to us a thoughtless force. The conscious mind, that which we proudly call 'reason', has its roots in this physical mind of matter as well and conditions us with its fantastic hypnotic power in everything we think and do. This physical mind has mentalized our cells as well—that is, it acts like a hypnotic prison warden that doesn't allow us to think, and let alone act, beyond a narrow and strictly limited domain. It is not only our own mentalization but also that of our parents, grandparents, friends, partners, doctors, professors, etc. That's why this clinging at mentalized conception has become such a great obstacle.

And yet, according to Aurobindo, the secret is that there is a supernal Light buried in that apparent powerlessness and immobility of matter. It is wrapped in a series of superimposed webs of intellectual, emotional, subconscient, and inconscient physical automatisms, which only the supramental power can unwrap.

But this can't be done if we are not willing to transform our ordinary outer ignorant consciousness, among others, the scientific sense-mind or physical mind which so firmly believes in the so-called 'physical laws'. All the indisputable truths that physicalism poses a priori are, from the supramental perspective, falsehood. We don't question material laws—for instance, those of the body which determine our health—because they have such a solid and concrete reality for us. They are so solid, so established, that they appear to be an absolutely undeniable fact. As absurd as it might sound, it is precisely these firm beliefs that condition us down into the body consciousness, in the mind of the cells, that we must learn to give up. There is an anxiety, a fear, an apprehension in our cells that can be translated into words as an atavism of being reasonable, being careful, being smart, taking precautions, etc. It is this whole mentalization of the cells that determine its behavior. It is a material mind that loves disasters and even attracts and creates them because it needs catastrophes and tragedies to grow in consciousness. Whatever is unconscious requires intense emotional shocks to wake up. Our physical mind permanently contemplates every possible disaster and, by doing so, opens the gates to all possible negative suggestions. This physical mind continuously conditions what the Enlightenment proudly called the 'power of reason'. It is a double-edged sword. On one side, it is a great power that sustains reason because it can, by logical inferences, predict, plan and organize the future. This is what allowed us to build all the sciences based on theoretical models which predict what will be observed or happen next. But it is when we mistake the physical mind for wisdom that things begin to go astray. For example, at the least little pain, we immediately think with a defeatist and automatic thought: 'Oh! Is it cancer?' It is this automatic and almost hypnotic thinking of our physical mind that nourishes our fears and anxieties, and that is closely connected to our physical illnesses. If we could also free ourselves only a bit from both, this would immediately have positive physical consequences. But our unaware thoughts, beliefs, conditioning, and fears prevent us from going beyond. Even doctors, with their thoughts, beliefs, and fears, crystallize the illness and make it real, concrete, hard (and then get the credit for treating it, when they can). The problem, however, is that there is a whole world of waves of suggestions that condition and hypnotize us and that we take as 'our thought', 'our reason', 'our common sense'. *"It's like a dog that has been beaten so often that it expects nothing but beatings,"* Mother used to say.

This cellular mentalization is so dry and empty of the true vibration, something so hardwired and firmly established in the subconscious that determines our thought processes. It is rooted so deep down in the unconscious and nescience of matter that only a supreme Power can change it. Then one discovers that the mind of the cells is conditioned not only by our body consciousness but also by a collective pessimism, doubt, or defeatist imagination anchored in our subconscious mind and in the cells of the body of everyone. This collective mental and physical consciousness is the real obstacle barring the way to the supramental physical transformation.

But this state of affairs in which the world finds itself is a falsehood; it isn't true. It is in our subconscious, unconscious, and nescient layers of our mind and body, but it is not true. It is a program, a habit, a mechanical repetition that isn't inevitable or that needs to remain there for eternity. It was an evolutionary mechanism that served its purpose to allow the sleeping God to emerge from matter by an external mechanistic process of natural selection and become conscious of itself. But, once the real Spirit in matter begins to awaken, these blind mechanical laws have fulfilled their aim and must be set aside if we want to progress further. It is when we stop believing in a mechanistic and inevitable fate and the inevitability of physical laws that the subconscious and unconscious abyss in matter can begin to be subjected to a process of transmutation.

The good news is that we are actually in a transitional evolutionary phase in which consciousness will be allowed to avoid the harsh path of the lower-Nature, which enacted the growth from below, by the external action of brute and blind physical forces through the geologic ages. If we assent and participate, we can change and further evolve by the action of a supramental Force, which is acting from above, descending downward deep into the inconscient, awakening not only the spirit in us but also every material fiber of our body.

This dark inconscient in the cells, this nescient abyss of ignorance, is not a cruel joke of an evil godhead. Quite the contrary, it contains the very secret of the reason of existence of matter itself. While Sri Aurobindo described *"the dumb occult consciousness ... which operates in the cells"* [266](ch. XIX), he added an important note, namely, that *"Almighty powers are shut in Nature's cells."* [267]*(Bk. IV, canto III)* In a famous passage in her diary, Mirra explains that *"at the very bottom of the inconscience most hard and rigid and narrow and stifling, I struck upon an almighty spring that cast me up forthwith into a formless, limitless Vast, generator of all creation."* [285] (Vol. 1, Nov. 8, 1958)

When the supramental force descends deep into the core of matter, one discovers the same supramental vibration below that which one had also realized above. In the pitch-black bottom of nescience, one finds the infinite golden power of omniscience. It makes the cells tremble with such power that the whole cellular life becomes one solid mass, a single block of vibrations which, however, only a transformed body can hold without bursting and collapsing under its action. It is something that takes hold of the body and which she described as an 'extraordinary vibration' or 'pulsation in the cells', like a 'strong mass' that is 'much more solid than matter'.

Therefore, one must learn to widen not only the inner consciousness but even that of the cells to make them acquire plasticity capable of containing that original Force. Her body was living a process in which one must be ready to go through death and conquer an atavist mechanical process of dissolution. It is a transformation that struggles with the subconscious of the body, crammed with defeatism and almost immutable habits, and on which only a supramental vibration can work. Something new is on the way to being realized during these times of transition.

But as the amphibian had first to develop lungs in order to step permanently out of the water, so must we acquire a new way of 'breathing' that will enable us to enter another state of consciousness and physicality. This means we have to de-program ourselves if we want to step out of what Satprem called the 'thinking fishbowl'. What we are dealing with here lies well beyond the cherished reason of the Enlightenment or any scientific analysis. It is something totally new and unexpected for science and also ignored by all spiritualities—a supramental consciousness that can, in an instant, 'derealize' everything painful and miserable, such as disease, cold, pain, or even death, into something blissful and miraculously wonderful. The supramental truth-knowledge does not admit any falsehood, suffering, or illness. It 'derealizes' it. But this 'derealization' dissolves falsehood not by external means or some sci-fi hyper-technology but by a divine Force that acts on and infiltrates matter by the vibration of Truth from within, canceling the falsehood. The purity of the Supermind is that of the Absolute. It doesn't tolerate even the tiniest bit of falsehood. Its 'divine intolerance' prevents it from mingling with our present state of consciousness that is still largely under the spell of ignorance. The supramental action is neither a material action nor an ordinary mystic transformation that the sages spoke about but, rather, an entirely new third position of consciousness that sees things from an entirely different perspective—the perspective of divine materialism. It is, at present, working at millennia-old habits, forces of division and polarization in the collective subconscious, against the resistances of humankind, belief systems, stagnant religions, and last but not least, the dogmas of a rationalistic naturalism that is delaying the evolution of consciousness on Earth.

When one steps out of the fishbowl, one realizes the indescribable glory and the fantastic eternal and limitless vast, absolute Light in which the next species, the gnostic being, will live materially here on Earth, not in some distant heaven. It will be a deathless *overlife*, with no need for food and sleep, basing its blissful physical existence on the true consciousness of what Mother described as 'the vibration of love in the body', 'the very essence of creation' which one finds in the 'inner silence and eternal peace in the cells' and where one simply and naturally is. In that physical state, the body experiences the formidable pulsations of the eternal, stupendous Love where *"all the results of the Falsehood had disappeared: Death was an illusion, sickness was an illusion, ignorance was an illusion–something that had no reality, no existence.... Only Love, and Love, and Love, and Love–immense, formidable, stupendous, carrying everything."* [285] (Apr. 16, 1970)

It is the second evolutionary transition, no longer from matter to life but from life to a life divine.

It is worth noting how the descriptions of the supramental experiences of the consciousness of the cells are reminiscent of some quantum mechanical descriptions of matter. Some scientists are getting close to the idea that in biological cells, there may be a place for quantum phenomena. In fact, many experiences described in the Agenda also point out the 'holistic ubiquity' of cellular processes at the subtle level, just as entangled particles in quantum mechanics do. There is, of course, nothing that shows us that we are speaking of the same phenomena, but perhaps, the analogy might not be entirely coincidental. The quantum world may be an interface between matter and a subtler realm.

At any rate, the mystic but very concrete and material experiences described in the Agenda go far beyond our rational human understanding. It is like the pre-human hominid trying to understand how the Sun's nuclear fusion processes work.

Interestingly, this new 'supramental vibration' turns out to be contagious. It is spiritually, mentally, and physically contagious and works behind the veil to accelerate evolution on Earth. The cellular state of consciousness in one body—that is, the new consciousness in matter—can, by contagion, affect the consciousness of the cells of another body. It turns out that cells are not independent material structures isolated from the rest of the world but are interdependent units through a mechanism of cellular communication that acts well beyond a single body.

But the world is still not ready. The inherent sense and nature of the Supermind is its character of Unity and Oneness. No one and nothing can be left behind. For a complete transformation, all the world must participate and transform. The distinctive character of integral yoga is that it aims at a collective transformation. But precisely the collective is still lagging behind. No one can do all the work alone. The structure of the world still represents a brake; a change in the terrestrial consciousness must take place. Will we have to go through one catastrophe after another or even toward a complete failure of the mind before the human species will assent?

In the vision of Sri Aurobindo, there is a Consciousness, the Supermind, the Divine, God, or whatever we may call it, that is preparing for a transition that will go beyond this human mental world. All the accidents, illnesses, quarrels, and conflicts are becoming more acute to the extreme because it is that Consciousness that *"puts a pressure for all that resists in someone's nature to come to the surface and manifest, and so the ridiculous or wrong side of the thing becomes conspicuous, and it has either to go or to...."* [285] (Aug. 23, 1969). Or things will break down, and we will again have to confront crises, catastrophes, meltdowns, etc. One has only to take a look at the news.

Sri Aurobindo went into seclusion to try this difficult metamorphosis that almost nobody had ever tried before (except some Vedic rishis or some very rare yogis who left us little anecdotal information). At the end of an intense sadhana, the bitter conclusion was: *"The earth is not ready."* He understood that the time still hadn't come. He said to his spiritual partner: *"We can't both remain upon earth, one must go."* She answered: *"I'm ready, I'll go,"* but Sri Aurobindo replied: *"No, you can't go, your body is better than mine, you can undergo the transformation better than I can do."* [285]*(July 26, 1969)* He passed away in 1950.

Later, Mother repeated the same experiment. All the experiences and her process of transformation have been transcribed by Satprem in her diary, the Mother's Agenda, a detailed account in thousands of pages. The steps of this physical supramental metamorphosis, covering the years from 1951 to 1973, were documented for the first time. But she, like Sri Aurobindo before her, reported that there is a problem, an impossibility of transformation: *"The earth is still not ready. This is the problem."*

Ready in what sense? What does that mean exactly? In her words:

" ... the whole makeup of the world still seems to act as a brake, there still seems to be something... And that 'something' is what this Consciousness is working on. For it to be established, a change in the earth consciousness must take place." - "Unfortunately, there's the whole influence of the outer world which makes it difficult for the body to be constantly like that and makes it tend to fall back into the ordinary way. That's why it can't settle in permanently."- "Life could be so marvelously simple and beautiful .... Man has really made it idiotic." [285]*(April 9 and 30, 1969)*

She wanted to try this transformation against all the evidence and all the terrestrial atmosphere that surrounded her. But she didn't find the way out. Perhaps there is no possible way. The transformation turned out to be impossible unless at least a small part of the terrestrial consciousness awakened. Nobody can come out of it alone. At the age of 95, on the 17th of November of 1973, she passed away too.

So, was it only an illusion, wishful thinking? Is such a transformation impossible, a simple utopia that tries to satisfy our hope in eternal life and power beyond death?

Many other questions remain open. But this negative note also gives us a different and wider understanding that not only allows us to better understand ourselves, the world, and the events taking place in a much deeper and wider context but also points us toward a new awareness of how things can and should be done. Mind alone won't do it. Reason is a tool, a limited transitory cognition that won't be able to deal with the challenges of real material and spiritual progress. The cult of reason must come to an end, at least as it is practiced in its present form. The rationalistic physicalism which refuses spiritual progress is ending. The question is, rather, if we are willing to participate in making this transition develop more harmoniously or if we choose the path of refusal and resistance against this change and, therefore, undergo a painful and at times catastrophic transformation—or, worse, extinction?

d. The Gnostic Being and the Life Divine

Following Aurobindo, the final aim of the evolutionary process, or at least the most distant aim actually visible, is the realization of a Superconscience in matter—a Superconscience whose intimate nature and character will be the unity in diversity where the individual consciousness will coalesce into that of the collective and yet maintain its individuality and divine personality, which will exist in a superhuman supramentalised form and live in a state of overlife which knows through a Knowledge by identity. It will be a new species, beyond the human, that of the gnostic being. T. de Chardin's Omega Point is already palpable here. The goal of the evolutionary impulse guided by spiritual energy from within will lead the noosphere to the *'planetization'* of consciousness. But Chardin's conception was a mental or even higher-mind contact with a deeper truth that was seen with a supermental vision by Sri Aurobindo decades earlier. Again, there is a difference between spiritual philosophy and a sight that sees things directly and without logical inferences.

What can we say about this future species which will surpass the human form and mind? Not much. Our ancestors, such as the Homo erectus, the extinct pre-human species from the Pleistocene, could hardly imagine what the mental Homo sapiens would be like. Nevertheless, someone who ascends the ladder of consciousness to the Supermind can give us an intuitive understanding of what the fundamental principles and aspects of this overlife will be.

The next evolutionary step will be a full emergence of the Spirit and its Light-descent into life and matter with a full replacement or transformation and uplifting of the mind, vital, and physical. This supreme Consciousness insists on the sense of oneness, a spontaneous and inherent unity fulfilled in diversity, which will create, inwardly and outwardly, a harmonic unity. There will be a new form of life ruled by a new power in Nature, a divine life that will penetrate every cell of the body. It will be a species with complete self-knowledge, capable of the perfect delight of self-expression in Nature which will be fully aware of its individual and cosmic personality and impersonality—something freed from the bonds of an untransformed subconscience, inconscience, and nescience that is actually still at the root of the present life. It will be a Life of the Spirit that manifests itself fully in matter, expressing a cosmic joy, a transcendent beatitude, the original and universal delight of existence. Its action will not be guided by the trials and errors of mind but by an instantaneous unfiltered divine knowledge, the divine gnosis. There will be no contradiction between materiality and spirituality, the inner and outer lives, the universalized individuality and the world. Falsehood, disharmony, half-truths, or any form of ignorance, pain, and suffering will be 'derealized' by the supramental Light. Illness, sickness, and physical, emotional, or mental suffering will be replaced by the supreme Love, Beauty, Harmony, and perfect effortless and flawless action of a Consciousness-Force.

Death will be recognized for its service and meaning but also for its falsehood, which will have been eradicated from the cells. The gnostic being will be immortal, physically and cognitively perfect, an image of the Spirit in physical form. It will live in the eternity of that past, present, and future and the unity of his individuality as the same unity in diversity of a gnostic collective expressing an innumerable Oneness.

Spirit will have made itself matter, a materialization of the Ananda—that is, the spiritual bliss of the Absolute—with peace and ecstasy that will inundate the body's cells and tissues. It is a gnostic vision that reconciles and fuses all differences and contradictions into a universal beauty and glory of a secret creation. The mechanical and automatic subconscient, inconscient, and nescient will be transformed into perfect divine superconscious automatism, which will replace the mechanical workings of Nature. Truth-Knowledge, Truth-Sight, Truth-Feeling, Truth-Will, Truth-Sense, and Truth-Dynamism will be the spontaneous and normal states of consciousness and the inherent operational activity of the gnostic being. There will be no separative and divisive assertion or imposition by one gnostic being onto another. A Truth-Vision of perfect unity in free diversity will form a great harmonized variation of a gnostic luminous collectivity where everyone has a complete consciousness of oneself and others by direct inner contact and interchange communicating inwardly and directly with each other. Everything will be done for the Divine, and for the Divine in all, guided by a complete universality, unity, mutuality, and harmony in oneness with all gnostic, human, animal, and plant life. A divine Will expressive of a free and spontaneous self-unfolding of the Spirit. A divine Life upon Earth.

"The gnostic life will exist and act for the Divine in itself and in the world, for the Divine in all; the increasing possession of the individual being and the world by the Divine Presence, Light, Power, Love, Delight, Beauty will be the sense of life to the gnostic being." – "Love will be for him the contact, meeting, union of self with self, of spirit with spirit, a unification of being, a power and joy and intimacy and closeness of soul to soul, of the One to the One, a joy of identity and the consequences of a diverse identity. It is this joy of an intimate self-revealing diversity of the One, the multitudinous union of the One and a happy interaction in the identity, that will be for him the full revealed sense of life. Creation aesthetic or dynamic, mental creation, life creation, material creation will have for him the same sense. It will be the creation of significant forms of the Eternal Force, Light, Beauty, Reality,—the beauty and truth of its forms and bodies, the beauty and truth of its powers and qualities, the beauty and truth of its spirit, its formless beauty of self and essence." [266](Bk II, pt. II, ch. 27)

II. Towards an Integral Cosmology

The previous chapter was an introduction to the spiritual cosmology and vision of Sri Aurobindo. It is a spiritual framework that, in its main traits, was outlined already in the 1920s. His real purpose, however, came to the surface only after his departure and is best recognized in the accounts of cell yoga that were described in the Agenda, thanks to Satprem's transcriptions, that appeared only in the late 1970s. Until then, the full dimension of their work hadn't come to light.

Meanwhile, the technological progress and especially the new discoveries of science, particularly in biology, evolution, consciousness studies, and quantum physics, were quite dramatic. These can hardly be ignored and necessitate a new way of seeing the world and ourselves. But, despite this progress, the fundamental questions that ask for the deeper meaning of things, for the essence of life and our place in the universe, remain unanswered. What is life? What are we doing here? What is consciousness? Why is there something rather than nothing? These and many other existential enigmas continue to be no-progress quests in the frame of a strictly naturalistic paradigm. As we highlighted at the beginning, this failure is the reason for a revived academic interest, especially in the philosophy of mind but also to some degree in other sciences, in theories like panpsychism and panentheism in the light of modern scientific knowledge, which led to the development of new theoretical frameworks such as cosmopsychism, analytic idealism, and panspiritism, just to mention some.

What we would like to do in this chapter is show that an integral cosmology in line with the insights of Aurobindo, but expanded to the new findings of science, can deliver an even more coherent spiritual framework inside which several of the issues and shortcomings that vitiated the previous cosmologies can find their natural accommodation. This extended philosophical and spiritual understanding of reality, which includes matter and spirit in a single and unique view, is more comprehensive and, we contend, more in line with modern science.

First, we will focus on some aspects that are most relevant to Aurobindo's supramental vision of reality. Then, these will be applied to our current understanding of the universe, its laws, life, matter, and space-time from a scientific point of view. Special emphasis will be placed on aspects of teleological nature and final causes in evolution to see how they might furnish an answer to longstanding questions which previously seemed to be irreconcilable with both physical and trans-physical conceptions.

1. Regaining Multidimensionality

One might call physicalism—that is, material monism—and the current forms of rationalistic and materialistic naturalism a 'one-dimensional cosmology'. Physicalism sees only matter and the law of physics as the sole determinants of everything that exists, from elementary particles to human consciousness. Similarly, dualism, as the word already suggests, can be seen as a 'two-dimensional cosmology' in which mind and matter have their distinct reality. Even being completely different ontologies from the simplistic Cartesian conceptions, panpsychism, cosmopsychism, and analytic idealism don't go much further than a one- or two-dimensional perspective because they conflate everything to mentality and consciousness or, worse, to only mentality (therefrom the term *'mental monism'*). Meanwhile, an *'integral cosmology'* that becomes aware of the more fine-grained gradations of consciousness and reality, as articulated by the planes and parts of being that we discussed in Pt.III-I.4, expresses a 'multidimensional cosmology'. This grand vision can accommodate aspects of reality, from the individual to the universe as a whole, which was previously mysterious or led to inconsistencies whose origin remained obscure. It is not only able to solve the problems, apparent contradictions, or any sorts of internal inconsistencies of the previous conceptual frameworks but can even explain why and wherefrom these arise.

Fig. 124 shows another figurative metaphor that the Austrian psychologist Viktor Frankl used to point out the limits of reductionism in psychology. On the footsteps of Plato's cave allegory, it illustrates what happens when we apply a 'dimensional reduction': Three very different 3D objects, such as a cylinder, a cone, and a sphere, once projected–that is, reduced–to their lower 2D shadows, mislead us to believe that they are one and the same object (the disk-shadows).

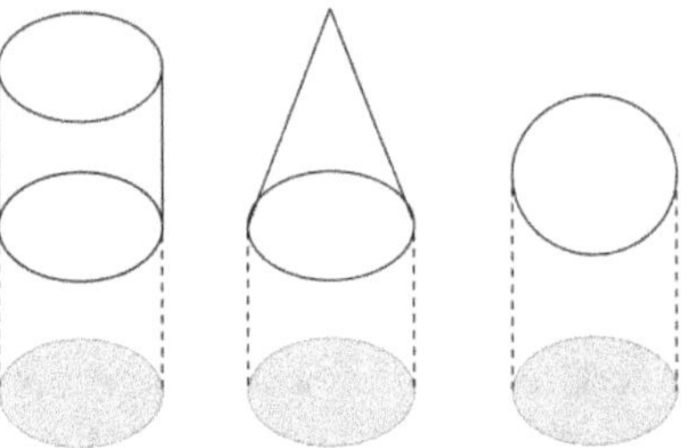

Fig. 124 A visual metaphor for the dangers of reductionism.

Analogously, reducing the human being, with all its deeper qualitative and psychological dimensions, into a lower-dimensional theoretical framework made exclusively of material processes leads to contradictory results and a misleading belief system about human nature. Here, we can go further and apply the same metaphor to highlight how sticking to a lower-dimensional philosophy, eventually by invoking principles of parsimony, or inappropriate applications of Occam's razor, we dangerously cage ourselves into an unnecessarily superficial belief system. We must be open to a higher dimensionality to apprehend and comprehend reality in all its integrality.

The conventional philosophy of mind and modern consciousness studies insist on cramming a reality that is inherently multidimensional into the limited mental 'flatland' conception. The idealist indeed goes a step further, some even adopting non-dual teachings as the basis of their theoretical framework and recognizing the centrality of a transcendent Self (Satchitananda, Brahman, Mind at Large, Emptiness, a witness Purusha or however we might name it) as the sole existence, with matter and mind having different reciprocal ontological statuses (inside a panpsychist, cosmopsychist, cosmoidealistic, or any other mental monist perspective). This is, without a doubt, a great leap forward compared to a strict and inflexible unidimensional physicalism that denies, from the outset, any other dimension. Nevertheless, these more recent theoretical frameworks of the Western philosophy of mind that embraces the Eastern non-dual Indian or Buddhist tradition limit themselves mainly to a two-dimensionality that doesn't go beyond the Self-ego dichotomy and continue to ignore the fact that already the millennia-old traditional Indian Vedic philosophy described a more sophisticated five-sheet cosmology and, most importantly, they lack a coherent evolutionary cosmology. Despite the fact that the present intellectual, cultural, academic environment is widely conditioned by an evolutionary worldview, even the most progressive scientists and philosophers of mind open to spiritual perspectives nevertheless look upon an evolutionary spiritual emergentism with skepticism. Philosophers of mind who embrace these simplified trans-physical cosmologies will inevitably have to cope with dissonances and incompatibilities that arise naturally, especially when we invoke principles of parsimony and cut out, with Occam's razor, the dimensions of a multidimensional reality reducing it to a flatland construct.

Fortunately, Nature couldn't care less about our anthropomorphic desires for principles of moderation and prudence and allows itself to manifest as a multidimensional realm whose richness, non-linear complexity, and variety the human linear flatland thinking finds itself inevitably unable to grasp. The settings of the modern philosophy of mind are still too low-dimensional models based on a too 'coarse-grained vision' of consciousness, the mind, and the universe. Inevitably, they can't make coherent sense out of it. This was the original motivation that stood behind our warning to not coalesce mind with consciousness and the emphasis on the potential risks and dangers of an uncritical and a priori application of principles of parsimony. By doing so, one misinterprets the part for the whole and misconstrues the latter with the limited properties of the former.

Things become more intelligible once we allow ourselves to consider that the physical universe represents only the tip of a much larger spectrum of intertwined ascending and descending planes of consciousness and substances that express themselves in hierarchical orders of manifesting realities determined by forces and (more or less subtle, plastic and conscious) 'stuff' of which matter is only the most exterior manifestation, the last rung of a descending ladder. Though interdependent and mutually interpenetrating, all these domains can be considered a universe on their own, of which our sense-mind is not aware but which, depending on our inner and spiritual opening, can nevertheless be intuited more or less vaguely and eventually be fully realized by ascending to higher states of cognition and psychic development. These planes of existence, or 'worlds', existed before the material universe. Matter is derived from these other planes of which an original Consciousness is a more or less involuted expression and which, whether we are aware of it or not, act and influence us mentally, emotionally, and physically.

2. From the One to the Many and the Nature of Mind

The question arises, then: How did a transcendent timeless and spaceless impersonal Absolute which is supposed to be the non-dual featureless and perfect Unity–the Nirguna Brahman–become a multifold polarity of qualities–the Saguna Brahman–manifesting itself as an extremely fragmented and particularized cosmos with material objects made of atoms, molecules and particles and with the appearance in the evolutionary process of our individualized consciousness, minds, vitals and bodies? Or, to rephrase it with the question of the decombination problem, how can the one Subject become many subjects and yet retain its own subjectivity? To answer this question, let us first elucidate a couple of other aspects of the integral cosmology of Sri Aurobindo.

By a direct experience, a new 'seeing' that goes beyond the sense-mind, the mind, and the intellect, one can become the witness of an immanent cosmic consciousness that labors behind the ignorance of matter, life, and mind. We have to draw back behind the veil of our own surface ignorance and come into contact with the Idea and Will that stand behind the appearances. However, even if we are still not able to ascend to higher states of consciousness, we might nonetheless be able to look further, at least with an inner intuitive perception of reality, and get an inner feeling of how every process in our universe is an event that occurs in a fundamental, indivisible Reality, an ever-changeless and, at the same time, ever-changing infinite Brahman. That's essentially what a natural philosopher like Goethe did.

From the point of view of an integral cosmology, as pointed out previously, the human mind is not a form of consciousness and cognition 'created' or 'generated' by an aggregation of matter resulting from evolution in time. Rather, it is the localized manifestation of a universal principle already inherent in the Creation since ever. From the supramental and overmental perspective, the individual mind is an exclusive concentration of the cosmic Mind. This cosmic or 'universal Mind' is a derivative power of the overmind. It is this cosmic Mind that, with its power of division, discrimination, and separation, creates particles and orders the cosmos with its material and physical laws. It is the first original power of distinction that pinpoints the Infinite into a finite object and makes our universe appear in its atomized and particularized form in infinitesimal points. Mentality is, indeed, one of the cosmic aspects that pervade and mentalize the entire universe. This is what makes the material universe appear to be amenable to mental cognition because our mind is also a derivative cognitive power of the overmind. The material universe has been organized according to mental principles originating from a cosmic Mind from the outset. It should therefore come as no surprise if our mental sciences can decipher a good amount of the cosmos' workings: When we look at the world with the mind, we see its mental aspect. The rationality and intelligibility of the world are a reflection of the immanent cosmic Mind that lies behind all the dynamical processes and the play of forces in space and time of the material universe, and with our mind being a localized reflection of it as well. This missing link led Einstein to say that *"the eternal mystery of the world is its comprehensiblity"* and *"the fact that it is comprehensible is a miracle."* [287] The paradox only arises if we assume the world to be soulless, purposeless, without meaning, and especially mindless.

It is here where things intersect with the mental monism of philosophical idealism. In a certain sense, the philosophical idealist who declares that all is mental has a point. For example, the Mind at Large of the analytic idealist is, indeed, a vague and intuitive perception of the existence of this cosmic Mind and the notion of the dissociated alter is, indeed, something in line with the notion of integral cosmology of the 'Infinite's self-concentration'. But the mental aspect of reality is only a superficial appearance of an immanent overmind that goes well beyond mentality alone. Declaring all the world being mental reflects a typical mental habit itself: Once it successfully realizes a partial truth, it rushes into extending and generalizing it to everything it can conceive of as the final truth. In this sense, the idealist has fallen into the same trap as the materialist. The materialist sees only matter with his physical mind and tries to reframe all of existence with matter, while the idealist rises a bit above, eventually rising to the higher-mind, and refashions the world with the mind. Both seem to not be aware of how the world appears to be only physical or only mental because we are still only physical and mental beings. If we rise above the mental, its higher mental, illumined mental, intuitional, overmental, and finally supramental[90] aspects would become obvious and self-evident facts as well.

But the cosmic Mind is not yet (or no longer) the overmind. In Sri Aurobindo's integral model, the cosmic (or universal Mind or Mind at Large) is the delegate in our spatial and temporal manifestation of the overmind,

---

[90] Of course, one could conjecture states of consciousness even higher than the supramental, but we leave this to posterity.

acting creatively with novelty in the physical and temporal manifestation. It is a notion of creativity of and in Nature that Henri Bergson would have probably welcomed and something he became automatically and subliminally aware of once freed from the limiting materialistic straitjacket.

In turn, the overmind is the delegate of Supermind, with both preexisting the appearance of matter and mind. The overmind links the planes of the unitarian supramental Truth-Consciousness whose inherent nature is always that of a perfect Unity and Oneness with the planes of polarity and multiplicity. Overmind is the in-between mediator between Unity and division, between the one-Idea and its expression in infinite possibilities. It is that which breaks down the transcendental Unity into the phenomenal multiplicity. At the same time, it has the unitary vision of the Supermind and the separative vision of the mind. Precisely in this reside the source and origin of the world of polarities, which it creates by the primal act of separation, division, and atomization from an original lost Unity characterizing the material domain to which we are accustomed, and yet can put everything in its right relations despite an enormous complexity of particularities.

The overmind allows for self-determination of the Infinite by a multiple exclusive self-concentration, an individual specialization, and limitation of a common universality which, however, is not a real separation and division. It is the indivisible Absolute that demarcates itself in itself becoming many, and nonetheless retains its primal Unity.

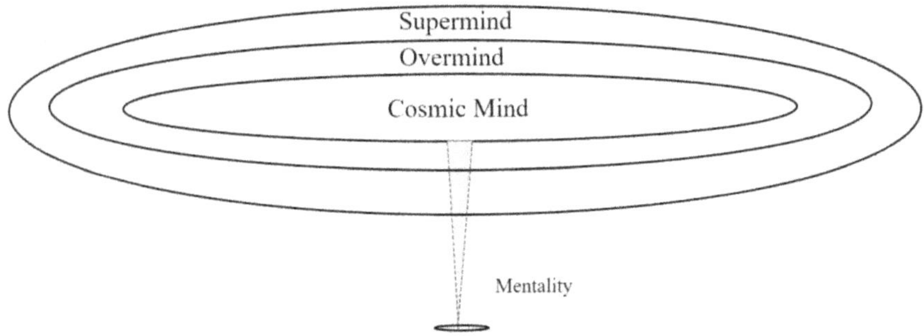

Fig. 125 Mind as a derivative consciousness of the cosmic Mind.

This cosmic Mind is in direct contact with the overmental and, indirectly, behind layers of existence, with the supramental domains. Nonetheless, reason can understand what its action in the form of an 'infinitesimal and extreme fragmentation of the Infinite' means: the One becomes the infinitely many and yet still is without a second.[91]

---

[91] Obviously, a meaningless and self-contradicting play on words for a reductionist analytic mind. Mind conceives only of one or many, tertium non datur. But this becomes a natural and self-evident fact for a cognition established on higher levels of consciousness.

The primary function of our mind is to distinguish, discern, separate, and divide because this is the cognitive function that is inherited by the original principle of division of the cosmic Mind from which our intellect is a minor representative of. Or, to put it in other words, our mind is a minor, reduced, and localized form of cosmic Mind in an organic embodiment and mimics or repeats its functions by a diminished power. The mental ability to discern, separate, and distinguish with our sense-mind, say, one object from another or a detail, property or quality from another, allows us to have a concept of quantity, number, and a qualitative spectrum. Our mental and sensory ability to locate these objects in a realm of separate points related by distances leads us to the notion of space and structure. Similarly, our mental and sensory ability to locate events in a realm of temporally separated instants related by time intervals leads us to the notion of change. It should therefore come as no surprise if *"the book of Nature is written in the language of mathematics,"* as Galileo used to say. Mathematics is precisely the study of quantity, number, structure, space, and change; that is, at the bottom, what number theory, arithmetic, algebra, geometry, and calculus are about. This is the reason for the apparently *"unreasonable effectiveness of mathematics in the natural sciences"* that physicist Eugene Wigner famously wrote in his article in 1959, which triggered an endless debate that continues today [288]. What appears to us an 'unreasonable miracle' becomes something natural once we recognize how our individualized minds, being particularizations of a universal Mind, work by its same principles of division and fragmentation and through which we look back in its cosmic physical appearance which has been originated by the cosmic Mind itself.[92] This infinitesimal fragmentation coming from the action of cosmic Mind which is the cause of atomic existence leads to the inconscient ocean we perceive as the physical universe and where the One is concealed by a fragmenting separative existence and consciousness. This inconscience works due its mechanical necessity, an automatism of purpose, with a principle of mathematical architecture, of design, of effective arrangement of quantity and numbers, as a base for quality and property.

The same fits for a cosmic Life, or a universal Life that already contains life as an inherent cosmic reality, rather than being an emergent purely material epiphenomenon of matter. As the action of cosmic Mind is the cause of atomic existence, so is the action of cosmic Life the cause of the individual life existence. It is not the other way around. For example, once we can see things from this perspective, we realize that the brain is not the origin of our mental activity but is the creation, instrument, and channel of a universal Mind in the

---

[92] For the more technically inclined reader in mathematical physics, we may also note the curious fact that one can rewrite all the laws of physics with a local measure of information, called *'Fisher information'*, which can be interpreted as the ability to distinguish, differentiate and separate. [386]

particularized expression of an individualized earthly being. Our brain and physical body are a derivative physical expression and process reflecting an original preexistent and immanent cosmic Mind and cosmic Life, respectively.

But, because the cosmic Mind is a derivative descent of the overmind, contrary to the human mind, it retains much of the original sense of unitarity and oneness that characterize the Supermind. It carries in itself a direct cognition of cosmic Truth. It comprehends the totality of things and their integral principle while acting in a cosmos of polarities and multiplicities where the personal and impersonal aspects are only two facets of manifestation of the same original Consciousness that transcends principles of personality or impersonality. For the overmind, the Divine can assume a personal or impersonal existence. While for our mind, the notion of a personal God like that of the monotheistic Abrahamitic religions, the polytheistic Hindu pantheon, or the pantheistic or panentheistic conception of a Spinozian God are all mutually exclusive and irreconcilable conceptions of the Divinity. The overmind recognizes them all as the different plays and appearances of the very same Entity. They appear contradictory only to our limited mental separative and exclusive cognition.

"Thus the mental reason sees Person and the Impersonal as opposites: it conceives an impersonal Existence in which person and personality are fictions of the Ignorance or temporary constructions; or, on the contrary, it can see Person as the primary reality and the impersonal as a mental abstraction or only stuff or means of manifestation. To the overmind intelligence these are separable Powers of the one Existence which can pursue their independent self-affirmation and can also unite together their different modes of action, creating both in their independence and in their union different states of consciousness and being which can be all of them valid and all capable of coexistence. A purely impersonal existence and consciousness is true and possible, but also an entirely personal consciousness and existence; the Impersonal Divine, Nirguna Brahman, and the Personal Divine, Saguna Brahman, are here equal and coexistent aspects of the Eternal. Impersonality can manifest with person subordinated to it as a mode of expression; but, equally, Person can be the reality with impersonality as a mode of its nature: both aspects of manifestation face each other in the infinite variety of conscious Existence. What to the mental reason are irreconcilable differences present themselves to the overmind intelligence as coexistent correlatives; what to the mental reason are contraries are to the overmind intelligence complementaries." [266](Ch. XXVIII)

That what appears to our mind as the apparently irreconcilable poles of personality and impersonality are the two faces of the same Reality. Contradictions appear only because of a limited and narrow conscious perception whose inherent nature is to divide, separate, and fragment. Once this fragmentation is complete and one seizes the world only by a fragmenting cognition, then, inevitably, one wonders where the unity has gone and, eventually, doesn't even admit it, branding it as a religious hallucination.

Therefore, for the supramental Consciousness, the One and Many, form and formless, finite and infinite are not opposites or mutually exclusive but complements of each other, two faces of the same and one Reality. From the standpoint of the supramental vision, *"the Absolute is beyond the distinction of unity and multiplicity, and yet it is the One and the innumerable Many in all the universes."* – These are *"to the dimensional mind irreconcilable opposites, but to the constant vision and experience of the supramental Truth-consciousness they are so simply and inevitably the intrinsic nature of each other that even to think of them as contraries is an unimaginable violence."* [265](Pt. II, Ch. I)–*"Supramental Nature sees everything from the standpoint of oneness and regards all things, even the greatest multiplicity and diversity, even what are to the mind the strongest contradictions, in the light of that oneness; its will, ideas, feelings, sense are made of the stuff of oneness, its actions proceed upon that basis."* [265](Pt. IV, Ch. XIII)

It is about a Consciousness that created a multitude of figures of itself and unfolded itself in a plurality of beings—the many results from a self-multiplication of the One by multiple exclusive concentration into forms of being. The Spirit founded upon the Knowledge by identity knows all as itself and in itself by a self-view and simultaneous self-vision, inhabiting each and all, knowing all things and beings as its individualized being of an all-containing existence, and yet, without getting imprisoned in the objects and beings in which it dwells. It is a double identity in which a large universal Identity contains a sea of smaller identities. The seer who has realized this Knowledge by identity has an intimate spiritual vision in which one sees, feels, and contacts all as oneself.

3. Conscious Force, Conscious Will and the Nature of Matter

Force and energy have a precise definition in physics[93], but definitions, as rigorous as they might be, don't tell us much about the nature and essence of what they define. The ontology remains obscure. Physics is about how processes occur but tells us nothing about what the essence and ultimate reality of these processes are. Physicists study what forces and energies do but don't care about philosophical ruminations that ask what these forces and energies really are. And it can't be otherwise if we accept that we are looking at things from a limited sense-mind consciousness. For example, take electromagnetic

---

[93] Simply put, force is defined as the amount of change in time of the momentum of a body and energy as the amount of physical work something can perform.

or gravitational forces. We can describe their action quite well: Once given the initial state and the boundary conditions, we can predict how they act dynamically on particles or other material objects. We can describe their dynamical evolution in time and space with remarkable accuracy, resorting to some fundamental principles. But all these laws and principles don't tell us much about what forces like electricity and gravity are in themselves. We know that they are just there and work with it.

From the view of a supramental consciousness, things acquire another significance. Following this integral vision, the source of space and time is a timeless, immutable Absolute that nevertheless creates the illusion of motion and mutability in itself without really changing—something which, as the apparent irreconcilability of personality and impersonality, for the human mind is only a wordplay, a contradiction in itself, an absurd impossibility. Because what our sense-mind realizes are only mutable relations of forms of external shocks of forces, it is unable to view all things in a multiple unity, as a Whole, as the Supermind does.

But this might not sound so distant and alien if we look at how modern physics indeed represents forces and energy. All physics could be reduced to force and space-time; there is nothing else. Four fundamental forces are known: the electromagnetic, the gravitational, and two nuclear forces. Force is an action that produces changes in time or keeps aggregates like particles, atoms, and molecules tightly together to give form to things[94], while energy is a measure of the potentiality to bring about this change. Its workings can, loosely speaking, be twofold: an action of change or, to the contrary, a tension responsible for the permanence and stability of material structures. Even matter presents itself solely by exchange of force. The solidity of material objects we perceive by contact—that is, by sensorial perception—is the expression of a force. In quantum physics, this is described by a probability field that tells us how much probability there is that in a given location in space and at a given time, a particle will be found or, more precisely, that an interaction between a force field with another force field will occur. Even atoms and molecules are described by *'probability clouds'* or *'quantum orbitals'*, not as solid objects. Modern quantum field theory describes elementary particles as vibrating field excitations of a universal and fundamental quantum field.[95] But these field excitations representing particles have no causal power unless they interact with other fields. Recall that, as we

---

[94] The weak nuclear force is a bit different 'force' inasmuch it transforms a particle type into another particle type and thereby is responsible for nuclear decays.

[95] Something very reminiscent of the Vedic image of the 'Cosmic Vibration' or the 'vibrating Prakriti' sustaining all things. The creative vibratory power projected out of Spirit that creates the ripples in subtle matter and that the universally known Sanskrit root word or seed-sound 'AUM' (or 'OM') symbolizes.

have elucidated in Pt.II-I.5&6, it is now accepted among quantum and particle physicists that the naïve notion of material particles in the classical sense of Newtonian physics, modeled as little chunks of solid matter and visualized as tiny marbles colliding with each other, has no meaning whatsoever. There are only abstract point-like force centers in space and time–that is, force fields–mutually interacting with each other. Matter is a 'compressed' force field localized in space and time that responds by contact between force and force. But physicists do not know what these quantum fields are, other than that they can be represented by abstract mathematical functions defined on a space-time manifold and that tell us about the excitation of a quantum field or the probability of observing the force in action, without telling us anything about the ontology of the force itself. It remains, therefore, to determine what more generally force and matter are from the perspective of higher levels of consciousness.

In this view also, elementary particles are the creation of an all-dividing mind. Matter is a self-expression, or rather a 'self-compression' of a Spirit which has precipitated itself into material existence by a 'self-limitation'. In Sri Aurobindo's words: *"Thus not any eternal and original law of eternal and original Matter, but the nature of the action of cosmic Mind is the cause of atomic existence. Matter is a creation, and for its creation the infinitesimal, an extreme fragmentation of the Infinite, was needed as the starting-point or basis."* [266](ch. XXV)

A cosmic Mind which dwells in its self-created atomic existence and which, by a creative process of associations of these infinitesimal points, carries out by a play of forces the Real-Idea it receives from above.

In the integral cosmology of Sri Aurobindo, force is a 'multiple self-forgetful exclusive concentration' of a separative active consciousness, a particularizing action by a separative movement of Prakriti—that is, Nature. Matter's atomic existence is also such concentration of consciousness, a 'self-limitation', or 'self-compression' of the Spirit in 'energy-substance', but passive, absorbed, and forgetful of itself. That modern physics, thanks to Einstein, also discovered a *'mass-energy equivalence'*, famously represented by his formula $E=mc^2$ (m is the mass, c is the speed of light, E is energy) is no coincidence. Matter is a form of compressed energy that, as is well known, can be liberated by a nuclear reaction.

This is also reminiscent of the quantum Zeno effect we discussed in Pt.II-I.5. Repeated and fast measurements slow down a quantum system's time evolution and, if fast enough that we can consider the observation as continuous, the system is 'frozen' in a specific physical state. Here, instead, it is the cosmic Mind that 'observes' by multiple exclusive concentrations something in itself and 'freezes' it into multiple material particles. If this should be considered only a coincidental analogy or it has a deeper meaning remains to be seen. But there is no contradiction or competition with the modern scientific view of things. One only goes a step further and asks for the inherent nature and essence of what a force ultimately is.

From the integral perspective, force is a Will inherent in existence as a potentiality of rest and movement in a fundamentally immutable Absolute. Will and consciousness are seen not as derivatives of forces. Rather, the opposite is true: What we perceive as force, action, and energy is the expression of an involved and immanent Will and Consciousness in a temporal and spatial reality. Schopenhauer had the right higher-mind perception: Indeed, there is a World-Will. But the 'integral seer' does not speculate, philosophize, and discover truths by process of reasoning or inference; rather, the seer 'sees' the things-in-themselves, as they are, without any thought at all, through a Knowledge by identity.

From the highest standpoint, change, mediated by energies and forces, is a manifestation of one of the infinite potentialities of this supreme Consciousness. It is a Consciousness that is a Will manifesting in the phenomenal universe expressing itself as a force. The clarity of the supramental vision reveals the universal clash of matter and forces as the workings of a Consciousness-Force and Consciousness-Will, a 'Force of existence'. The identity of Force and Will arises because *the Energy that creates the world can be nothing else than a Will, and Will is only consciousness applying itself to a work and a result."* [266](Bk. 1, ch. I)

Forces are one appearance of the power of Satchitananda–that is, Existence-Consciousness-Bliss–in the temporal manifestation and, thereby, reveals its biune aspect of Self and Power, emerging as a dual Person, '*Ishwara-Shakti*', the divine Self and Creator of the universe.

In this evolutionary integral cosmology, the Supermind is still not directly active. It maintains itself behind the veil of appearances but, even though concealed, is nevertheless indirectly at work in every form and force by the overmental concentration and particularization that pinpoints itself in itself. All the processes in our cosmos are self-presentations of It in itself, and physical forces are a Will, which is also a Knowledge of a Consciousness-Force acting in space and time by an exclusive concentration. For example, the scattering process between two electrons, and that in quantum field theory is represented as the interaction between two force fields, is the dynamic manifestation of an involved and self-forgetful exclusive concentration of a cosmic consciousness in itself—that is, an inconscient and mechanical will acting in a localized space-time region—something that, however, is the action of a Will-Force in its origin. The mental relations we build from the play of forces is that which we call 'form', and all the phenomenal world we experience is a cosmic multitude of sensations resulting from a form of force responding to another form of force.

It is in this sense that we must posit Consciousness at the origin of all things. Consciousness is the uncaused and irreducible ontological primitive. Consciousness-Force, Consciousness-Will, and Consciousness-Knowledge are its dynamic spatio-temporal aspect we perceive as action in time and space.

4. The Supramental Vision of Space and Time

Modern theoretical physics is searching for a unified theory of matter, forces, and space-time that aims at merging quantum mechanics and general relativity, in a single and unique description, into a *'theory of quantum gravity'*, popularly known as the 'theory of everything'[96]. For over half a century, generations of physicists have tried hard to find such a theory but, so far, with no tangible results that could be verified experimentally. There is now a growing consensus that one of the reasons for this failure is that physics always took it for granted that space and time—or, more generally, space-time as described in general relativity—is fundamental and for our normal human understanding, something that presents itself as a self-evident given and obvious reality that only philosophers felt was worth further investigating. Until recently, physicists felt it unnecessary to question what space and time are; they just posited it as a self-explanatory fact. After all, that's what physics always did (with some nuances introduced by Einstein's relativity), and it worked egregiously for more than three centuries. This lack of further philosophical curiosity was also encouraged by the fact that space and time lend themselves quite well to a mathematical formalization. Space-time can be described very efficiently, relying on a four-dimensional coordinate system (three spatial dimensions plus one temporal parameter), eventually extending it to a non-Euclidian geometry, as done in general relativity.

However, it is now felt that, at least at the most fundamental level investigated by particle physicists, our underlying assumption with respect to space and time must be reconsidered. For example, it can be shown that the *'Schrödinger equation'*—one of the most fundamental differential equations of quantum mechanics that allows us to describe the dynamical evolution in space and time of particles or larger systems, such as atoms and molecules— when applied to the entire universe as a whole becomes the so-called *'Wheeler-DeWitt equation'*, which has no temporal dependency. Applying the known rules of quantum physics and relativity , not to a limited system inside the

---

[96] The expression 'theory of everything' may suggest that it will explain everything. This is a very popular misinterpretation promoted by the media. A theory of quantum gravity will eventually reformulate the theory of general relativity and quantum mechanics in an extended theoretical framework and express all the physical forces known as a single 'super-force'. But this won't explain to us how a cell functions or how a brain works, let alone tell us anything about human psychology or the deepest essence and nature of our spiritual existence.

universe, but taking the entire universe itself as a unique quantum system, then we are forced to consider it as a timeless object. This seemed at first to make no sense. We all know very well, through our everyday experience, that in our universe, things change in time. The quantum system we call the 'universe' does not look and feel timeless at all; it appears to us as being subject to change. Meanwhile, the Wheeler-DeWitt equation seems to suggest to us that our dynamically changing universe is ultimately something that deep down is in static immobility, contradicting any mental conception and sense-mind experience to which we are accustomed.

Many other inconsistencies and divergences arise in modern theoretical physics once we try to unite quantum mechanics with general relativity. Nowadays, physicists still do not know the deep and fundamental message that Nature is trying to convey to us. However, after decades of failures in framing a theory of quantum gravity, the question about the nature of space and time gained new momentum. The lack of progress in building a theory of quantum gravity has led physicists to question whether space and time might not be themselves emergent properties, such as colors, tastes or the wetness of water and other qualities of the macroscopical world we perceive with our senses and that are emergent properties of physical phenomena at a microscopic scale. But nobody knows for sure what to look for, or where, and in what direction physics should proceed. Physics is itself an intellectual construct that stands upon the pillars of the notion of space and time. Physics without space and time is like math without numbers and functions. If we take away these fundamental concepts, one must wonder what physics is supposed to be about.

It is here that the vaster and more comprehensive vision of the universal process in the frame of a supramental gnosis can become of interest and reconcile the apparent opposites. Again, that what from a mental perspective seems utterly irreconcilable, such as the personal and impersonal aspect of a transcendent Consciousness, or its unity and multiplicity, looks equally nonsensical to conceive of a spaceless and timeless Entity being also spatial and temporal.

This is not so from the gnostic heights, where one no longer sees or conceives of physical reality as an unfoldment of time-successions but, rather, as the action of a timeless and changeless eternal Being. A causeless, eternal Absolute—a changeless and 'atemporal' Infinite—that does not create but manifests itself in itself through self-determinations in time and space, and nevertheless is above that very same time and space, what Aurobindo used to call an 'eternity in motion status'. From this perspective, space and time appear neither as separate entities nor as separable aspects of the Infinite but as self-extensions.

Space is an 'objective self-extension' of the Brahman, and time, its 'subjective self-extension'. It is objective in the sense that this infinite Consciousness steps back from itself, so to speak, and thereby becomes an object of experience of itself. It is subjective in the sense of a qualitative experience of a supra-cosmic Consciousness of itself.[97] *"Space and Time are our names for this self-extension of the one Reality." – "Space would be Brahman extended for the holding together of forms and objects; Time would be Brahman self-extended for the deployment of the movement of self-power carrying forms and objects; the two would then be a dual aspect of one and the same self-extension of the cosmic Eternal."* [266] (*Bk. 2, ch. II*).

In the theory of general relativity, space and time are represented as a unique space-time manifold. In general relativity, as in quantum mechanics, time is just a parameter and codes for the flow of a 'space-time slice' through and along a larger four-dimensional 'space-time block', and where past, present, and future are represented as a unique and eternally present block. This is a philosophical perspective called '*eternalism*' and is sometimes also referred to as the '*block time*' or '*block universe*'. In this model, every moment is eternally present, and the present moment is only a selection that picks out a time-slice. This understanding seems to us so counterintuitive because we identify our individualized mind as the selecting agent, while from the supramental perspective, one realizes that the 'Selector' is the universal consciousness itself.

However, if time can be considered just another geometrical dimension in which nature and essence can be placed on an equal footing with the three spatial ones is not so obvious. While space appears to us as an objective property of the external physical world, there is something subjective and non-geometrical that we intuitively experience about time. We can measure the extension of something—say, the length of the side of a table—with a measuring rod or a tape measure, but while our common language seems to suggest otherwise, time can't be 'measured' in the same sense as we do with material objects. Time is something we always 'measure' with a clock only insofar as it counts the ticks of a regular physical phenomenon, but we don't apprehend time as an object of cognition itself. In the physical world, what we call 'time' cannot be objectively measured as if it were an object 'out there', say, like a table or rock. You don't find time in a specific place and then measure it somewhere. Measuring time means measuring the degree of change of a physical process—usually, the clock hands or display. When we measure the size of a physical object, we see the object, but when we measure an interval of time, we don't 'see time'. We don't even know what that means. Rather, we still look at a physical object and its change in time. Therefore, space is 'objective' in the sense that we relate it to an extension of and between objects we perceive out of us, but time remains something we can't entirely

---

[97] To physicists, this may be reminiscent of John Archibald's Wheeler *"it from bit"*, which symbolizes the idea that all things physical are information-theoretic in origin because, ultimately, a physical experiment and observation is a part of the universe observing itself.

detach from an internal subjective experience. As we have seen in Pt.II-I.2, there is nothing really objective; at best things are inter-subjective. However, as with the distinction between primary and secondary qualities, we can set time apart as a predominantly subjective phenomenon compared to the intersubjectivity of space and matter. Space is something we perceive outside us, while we perceive time as something intrinsically correlated to our subjectivity inside us. And, as a twofold character of the individualized objectivity vs. subjectivity exists, a twofold character also exists in the form of cosmic objectivity vs. cosmic subjectivity. The supreme real-Subject, the Divine, projects and apprehends itself as a two-fold extension of itself: It becomes a subjective extension we call 'time' but identifies itself also by a projection of itself as an objective extension, which we call 'space'. Space and time are the two aspects of the same Self which extends itself in two modes or states of being.

That time and space must have something to do with consciousness is not really surprising. We know well how they can be experienced in different manners according to our state of consciousness. For example, the very different perceptions of space and time we have between the waking or sleeping state (with or without dreams) are a vivid example of this. But, as usual, our anthropomorphizing rationalistic and materialistic mentality convinces us that the 'real' time is that which we perceive in our waking state, especially that which measures the change of physical process. However, by rising to higher states of consciousness, we discover how our ordinary waking state experience of space and time is illusionary as well. Space and time are two different but complementary conditions of the Spirit, which appear according to the state of consciousness of the individualized spirit that perceives or conceives of it at a specific plane of existence. Time and space are relative but real in themselves or, at least, 'real' in the sense that any other extension, projection, or concentration of the original Spirit is, such as matter and physical forces.

The fact that, from the view of the supramental consciousness, time is a subjective extension, and space is an objective extension of a transcendent Entity does not imply that it sees them as our dividing sense-mind does. Again, the separative activity of mind divides and distinguishes this temporal subjective self-extension into separate instants and the spatial objective self-extension into separate positions or points in a spatial coordinate system. There is always a timeless Infinite that may condition itself in a spatial and temporal context, but it is our mind that creates a mental division of the pure unconditioned, featureless, time-less, and space-less indivisible Brahman. The separation in space and the successions in time are figures in a mental or still non-supramental experience.

The Spirit sees itself always as a Whole, a One without a second, as an undifferentiated and unaffected Oneness. This holds not only spatially but also temporally. If the succession in time is a mental figment, how does time appear from the heights of a supramental consciousness? At that point, one enters the *'supramental time-vision'* (*'trikaladrishti'*—from Sanskrit: the 'threefold vision'). This infinite time consciousness is *"founded first on its eternal identity beyond the changes of time, secondly on a simultaneous eternity of Time in which past, present and future exists together for ever in the self-knowledge and self-power of the Eternal, thirdly, in a total view of the three times as one movement singly and indivisibly seen even in their succession of stages, periods, cycles, last– and that only in the instrumental consciousness– in the step by step evolution of the moments."* – Consciousness moves between *"Two states and powers of existence, that of the timeless Infinite and that of the Infinite deploying in itself and organizing all things in time."* – *"With the knowledge founded on the supramental identity and vision ... the two are only coexistent and concurrent status and movement of the same Truth of the Infinite."* – The unified infinite time consciousness of the timeless Infinite *"maintains in itself at once in a vision of totalities and of particularities, of mobile succession or moment sight and of total stabilising vision or abiding whole sight what appears to us as the past of things, their present and their future."* [265] (Pt. IV, Ch. XXV)

An eternal time consciousness that has a total and unified vision of actualities. Time, as we know it, from our limited mind-consciousness, is a mental artifact, an artificial shadow of a much deeper truth of the phenomenal universe.

The experience of the past, present, and future that is seen as an inseparable and indivisible whole is obviously something that goes far beyond the comprehension of reason. The analytic mind, whose inherent nature is to separate and divide, can't follow. We can have an intuitive inner feeling of these things only if we ascend to a state of consciousness where all this might become not only comprehensible but also a self-evident fact. Nevertheless, it could be a good practice to remember that what we see is not what is. Physics is slowly but steadily discovering that the cosmos as we experience it is an illusion, and yet it is the projection of a very concrete Reality.

5. Creative Randomness, Evolution, and Fate

The integral cosmology described so far has, obviously, inescapable consequences for how we understand the world. An important aspect, one which is relevant for our reflections, that tries to bridge science with spirituality, is its inherently teleological character. Of course, a cosmology that bases itself on an ever-existent and supreme Consciousness, which works out by a Consciousness-Will and a Consciousness-Force all events in time and space, is inherently a cosmology to which we could ascribe an aim, a purpose, a télos entailing a finalistic conception standing behind all the cosmos. The question is: if and how can this consistently be integrated into the conceptual framework of modern science, which, however, is concerned with a strictly mechanical universe where any final causes are denied from the outset?

One option is to continue to deny it, devalue every human experience that is not exclusively based on a materialistic sense-mind analytic thinking, ignoring all the inconsistencies and contradictions arising from it, and brand any metaphysical conception of reality as empty babble or, at best, not more than poetry, eventually resorting to Occam's razor or the usual progressive-quest argumentation to permanently justify this attitude.

However, if you have read so far, you might be willing to consider a second option, namely, to see if and how a vision of things that goes beyond sense-based mental conceptions is compatible with the findings of modern science, and also solves seemingly irreconcilable divergences or apparent inconsistencies. We will discuss if and how what we have learned about an integral vision arising from the standpoint of a higher level of consciousness is in line with present scientific knowledge and if this can eventually also shed light on several issues that otherwise remain inexplicable, still obscure or half-lit phenomena.

However, to do so, it might be useful to first ground ourselves by reevaluating some very elementary but, at times, quite abstract concepts and misunderstandings surrounding it, and that are pervasive in modern science.

a. Reconciling Classical Indeterminism with Will and Purpose

Too often, we use terms, concepts, and meanings in specific contexts without knowing what they really mean. In particular, one often sees people extrapolating from the notion of randomness or 'pure chance' of an event or phenomenon and jumping to conclusions that entail a philosophical statement, such as proof of a lack of aim, will, or purpose behind that phenomenon. This line of reasoning is particularly evident in biologists who, starting from the fact that mutations are 'random', jump to conclusions when it comes to the question of purpose and intelligent design in evolution, or physicists who point to the quantum world being 'purely random' as proof that there can't be any connections between QP and consciousness.

These extrapolations aren't correct and, while less relevant in a pragmatic and mathematical domain, they are, however, misleading in a philosophical context in which we ask for the deeper meaning of things.

So, what is randomness? What is chance? What is coincidence?

Let us first take a view from the standpoint of our everyday experience and the non-quantum classical physics[98]. The perspective from quantum theory will be addresses in the next section.

---

[98] Here, with 'classical physics', we mean the Newtonian and also relativistic physics, but not phenomena ruled by quantum physics.

According to the online Cambridge dictionary, random is something *"happening, done or chosen by chance rather than according to a plan"*– *"without choosing intentionally."* Or also something that is *"informal, strange or unusual."* Chance is *"the level of possibility that something will happen"* and a coincidence is *"an occasion when two or more similar things happen at the same time, especially in a way that is unlikely and surprising."* The Oxford English Dictionary instead defines 'random' as anything *"having no definite aim or purpose; not sent or guided in a particular direction; made, done, occurring, etc., without method or conscious choice."*

Wherever you look for definitions, you will get to know that randomness is *"something proceeding, made, or occurring without a definite aim, purpose, reason, or pattern, method or adherence to a prior arrangement; not following any prearranged order and lacking any definite plan"* or as something *"accidental,* incidental, *aimless, designless, fortuitous, purposeless, unaimed, unplanned, unpremeditated, unintentional, unforeseen."* Chance is *"the absence of any cause of events that can be predicted, understood, or controlled; not planned or expected; accidental; by chance, without plan or intent."* Meanwhile, a coincidence is *"a striking occurrence of two or more events at one time apparently by mere chance."*

The problem is that these are only vaguely defined concepts, which may be fine as long as we use them in our common language and ordinary understanding, but so defined, have no significance in science.

So, what is randomness, chance, and coincidence in science or, at least, in physics and mathematics?

Loosely speaking, what we usually mean by 'random events' is a set of outcomes that are unpredictable and to which a probability is assigned. The concept of probability dates back to the 17th century with Blaise Pascal (1623–1662), Pierre de Fermat (1601–1665), and Christiaan Huygens (1629–1695). In 1657, the latter introduced the concept of *'expected value'*. Since then, the theory of probability developed quickly into a full-fledged (sometimes quite complicated and abstract) mathematical theory, and which serves well for several practical purposes, as it is practiced in finance, surveys, medicine, game theory, etc. In physics, using statistical mechanics, one can use probability functions to make meaningful statistical calculations and predictions about the microscopic properties of gases (physicists speak of the *'molecular chaos'* of the random thermal dynamics of molecules in a gas) considering macroscopic state variables, such as the pressure, volume and temperature., etc. Similar considerations hold for other natural phenomena and physical systems difficult to predict, such as the weather, earthquakes, the chaotic orbits of stars in a stellar cluster, etc., and that are more generally labeled with *'deterministic chaos'*. These are deterministic phenomena that can be described with classical physics alone, and yet are chaotic and unpredictable.

Thus, in all these scientific contexts, probability and randomness are only measures of unpredictability. Going from there to speculative theories about will, aim, or purpose is an unwarranted extrapolation. A scientific statement that starts from the more or less abstract and loose notion of randomness, connecting it naively to a lack of purpose, meaning, will, or any kind of intentionality, is misplaced. Using unpredictability as a measure of intentionality is bad science and philosophy, especially when it comes to deeper philosophical questions of meaning.

But one might object and ask what aim or purpose could there be in a molecule zigzagging in a gas, in today's sunny weather, or in a specific stellar orbit in a globular cluster? Obviously, undirected phenomena can eventually produce random outcomes. The question is if the inverse is also true: are random phenomena an obvious sign of a lack of purposefulness?

First of all, one important point we must always keep in mind is that randomness is in the eye of the beholder.

For instance, think of the tossing of a dice. Is it random? The question might sound silly because everyone knows that an unbiased dice is always considered to display random faces between one and six. But, more careful reasoning reveals that the answer depends on how we observe it.

In fact, suppose you have all the physical parameters about the dice (form, weight, elasticity, etc.) and know with very high precision how it has been thrown on a table (about which you know all the physical parameters too, for example, its minimal inclination, roughness, hardness, and so on), and have a supercomputer that has been fed with some program knowing everything about the initial and boundary conditions of the dice and the laws of physics determining the dynamics of solid bodies, such as elastic or inelastic collisions, etc. Such a computer could, at least in principle, predict what will happen. A dice is subjected to a very complicated action of forces influencing its dynamics; however, if all is known about the physical state of the dice, its initial conditions, and the forces acting upon it, we could predict how it will roll on the table and, finally, predict which number comes up.

Would we then still label this process as 'random'?

This example makes it clear how the idea of connecting a measure of predictability to a property inherent in the process is dubious: what appears to be totally unpredictable in some experimental context might well turn out to be predictable in another one. Randomness is not a property of things, it is a measure of our lack of knowledge. In classical physics we don't perceive a contradiction between determinism and randomness, that's why one speaks of *'deterministic randomness.'* At the bottom, at least in classical physics, what randomness measures is only our ignorance.[99]

---

[99] That's also why the notions of randomness, order, and disorder are closely related to concept of entropy.

We can illustrate this further with a very simple example that, if you are a programmer with some background knowledge, you can perform on your own computer. In Fig. 126 left, you can see an image that seems to be totally random. It is a picture with only 128 x 128 pixels encoded in 128 colors (grayscales). Each pixel corresponds to a number between 0 and 127. There seems to be no structure in the image nor the geometry or color (grayscale) distribution.

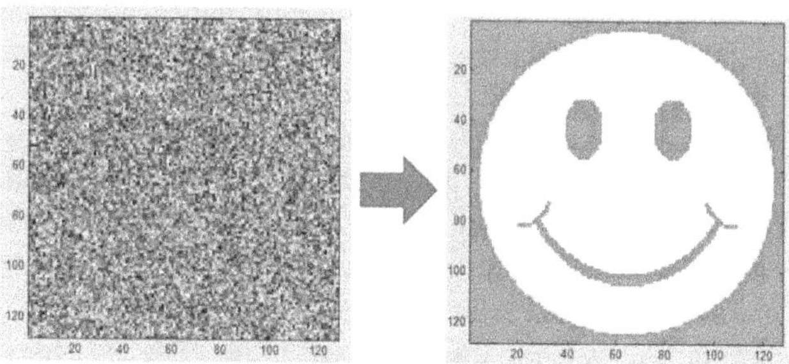

Fig. 126 Left: Encrypted data. Right: Decrypted smiley.

The overall distribution shows no signs of organization, and we can label it as almost pure 'noise' with some statistical fluctuations here and there. When one looks at the digital code, at the occurrence of the ones and zeros appearing in the binary file, and calculates the probability that binary number one or zero will show up, one finds that there is an almost perfect 50% chance each of getting a one or a zero–that is, the ones and zeros appear to be statistically distributed with equal weight. It is then fair to say that this is a pretty much random data sequence.

From this, can we conclude by whatever statistical analysis that the image doesn't contain any meaningful information and could not be authored by an intelligent agent? Can we assert with certainty that this image doesn't convey a message or had a purpose?

This apparently random data of Fig. 126 left is nothing other than the encryption of the smiley of the right picture. Both images represent the same informational content, but one was encrypted with an encryption algorithm.

As you know, cryptography is a science that has a wide range of applications. Nowadays, it is common to shop on the internet with a credit card. Your credit card data is always encrypted according to a given standard–that is, a mathematical scheme—which makes the sensitive information unrecognizable, apparently random, and meaningless; only then is the data sent over the web and decrypted back by a server on the other side. This is an operation that happens millions of times daily online. Decryption is making sense and letting something acquire meaning from an almost perfectly random sequence of data to which we usually ascribe no sense and purpose at all. Encryption technology has been created just for that reason: to transform sensitive data into a stream of data and events optimized to appear random. What was previously seen as just white noise without any regularity, let alone meaningful structure—that is, a stream of apparently purely coincidental data dominated by chance, without meaning and purpose—once decoded, suddenly makes sense.

So, what makes the difference between this image and the previous one is not a meaning, a purpose, or whatever aim, and even not the notion of 'randomness' or lack of randomness in itself, but that in the first case we didn't have a clue about the encoding and decoding scheme, or more loosely speaking, about the structure, grammar, rules, and laws of the 'language' in use. In the second case, we have a decryption code that allows us to 'de-randomize' the message. What might be a very ordered and meaningful structure in one representation and language can be very disordered in another. Data and facts can be encoded or not, but the meaning, sense, or purpose is in the mind of whoever produces the data and can be received only by another mind. Again, randomness is not a thing inherent in objects. One person's noise is another person's data.

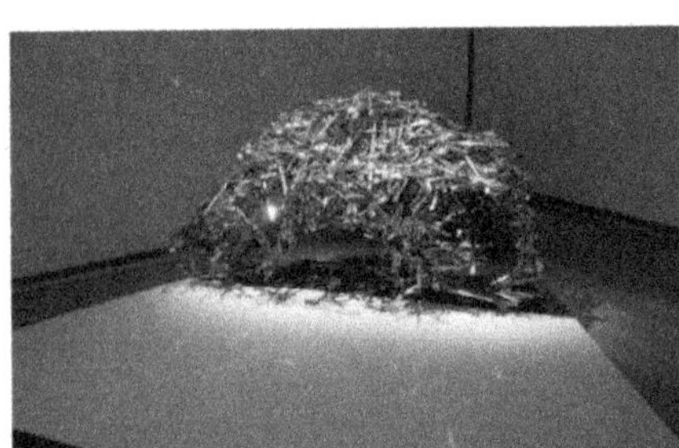

Fig. 127 Shadow art of Shigeo Fukuda, "Lunch with a helmet on", 1987.

Yet another example of how meaning vs. randomness are in the eye of the beholder is illustrated in Fig. 127. It vividly shows how a pile of junk made from hundreds of forks, knives, and spoons, and that, at first sight, we would call assembled 'randomly' (Fig. 127 left), reveals its real meaning and intent only once light is cast onto it at the correct angle (Fig. 127 right). What to one person might look like a coincidental and meaningless assemblage appears to someone else as what it really is: a sculpture created by the Japanese artist Shigeo Fukuda that projects the silhouette of a motorcycle on the floor.

Another context where we frequently fall in this randomness→purposeless fallacy is in the relation between the micro and macrocosm.

What appears to be noisy at a small scale might well turn out to be a highly ordered structure at a larger scale. If one looks through a microscope at the surface of a well levigated table, one will nevertheless observe small, microscopic random ripples. A bacterium living in this microscopic environment might consider it a structure that has formed through purely random natural processes.

The flickering of a single pixel on a monitor may appear completely random, and focusing on it only, without being aware that we need to capture the image as a whole, cannot reveal to us that, despite its random appearances, it nevertheless is part of a purposeful and willed intelligent creation: a movie of a moviemaker. Micro-randomness must not necessarily lead to macro-randomness.

Notice how the temporal dimension can also play a decisive role: shortly after the Big Bang the universe was a random patch of hydrogen clouds without structure and apparent order but, after 13.8 billion years, galaxies, solar systems, planets and even life formed. Something random today may reveal its self-organizing structure tomorrow.

Therefore, seeing things only from one of the (spatial or temporal) micro- or macro-perspectives and excluding the other might lead us astray. Contrary to common belief, randomness is not a measure or a 'litmus test' that can tell us something about final causes, the existence of a conscious agent, or the impossibility of the emergence of an order in things.

This obviously didn't tell you something that is new so far. Everyone who thinks about it can easily arrive at such conclusions. However, much too frequently, even top scientists forget about all that and continue to connect randomness with the idea of purposelessness, aimlessness, lack of a goal or intelligent causation, etc. When they are confronted not with human computer-generated images or data streams, but with a sequence of natural phenomena, they instinctively and subconsciously immediately switch back to the opposite point of view, making metaphysical statements about the lack of an aim, purpose, or meaning of life, evolution, the universe, etc.

Moreover, there are also things that are numerically random, while they are not at all from a more abstract perspective. For example, consider the sequence of decimal digits of the irrational number that is $\pi = 3.14...$ If you try to find a pattern, a regularity, a structure in the occurrence of the digits

$$\pi = 3.1415926535\ 8979323846$$
$$2643383279\ 5028841971$$
$$6939937510\ 5820974944$$
$$5923078164\ 0628620899$$
$$8628034825\ 3421170679$$
$$8214808651\ 3282306647\ ...$$

Fig. 128 The first 120 decimal digits of Pi.

of π, you will find none. According to the law of large numbers, the statistical distribution seems to appear completely random. Every digit, from 0 to 9, has an equal chance of showing up. And yet, this constant Pi has a precise meaning. For example, as is well known, it defines the ratio of a circle's circumference to its diameter and is ubiquitous in all mathematical and physical theories describing Nature. It is a random sequence of digits which, nevertheless, has a precise meaning and tells us something very fundamental about the geometry of our universe.

So, are the decimal numbers of π random or not? It is a matter of convenience. In some contexts, you might prefer to use the standard random notion based on the frequentist definition of randomness as a sequence of numbers having an equal probability of showing up, while in other contexts, you might say that it is not random since it signifies something very precise, in a certain sense also very 'concrete.' If we would receive radio signals from an extraterrestrial civilization sending this sequence of numbers, we would immediately recognize it as the product of an intelligent agent, and people would hardly speak of a random signal.

Thus, randomness is in the eyes of the beholder, a human-made notion. It is not an intrinsic property of either mathematics or any object or phenomenon in Nature. It depends on how we see things, and on what we know.

Moreover, keep always in mind that probabilities in science are always based on a third-person perspective–that is, they are based on the knowledge of an outside observer. For example, I might know what I'm going to do in the next five seconds but, if I don't tell anyone, an external observer could at best only guess using some statistical analysis to predict what I'm going to do. From the perspective of the acting conscious agent, the will to act might have been decided and is coming about with certainty–that is, with probability p=1. While from the perspective of another observer that necessarily must take an external point of view, a degree of uncertainty remains that inevitably forces one to resort to probability distributions. This distinction is essential when we deal with ontologies like panpsychism or cosmopsychism, and where an inner volitional factor is assumed to exist, say in a particle, a cell, or a cosmic consciousness.

Nevertheless, one frequently hears people making distinctions between only apparently random phenomena, so-called *'pseudo-random processes'* and 'truly' random processes, or between *'deterministic'* and *'non-deterministic processes'*. Is this distinction warranted? Can you come up with a single example of a 'truly' random process? As we have seen, there is no means to do that. The distinction is only intuitive, colloquial, not scientific. For example, you will hear that coin or dice tossing is a 'truly' random and non-deterministic process. But we have already seen that it is easy to show, with a thought experiment, that this is not at all the case. The implicit connection we make between a truly random event and the notion of unpredictability is unwarranted.

Realize also that the so-called *'random number generators'* on your computer are usually only very complicated deterministic programs that create sequences of pseudo-random numbers—that is, generated by a deterministic algorithm. The algorithm is designed to be so complicated that its output appears to be a random sequence to anyone who doesn't know the algorithm. When we talk about the random generators of a normal computer, we mean a program that 'hides' its variables. If we can know these *'hidden variables'* and the algorithm that stands behind them, then we would be able to predict the outcomes, at least in principle. This is also the conception of randomness in classical statistical mechanics.

This kind of classical indeterminism is governed by classical physics, and is only apparently indeterministic, a deterministic chaos. Because it arises only due to our lack of knowledge of all the microscopic details, and that determine the evolution in time of the system we are observing–say, for example, due to imprecise measurements. Classical indeterminism is just a complicated and messy phenomenon we can take advantage of, like playing dice that, however, deep down, is still the everyday causal realism, the Laplacian classical physical causal determinism.

At the end of the story, we rediscover what was obvious from the outset and that (more or less subliminally) everyone knows but that even well-trained scientists and philosophers like to forget: At least, in the classical (non-quantum) world, if randomness means something at all, it is subjective unpredictability–that is, only lack of knowledge, just our ignorance.

b. What is Quantum Indeterminism?

There might be an exception, however. In fact, what about quantum physics?

Here, we repeatedly said that phenomena in quantum physics are indeterministic, random, uncertain, fuzzy, or whatever kind of label you might use to describe them, and that this uncertainty is not due to our ignorance but, rather, is 'intrinsic'. Quantum mechanics is considered 'inherently' indeterministic or 'inherently' random, or just 'truly random'.

But what does this mean exactly?

Saying that in QM phenomena are quantum indeterministic, or just 'truly random', means that the theory that predicts the observations is a complete description of reality and has no hidden variables.

The question then is, what is a 'hidden variable', what kind of 'reality' are we talking about, what all this allows us to conclude, and, more importantly, what kind of inferences we are *not* allowed to make?

Let us consider the first question. In general, by hidden variables, one means the hidden information about the unknown physical quantities that determine the observed phenomena and that are in principle knowable but, for practical reasons, almost impossible to establish. Consider again the example of dice tossing. All the physical quantities, such as its exact form, weight, elasticity, etc., and the inclination, roughness or hardness of the table, are attributes of a physical reality that can be known only with limited precision and sometimes are simply impossible to be measured everywhere. Think of all the microscopic details of the surface texture of the table: These can be known in principle but in practice, they are 'hidden variables' that, however, can determine a macroscopic outcome. Or, again, think of the molecular chaos a molecule goes through in a gas. This is also a process that, according to the classical causal determinism, in principle, Laplace's almighty demon could predict: If all the physical quantities involved, namely the exact mass, position, and momentum of every gas molecule are known with infinite precision, we could calculate the movement of each of them in time and how they scatter with each other, just as we imagine with our causal deterministic realism the scattering of billiard balls. But since there are hundreds of billions of billions of entities involved, and which positions and speed we certainly cannot all measure instantaneously, we are forced to give up any hope to know these hidden variables and must resort to statistical considerations. We can speak of a probability that the particles involved have an average energy or average momentum, but we will never be able to make any specific statement about the behavior of a single particle at a specific time. Because of the ignorance about the hidden variables, here statistics and probability functions reign and are all that can be said.

Hence, a hidden variable is an unobservable but real entity, and that could influence the measurement outcome of a physical quantity or, more formally, it is an extra parameter in a probability distribution. An unknown (but in principle knowable) hidden cause that determines an observable effect making it appear unpredictable–that is, 'random.'

What is a complete theory without hidden variables then?

In Pt.II-I.5&6 we pointed out the probabilistic and statistical nature of quantum physics. The question at this point is whether in QP, this apparently ineliminable uncertainty, indeterminacy, and unpredictability of quantum phenomena (think of Heisenberg's uncertainty principle, the random measurement of spin states, the collapse of the wave function, etc.) is also due to our ignorance about underlying hidden variables? For example, where the particle will displace itself on the screen in the double-slit experiment or where it will be found according to Heisenberg's uncertainty principle is a matter of pure chance—a random event.

People conjectured that the statistical nature of the quantum world perhaps arises like in thermodynamics–that is, we can't describe and predict the position and speed of every particle due to our ignorance about all the details and properties at the most microscopic level of the quantum. Perhaps, also in QM, there might exist an underlying physical reality yet to be discovered, and with its physical quantities and the properties of the elementary particles that are so tiny and hidden from our measurement devices that might be responsible for the probabilistic and unpredictable nature of the quantum world.

However, from several theoretical considerations and experimental verifications, we know that there can be no hidden variables in QM, at least not in a locally realistic theory. From experimental evidence, we know that quantum physics violates the so-called *'Bell's inequalities'*, a mathematical theorem that sets boundaries on our conceptions of reality and that tells us that we must abandon either locality–meaning that particles can interact with each other only in a limited range because nothing can be faster than light–or the deterministic causal realism or, probably both. If this is the case, it seems to imply that we have to deal with a truly random domain where things are inherently not predictable, not even in principle, simply because there are no hidden variables at all. Because, if a theory is complete, has no hidden variables, but, nevertheless, makes only probabilistic predictions, how else could we interpret this fact?

In this sense, a statistical theory without hidden variables can be called 'complete' when there is nothing more to know and to say about its structure and predictive power. The indeterminacy in QM is not due to our lack of knowledge but is inherent in Nature. The best we can do is resort to probabilistic predictions, but never, ever will we predict the single quantum phenomenon as we could in principle do with the outcome of the dice tossing or decrypting an image or by knowing the algorithm of the deterministic random number generator because, in the quantum world, there is no algorithm in the first place. That's also the reason why in some experiments in which you want to be absolutely certain that you are using a truly random generator, you use so-called '*quantum random generators*' (for example, by reading out the quantum fluctuations of the vacuum.)

In the microscopic domain ruled by the laws of QP, a 'quantum state' is represented by its wave function. The wave function is considered to furnish the full description of the system, there are no other hidden variables or parameters that determine its formal structure. The wave function is a mathematical function which sole purpose is to allow us to calculate the probability of a possible measurement outcome. It tells us how likely the occurrence of an event is (an atomic decay, an energy transition, etc.), or the probability to measure a particle's property, such as its position, spin, momentum, etc. For a theory that has no hidden variables, the wave function is a formal complete description of every quantum state; it is all there is and that can be said. The collapse of the wave function to a specific outcome from a set of eligible outcomes occurs for no underlying reason at all.

This also implies a different temporal conditioning of physical phenomena. A theory with all known variables (or with hidden variables that we can know in principle), like in classical physics, once given the initial and boundary conditions of a system at a certain time t_0, would be able to predict the exact state of the system at a successive instant t_1. A single cause leads to only one possible effect, and each classical state is uniquely determined by its past condition.

Conversely, a statistical theory described by probabilities, and without hidden variables, like in QM, describes a reality where physical phenomena are not determined entirely by their past conditions. Even if all the properties of the system, its initial state, and boundary conditions are known with infinite precision, the future state can be described only in terms of the probabilities of what we will observe. But the single event we observe now does not depend on the past state of the system, and a single future outcome will not depend on the present conditions. There are aspects of the present reality with no causal dependence between its past and future. In other words, there could be observable effects that have no antecedent causes. A quantum theory without hidden variables is an 'acausal ontology' where events just happen and must not necessarily be what they are because of preceding causal relations with past events. The state of a system is not a function of the past.

In the quantum domain, one can have effects literally without a cause. The exact time at which a specific atom decays is causeless in the sense that it is not predetermined by its previous physical state. It is only when we measure a sufficient number of decays that we recognize a statistical law. The fringe that the single-photon or electron will hit on the screen of the double-slit experiment is not determined by something in advance or by the influence of external boundary conditions; it just happens at the instant of the measurement even though, after a large number of particles is detected, an overall statistical distribution appears—namely, the interference fringes.

Or, conversely, but equivalently, we can say that in QM causes make effects probable but not certain. A specific physical phenomenon at time t_0 causes an effect at time t_1 only with a specific likelihood, but not with certainty. And there are no hidden physical causes that make effects more or less probable. We could say that in quantum physics, the events dictated by quantum random and purely probabilistic laws have no inherent and hidden causes. These events just happen and are 'causeless', appearing to us as an undirected sequence of phenomena that are constrained only by probability laws. The single event is a potentiality out of many (or infinitely many) possible ones but no physical selecting mechanism exists.

This casts doubts on one of the most fundamental assumptions of everyday experience. Spinoza's and Leibniz's principle of sufficient reason which states that everything must have a reason, cause, or ground, is questionable in this domain. For our common sense macroscopic intuition and experience, there must always be a reason for whatever happens. But for a complete theory without hidden variables, like in QM, there is no 'reason' for an electron to 'choose' spin up over spin down when both are equally likely and equally possible measurement outcomes (recall Fig. 67). Such an ontology is ruled by a fundamental quantum indeterminacy, with 'fundamental' meaning that an event is unpredictable–what we colloquially call 'random'–because it emerges as the manifestations of one of the many possible potentialities inherent in existence independent from past conditions and is not reducible to a mechanistic process in which causal relations are fixed in time. These potentialities that are not functions of the past that can be described with statistical theories only and that, contrary to their classical counterparts, cannot resort to computable mechanistic models and sense-mind representations and models of reality, because there is no underlaying ontology that can be framed in our naïve sense-mind realism. In QM every physical event has a 'potentiality', 'propensity', 'aptitude', 'disposition' to manifest, but must not. Outcomes occur for no reason at all, and there is no underlying hidden reality determining this 'choice'. A 'cause-less' and 'reason-less' phenomenality.

Thus, there is an important distinction between the concepts of probability and randomness in classical and quantum physics. Both rest on statistical analysis but are inherently very different and must be distinguished.

That's why Einstein, who refused to give up the local and deterministic realism, famously said: *"God, doesn't play dice."* Bohr's answer became equally famous: *"Stop telling God what to do with his dice."*

However, most physicists accept this sort of 'ex-nihilo' or 'out-of-nowhere magic' as it is and without questioning it further. After all, it led to powerful insights and theories that were experimentally verified.

Fig. 129 Does God play dice?

c. Reconciling Quantum Indeterminism with Will and Purpose

If this genuine indeterminism is a feature of the world, how can there be any place for free will? If, on the one hand, everything is ruled by strictly indeterministic microscopic processes, where everything seems to be ruled by a blind chance, by pure coincidences, just random events, how can any form of volition–that is, a consciousness applying itself to a goal-directed work and a result–be conceivably possible? How can chance be the cause for our actions? How can indeterminism be the cause of our freedom? Since will implies control, how can random physical events cause such a control? Can we exclude any teleological argument in quantum physics by resorting to its random character?

These questions led British-Australian philosopher J. J. C. Smart to frame the '*standard argument against free will*' in 1961 [289].

The first objection against free will is the '*determinism objection*': If all events are caused deterministically, then all our choices must be predetermined, and there can be no free will.

The second is the '*randomness objection*': We can be free only if our choices are free, and these are free only if they are neither deterministically caused nor necessitated by antecedent events. But, a free choice that is not causally determined by antecedent events is a random occurrence and, therefore, cannot be regarded as a free action as well.

In other words, it doesn't matter if our actions are caused deterministically or non-deterministically, either way, we can't have any control over it.

The random objection is also the typical argument of the scientist speaking against quantum randomness as having anything to do with consciousness and (more or less free) will, mind or brain states. It assumes (subconsciously) the naïve conception of randomness and chance as being synonymous with an undirected phenomenon, an indicator of the lack of any conscious agency. It is just unpredictable 'noise' that has no meaning and, thereby, can't be considered a source of volition or creative power. But, we have already shown how any extrapolation from randomness to a supposed lack of will, purpose, or aim is misplaced.

However, this reasoning also hides another unaware assumption. Whoever invokes these standard objections assumes them to be exhaustive of all possibilities. There can be only determinism or indeterminism, what else?

This ignores the third possibility that great minds such as Plotinus, Spinoza, Descartes, St. Augustine, Schelling, Schopenhauer, Hegel, Whitehead, Husserl, and others contemplated: The '*causa sui*'–that is, something that is generated within itself, is self-caused, or cause of itself and independent of any other ground, yet containing within itself a sufficient explanation of its own being and source of every self-creative process. It is neither determined, nor undetermined, but self-determined.

Spinoza begins his magnum opus, "Ethics", stating: *"By cause of itself (causa sui) I understand that whose essence involves existence; or, that whose nature cannot be conceived except as existing"* [290](I, Def. I.) For Spinoza the causa sui was the very notion of Substance, or God, as the first cause of all things, and also the cause of Itself. This Substance, however, was not meant by Spinoza as a resting being, but as an unconditional and temporally power of action, a creative and dynamic force, yet not subject to any temporal change. *"That thing is said to be free (libera) which exists by the mere necessity of its own nature and is determined to act by itself alone"* [290] (I, Def. XII.)

Whitehead claimed that each actual entities, the *atomic occasions of experiences,* are self-creative or causa sui: *"Self-realization is the ultimate fact of facts. An actuality is self-realizing, and whatever is self-realizing is an actuality."* [235] For Whitehead, each actual entity has a drive, as it were, to realize its potentialities, it creates its own identity and strives for self-actualization. It achieves its subjective aim, and becomes actual: Every process is active in and of itself and reveals a drive to realize its potentialities.

Aurobindo did not call it 'causa sui', but held a very similar vision of original spiritual self-determinations of the Infinite.

"It is perfectly understandable that the Absolute is and must be indeterminable in the sense that it cannot be limited by any determination or any sum of possible determinations, but not in the sense that it is incapable of self-determination. The Supreme Existence cannot be incapable of creating

true self-determinations of its being, incapable of upholding a real self-creation or manifestation in its self-existent infinite. Overmind, then, gives us no final and positive solution; it is in a supramental cognition beyond it that we are left to seek for an answer. A Supramental Truth-consciousness is at once the self-awareness of the Infinite and Eternal and a power of self-determination inherent in that self-awareness; the first is its foundation and status, the second is its power of being, the dynamis of its self-existence." [266] Bk.II, Pt.I, ch.I, Pg. 327

"These [trinity of Self as subject, Self as object and self-awareness holding together Self as subject-object] and other primal powers and aspects assume their status among the fundamental spiritual self-determinations of the Infinite; all others are determinates of the fundamental spiritual determinates, significant relations, significant powers, significant forms of being, consciousness, force, delight,—energies, conditions, ways, lines of the truth-process of the Consciousness-Force of the Eternal, imperatives, possibilities, actualities of its manifestation." [266] Bk.II, Pt.I, ch.I, *Pg. 329.*

Even the Infinite, the Divine–that is, God–can be truly free if, and only if, it is self-determining, and doesn't need reasons or antecedent causes to be or to cause something.

The question then is: Could we, at least in principle, reinterpret quantum indeterminism in the frame of a self-determining or 'self-causal' ontology? Something along the lines of Leibnitz's panpsychist monadology where each Monad is an actual entity having not just a passive proto-conscious atomic occasion of experience but also being an active proto-volitional self-determining entity? Quantum fields not just having but being minutest causa sui propensities for volition?

This would be a point of view in which we don't ask how free will relates to causation, or a lack of causation, with all its paradoxical implications with regards to an apparent incompatibility with determinism and indeterminism, but, rather, is a paradigm where we posit causation itself as an aspect of a volitional power. It is a microscopic view that contrasts and even replaces what philosophers call an *'event-causation'* (every event causes another event–that is, the standpoint of the mechanistic Laplacian clockwork), with an *'agent-causation'* ('being' is an agent that causes an event). A paradigm that posits consciousness, will and agent-causation as being ontological fundamental primitives.

It is in this sense that quantum indeterminism might be an entry point that provides an opening for free will and an ontology where, not matter, but consciousness is fundamental, and where the standard argument against free will fails to be valid. Because will isn't a hidden variable. Will is an aspect of consciousness, it cannot be just a hidden parameter in a mathematical expression or some unknown gear mechanism determining a probability distribution. Free will is the cause of itself and doesn't need anything beyond itself or a 'reason' to will.

Nevertheless, one possible objection against this 'quantum agent-causality' is the following (for example see [291]): While we might concede quantum randomness making room for agent-causation freedom, we shouldn't forget that quantum mechanics assigns probabilities to each of these possible 'choices' according to strict probability laws algorithmically determined by the laws of QM[100]. The single measurement outcome is unpredictable and indeterministic, but the probability function is determined. Even though the so-obtained probability distribution does not say which event will become actual, nevertheless the overall behavior of the system is constrained by these quantum laws. For example, the photon diffracted at the double slit will have to hit one of the interference fringes, it can't displace itself 'at will' on a dark band (recall Fig. 65). An atomic decay can happen at any moment, but the large number of decays can't distribute themselves 'at will', they must conform to a half-lifetime exponential decay law. Thus, the saying goes, that agent causation could not peacefully coexist with quantum mechanics because this makes it a slave to quantum-mechanical probabilities. Quantum mechanics does not help the free will theorists.

But careful analysis of this reveals the fallacy. Probabilistic laws of Nature do not require, for any finite number of trials, any precise distribution of outcomes. Several highly unlikely outcomes may occur, without violating the overall statistical law. A conscious agent may choose a prescribed distribution law of outcomes but there is no prescribed order to do that. An agent can freely choose how to approximate a target distribution in each case. It is only constrained to match the distribution law after a large number of 'choices'.

Thus, free will is perfectly compatible with quantum indetermination as long as it does not cause any observable deviations from quantum laws.

One might still object that, nevertheless, we are partially unfree since, in the end, we must make our choices so that these conform to a certain probability distribution. We have only relatively small freedom to choose until discernible statistical deviations from the prescribed law begin to appear.

This is only partially true. Our will is obviously constrained by physicality. To be a physical living being our bodily functions must work according to the laws of physics, chemistry, and biology. I can't stop my aging by my will-power. Whatever I will, I can't fly by flapping my hands, as I may do in a dream. Physical laws, quantum and classical, constrain our free will. Nonetheless, there remains an enormous creative potential to be expressed. To clarify why there isn't any conflict between an agent-causation theory and QM, let us use a useful analogy, originally proposed by F. Faggin [292], which

---

[100] For the technical minded: The solution of differential equations, such as the Schrödinger equation and the Born rule.

shows how whatever probability distribution can be taken neither as a signature for lack of agency nor as a violation of natural laws.

Consider the probabilistic distribution of the alphabet's letter in a written text. Statistical analysis shows that, given a sufficiently long text, regardless of the language and of who writes, the probability distribution describing the occurrence of each letter will always be the same. For example, in the English language, no matter who is writing and no matter what he/she is writing, the most probable occurrence will always be the letter "e", appearing with a probability of about 12.7% (see *Fig. 130* top).

Without the knowledge of the language, the alphabet, the syntax, and most of all, the semantic content of the words, one would see only the occurrence of an unpredictable sequence of symbols and may conclude that no conscious volition and willful semantic agent stands behind the occurrence of these randomly appearing letters. Of course, though one does not know the English language, nobody would seriously believe that. But this is only because we ourselves are conscious agents knowing from the outset that, say, behind an English classic like "The Lord of the Flies" stood another conscious agent, namely, Charles Dickens.

The point is that we are not allowed to extrapolate from this unpredictability the absence of an immanent teleology. Though this probability distribution of the occurrence of the letters in English is not white noise–that is, an equal probability of appearance for all letters–nonetheless, the occurrence of a single letter remains completely unpredictable. It can be only characterized by an expectation value out of a probability distribution. From the statistical point of view, this is exactly the same notion of probability distribution, expectation value, unpredictability, and randomness of quantum mechanics. Conceptually, there is no difference.

To illustrate this further, consider *Fig. 130* bottom. It shows the emission line spectra in the visible light of an atom (Neon). It is a snapshot resulting from an exposure that collects the flux of photons in a time interval. The horizontal position of the line represents the wavelength of the photons and its intensity (the width of the line) the number of photons collected in that time interval. This also dictates the probability that one will observe the single photon hit the specific line every time the atom emits a photon. Which photon will hit next which line is an exclusively quantum process; one can speak only of the probability with which the next photon will appear on one or the other spectral line, but no matter how much we know about the quantum system, the single event remains intrinsically unpredictable.

Obviously, this is not to say that language has something to do with quantum physics, but this analogy highlights how there is no conceptual difference between the two statistical tests: the statistics of the occurrence of the alphabet's letters in a text and the occurrence of that of the photon's wavelength by an atomic emission. If the former is not a 'litmus test' that does allow for any extrapolations for or against teleological speculations, the latter is not either. Arguments based on quantum randomness (or unpredictability) supposedly demonstrating a lack of will, teleology, final causes, unguided evolution, agency, mind, or consciousness are flawed logical inferences.

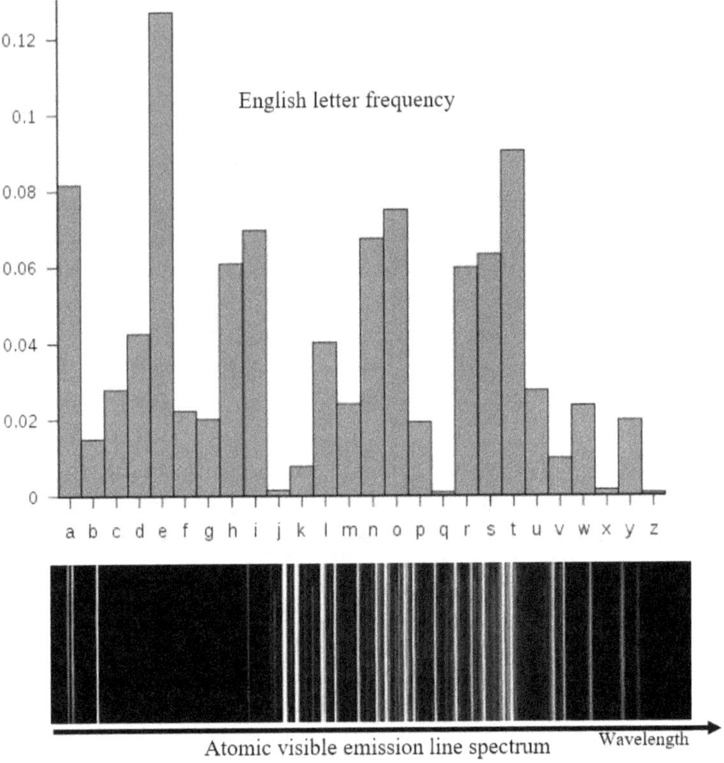

Fig. 130 Top: Probability distribution of the English alphabet.
Bottom: Atomic spectral emission lines.

In conclusion, self-causality has been misinterpreted as a 'reasonless acausality' or, what popularly people call 'true randomness', while it is precisely that which allows for volitional causal agency to enter at the quantum level resolving the puzzling aspect of indeterminacy of the quantum laws. The best free will *is* self-caused will. Self-determination is the true meaning of what we call 'freedom.' The strive for freedom *is* the strive for self-actualization that enacts a self-realization. A quantum phenomenon ruled by laws without hidden variables is not just indeterminate, because indeterminism only implies unpredictability. Quantum indeterminism differs from classical indeterminism inasmuch that it is not only unpredictable for an outside observer but is also self-determined from the inside because there is nothing else than itself as a possible causal source or agent and that is not dependent on prior events. This is not in contradiction with the laws of QM. Quite the opposite, a theory that is complete and without hidden variables even requires self-determinism. QT even demands some sort of primitively volitional capacity on the part of a quantum system. We could posit it per definition: A theory that is complete and without hidden variables *is* a theory of self-causation.

For a more technical assessment of this chapter see also my articles [293], [294].

d. The Conscious Universe

> *"Randomness is our inability to perceive God's patterned encryption. It is a boundary where our knowledge ends and our ignorance begins."* - Robert Edward Grant

If consciousness is fundamental–that is, an ontological primitive that precedes physicality–then it is questionable whether it obeys our conventional understanding of causality. The principles of cause and effect that, from the standpoint of our mental consciousness we take for granted, and where we see a cause temporally preceding the effect, make no sense from the standpoint of the supramental gnosis where past, present and future are seen in a one and unified threefold vision. A transcendent Consciousness needs no antecedent causes or 'reasons' to be and to do something in its manifestation in the present moment, as it is trans-temporal in the first place. Cause and effect exist simultaneously in the Will. The mind figures processes in relations and successions in the form of a temporal causality which, however, is only a means of interpreting the manifestation of an immanent creative Power that stands beyond the appearances and mental conceptions of causality.

One might even wonder whether Schopenhauer's intuition, which identifies causality and will, might have something to do with the quantum randomness that determines when and where a detector clicks. Or, might what Sri Aurobindo described as the Consciousness-Force and Consciousness-Will acting in and between matter determining the events and the history of the Cosmos, also work at the quantum level?

We have seen that there is no logical reason or physical law that prevents us from conjecturing that quantum randomness might be, indeed, the expression of a selecting Will, purpose, a directional volition intrinsic in matter, space-time, or both. This would, at least in part, be in line with Schopenhauer's vision of the 'World as Will' or a Spinozian universal 'Substance', a Natura naturans at work or Aurobindo's Consciousness-Force and Consciousness-Will.

The universal quantum field of quantum field theory indeed offers a substratum in the natural world where

Fig. 131 The universal random quantum field: "The World as Will"?

teleological action and free will may well find their reason to be. It would not be the free will of the individual but of the cosmic consciousness. Taking this point of view would present Einstein's famous statement that *"God, doesn't play dice"* in a different light. Indeed, if quantum indeterminism *is* one of the manifestations of God's will on the physical plane, it is hard to believe that we, as limited cognizant beings, could interpret it other than being 'random' and see God playing dice even if he doesn't.

It is by taking this perspective that the Anglo-American theoretical physicist and mathematician Freeman Dyson conjectured: *"Atoms in the laboratory are weird stuff, behaving like active agents rather than inert substances. They make unpredictable choices between alternative possibilities according to the laws of quantum mechanics. It appears that mind, as manifested by the capacity to make choices, is to some extent inherent in every atom. The universe as a whole is also weird, with laws of nature that make it hospitable to the growth of mind."* [295]

The main difficulty we frequently have, however, is to switch our view from a micro-to a macro-causality, and back. As we have shown with the case of the white noise of an encrypted picture, or the flickering of a single pixel on a monitor, etc., also the unpredictability of microscopic quantum events doesn't disprove its potential efficacy in determining a goal-directed, volitional, and purposeful macroscopic physical phenomena. The fact that we observe a totally random series of quantum phenomena at a microscopic scale doesn't imply that these couldn't be part of a larger design to create ordered structures at a mesoscopic or macroscopic scale. If something appears locally as undirected, chaotic, purposeless, and meaningless, this is no good reason to assume it can't instantiate a directed, ordered, purposeful and meaningful dynamical structure globally. From this standpoint, will is not caused by randomness, rather, will in action couldn't appear other than as random.

e. Quantum biology and Theories of Quantum Consciousness

This brings us to the possible links between biology and quantum physics—a possibility that has recently been investigated by the science of *'quantum biology'*, though it dates back to the inception of quantum physics itself when physicist Pascual Jordan first conjectured about the possible quantum effects in living cells in the 1930s [296]. Jordan was the first to speculate that the unpredictability of life, our behavior, the mind's apparent indeterminacy could be reconducted to a macroscopic amplification of the chance and unpredictability of the quantum world that is somehow scaled up inside living organisms. He called this the *'amplifier theory'*. The idea is that microscopic quantum indeterminism could also influence macroscopic biology by amplifying the effects of Heisenberg's uncertainty principle and, thereby, could influence the entire organism.

Nowadays, this amplification theory could be compared to the famous *'butterfly effect'* in non-linear dynamical systems in which only a tiny difference in the initial conditions of a physical system can cause entirely different behaviors and outcomes in the long term. The typical analogy people use to exemplify this is in weather forecasting and atmosphere dynamics. Say, for example, that a butterfly flaps its wing in one part of the world today; this would determine, in the long term, a completely different weather condition on the other side of the world than that what it would have been if the butterfly had not flapped its wings. This is a well-known extreme sensitivity to the slight modification of the initial conditions of non-linear systems—that is, it is an amplification effect. It is without question that the brain is a strongly non-linear system that inevitably must be sensitive to 'neural butterfly effects' as well. The question is: Does this sensitivity go so far as to also incorporate the quantum effects?

If so, this could be the physical entrance door of an unphysical mind into the physical brain, a possible connection between quantum indeterminism and free will, and quantum randomness as a manifestation of will and intentionality that can influence macroscopic events departing from microscopic ones. But while this 'bottom-up quantum télos' would be unrecognizable at the level of organic life, it ultimately could drive its evolution, such as determining genetic mutations, the behavior of single cells, and the cerebral activity of any living organism. It would be the action of a 'universal quantum mind' that drives life from the inside out rather than being determined only by the external factors of the environment.

The problem, however, is that the human mind is not so wise as the universal one and can fall not only into error but also into perversion. What made Jordan's theory fall into disgrace was Jordan himself; he embraced Nazi ideology and even tried to connect his theory to supposed principles of life with the necessity of dictatorial leaders like the 'Führer'—that is, Hitler. Obviously, and in part understandably, all his physical quantum biological ideas were soon dismissed and ignored to date. Nevertheless, we believe that Jordan's research remains a direction worth being explored.

For a long time, this idea was left in the background and was, at best, only addressed by physicists in their popular science books. Dyson openly stated in 1979: *"I think our consciousness is not just a passive epiphenomenon carried along by the chemical events in our brains, but is an active agent forcing the molecular complexes to make choices between one quantum state and another. In other words, mind is already inherent in every electron, and the processes of human consciousness differ only in degree but not in kind from the processes of choice between quantum states which we call 'chance' when they are made by electrons."* [297]

In fact, similar ideas were developed later by Australian neurophysiologist and philosopher Nobel laureate John Eccles, who suggested that quantum theory plays a role in the workings of the brain as an alternative to a rigid deterministic neurophysiology. In 1994, in his book "How the Self Controls Its Brain," Eccles claimed that quantum theory enters brain dynamics in connection with cerebral exocytosis—that is, the transport of vesicles filled with neurotransmitters from a nerve terminal into a synapse. Exocytosis is triggered by an action potential build-up caused by the flow of calcium ions through ion channels into the nerve cells. These ion channels are the size of a nanometer and, thereby, are small enough to be subjected to quantum random events. If so, quantum effects are amplified through the exocytosis processes that determine the workings of the 80–100 billion neurons of all the brain circuitry.

Much more popular became the model proposed in 1994 by theoretical physicist and 2020 Nobel laureate Roger Penrose together with anesthesiologist Stuart Hameroff, who collaborated to produce the theory known as '*Orchestrated Objective Reduction*' (Orch-OR)([298], [299]). Penrose proposed that the quantum wave function collapse can occur in isolation, called 'objective reduction', and Hameroff suggested that microtubules—that is, the polymers that form part of the cellular cytoskeleton for the cell's structure and shape—could be a suitably small structure to make this objective reduction, together with other quantum effects, become effective. However, some of the Orch-OR model predictions have been falsified, and it is questionable whether it will survive as a tenable speculation in the future.

Less known but noteworthy is Karl Pribram's '*holonomic brain theory*' developed in 1999 in collaboration with David Bohm where the brain is conceived as a holographic storage network [300]. Pribram suggested that memory is encoded in the brain in the form of electric wave oscillations in dendritic networks like a quantum holographic memory. A theory that, in principle, could account for the non-local aspect of human's memory storage.

At about the same time, the Italian physicist of the University of Salerno Giuseppe Vitiello, followed a similar path developing a quantum field theory of many body systems which conceives the brain as a dissipative condensed matter system exhibiting ordered patterns maintained by long-range correlations and which could also account for the neural mechanisms of memory formation ([301], [302]). While Henry Stapp proposed that the wave function collapse is a conscious selection process, and that the quantum Zeno effect could work as a possible mechanism against quantum decoherence processes [303].

Others speculate about the possible connection between a quantum field and neural activity. These include Tim Palmer, a physicist at the University of Oxford, who suggests that the brain's sensitivity to thermal and quantum noise can be a factor for creativity [304]. Worth mentioning might also be Joachim Keppler's theory of a phenomenological '*ubiquitous field of consciousness*'—

that is, the universal quantum field—which modulates and determines the brain's state by a quantum mechanism leading to coherently oscillating neural cells tapping into this universal field of consciousness [305]. G. M. D'ariano and Federico Faggin postulate a *'quantum informational panpsychism'*, positing quantum entanglement as building up thoughts and qualia, with experience being a fundamental feature of information [306].

All these attempts to frame a connection between QP and consciousness, seem to reverberate the more metaphysical thoughts of Ervin László, a Hungarian philosopher of science, composer, and peace activist who proposed an evolutionary philosophy based on the system theory of quantum consciousness, positing that all the universe is an informational 'akashic field'.

So far, however, any theory suggesting links between quantum physics and consciousness remains purely speculative and highly controversial. The main objection of the skeptics is that thermal decoherence—that is, the effects of the heat coming from the environment on these supposed quantum processes— would immediately destroy any quantum coherence, such as entanglement or superposition of particles inside the cells, and which are notoriously much too hot compared to the temperatures needed to also sustain the simplest form of quantum effects.

Note, however, that Jordan's, Eccle's, Stapp's, and Palmer's models do not require effects of quantum coherence and don't rely on particular exotic quantum states of matter; rather, they are based on quantum random indeterminism alone.

One could also ask if noise, independent of the question of whether it is quantum or classical deterministic randomness, could be a source of innovation and creativity. At much larger scales than those occurring at the microscopic quantum levels, it is now known that noise is, indeed, a novel creative process. Modern developmental biology is recognizing how, besides genetic and environmental factors, the fluctuations in the biological process determined by the molecular jitter due to Brownian motion—that is, the molecules bouncing around inside due to thermal effects—are an important factor in determining diversity and variation in the phenotypic traits of a species, such as its shape, size, color, and even behavior. Genetically identical organisms living in the same environment can develop very differently. For example, it turns out that the synthesis of proteins, as well as their folding, their function, or how they assemble, also depend on how they have been randomly bound, pulled apart, or diffused inside the cell. Diversity is fostered by stochastic developmental events (see [307], [308], [309], [310]).

Denis and Raymond Noble even propose that stochasticity could be the source of free will. The unpredictability of our choices and the novelty in natural phenomena may be grounded in stochastic phenomena that a deterministic view of life can't explain. *"Harnessing stochastic and/or chaotic processes is essential to the ability of organisms to have agency and to make choices* [311].

Therefore, at least from the logical point of view, biological diversity, conscious creativity, teleology, and intrinsic noise are not mutually exclusive. On the contrary, what we call 'noise', 'randomness' and 'chance' might well be the backdoor of a supra-physical intelligent agency. When it comes to the connection between randomness and meaning or purpose, any claim for or against teleology in Nature is an unwarranted extrapolation.

f. Randomness, Purpose, and Design in Evolution

> *"Teleology is like a mistress to the biologist; he dare not be seen with her in public but cannot live without her."* – J. B. S. Haldane

The connection between the apparently random events in the microcosm with the structure of the macrocosm is frequently ignored also in the domain of evolutionary biology. Things can go also the other way around, namely that an apparently disordered macrocosm may have structure in the microcosm. Failing to recognize this, is what I call it the 'ant fallacy': In the eyes of a tiny ant, what is the world outside of its anthill if not an unintelligible stream of random 'coincidences' and 'accidental' events?

The typical context in which this fallacy is applied is in biology and Darwinian evolution. An example where scientists often jump to conclusions, can be found in the so called *'argument from poor design'*, the common misconception that, since biological systems are the result of 'random chance', as a consequence, they are 'suboptimal' and 'non-robust' having 'flaws' and a 'poor design', in other words: organisms are 'imperfect.'

For example, the human genome contains 'genetic disorders'–that is, 'mistakes'–that can lead to genetic diseases. On our eye retina we have a blind spot of which we are unaware only because the brain integrates the information from the surroundings and reconstructs the images without obscuration of the visual field. The appendix in our intestine has no known function and can be the cause of several diseases.

A very common objection against any form of theistic evolution is that which points out the existence in every organism of morphological traits and 'fossil' genetic information which was once useful for a specific function and purpose but ceased to have any functional necessity in the presently existing species. For example, some aquatic mammals retain morphological traces of limbs which once upon a time served for locomotion on land but now have lost their function. This supposedly proves that evolution is incompatible with an 'intelligent design' because there is no reason for the 'designer' to endow a living organism with unnecessary and useless functions (such as limbs in an aquatic environment). As an analogy, think about how humans design cars. It was once normal to start the engine with a starting handle, but newer (that is, more 'evolved') car versions replaced it with the nowadays familiar starting key. Modern cars obviously do not retain such an old device that necessitated such an uncomfortable mechanical maneuver. Instead, evolution proceeds by a process of accumulation and modification, which often retains old traits with obsolete, unnecessary functions or even useless information in the genome.

These were only few examples of 'flaws' in biological organisms that no one would deliberately engineer. Why should a designer purposefully plan these sub-optimal or even dysfunctional organs? The orthodox Darwinist believes that these can be explained, and are even expected, in the frame of a theory that postulates its existence as the remanences of a long and randomly-selective evolutionary process.

This is the sort of metaphysical argument one hears frequently coming from modern biologists. There are different possible answers to this line of reasoning. Let us take up some that look at things from a more comprehensive and integral view.

First of all, what is supposed to be the criterion of 'good design' or 'perfection'?

In biology, a system is considered to be 'optimal' when it maximizes or minimizes some functions, such as survival and reproductive abilities, under given constraints and adapts best to the environment. But these are human-defined criteria. If there is an aim or purpose in evolution, we have no idea what it possibly could be and how it is supposed to be implemented in the grand-design. Maximizing functions or reproductive and survival skills might be the means, not the goal. To say whether an aspect in biology is 'poorly designed' implicitly anthropomorphizes the designer and assumes that we are able to know what the designer aims at and what the designer's thoughts are. And since the 'mind of God' does not agree with the human mind, we proudly conclude there could be no God in the first place. Even the worst medieval anthropocentric conception didn't go so far!

One could also wonder if and how a human-designed self-reproducing machine—something we are light years away from being able to build anyway—would still be able to renew itself without dragging behind something of its evolutionary past. If reproduction were not precise enough, the organism would be born faulty or even die before birth, and speciation wouldn't be possible in the first place. And because reproduction must achieve a threshold level of exactitude, it is not surprising that it also copies the old and potentially unnecessary traits and information. Would such precision be possible without a designer?

Moreover, taking a strictly functionalist point of view, that is so exclusively focused on fitness criteria, separates a priori between the organism and the environment, failing to recognize how the adaptation of the single organism to the environment is only one of the many internal functions of a totality. This totality is a '*Kantian Whole*'[101]—that is, a system where the parts exist for and by means of the whole and the whole exists for and by means of the parts. For example, an organ like the heart, with its properties and its function, can't be separated and abstracted from the organism as a whole. An organ like the heart serves the purpose of pumping blood for the best survival of the organism in its entirety. It is not designed to optimally fit in our chest alone, and its activity is not aimed only at maintaining its own robustness or fitness, even though it might have to satisfy certain criteria of adaptation to the chest and tune its activity to avoid functional failure. But no biologist would speak of the 'optimal adaptation' of the heart separated and abstracted from the overall body's functions.

This fallacious line of reasoning is precisely what we do when we think of the adaptation of the organism in relation to the biosphere or the supposed DNA transcription 'errors' without being aware that we have lost sight of the cell's workings in its totality.

Seeing comprehensively necessitates, not only a spatial, but also a temporal comprehension of the wholeness of Nature. For example, if 65 million years ago, a gigantic asteroid or a comet had not crossed the Earth's orbit at that precise time in that precise place (of course, by 'pure random chance'), it would not have caused the extinction of dinosaurs and then humankind would possibly not exist. Was that a 'mistake'? From the limited perspective of the dinosaurs, it was for sure. From our perspective, as humans, it was a decisive event necessary for our existence. Indeed, suddenly, you won't find anymore an astronomer or a biologist labelling an asteroid impact as an 'error'.

Since we are unaware of our inability to follow the complex dynamics of the whole, we assume it to making 'errors' and 'mistakes'. Concepts such as 'optimal', 'good or bad design', 'imperfection', etc., arise only because of an anthropomorphic understanding of how the world is supposed to be according to our human standards and desires. Since some processes didn't turn out as expected from our limited sense-mind understanding of the world, which is unable to consider all relevant constraints and can't see the whole complexity of Nature holistically, we conclude that it couldn't be other than a 'bad coincidence' or an 'error' without meaning.

The human mind is a far too low-level form of cognition that can't recognize and capture the complexity of the system as a whole. From a higher perspective, there is neither an organism nor an environment; there is only a single whole where the mutual interplay of its parts is the expression of the whole itself. This separation appears only at the level of our separative mind where the lower-Nature appears to us as a clash of competing entities which sum builds up a whole made of separate parts glued together.

---

[101] A term and concept we borrowed from Stuart Kauffman. [381]

Thus, to derive from these metaphysical inferences about the meaning and the purpose of our existence is, again, an unwarranted and inconsistent logical move, no more and no less than interpreting Fig. 126 left as being without any information, meaning, or purpose. What we interpret as 'mistake' may well be a complex conjunction of purposeful events that we, with our limited cognition, can't disentangle. Randomness and chance are the obvious expressions of complexity. However, complexity isn't an argument against teleology.

While, in the frame of an integral evolution, this line of reasoning is not only an invalid objection but even requires such an apparently redundant evolutionary processes based on an accumulation of old and new traits governed by trial and error. Apara-Prakriti is called a 'Nature of ignorance' precisely because an original omniscient Spirit has plunged itself into oblivion and forgetfulness of itself in the material domain. It is finding back itself step by step in this state of ignorance by a process and method of self-finding which a large degree must be mechanistic and imperfect and which cannot and even does not need to redesign the species from scratch but works from what it has, from the organism as it is in its present state of evolution—that is, from the near to the far. If some useless appendixes are conserved by reproduction during the evolutionary journey, this is no issue.

Overall, these sorts of objections are far from conclusive, other than showing (quite conclusively) once again the hidden assumptions and the bias with which modern materialistic naturalism works.

g. The Unsolved Mysteries of Deterministic Evolution

"We as computer scientists don't know any algorithms that you would want to run for a billion years and would still do something interesting."
Jeff Clune [312]

The physicalist narrative tells us that there is no goal, purpose, or direction in evolution. Our life is a meaningless and purposeless flicker in an almost empty and chilling universe which is just what it is. Even the question about meaning is a meaningless question we should stop asking. So often, we get to know that evolution is a directionless, unguided process in which chance and randomness play a central role. The first statement tells us that, among other evolutionary mechanisms, the mutations that affect evolution—that is, the change in the sequence of an organism's DNA—are 'random'. The second statement therefore concludes that evolution is blind, undirected, purposeless, and meaningless; it is just an accidental walk in time and Earth's natural history. While the former is a scientific statement that has an operational validity, the second, as we have seen, is an unwarranted extrapolation motivated to sustain an ideological belief system. Evolutionary sciences are frequently invoked as evidence against finalistic conceptions of a goal-oriented Nature and, implicitly, against any form of theism.

However, upon an honest and closer inspection, people are increasingly realizing how unconvincing and meaningless this reasoning is as well. This pervading neo-Darwinian belief that the immensely complex evolution that gave form to life could be explained away by a totally unconscious blind undirect processes without an agency is becoming increasingly untenable under the pressure of the new scientific evidence.

Keeping out of the equation any teleological temptation in scientific evolution is, perhaps, the right thing to do from a pragmatic point of view, which is not interested in philosophical, existential, or spiritual quests. In the present stage of science's development, which isn't sufficiently self-aware and still can't tip into the higher-mind seeing, a dry attitude that ruthlessly expunges any temptation which asks for why things are, beyond considering how things work, might have been necessary primary school homework for the analytic mind which, otherwise, confuses and mishmashes deeper ideas, ending up in a dogmatic belief system itself. A disinterested and healthy agnosticism would then be the most appropriate attitude. In this context, even disbelief and doubt would be a genuine form of scepticism that could, and even should, have its place in a rational discourse.

Unfortunately, some neo-Darwinists, who self-proclaim themselves 'skeptics', have embraced the opposite of the above attitude and embarked on a sacred life-mission promoting neo-Darwinian evolution as a scientific argument against any form of finalism, becoming a sort of militant atheistic movement which crusades against any alternative formulation.

From the opposite side, the reaction to this materialistic culture; expressed itself in a philosophically dubious form of religious and fundamentalist creationism which only strengthened the position of these 'skeptics'. In modern times creationism evolved to a more elegant form of the theistic concept of evolution, represented mainly by the movement of *'intelligent design'*. In the author's eyes, while the ID movement is a step in the right direction compared to any unsustainable forms of young Earth creationism that stick to literal biblical interpretations, its pretention to be a scientific non-philosophical movement appears questionable. It denies having any religious or political affiliation, but it is quite clear from the statements of its supporters how it still struggles to get rid of an exclusively Christian-cantered and anthropocentric worldview and shows to be unwilling to open itself towards a culturally more inclusive and comprehensive spiritual understanding of life and, worse, adds to its arguments political (mainly right-wing) nuances that should be kept out of any serious scientific and philosophical discussion.

Both of these antipodal positions, the skeptical and the ID proponents, paradoxically have the refusal to step out of their self-imposed limitations in common. While, from the scientific perspective, the only honest standpoint is agnosticism. As long we are still not able to ascend higher states of consciousness, the question of whether the material universe might nevertheless be governed by a hidden determining Will of a cosmic Mind or Consciousness or a Creator; remains forever a matter of personal and subjective choice.

On one hand, everything seems to depend on a self-organizing dynamic chance, an unaccountable freak and fantasy of the cosmic phenomenon we call 'Nature'. To the mind, everything appears to be an inconscient force that acts and creates at random and nevertheless has given birth to an ever-increasing complexity and to conscious beings. In fact, the preferred argument of scientific Darwinism against teleological temptations is the *'blind watchmaker'* analogy that states how by natural selection, random mutations, genetic drift, and genetic flow which, together determine adaptation, the survival of the fittest, induced by chance events dictated by the environment and other external constraints or biochemical factors, etc.—then, by a cumulative and selective mechanism that lasted for billions of years, everything can be explained and any necessity for a divine intervention becomes unnecessary. It is a non-necessary hypothesis that must be cut off with Occam's razor.

On the other hand, despite the optimistic claims and the hype in the media, science is very far from having proven that these mechanisms are self-sufficient explanations. Whatever science tells us, the tension between the idea of a blind deterministic universal clockwork and a Creation guided by an inward intelligent cosmic consciousness never ceased and didn't find a definite resolution. The point is that mind alone can never argue, reflect and finally establish with certainty the final word in these matters. Ultimately, the limited individualized human mind can only speculate, not prove conclusively. It can only debate endlessly on these issues because there is something in its very nature that inherently lacks the power of comprehension.

In fact, the explanatory omnipotence of the principles of Darwinian evolution as being self-sufficient remains wishful speculation. It is true that these natural mechanisms exist and play a role in forming and molding the species, but biology is far from proving that this alone is a sufficient basis to explain the existence of complex and conscious organisms. Natural selection is certainly a part of the solution at some level, but we are light years away from proving it as being sufficient to explain the entire evolutionary process in all its complexity, including the existence of sentient beings. The laws of evolution are best documented for the microbiological domain, but already in this domain, the complexity of the simplest organisms (such as the cellular organisms we have discussed in Pt.I-IV.3&4) is far beyond our comprehension. The idea that complex organisms, like us, can emerge due only to a blind natural selection and by random mutations alone is just a conjecture, at best a working hypothesis that remains unproven. To proclaim natural selection and the known physical principles of evolution to be able to explain everything we observe is an extrapolation of the mind which, once it discovers a truth, immediately wants to extend it to everything. But there is no proof whatsoever that these principles are self-sufficient to bring what we see into existence. And, resorting to more abstract ideas such as the 'evolutionary advantage' is unconvincing because, by that, one can explain everything and the contrary of everything.

Moreover, there is no notion of 'progress' in Darwinian evolution in the sense of greater complexity and intelligence 'becoming better' or improving, let alone an intentional progress towards a goal. It is only about reproduction and survival. Darwinism is a theory of accumulation—that is, it posits that life developed organisms by summing up random favorable traits. And what we often fail to notice is that Darwinian natural selection 'selects' and does not 'create', meaning that it removes or 'filters' out those already existing traits or mutations that confer more fitness. It sorts out the harmful from the beneficial effects on reproduction and survival, but natural selection does not generate anything. It only sifts out what was already present. Biologists know well that it may eventually also lead to a loss of complexity or 'suboptimal designs' if the environment demands it (the paradigmatic example is that of species that lost their sight after moving into dark caves or the case of our own eye's blind spots).

Therefore, that those incredibly complex organisms—some, such as us, bestowed with intelligence—could form only by the process of accumulation and sifting is far from obvious and certainly not a proven fact. We, as humans, did certainly not build relatively simple machines, like computers or cars, by amassing things and removing impurities. What are the logic and rationale that should make us believe that Nature did instead accumulate and sieve its way through to the functionally complex organisms we are? What is the rational, scientifically-based reason or evidence that stands behind this assumption other than a hypothesis based on a theoretical extrapolation that would like that to be true?

At any rate, natural selection and mutations are processes occurring in an already formed biological cell. Before the existence of the first cell, there could be no Darwinian evolution. How then could Nature build such a complex aggregate like a cell before there could be any evolution to begin with? There is an explanatory gap, because also the most primitive self-reproducing cell must already have been a very complex structure that could not appear by pure chance. In the origin of life theories appears a sudden 'jump' or an evolutionary 'saltation' that brought into existence the first cell, and that could not be determined by the known Darwinian evolutionary processes. It is now becoming clear that there exists a minimal trade-off between the replication accuracy and survival probability of a cell. If its replication fidelity and/or survival probability aren't both about 60%, life would quickly die out. In a 'pre-evolutionary era' where the Darwinian processes still could not take place, there must have been another 'natural law' or evolutionary 'driving force' that allowed for this 'primordial saltation' to lead to the formation of the first cell with a minimal survival and replication efficiency from the very beginning. Thus, Darwinian evolution alone alone can't be the whole story of life (for more on this see [313].)

Please note that we aren't implying that the modern scientific concepts of biology and evolution are wrong or unproven. There is plenty of evidence that evolution is a real phenomenon and how these mechanisms, such as natural selection, adaptation, and random mutations, are real (just search on the web for "evidence for evolution," and you will find tons of it). However, the point is that, when it comes to teleological questions, all these known mechanisms are far from being sufficient to be able to explain what they claim to be able to do. That a mechanical determinism alone is what rules Nature is given for granted while, in truth, it remains just an unaware metaphysical assumption far from being proven.

It is also true that this doesn't imply that a relatively complex structure always needs a teleological drive to come into existence. Structures that emerge due to a '*spontaneous self-organizing process*' exist. From an initially chaotic disorder, a spontaneous well-ordered object can emerge without any intentional design.

For example, think of how beautiful symmetrical crystals and snowflakes can form. Other examples could be the formation of thermal convection cells in the atmosphere or the oceans, tornado clouds, the more or less ordered structures emerging from phase transitions in solids or liquids due to symmetry breaking and, in biology, the spontaneous folding of proteins, patterns on animal furs and butterfly wings or the self-assembly of the cell membranes. In physics, these patterned structured formations can be explained by dissipative processes, which

Fig. 132 A snowflake: a non-random and well-formed structure without 'design'.

can locally decrease the entropy—that is, the local disorder—to form more ordered structures without violating the second principle of thermodynamics which imposes an ever-increasing global entropy in the universe. These can also be dynamical systems that can acquire a variety of conformational states (in physical terms, these correspond to the local potential minima), each with its specific arrangement and adapting under environmental perturbations. Think of a marble rolling on a surface with peaks and basins: It will find its natural local state of equilibrium in one of these basins. The same applies to these self-organized states, which can have different robustness to the external perturbation and reconfigure themselves if a specific threshold is achieved. Networks made of several interacting system components can be spontaneous, self-organizing dynamical entities (the paradigmatic example is a neural network).

There is no doubt that evolution exploits self-organization to form stable structures and organisms. However, things can't be simple as that. Spontaneous self-organization in non-linear complex system, eventually augmented by processes of natural selection and genetic drifts, can't explain higher-level organizations such as the emergence of the first cell. This widespread but misplaced belief rests

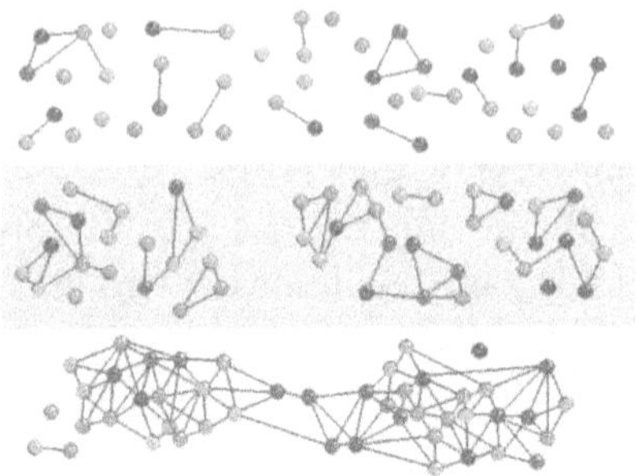

Fig. 133 Network self-organization stages.

on a too optimistic extrapolation of the physical law of thermodynamics according to which the spontaneous formation of ordered structures is possible by reducing the local entropy to the expense of the global environment. And extrapolation that forgets, however, how this can only be done by lowering also the energy state of the emerging structure. For example, the snowflake is colder–that is, in a lower energy state–than the water vapor it originated from. This entails that we have, indeed, a nicely ordered structure, but with less energy available to do work, what in thermodynamics is called the '*free energy*' (lower entropy → less free energy.) The process may lead to a nicely ordered structure that, however, is functionally unable to do anything. Whereas, life goes in the opposite direction: It also decreases the local entropy to form, say a cell, but this cell is in a higher energy state than its surroundings (lower entropy → more free energy). Otherwise, it wouldn't have free energy available to run the extremely complicated machine that a metabolic system is, with all its life-sustaining chemical reactions necessary for the vast amount of cellular processes, and to grow, reproduce, maintain structures, and respond to environmental stimuli. To state that a structure self-organized itself in a primordial soup and also raised its free energy, is a statement that contradicts the laws of physics. Adding natural selection or random mutations into the mixture doesn't help, they can't violate these laws either. So, obviously, much more is needed than such a simplistic mechanism.

Moreover, the non-organic spontaneous, self-organizing structures are always composed of the same substance. In comparison to living cells, snowflakes or convection cells are extremely simple structures made by only one substance—namely, ice and water, respectively (or, more generally, by a single solid and fluid substance). Spontaneous self-organization arising from the interaction of different materials has never been observed. Because in the organic world this is, instead, a common phenomenon all over the place, it is hard to believe that such a complicated being like a living cell could arise just due to a self-organizing process that spontaneously assembles a functional and even self-reproducing structure. The combination of both processes—that is, spontaneous self-organization and Darwinian evolution put together—can't be the whole story.

A fact that indicates how the current evolutionary process of natural selection, adaptation, fitness, improvements with a variation augmented by random events, and self-organizing processes can't be the whole story is also suggested by the limited success of evolutionary algorithms. Evolution, as conceived by current neo-Darwinian theories, is essentially a computational process that can be encoded in a computer algorithm. There was a time when so-called *'genetic algorithms'* seemed to be the holy grail of AI. Indeed, by encoding some fundamental evolutionary rules, it was possible to build algorithms that are used to find optimized solutions to a particular problem or to evolve an algorithm or a program for solving a particular class of problems. But as is so often the case, it turned to be an overhyped branch of IT that did not meet the expectations and is nowadays out of fashion. While the results initially seemed impressive, after about three decades of research, the power of evolutionary computing remains limited. Computer simulations of biological evolution do not produce any complex artifacts, let alone any automaton having a resemblance of real single celled organisms. A recent analysis of this state of affairs showed that the problem can't be downplayed to the lack of computing resources alone [314]. Because running evolutionary algorithms longer does not only fail to produce better results but even leads to progressively diminishing results. No one has even succeeded at evolving nontrivial software from scratch, unless human assistance intervenes and guides the overall developing process. The expectation to develop an evolutionary software able to design an engine, a house or plane without human guidance—that is, without added information coming from an implicitly embedded intelligent agency—turned out to be delusional. Computer simulations mimicking the principles of Darwinian evolution fail to reproduce Nature's real evolution.

Again, this does not mean that modern evolutionary concepts are wrong, but it suggests that evolution reduced to an algorithm reflects a much too superficial understanding of what is really going on. A theory of evolution conceived as a random deterministic selective algorithm only works if it is also augmented by an intelligent agent that introduces directives, implicit conditions, and necessary information a priori. Otherwise, without this external guidance imposed on the process from the outset—that is, without a teleological input—it continues to fail and produce sterile results.

Then, unanswered remains the question of why our universe seems to have implanted in itself from the beginning just the right physical laws from which randomness can profit so successfully? Because, also, the present scientific evolutionary theories rest on the unexplained fact that the laws of physics are just what they are. Without the predetermined background of the physical laws as they are and the biological principles and guiding laws of Darwinian evolution, chance alone would lead to chaos with an infinite increase of entropy globally and locally. Natural selection with randomness as its source is possible only because the laws of physics allow for it. These laws and principles by which our universe abides are far from being self-evident and obvious. Science simply posits it because it observes it. They are just 'out there,' and we work with it in our theories but without further reflection and questioning, and believe the clockwork of the blind watchmaker is doing the rest.

There is another aspect that makes a strict materialistic worldview untenable, and that is rarely recognized. We should become aware of the fact that the inter-dependence between the micro-scale and macro-scale structure is much greater than usually assumed. The microphysical state of every neuron in a brain depends strongly on the macro-physical state of the environment in which an organism lives. For example, seeing coincidentally an apple falling from a tree may trigger in someone's brain circuitry the desire to eat an apple and substantially change one's own personal future thoughts, attention and actions. This is not something happening only with an active living organism immersed in an environment but also with very simple chemical reactions. Just think about how any physical and chemical process depends on the external pressure or temperature of the environment. These are examples of a *'downward causation'* from the macro- to the micro-cosmos that shapes the underlying micro-physical processes.

This interaction also goes the other way around. For example, humans are the most obvious case of how a species can affect the ecological balance of an entire planet. We have destroyed substantial amounts of the natural environment and polluted air and water; one can safely see this as an example of *'upward causation'*–that is, from the microscopic brain processes of a living organism that have shaped the macroscopic surroundings. This is firing back now by another downward causation: the macroscopic state of the environment. The effects of climate change, which forces us to reconsider all our industrial policies and living standards, are only the most visible example.

There are not only two but many layers of upward and downward causation. The shape and state of an environment at a certain scale also depends on the higher-level scales. For example, the type of climate and its cycles are strongly connected to the Earth's distance from the Sun and its axis tilting, which, in turn, were determined by the cosmological history of the universe as a whole.

The point we want to make here is that the macro as a micro-physical state of everything in the universe is itself highly dependent on a multilayered upward and downward causation flow in an immensely complex network of causes and effects. It is not sufficient to know in every detail the state of a brain to be able to predict its actions. One must also take into account the whole of the environment and, for a long-term prediction, the whole universe. It is not sufficient to know the initial state of each particle of a system to predict its dynamical evolution by solving differential equations. One must also know the external constraints determined by all the surroundings. [102] As we have already seen so many times, one can't abstract the part from the whole and must consider the whole as more than the sum of its parts, not only in its forms and shapes but also in its dynamic and temporal evolution. And besides the unpredictability of a non-linear dynamical system and the indeterministic aspects of quantum physics, this adds yet another layer of complexity that makes any Laplacian dream appear as an even more distant chimera.

At any rate, if we believe that the entire universe is only a giant physical causal deterministic machine, with life a meaningless accident, sort of what Laplace conjectured, then everything in the universe that exists now, in the present moment, was causally predetermined by the state of the universe in the preexisting instant. Similarly, every state of the universe of the preexisting instant was predetermined by the causes of the instant before, and so on, until we reach back to the times immediately after the Big Bang when the physical universe was nothing more than an extremely hot and dense plasma without any order and structure.

If so, one must conclude that our present physical brain states were initially inherent and already encoded in that initial state of the universe from the outset. This naturally raises the question of whether all the works of Shakespeare, the art of Leonardo da Vinci, the music of Mozart or Bach, the genius of Galileo, Newton, and Einstein, and the content of all the books that have been written by humanity were already predetermined, concealed and encoded in every detail in that initial primordial almost perfectly uniform and boring soup? If one truly believes in such an extravagant idea, one must then explain: How did all that get there? Who or what encoded all that into the initial state of the universe? Is just an immensely improbable coincidence becoming real?

The paradox is that in using a strictly deterministic conception of reality to explain the existence of complex structures such as life, the brain, and its creations, we are forced to posit its existence in potentiality from the outset. The logic becomes circular because to explain the existence of something, we must assume its existence already in latency. Such a logically recursive inconsistency is typical of the limits of the analytic mind, which can think only of a deterministic causality conceived in terms of separated spatial and temporal events and distinct particles whose true nature and essence it knows nothing about.

---

[102] For a more rigorous and in-depth analysis of these aspects see [388], [387] and [389].

h. What is Life?

What about the vital force, the élan vital of Bergson? This was no new hypothesis, as already Plato and Aristotle held that there is a force immanent to the organisms beyond the reach of our sense-mind perceptions, which is nevertheless visible as something making life fundamentally distinct from non-life. In the Middle Ages, the belief in essential life forces was quite common. The French biologist Jean-Baptiste Lamarck (1744-1829), better known for his foundational contributions to evolutionary biology, postulated the existence of a life-force standing behind evolution and responsible for the increasing complexity of organisms. Also, the French chemist Louis Pasteur (1822–1895), known for his contributions to microbial biology and as the developer of vaccines, postulated that fermentation is a *'vital action'*. He was inspired by F. Redi's experiment that disproved spontaneous generation from rotten meat (recall Pt.I-IV.1) taking this as evidence that life must always originate from life—that is, there can't be, not even in principle, a spontaneous generation of life from non-living matter. Pasteur contended that it is a vital action that makes life inherently different and special compared to non-living matter.

But the success of the purely materialistic approach of biology led to a quick demise of vitalism. It was no longer a matter of debate and was almost forgotten.

Nonetheless, until now, nobody was able to hit the final nail in the coffin answering Plato's and Aristotle's doubts, let alone generate life in a test tube. As a side note it is worth recalling that, so far, all origin-of-life experiments have all failed to produce life from non-life. The famous *'Miller-Urey experiment'* done in the 1950's, and that bombarded gases with electrical discharges producing amino acids, was claimed to be a big success in reproducing the first 'building blocks of life'. It turned out, however, that it was a flawed experiment, since the gases used did not represent the composition of the Earth's primordial atmosphere. In fact, no experiment realistically simulating early Earth conditions has ever even produced any amino acids. Furthermore, the real 'building blocks' of life are not amino acids but proteins, and the simplest living thing requires more than 100 different types of proteins and that no one has ever created in a laboratory experiment designed to simulate early Earth conditions.

Anyway, biologists and scientists tried hard throughout generations to define what should be considered a living organism. Famous is Erwin Schrödinger's book "What is Life?" which precisely tried to address this issue but without any convincing final answer. Nowadays, biology identifies as 'living', loosely speaking, anything capable of metabolizing, eating, excreting, maintaining homeostasis, growing, adapting to the environment, reproducing, and evolving. However, there is no consensus regarding a universal definition of life because, for example, the demarcation line between what is living and non-living remains unclear. NASA adopted Carl Sagan's original proposal to define life as a "self-sustaining chemical system capable of Darwinian evolution" [315]. But one could make several counterexamples of something non-living and nevertheless being able to satisfy several of the above criteria. Viruses have no metabolism and remain inactive without reproducing as long as they do not encounter a cell. And yet, they are capable of Darwinian evolution. Whether or not a virus must be considered a life form is a matter of debate. In a sense, cars also 'eat', 'metabolize' and 'excrete', but nobody would recognize in them anything nearly resembling life. One might object that cars do not reproduce. Computer programs can simulate 'artificial organisms'—that is, digital cellular automatons —fitting the above definition, though only at a much simpler level of complexity than that of a real living cell. However, these cellular automatons do not exist other than as a stream of bits and bytes in codified software in computer chips. Are these living beings? Even though such a futuristic technology will probably remain a sci-fi scenario for a long time to come, we could, at least in principle, imagine that sooner or later, we will be able to create self-reproducing machines. Would these then become alive? To highlight how reproduction can't be the decisive aspect individuating life while being a common trait of all living beings, think of the opposite case: Would anybody consider a sterile man or infertile woman as 'non-living'?

Therefore, only a little bit of critical reflection makes it clear how any definition of life based on such exclusively material functionalism always fails to capture something of what it is trying to define. In life, we see consciousness, will, desire, motivation, and goal-driven behaviors. The physicalist tried hard for centuries to reduce these psychological properties to functional descriptions of material processes. Could it be that we should give a chance to the hypothesis that things might work the other way around? That consciousness, will, goal-directedness, etc., are fundamental principles that impel the organization and function of living matter itself?

For some reason, there is something undefinable, ineffable and intangible in life that escapes a rigorous and universal scientific definition inside a purely reductionist and physicalist paradigm. Why is it so hard to define something whose existence is undeniable and whose distinctiveness is so evident? Perhaps for the same reason why consciousness escapes any definition? Might life be something unphysical such as consciousness or mind?

Because of these philosophical issues, on the line of those who eliminate consciousness as an 'illusion', others, such as molecular biologist Andrew Ellington, simply declare, *"There is a more obvious conclusion to be drawn from our failure to define life: there is no such thing as life. Life is a term for poets, not scientists. There are only replicators with different degrees of complexity. PS: many of you are closet vitalists."* [316] This speaks volumes about the hypnotic power of the physical mind and how desperate physicalism must be.

The timid attempt of Bergson to revitalize vitalism at the beginning of the 20th century met with indifference or ridicule. The British biologist Julian Huxley sarcastically rejected Bergson's idea comparing the élan vital hypothesis to that of an 'élan locomotif' ('locomotive driving force') to explain the operation of a railway engine. The argument of Huxley, however, missed Bergson's point: The question is not really what stands behind the motion force of organisms, which can obviously be explained away by mere physical electro-chemical reactions. A simple electrical impulse can lead to a muscle contraction in a dead body, as the Italian physicist Alessandro Volta, well known as the inventor of the electric battery, first observed by applying an electric current to the amputated legs of a frog. The question is, where does the creative impulse of novelty and the urge to reach an ever-increasing complexity and variety of forms we observe in the evolution of life come from? Why does every living being act as if it has agency, aim, and purpose? What is and from where comes that will to action, will to reproduce, will to grow, will to expand, will to know of living organisms? There is no sign that the élan locomotif—namely, the steam's heat transformed into mechanical energy—imparts any creative force to the engine, that will generate new engines or lead it to a novel behavior with agency and purpose other than endlessly spinning the flywheel.

Until recently, biology answered Bergson's élan vital hypothesis more elegantly by pointing out that there is no life force because every morphological organization and its development can be explained in terms of genetic expressions and their mutation in time determined by the interaction with the environment. This is yet another ad hoc hypothesis without substantial proof, but that could, in principle, save appearances.

However, it is now well known that the idea which was dominant throughout the 20th century, according to which all our morphological traits are encoded in the DNA molecule, was flawed. The complexity of an organism does not scale with the number of genes in the DNA code (a soybean plant has more than twice the genes of a human!). The DNA's functionality and causal power are much more limited than the popular belief imagines. It serves as a template for protein synthesis, but it does not specify how they will get together to create a fully developed organism. The folding of proteins is not entirely dictated by the DNA; it is also determined by the environment where these proteins are created. The instructions to form an organism arise from an extremely complex network of interactions and relations between components such as a myriad of cell signaling factors, enzymes, other proteins, amino acids, vitamins, minerals, etc. Genes are a passive database to build proteins but they neither design nor control the morphological structure of a living being. We now know that the DNA codes for the ingredients and not for the recipe. It is even not a "self-replicating" molecule, as we are sometimes told. The DNA is replicated by cellular machinery, no more and no less than a sheet of paper with words on it is copied by a photocopying machine. Nobody would say that the piece of paper has 'replicated itself.' As the study of epigenetics has further shown, the heritable phenotype changes are possible without altering the DNA sequence involving the gene activity and expression, which, among other things, may result from external or environmental factors. There are complicated non-genetic factors that cause the organism's genes to behave differently.

Of course, physical traits like hair, skin and eye color, facial features, and so on depend on the inherited gene pool. However, none of these traits depend on a single gene and, rather, are the result of several streams of chemical synthesis controlled by these regulatory networks whose dynamics control the genetic transcription. In most cases, even this results in a statistical outcome, not a certain and predetermined fact. More evolved functions depend upon even vaster regulatory networks and thousands of genes. Genes are only one of the players in an incredibly complex process inside the whole cell. Moreover, it is now known that genetically identical individuals do not grow identically even if they are subjected to the same environment, while large genetic variations can lead to the same phenotypical trait via multiple alternative pathways (as an analogy, think of how one can take several paths to get from point A to point B) [317]. It is now also known that cells and organisms can alter their own DNA, rewriting their genome throughout life [318]. The DNA in the cell's nucleus contains only part of the information. Another example comes from 'ontogenesis' –that is, the embryo's development. It turns out that the ontogenetic information is also carried by the cell's membrane, independently from the DNA molecule. It is determined by complicated bioelectric phenomena and a sugar code on the cell's surface. The embryo's development would be impossible based on the information contained in the DNA alone [319].

As Stanford University neurologist Robert Sapolsky reminds us, we humans share 98.9% of our DNA with chimps [320]. The remaining 1.1% account for differences in olfactory receptors, size of the pelvic arch that allows us to walk upright, fur, and differences in the immune system. But nothing in the genome determines a cognitive difference except for genes coding for a higher number of rounds of cell division during fetal brain development, leading to trice as many neurons in the human brain as in a chimp brain. This makes it clear that the difference in cognitive abilities between humans and chimps is not stored in the DNA. The genes encode only for a quantity which, much later, enables a quality, but there is nothing in the genome that determines that quality. The latter must somehow be acquired with experience by our human neocortex, inside which reside the main cerebral differences between our brain and those of other animals.

Curiously, however, the entire neocortex seems to be connected to motor tasks. Here, cells connect with areas of the old brain related to the generation of movement. This is not what one would expect from an organ supposedly dedicated to the display of complex higher-order cognitive functions. Moreover, while one would expect, for the same reason, great diversity and variation throughout the neocortex, one finds instead that it is very similar all around. The variations between areas are small. One of the main distinctions between our neocortex and the rest of the brain are the hundreds of thousands of *'cortical columns'* —that is, a particular modular structure of columns of neurons—but its functional role remains unclear [321]. Of course, from brain scans, we know that the neocortex lights up when language and semantic processing (in particular, in the Broca's area) occur and reasoning, thought, or emotional events take place. However, all these anatomical specifications and localizations don't tell us much about how and why these particular neuronal geometries should generate such cognitive skills and phenomenal experiences and, for the reasons we have amply elaborated on in Pt.I, sufficient facts lead us to doubt that these neuronal structures supposedly do this in the first place. But, setting aside the hard problem of consciousness and its neural correlates, what is nowadays certain is that nothing in our genes codes for these skills, other than saying "multiplicate chimp neurons times three".

Recall that we pointed out in Pt.I-IV.2, how cognitive higher-level functions can't be a matter of brain size since other animals have more neurons and/or brain mass and volume. How can we reconcile these facts with the belief that it is our brain that makes us what we are? To save the brain-centered (and anthropocentric) paradigm, we must find something that distinguishes us from animals, enthroning us on the top of the evolutionary hierarchy. And guess what…if you try hard and search long enough for something that confirms your bias, you will always find it. In fact, it turns out that humans are the ones who possess the largest (in nr. of neurons) cerebral cortex: about 16.3 billion neurons vs. the nine billion of a gorilla and the six billion of the chimp.

Moreover, it is now known that organisms themselves can also facilitate their own evolution beyond mere accidents. Cells can rearrange and restructure their DNA with mobile genetic elements. For example, retroviruses infecting cells can insert their genetic material, copying and spreading it throughout the genome. These mobile pieces of DNA allow organisms to evolve beyond a process based on mere natural selection and build up the 'non-coding' DNA (once labeled as 'junk DNA') that does not code for protein synthesis but has regulatory functions for genome expression and even capabilities of self-restructuring. Cells have ways of changing their genomes and the capacities to regulate these changes. These could also be determined by environmental factors: Evolutionary changes can be caused by ecological ones, leading to the creation of new phenotypes without necessarily implying a genotype change. Epigenetic changes can determine how an organism reads its DNA, forming different tissues out of the same genome. The DNA regions that are active and expressed are determined by an epigenetic regulation that alters the way the

DNA is read, and that can change over time. The same region of DNA can be read in different ways—that is, it can encode structurally different proteins. These variations are not directed by a blind selective or random process but by the organisms themselves, presumably when under environmental stress.

The idea that genetic variation and natural selection account for all the evolutionary changes in a passive living organism turned out to be quite an inaccurate oversimplification. The materialist's idea that our brain structure and function–and with it, our cognitive skills, our personality and all what we are in terms of a psychological identity–as being entirely predetermined by the genetic code, revealed itself a much too simplistic belief. Even though twin research has detected behavioural affinities, it is now known that identical (monozygotic) twins can nonetheless develop very different personalities, preferences and psychological traits, despite being genetically almost identical. One might argue that the differences arise later due to dissimilar life-experiences that quickly modify such a complex circuitry as the human brain, even if identical twins live in the same family-environment. However, already with very simple organisms in vitro, it can be shown that brain connectivity changes dramatically almost immediately after birth. For example, the 300 neuron sized brains of different Caenorhabditis elegans, the roundworms we already encountered in Pt.I- IV.2, display substantial differences in chemical synaptic connectivity immediately after their postnatal development making their brains unique, despite being 'isogenetic'–that is, having exactly the same genome–and in spite of living in controlled identical environmental conditions [322]. The popular myth that, to some degree, conditioned also the academic belief system, which correlates genes with brain and personality identity has no scientific foundation. Genes do less than we thought.

Other processes, such as symbiogenesis—that is, cell fusion—contribute to the evolutionary walk independent from selective and genomic aspects as well. The paradigmatic example has become the origin of mitochondria and plausibly other cellular organelles, resulting from an endosymbiotic association of a bacteria with an archaea cell.

Let's not forget, as we already outlined in Pt.I-IV.4, it is now clear that single-celled organisms have a degree of sensing and information processing of their surroundings that can hardly be explained inside the orthodox paradigm. A form of 'basal cognition' in cells and plants exists that previously was thought to be possible only in organisms with a brain, or at least with a nervous system. This basal cognitive behavior in response to environmental stimuli shapes the evolutionary trajectory that, again, cannot be explained by random mutations and natural selection alone. Meanwhile, mounting evidence challenges the prevailing paradigm that mutations are a directionless force in evolution. Genetic mutations occur with non-random frequency. Somehow Nature knows, how to reduce deleterious mutations [323], while making appear more frequently those having a protective function [324], and this independently from randomly selecting processes.

Overall, the gene-centric model that considered the DNA as something that can 'selfishly' act alone by a fixed bottom-up control, like a computer with a ROM memory, turned out to be grossly misleading. The DNA is more a database, a sort of RAM memory where the CPU and the program must reside elsewhere. It is mostly a passive tool that cells can use to change themselves. We are nowhere near 'explaining' life, as genes 'explain' life no more and no less than the words in a dictionary 'explain' literature.

These new findings have now also been accepted by mainstream molecular and evolutionary biology. What, nevertheless, remains a matter of controversy is where the emphasis must be laid. The Neo-Darwinian orthodoxy defends the 'modern synthesis'–that is, the mid-20th-century formulation of evolutionary theory that complemented the classical Darwinian selection theory with genetics, positing these as the two primary evolutionary driving forces. Its 21st century version also accepts and incorporates the above-mentioned factors, such as epigenetics, symbiogenesis, the cell's basal cognition, etc., but regards these as secondary engines of evolution. Other academic movements incarnated especially in the so-called 'extended evolutionary synthesis' [325] or 'The Third Way' [326], consider the gene-centric model of the modern synthesis as outdated and tends to highlight the non-genetic mechanisms as having equal or even more importance as compared to genetic factors in determining the evolutionary change [327]. The debate is far from being settled, but we feel that it is clear that biology is on the verge of a major paradigm shift.

Thus, the popularly conceived gene as a blueprint encoded in the DNA determining the organism's structure, its development, its variations up to every physical trait, and even with all our psychological inclinations does not exist [328]. By reading the DNA, one can't determine the shape and size of an organ of a creature; it doesn't even tell us if there is a particular organ at all because it is not the DNA molecule that determines it. We are not nearly as determined by our genes as once thought. The hope that by manipulating one or a few genes, we could control our intelligence or social behavior turned out a vain chimera. As British psychologist Ken Richardson summarizes: *"So the hype now pouring out of the mass media is popularizing what has been lurking in the science all along: a gene-god as an entity with almost supernatural powers. Today it's the gene that, in the words of the Anglican hymn, 'makes us high and lowly and orders our estate."* [329] Or, as biologist Denis Noble says: *"It's time to admit that genes are not the blueprint for life."* [330] And, together with his brother Raymond, asserts that there are good scientific reasons to argue that life is *"purposefully creative,"* that it *"exists to exist,"* because *"function necessarily implies purpose."* [331]

This neurocentric 'gene-god' myth has been kept alive for more than a century and still survives because it resonates with the linear narrative of the physical mind, which likes to think in units, particles, molecules, and single genes doing everything. We didn't doubt this self-induced narrative until compelling evidence to the contrary emerged. But now we are so accustomed to that narrative; it continues to repeat the same false belief system in the mass media and, sometimes, in scientists as well.[103] But after more than a century of the dominance of the Darwinian and neo-Darwinian paradigm, the non-gradual forms of transition and its phenotypic novelty and complexity, continue to suffer of an explanatory deficit that seems to have no end. This is not to say that the current neo-Darwinian paradigm is wrong, but rather that something more is at work in the evolutionary process, and that refuses to be caged into a simplistic mechanism. Not only forces from the outside to the inside, but also forces from the inside-out determine the evolutionary journey.

Genetics, biophysics, chemistry, and biology explain the mechanical workings and metabolic dynamics of the material substrate but don't tell us anything about the volitional force, the coming into existence of a conscious subject, and the will that determines and moves it. Slowly but steadily, scientists are becoming aware that the machine metaphor in biology doesn't work [332].

Vitalism does not pretend to explain the nature of the mechanical principles and forces that set an organism in motion. Rather, it posits the existence of an intentional principle, a 'desire-force' and a 'will-force' inherent to life dynamics. Saying that this can be explained away by the energy of chemical reactions is like stating that the principles of thermodynamics describing an engine's workings also explain the motives, will, and intentions of the driver.

For those interested in an in-depth analysis of why current biological sciences did not falsify alternative vitalistic theories or paradigms that posit consciousness as an ontological primitive the author published a paper that shows how at least some forms of vitalism could be linked to some forms of idealism if we posit life and cognition as two distinct aspects of consciousness preeminent over matter. Contrary to a physicalist doctrine, these were and remain coherent worldviews and cannot be ruled out by modern science, as I have outlined more rigorously here [190].

The human mind is still not developed enough to think in non-linear and holistic terms beyond a few units. It instinctively needs a separative, dividing, and reductionistic conception of life that appeals to physicalism; otherwise, it feels lost and ignorant. Because, after all, the human mind is the individuated

---

[103] In physics the simplistic atomic model of Bohr that imagines marble-like electrons flying around the nucleus along well defined and discrete orbits, like planets in a solar system, is another example of a collective misconception that survives throughout the generations despite having been disproven almost a century ago.

expression of a fragmentary cosmic Mind and must inevitably see everything through the lenses of the sense-mind. Unless it rises beyond itself, it is constrained to do so forever. Science likes to believe that it is nearing the ultimate secrets of life, but we are light years away from having discovered a mechanistic explanation for Bergson's 'creativity' of evolution. The reasons why vital forces continue to be dismissed do not rely on any rational or scientific basis but must be found in an ideological mindset with all its historical and social background and, of course, because of the usual appeal to razors and principles of parsimony that deny it from the outset. We believe that the exclusion of vital forces from the picture might have been too premature and might soon find its way back into our world-constructs, at least in the metaphysical discussions as idealism, panpsychism, and various forms of cosmopsychism are already doing.

i. Coincidences, Incidents, and Fate

At this point, one might argue that, nonetheless, there must exist events in life in which we have no reason to believe there could be any aim or purpose. At least some phenomena or life experiences are neither planned nor have a goal.

Say, for instance, that a couple meets coincidentally and, knowing each other, falls in love and builds a family. Would that unplanned and apparently coincidental encounter have not taken place at exactly that time and at that precise location, they would have never known each other, and their lives would have taken completely different paths. Well, that's what, fortunately, happens in most cases because it is a necessary source of genetic variations, isn't it?

Or what about so-called 'incidents'? Say a drunk driver crashes into your car, you get injured and will have to spend the rest of your life in a wheelchair. By a tragic chain of 'coincidences', you were in the wrong place at the wrong time. What kind of aim or purpose could reasonably stand behind that blind destiny? It is just a matter of fate! And if that is God's will, as some religious-minded might believe, the natural question, then, is what kind of cruel deity this must be that allows for so much suffering?

Especially the unconvincing reply to the latter moral and ethical-theological question leads many to opt for an atheistic worldview or, in rare cases, to embrace a Spinozian theology that considers the Deity itself as not really a conscious moral and purposeful Entity but, rather, as a universal Being that is driven by its instincts and that does not really plan or have an aim, being itself subjected to the mechanical laws of the physical world in which it emerges and, only by that process, becomes slowly self-conscious. It is a pure World-Will or World-Spirit that must still learn to know itself and can't do anything other than follow its own mechanical and instinctual nature, which is destined to emerge by a trial-and-error mechanism and grow by a natural selection that harnesses randomness and chance.

The other option is that of the non-dualist who denies that there is any universal drama from the outset. Everything is an illusion and all is non-real, a bad nightmare that never occurred. If we could detach from the cosmic illusion, Maya, we would live in a transcendent bliss and anything happening in our lives would not touch us. In a sense, it never did, but we, who are still not awakened, only believe that it does. Things simply happen and are just as they are. It is all about transient and illusionary phenomena to which we attach a meaning and purpose they don't have.

From the integral perspective, these viewpoints all contain a partial truth but also a too limited understanding of the universal circumstances. The atheist has the right to refute the idea of a punishing God that purposefully subjects its creature to random suffering. The religious mind is right in perceiving that there must be a deeper truth that our limited knowledge can't grasp. The idealist or Spinoza-like philosopher rightly captures the mechanical and instinctual surface aspect of the Divinity, while the Vedantin recognizes that we are like children who tend to be scared of the snake's venom and don't realize that it is a rope.

But the integral perspective tries to integrate and, at the same time, go beyond all these viewpoints. If we realize that the universe as a whole is still the same Divine, the God that has projected itself in itself, then the moral aspect vanishes. We are that very same Divine. We are ourselves an emanation, a spark of the Divine itself, and every pain and suffering we feel is that of the Divine. There is no angry or cruel personal God that punishes its earthly creatures. We are an individualized aspect of the Divine that has subjected itself, by a planned and conscious choice, to plunge into the universal inconscient, which, again, is another aspect of itself as well. From the moral perspective, there is no contradiction: We are the very same Entity that has agreed to participate in the pains and pleasures of evolution. This also means that, while our surface awareness vehemently denies it, there must be something deep down in us that accepts all the turmoil of what is, more or less 'coincidentally', happening to us in our lives. Our inmost soul, the psychic being of which we are mostly unaware, accepts what happens because it knows that it will help it to progress and evolve.

Of course, this might sound outrageous. It implies that the soul of someone in a paralyzed body who must spend the rest of his or her life in a wheelchair has intentionally chosen to undergo this experience. Or that in all those innocent people who have been tortured, killed, or starved to death, there was something in their intimate inner nature that consented or tolerated it. Because there is something to learn, something which allows us to evolve, and grow from lifetime to lifetime, and which, otherwise, we would not have learned.

This point of view might be quite difficult to swallow. But from whatever kind of metaphysical angle you look at this philosophical issue, the point remains that nothing prevents us from seeing purpose in any circumstance and event. From the micro- or macro-physical phenomenon to our daily life experience, any correlation between words such 'randomness', 'chance' or 'coincidence' with a supposed lack of purpose, meaning, or aim in Nature's workings or in our lives is a void claim that has no logical and philosophical basis whatsoever. Claims that indicate randomness as a measure for the absence of intentionality, meaning, will, purpose or plan, optimal design, perfection, etc., are made only because we never think about what randomness is.

Of course, these reflections are also true for those who defend the opposite camp. If you believe that there is a purpose, a meaning, an intelligent design behind the universe and life, don't be afraid of randomness. All that has no metaphysical meaning. Trying to show that the Big Bang, the creation of the universe, evolution, or whatever natural process was not caused by what seems to *us* random, means applying the same fallacious and misplaced reasoning. Claims to distinguish between random and non-random phenomena in order to support a final cause in nature—and ultimately, the existence of an almighty Creator—amounts to the pretense of being able, as the tiny ant, to understand the incredible complexity that gave rise to these events. Or, if you prefer, it amounts to the pretense of understanding the mind or Supermind of that Creator. That sounds a bit arrogant, doesn't it?

6. Integral Teleology and the Evolution of Life from the Integral Perspective

a. Spirit Calls Nature: The Central Principle of Evolution

Because scientists are faced with the new discoveries of basal cognition, plant intelligence, the irreducible nature of consciousness, and with this so annoying aspect of evolution to be seemingly goal-directed, there are renewed research efforts aimed at showing how complex agency and goal-directedness could be explained away as something evolving from ancient mechanisms in a naturalistic framework developing contrived theories relying on complicated molecular mechanisms, dynamical systems, information theory, bioelectrical networks, etc.[104] We wish them good luck but also feel assured enough to predict that even these attempts, like all the past ones, will fail in doing so. The mystery and paradoxes of the origin of life will persist, and its goal-directedness won't be banned by resorting to ever more desperate attempts to

---

[104] See, for example M. Levin's speculations [375] or [374].

cram everything into a mechanistic worldview. The doors that will continue to allow for vitalistic and goal-oriented interpretations of Nature will remain wide open.

These paradoxes arise because the mind accepts only one standpoint—preferably, its own sense-mind point of view—and has the inherent tendency to exclude all others. It is not just a matter of choice; it is in the very nature of the mental to opt for one or the other truth because of its separative and distinction-based cognition. Mind always struggles to combine diverging truths that it sees as incompatible and contradictory. That might also explain why the good old Hegelian dialectic based on a thesis, antithesis, and synthesis is so praised and held in high regard by everyone but, de facto, practiced by almost no one.

However, according to the integral vision, all the forms of knowledge that humanity has developed so far, be it that of the physicalist, the reductionist and deterministic perspective, that of the philosophically inclined mind, or that of the mystic trans-rational point of view, are different forms and fragments descended from a single integral truth-vision. Their apparent opposition arises because the mind considers one or the other paradigm as the only possible truth and excludes all the others as 'untrue'.

On the other hand, putting them together, trying to unite them, and find common ground will not lead us much further. The botched attempt to unite all standpoints in a unique vision usually ends in an unconvincing pseudo-synthesis because one tries to embrace them all uncritically without realizing what connects and transcends them all. Or worse, it falls into a blind relativism that reduces everything to one category to the exclusion of others (no good or evil, no better or worse, no progress or regress—everything is only just 'change'). Whereas, it would be more appropriate, and potentially more fruitful, to find a great central principle that is common to all of these seemingly divergent methods and understandings of reality and which finally reveals the common secret connecting them.

The first step to take in order to find this central principle is to recall that, of course, in the view of an integral cosmology, as in other spiritual practices also, the foundation of all the cosmos isn't matter, space-time and the physical laws; rather, it posits consciousness as primary and the ultimate essence (Spinoza would speak of 'modes of substance') of all there is. In this integral conception, the very foundation of all existence is the Spirit, the Conscious Soul, Brahman, the Divine, the Absolute, the Satchitananda, or Existence-Consciousness-Bliss. This is not posited by an intellectual discourse or logical inference but becomes the self-evident truth of things by an experiential realization that transcends the limited analytic and intellectual separative physical consciousness.

The second step is to reconsider the nature of Nature. Because the word 'Nature' has acquired many meanings in our contemporary culture, its deeper significance in the present context requires further clarification. What it is not meant to be here is a simplistic natural identification from a merely biological and material perspective—that is, we are talking about something which goes beyond an image that sees life and the physical cosmos with its physical laws as a display of 'nature'. It is a nature (lower case 'n') we implicitly assume much too often as being separate from us as if we are not part of it as well. Needless to say, it is precisely this conceptual separation between humankind and Nature that stands behind the environmental issues we are facing today. Nature (capital 'N'), in its integrality, encompasses all the manifestation, from the most material to the most subtle planes of existence, from the stone to the individualized mind as the universal consciousness in its dynamic play, and of which the human is clearly part. Nature is here conceived in a much wider sense. In Aurobindo's words: as *"an active force of conscious being which realizes itself in its powers of self-experience, its powers of Knowledge, Will, self-delight, self-formation, with all their marvelous variations, inversions, conservations and conversions of, even perversions, is what we call 'Prakriti' or 'Nature', in ourselves as in the cosmos. All Nature is simply the Seer-Will, the Knowledge-Force, and the Conscious Being at work to evolve in force and form all the inevitable truth of the Idea into which it has originally thrown itself."* [333]

In an integral view, Nature—that is, Prakriti—is the other pole of existence of the Witness-Consciousness, Purusha. It is neither separate nor other than the Divine; rather, it is the Divine itself in a manifestation of dualities. Nature is the Absolute that presents itself in itself and, on the lower hemisphere of conscious existence, seemingly other than itself. And yet we feel and know that in Nature there is something which inherits a fundamental Unity.

On one side, we have the One, the pure and undifferentiated Purusha, the Self-conscious and Self-existent omniscient and omnipotent Satchitananda. On the other side is an aspect of Nature with its temporal, spatial, and material manifestation in an evolutionary process in diversity and multiplicity, which seems to have lost itself in an unconscious and mechanical universe. Therefore, Aurobindo invites us to reconsider and expand our conception of Nature. It is its relationship with the Spirit from which we have to start.

The Indian tradition of Vedanta conceives of the Spirit, or Purusha, as the ultimate truth, while it conceives of Prakriti—that is, Nature—as an illusion, the 'veil of Maya'. Every phenomenon, from the most material to our phenomenal perceptions, be it a sensory experience, an emotional state, or a fleeting mental thought, is an impermanent and transitory event without inherent reality, just a figment or illusion. It simply happens and is projected on the screen of an immutable and undisturbed Self. The Self is the sole and ultimate reality. This has also become the most accepted interpretation in Western spiritual circles inspired by this ancient Eastern doctrine.

Western science, and most of its philosophy, especially in its modern analytic and empiric form, instead went the other way around: It conceives of the Spirit or anything which can't be described in precise analytic and mental terms, or in the form of particles, forces, and space-time, as an illusion, a superstition we must get rid of by cutting it out with Occam's razor, and starts instead with Nature as the ultimate truth from which everything else must be derived, where consciousness appears as nothing other than an epiphenomenon of matter.

In between stand the philosophical idealists who recognize all the physical manifestation, the universe as a whole, as a universal Mind. The idealist understands the delusion of scientific objectivity intellectually and distances itself from a physicalist perspective but, depending on the school of thought, remains agnostic or cautious, if not even a skeptic, in embracing a spiritual or mystic worldview. Because everything is seen through the lens of the analytic mind, it concludes that everything is fundamentally mental in its nature and essence and limits itself to a monistic conception as well.

Instead, from the integral perspective, all three standpoints—the materialist's, the vedantin's, and the idealist's—are partially true and partially false, or, more precisely, are different narrow windows through which the very same reality is seen. They seem irreconcilable and mutually exclusive only because of the monistic and separative tendency of mind, which admits only matter, or only spirit, or only itself. Integrality cannot be realized because, in the background, the mind remains the primary cognition tool that compels us to make a choice. Whereas, if one looks at things from an integral perspective, one can not only look through these three windows simultaneously but, by rising above the mind, recognize them as three low-level entries which can be widened and surpassed if at least the conception of higher forms of cognition is allowed. An integral cosmology recognizes physicalism, idealism, and non-duality as the three forms of cognition reflecting the three steps of the descending (or ascending) ladder of existence. And the supramental vision not only reveals these three steps but embraces several other levels, as outlined in the planes and parts of being in Pt.III-I.4.

Therefore, the central principle that appears in this integral cosmology is the relation between the Transcendent, the Purusha —that is, the unmanifest aspect of the Divine or God—and Nature, the active Prakriti . On one side is a universal Consciousness-Soul which already contains in potentiality all the differences which remain unexpressed, while on the other side is a Nature-Soul which expresses these potentialities by and through the descending ladder of planes from the supermind and overmind down to the involuted forms of consciousness such as mind, life, matter, the subconscient, and the inconscient. One is the pole of a superconscious knowledge and pure existence but at rest, while the other pole is that of action, which works out the universal phenomenality by an expressive Will-Force or Will-Power in the cosmic play.

The pole of the Unmanifest beyond the temporal manifestation projected itself into the other pole of existence, by a vertical plunge from a higher to a lower hemisphere—that is, from an immanent to a manifestly dynamic and evolutionary Nature. Therefore, Nature is an instrument for the expression of the Divine consciousness but not something separated by it. It is the Power of the Spirit, the executive Consciousness-Force and cosmic Consciousness-Energy, or Shakti, and is the origin and cause of all dynamical activity in the temporal manifestation.

Recalling our overview in quantum physics, this might sound familiar because it resonates with D. Bohm's implicate and explicate order. Bohm might have sensed something and was trying to translate it into the language of particle physics. However, here we are not referring to a universe made only of particles interacting holistically. The twofold aspect of Nature goes way beyond that of a model made of chunks of matter interacting among themselves (however, holistically, non-locally, entangled, or whatever). Matter and particles are only the very last rung of a ladder that goes much deeper (or higher, if you prefer) than punctuated entities in space and time.

The higher and lower hemispheres are not just two independent blocks. The planes of consciousness of the higher cosmos, from the Supermind that connects the Unmanifest with the Manifest, the overmind, the intuitive mind, the illumined mind, and the higher-mind, capture to descending degrees the original unity of its underlying Origin, the Oneness of the unlimited Satchitananda. The lower hemisphere—that is, the cosmic Mind, life, matter, and sub-, in-, and nescient planes—are domains of division, separation, and limitation.

Sri Aurobindo, therefore, distinguishes between a *'higher-Nature'* (*'Para-Prakriti'* or *'Super-Nature', the 'Divine Maya'*) which acts by unification and by transcending limitations, working in and by a divine Force and Unity, and a *'lower-Nature'* (*'Apara-Prakriti'*) acting through division and limitations. One is the supra-conscious aspect of Nature which upholds the other mental, subconscious or inconscient aspect of itself. The higher-Nature reconciles, unites in oneness, and works on the lower-Nature life, mind, and matter to uplift it into its inherent light, force and joy. The higher-Nature has a wider self-guiding consciousness and a vaster force, while the lower-Nature has become more mechanical and deterministic, being apparently ruled by blind forces because it is an involved form of consciousness. This twofold aspect of Nature is also reflected in a twofold evolutionary process: an unfolding from below by a spirit in ignorance that ascends the vertical ladder of consciousness in a still unconscious or only semi-conscious hemisphere and a dynamic action from above ruled by a Spirit of knowledge descending the ladder from a supraconscious higher hemisphere.

In this larger vision, we discover the central principle of evolution: that of a Spirit that calls Nature. It is a Spirit that calls Nature to initiate an ascension towards it. A call of the Spirit onto Nature that can't be ignored forever and which is actually underway, especially in the human being, which is the first earthly species able to consciously participate by answering this call, making it possible that also Nature calls the Spirit.

In the phenomenal world of matter, this call comes from the Supermind's organizing Real-Idea by descending through the higher Nature into the lower-Nature realms of name and from. An original Real-Idea of undivided unity without contraries is captured by the universal overmind where the first step of fragmentation in polarities arises. This is reflected by the higher-Nature of the World of multiplicities and unity and Knowledge which is still fully aware of its divine Origin. The higher-Nature stands behind the workings of the lower-Nature of the World of Ignorance in the form of the cosmic Mind, cosmic Life, and cosmic Physical.

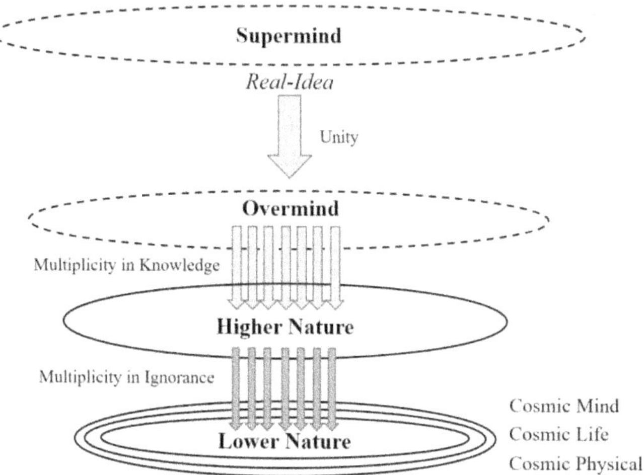

Fig. 134 The Real-Idea as the call of the Spirit
through the higher-Nature onto the lower-Nature.

The same central principle is reflected in the relationship between the individual and the cosmic. However, instead of a vertical system of planes, one has a horizontal hierarchy of parts of being, the concentric system we outlined in Pt.III-I.4d, which starts from the inmost central soul, the psychic being, surrounded by an inner mind, inner vital, and inner subtle physical, and that represent our true inner personality subliminally in contact with the cosmic planes.

The outer untransformed personality—the outer mind, the outer vital, and outer physical (the body), which are what we indicate as our' personality'—is the lower part of our being subjugated to the lower-Nature. A subjugation that our soul has accepted to elevate itself by climbing the scale of consciousness.

This multidimensional vertical and horizontal dynamic cosmology connecting the highest supraconscious with the lowest unconscious is simpler than it may look at first if we realize this central and uniting principle: the relation between an infinite Spirit emerging by a bottom-up and top-down evolutionary process in a finite Nature connected by planes and parts of varying degrees of unitarity (or of fragmentation) in its individuality and 'cosmicity'. It should be kept in mind that Spirit and Nature are two equally real poles of existence of the very same Reality. Thus, the aim and purpose of evolution are to make the lower-Nature increasingly conscious of its integral and unitary Origin using a supreme Spirit that calls on it through an intermediary higher-Nature.

Although life is outwardly not conscious of its purpose—that is, it is still guided by a lower-Nature dynamic—inwardly, its movements are determined by an active Will within, coming from a higher-Nature Force. It is about a concealed Consciousness and Wisdom in all things driven by a Will-Force that appears to the mind as a random and mechanical force. Physical force is a mechanical will forgetful of its spirit and goal. Life is a search, a longing in need of light to evolve beyond an inconscient and nescient 'Night-Nature' from which it came. It is a concealed action of the Spirit through inconscient forces and a subconscient life that is nevertheless a conscious power executing a grand-design whose aim and goal is that of linking the finite with the infinite. It is a Spirit operating in and through a twofold-Nature that works out a divine plan and intention in the cosmos. What appears to be a brute machine on the surface is the movement of an unseen God in things, a great working of a cosmic Power behind the veil. Creation is the manifestation of a creative Force impelled and guided by the creative Spirit's intelligence within. It is the Eternal that cast itself into time and the One organizing itself into the many. A double movement and action in which the Divine descends by a downward plunge in matter and the soul ascends, following its call. In this process, Earth, life, and all its creatures are urged from below, guided from within, and sustained from above. Aurobindo expresses this as a principle that also characterizes the yogic practice and summarizes it as follows: *"We see, then, what from the psychological point of view—and Yoga is nothing but practical psychology— is the conception of Nature from which we have to start. It is the self-fulfillment of the Purusha through his Energy. But the movement of Nature is twofold, higher and lower, or, as we may choose to term it, divine and undivine. The distinction exists indeed for practical purposes only; for there is nothing that is not divine, and in a larger view it is as meaningless, verbally, as the distinction between natural and supernatural, for all things that are natural. All things are in Nature and all things are in God. But, for practical purposes, there is a real distinction. The lower Nature, that which we know and are and must remain so long as the faith in us is not changed, acts through limitation and division, is of the nature of Ignorance and culminates in the life of the ego; but the higher Nature, that to which we aspire, acts by unification and transcendence of limitation, is of the nature of Knowledge and culminates in the life divine."* [265](Introd., Ch. V)

Once this principle and the integral vision resulting from it begin to take shape and become our working paradigm, the three aforementioned 'modes of knowing'—that of the physicalist, the idealist, and the non-dualist—appear in a new and much more comprehensive perspective.

The physicalist's mode of cognition observes and analyses the physical cosmic plane only with the outer mind and sensory means of the outer physical—the sense-mind. It is an externalized state of consciousness in which matter appears fundamental because there is no contact with the higher planes of cognition. Would we see the universe through the lens of the overmind, we would recognize the atomic existence of things which we call 'matter' as the extreme fragmentation of a universal Self of itself and in itself. However, the physicalist knows and experiences nothing of this principle of increasing division from Unity to multiplicity in nature and the origin of the atomicity of matter and takes the latter—that is, the endpoint—for the starting point. The materialist's perspective seizes the bottom-up evolutionary aspect of the lower-Nature but, being limited to a sense-mind form of cognition, which is unaware of the dynamic action of the spirit in the manifestation and the deeper nature of matter as a result of an involution, misses the top-down evolutionary aspect of the higher-Nature and fails to acknowledge the consciousness hiding behind the mechanical material processes, which to the mind inevitably appear as unpredictable–that is, random.

The philosophical idealist goes a step further and, by an inward turn and a mental exercise of self-knowledge, begins to doubt the absoluteness and objectivity that an externalized analytic reason suggests. The naive identification between noumena and phenomena is questioned. The separation between subject and object becomes thinner, and by crossing the boundary between mind and the higher-mind, one becomes aware of the purely phenomenal aspect of existence as inextricable from the subjective phenomenal experience. The scientific concept of objective truth unveils itself as an illusion, the centrality of phenomenal consciousness becomes an almost self-evident fact, and something from higher cognitive planes begins to instill the doubt that, at the bottom, perhaps everything is a form of consciousness or mind. The idealist's mode of cognition begins to intuit philosophically the possibility of a universal Mind and its workings as higher-Nature, though still missing the direct knowledge of it that only the mystic experience can furnish.

In fact, the mystic who steps back and, by a giant leap, transcends both the lower- and higher-Nature and any form of cognition altogether, identifying directly with the ultimate Source, where the mental individuality, the ego, is abolished like a 'drop in the ocean', becomes aware of the non-dual Conscious-Self, the transcendent pole of existence, and eventually also experiences the cosmic planes.

This is, indeed, an enormous progress in the evolution of human consciousness. While the physicalist and idealist still dwell on the mental, or at best, intuitive levels of awareness, the mystic does not capture reality by a philosophical theory or a series of logical reasonings and logical inferences. Rather, the mystic does so by a direct experiential approach, stepping out entirely from the manifestation.

Nevertheless, spiritual knowledge, in its traditional non-dual Vedantic version, also has its limitations. While it discovers the pole of the universal Conscious-Soul, it still sees the pole of the Nature-Soul through the lens of mind. The 'realized master', the 'liberated soul', the Buddha who lives in a conscious state of Nirvana, is not necessarily an individual who has realized a psychic and spiritual, let alone supramental transformation. If the psychic being still hasn't come to the front and the consciousness of the practitioner did not ascend towards higher spiritual, cognitive domains, the phenomenal universe and all its temporal display in space, time, and its clash of forces will still appear as a process without aim and purpose. The non-dual 'liberated soul' becomes aware of a previously unknown pole of existence but remains in the dark as to what connects the unmanifest Supreme with the manifestation and what their relation is. Therefrom originated the idea of the famous 'veil of Maya', the cosmic illusion which appears to be a distortion or, at any rate, something less real than the Brahman, which is perceived as the sole and ultimate reality.

It is no coincidence that the non-dual teachings did not develop an evolutionary concept. Evolution itself is part of the illusion; it just happens. For the Buddha-mind that looks at the manifestation from the eyes of the Absolute, everything is still a process, a being and becoming but, because the possibility of a supramental transformation, down to the most material tissue, isn't known, evolution inevitably remains an unnecessary hypothesis or irrelevant aspect for the spiritual realization, which should not distract us from our spiritual practice and discipline.

In this sense, all three modes of cognition capture only an aspect of reality and fail to be truly integral.

This integrality of existence is visible only to the seer by a supramental vision. According to Aurobindo's integral yoga , if we want to become aware of reality in its integrality, our being must also be subjected to an integral transformation. Only by transforming our mind, vital, and body, opening our modes of cognition to the knowledge by identity of the supramental heights, with our soul awakened and free, will the universe disclose us as it is. Maya will appear neither as an illusion or process without true substantiality, as some non-dual teachings suggest, nor as an objective and exclusively material reality, as the physicalist would like to believe, and not even as a play of an inherent universal mentality, as the idealist contends. Rather, it discloses itself as a dynamic evolutionary manifestation of a transcendent Being—that is,

Nature in its polar aspect—which is as real as the transcendent Being itself from which it comes. The universe appears as an indivisible and undifferentiated unity and totality, which expresses itself by a bottom-up Para-Prakriti action determining the unfoldment of Apara-Prakriti through an evolutionary process characterized by infinite variety and multiplicity. The One and the infinitely many, the objective and the subjective, the personal and the impersonal, Nature and Super-Nature—are all the different aspects and modes of the same Being. All are real, and none is an illusion.

Rising above the human mind, one begins to perceive the cosmos' aim and hidden final causes as an obvious state of affairs. It is a point that Aurobindo summed up as follows:

"The Truth-Consciousness is everywhere present in the universe as an ordering self-knowledge by which the One manifests the harmonies of its infinite potential multiplicity. Without this ordering self-knowledge the manifestation would be merely a shifting chaos, precisely because the potentiality is infinite—which by itself might lead only to a play of uncontrolled unbounded Chance. If there were only infinite potentiality without any law of guiding truth and harmonious self-vision, without any predetermining Idea in the very seed of things cast out for evolution, the world could be nothing but a teeming, amorphous, confused uncertainty." [266](LD Bk. 1, ch. XV)

The question of final causes, a purpose, and an aim in the creation and evolution of the cosmos cannot be solved at the rational scientific level, no more and no less than the variability in intensity and color of a single dot or pixel on a screen can tell us about the screenplay. Only by going beyond this limited vision comprehending the whole dynamical play of forces and events can we realize the meaning and deeper significance of the phenomena and then, literally, 'connect the dots'.

"All action, all mental, vital, physical activities in the world are the operation of a universal Energy, a Consciousness-Force which is the power of the Cosmic Spirit working out the cosmic and individual truth of things. But since this creative Consciousness assumes in Matter a mask of inconscience and puts on the surface appearance of a blind universal Force executing a plan or organisation of things without seeming to know what it is doing, the first result is kin to this appearance; it is the phenomenon of an inconscient physical individualisation, a creation not of beings but of objects." [266](Bk.2, Ch. XXVI)

"The principle of free variation of possibilities natural to an infinite Consciousness would be the explanation of the aspect of inconscient Chance of which we are aware in the workings of Nature—inconscient only in appearance and so appearing because of the complete involution in Matter, because of the veil with which the secret Consciousness has disguised its presence." [266](Bk.2, Ch. I)

The supramental vision reveals to us the root cause of the dilemma that plagues, on one side, the materialist's denial, which sees randomness as the characteristic sign of a lack of purpose but can't go beyond an ideological statement of facts, and, on the other side, the spiritualist who would like to recognize in the workings of Nature a final cause, but desperately tries to eliminate chance and randomness, unknowingly being the victim of the very same physicalist misunderstanding.

Indeed, there is a Consciousness-Force at the base of every universal event, but because it is involved and self-forgetful of itself in matter, it must appear as a mechanical and repetitive blind action. What we perceive as randomness and chance are the results of an involved conscious Will-Force which manifests as mechanical and automatic. On the other side, this bottom-up process is also conditioned by a top-down overmental and indirect supermental knowledge which contains all the variation, multiplicity, novelties, and creative powers of a supreme Consciousness.

Or, looking at the same cosmic process from another perspective, randomness and chance appear because of an evolutionary process of a divine Consciousness that plunged itself into the mechanical and inconscient matter and which can find itself back only through a bottom-up process of trial and error. It is about a reality that is manifesting and working itself out by the evolutionary process but which has lost its original supreme consciousness in matter. To the mind, this appears as a disordered and apparently incoherent amalgam of play of forces which it takes as proof of the absence of purpose, goal, and will. However, this is a limited view of the processes of Nature. A top-down action of a universal higher-Nature Consciousness-Force is necessary to direct the process in the lowest domains of the universal lower-Nature unconscious force. This allows us to see in a wider perspective Bergson's 'creative evolution'. If we look at Nature as being active by *both* the bottom-up *and* the top-down universal operations at the same time, then the contradictions and inconsistencies that appeared in the frame of the single approach disappear. If we become aware of both activities, or, at least, consider both points of view having an equal right to exist, there is no contradiction between the fact that we observe a cosmic chance, a huge combination of coincidences, a lucky play of blind material forces at work and a creative teleological goal-driven process which leads to the existence of complex living and consciousness organism. Mind and life are not a coincidental emergence of a process of natural selection but are the obvious consequences of the inherent cosmic twofold structure of Nature.

b. Unity in Diversity: The Universal Principle of Existence

Paradoxically, the reductionist approach inherently has something holistic itself. This becomes clear once we realize another important aspect of the integral cosmological teleology with its inherent multiplicity in unity with sameness and variation. What we observe everywhere in Nature is how it sticks together elementary constituents building the world in a hierarchical scheme. Reductionism conceptually organizes the world as being built up from the microscopic to the macroscopic cosmos by successive orders of organization playing with a multiplicity and variety of units that form new unities. This is, after all, the main motivation of the reductionist mind which sees everything from the bottom-up perspective. Broadly speaking, the physical plane is piled up as follows.

A variety of atoms emerges from the combination of a multiplicity of elementary subatomic particles. These form a systemic unity we call 'chemical elements'—a self-organizing complex system of greater size with atomic properties that weren't present in its constituents.

By the combination of a multiplicity of atoms, a new multiplicity, which is also a new systemic unity, forms what we call 'molecules'. A huge number of molecules manifest with chemical properties that were inexistent in the previous atomic order of organization.

Similarly, in combining a variety of molecules, which are organized into amino acids, these are then organized into protein molecules and complexes, which are again organized into organelles. Organelles are then organized into units of complex living entities that we call 'biological cells'. Then cells are organized into tissues, which are organized to form organs, which in turn, finally, are organized into living organisms.

At every step, a new structure emerges, manifesting unique properties. At each configurational level, a new multiplicity forms a new unity which becomes the basis for a multiplicity itself, and so on.

The rarely addressed question is: Why is this so? We tend to perceive this as a normal and almost obvious state of affairs. However, closer scrutiny reveals that it is not so obvious how this bottom-up process of aggregation could come into existence with this particular dichotomy of multiplicity in unity.

Imagine being present shortly after the Big Bang, when only a hot and homogeneous plasma of elementary particles formed in a universe still lacking any structure. There are still no stars, galaxies, or planets—just an incoherent gas of particles swirling throughout space. Even if you knew everything about the physics of elementary particles, would you have been able to predict the long-term outcome? You might have been able to predict the formation of hydrogen atoms, and perhaps a few other heavier elements, and eventually that they might somehow stick together, but you wouldn't be able to go much further than that—especially when it comes to completely novel creations with all the complex chemistry which led to the formation of unicellular and multicellular organisms and the emergence of complex sentient, conscient and intelligent beings. That would not only have been impossible to predict but, especially the modern scientific mind, would have branded these only as a phantasy, a sort of new-age woo, that must be expunged by resorting to Occam's razor. Fortunately, cosmic history went otherwise, and if we take the result of this bottom-up creation for granted nowadays, it is only because we know the outcome in hindsight.

If we ponder this mechanism without taking it for granted, it is not that clear why an association of sub-units must necessarily combine to create an emergent higher-order functional and unique, novel, and quite useful aggregate. By combining whatever kind of random units or entities, these hierarchical structures could not come into existence. You won't be able to create anything meaningful and useful just by taking whatever kind of building blocks are available and putting them together by whatever means. It is not obvious that sticking together chunks of something would lead to anything that could exhibit a plurality of new units with functional properties forming a new totality and, at the same time, a basis for another unity on the next hierarchical level.

What receives scarce attention is that all this was possible because of some pre-existing and built-in premises in the structure of the world. The building blocks must have very specific properties to lend themselves to an emergent bottom-up creative process expressing new things in this duality of multiplicity in unity.

For example, as is well-known, atoms are made of protons and neutrons — the electrically positive, neutrally charged particles in the atomic nucleus respectively—and electrons, the negatively charged particles which form the atom's shell. These could form all the variety of atoms and molecules and build up the material universe as we know it. However, if there were no neutrons, then only the hydrogen atom could form, and no other more complex atomic structure could come into existence. Without neutrons, all other atomic nuclei heavier than the hydrogen atom could not form (by their combination, protons would repel each other due to their positive electric charge, and any nuclei would disintegrate almost instantly). That is, neutrons are essential for the creation of multiplicity and the stability of matter. Without them, nothing new could form by a bottom-up approach.

On the other side, protons , neutrons, and electrons are the only material particles with lifetimes long enough to form anything worthy of mention.[105] What if there were one more stable particle? Say, the '*muon*' elementary particle, which has all the properties of the electron, with the only difference being that it is 207 times heavier? If a muon did not decay in a fraction of a second, as it fortunately does, but remained stable for the entire life of the universe like its lighter brother fortunately does, it would form 'muonic

---

[105] More precisely, neutrons are stable only if confined inside a nucleus. Moreover, also some types of neutrinos might be stable, but we won't bother the reader with further technical details which are not essential for the point we would like to make here.

atoms'—that is, the electrons of atoms would be replaced by muons, forming much smaller atomic shells. This would automatically form much smaller hydrogen atoms (and molecules with completely different chemistry) that would allow for nuclear fusion reactions even at room temperature. In turn, this would ignite such fast nuclear reactions that stars would burn out very quickly, and life would be unlikely to form in the first place. The emergent hierarchical build-up process would have been stopped very soon, resulting in a universe with almost no organizational structure other than, at best, inactive stars and dead rocky planets doing nothing.

Moreover, the coming into existence of these successive orders of organized complexity is possible only if the properties of the new emergent units lend themselves to a proper combination allowing the transition to the next organizational level.

For example, the protons, neutrons, and electrons are not enough to create a multiplicity of elements. There had to exist an inherent universal law that prevents the electrons from collapsing all on the nucleus occupying the same space in an undifferentiated manner. Otherwise, all atoms would collapse into electrically neutral and chemically inactive and uniform chunks of matter, and the creation of molecules would become impossible, preventing a further build-up process towards the next order of molecular complexity and multiplicity. Fortunately, Nature embedded a specific universal principle into the cosmos, namely the *'Pauli's exclusion principle'*, which avoids precisely that. Thanks to this inherent quantum principle, electrons can pile up in a manner such that atoms would not all be electrically neutral, thereby enabling the formation of chemical compounds

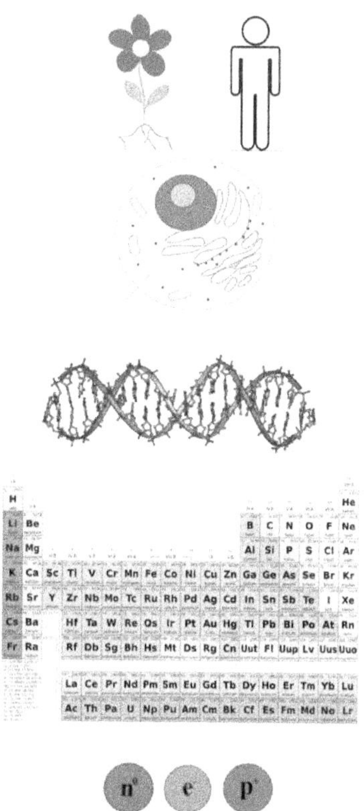

Fig. 135 *The bottom-up structure of matter and life: elementary particles, atoms, DNA molecule, cell and living organisms.*

according to a specific scheme constituting the well-known periodic table of elements—that is, an emergent creative layer of novelty.

As the last example, consider that we live in a three-dimensional universe. There is, however, no known physical law or principle that imposes precisely three spatial dimensions. That's why physicists and astronomers speculate about the existence of other universes that might have other than three dimensions. However, there is a very stringent reason we find ourselves living in this universe—because it can be easily shown that in a two-dimensional or higher-dimensional universe, stable atomic structures and stable planetary orbits would be unable to form and, again, everything would quickly collapse into giant black holes, leaving only a dead and uniform universe without structure and variety. If a universe must be built according to a bottom-up principle of diversity and unity, it must be three-dimensional.

Many other examples of this sort could be mentioned. Of course, what we are talking about is the famous '*fine-tuning*' of the universe that allows for the emergence of life (for a technical review, see [334]). However, an aspect that is seldomly appreciated and that we are pointing out here is that a 'selective-tuning' is also necessary to create a universe that a reductionist bottom-up model can describe. The creation of the universe based on a bottom-up approach was not at all a guaranteed outcome and became possible only because its primal constituents, the elementary subatomic particles, exist in a specific number and variety by possessing very specific physical properties and only because the universe has a specific structure and is governed by very specific laws enabling it to do so.

So, to make a very long and complicated (but quite fascinating) story short: The reductionist assumes something being self-evident, which is instead deeply mysterious. Scientists work with a reductionist view of the universe because they observe it as something already given by Nature, as a fait accompli, and thereby accept it without question. But looking at the world from a reductionist bottom-up approach doesn't explain much; it only describes it. The fact that things can be described by a reductionist approach where a multiplicity of entities with sameness and variation at each order of complexity becomes the basis for the following order of a new complexity having, again, an inherent variety in unity is not an obvious and inevitable aspect of reality. On the contrary, it appears to be a 'miracle' that avoided the opposite destiny, making the universe end in a clump of matter with only sameness and without much structure and variation.

From the integral cosmological perspective, this apparent mystery not only disappears but also asks for a principle of unity in diversity to be a necessarily inherent aspect of all reality. The fact that the universe so nicely follows laws of multiplicity and diversity in unity, from the elementary subatomic particle to the most complex biological organism, is an intrinsic aspect of the Transcendent, the Supermind, and its overmental manifestation. One and the many, commonality and diversity, sameness and variation, unity in diversity are the very essences of the manifest aspect of the Divinity in the first place. These are not derivative qualities emerging from some weird alchemic mixture but the first principles of the cosmos, which are already inherent in what is beyond the cosmos itself. That Nature lends itself to a partial description by a bottom-up reductive theoretical framework is because the material manifestation *is* an aspect of the supreme Spirit which inherently contains this dichotomy. Or, if you prefer, it is due to an immanent archetypal property of the cosmos, without which, speaking of the combination of units forming higher-order structures, would be something impossible to even imagine. Reductionism is not a human idea; it is a divine Real-Idea about how the world should be. It is an inherent feature that transcends physical reality, has been ideated in advance, and could not exist without its opposite, which is holism. At the bottom, reductionism is the alter ego of holism.

Moreover, out of these bottom-up processes, something new always emerges. The above-described processes of emergence pertain only to the physical domain. The emergence of something so utterly different and novel as life, mind, and consciousness is even more difficult to account for in the frame of a naturalistic paradigm. However, in an integral view, the emergence of novelty is also not a coincidental or fortuitous aspect that allows us to exist. It becomes, instead, something inherent in the very nature of things because it is the reflection of an inherent Will to emerge, the expression of a Spirit that evolves out of matter and emerges in and through it.

c. The Real-Idea and the 'Association of Infinitesimals'

> *"A little science estranges men from God,*
> *but much science leads them back to Him."*
> Louis Pasteur

In this integral perspective, things are not as simplistic as they are in the alternative monistic or dualistic cosmologies, but they can nevertheless furnish a much more powerful and consistent theory of consciousness, evolution, and the emergence of life. The universe is not just about matter and spirit, while an organism is not only body and soul and not even just a dissociated alter in a Mind at Large, let alone only a body without spirit and soul, as the materialist would like to believe. The evolution of life can't be seen as just a play of physical forces alone; it is determined by several other factors acting on other trans-physical and trans-rational planes of existence. There is not just matter, or only mind, or only matter and mind. Above, there is the Transcendent and the immanent in its impersonal and personal forms; there is matter, life and mind, and the planes above the mind, while below there is a subconscient, inconscient, and nescient basis which are interdependent and which mutually interact and compenetrate each other.

By also integrating the integral evolutionary twofold higher- and lower-Nature central principle, which contains in its bosom the principle of unity in diversity, the scientific perspective of *'abiogenesis'*—the process of the origin of life—and its subsequent evolution is turned upside down: The aggregation of matter does not 'produce' any consciousness, life, and mind but only furnishes a physical basis for the material manifestation of a universal

Consciousness, a universal Life and a universal Mind which express themselves in and through a multitude of individualized forms on the sub- and unconscious physical plane.

Of course, Aurobindo did not express particular scientific strong opinions on the matter of the evolution of life in terms of a scientific language. First, he was neither a scientist nor a philosopher; rather, he was a spiritual master who developed a practical psychology with no interest in developing scientific theories. Second, it goes without saying that in the 1920s, scientific knowledge was not the same as that of today. However, with this in mind, the spiritual framework of integral yoga not only can tell us something about life and its development from a trans-physical perspective transcending the material and mechanical understanding of the workings of Nature but may even be an interesting extension of the modern scientific view of evolution without conflicting with it.

In the integral cosmology, consciousness, as well as its individualized expression in sentient beings, does not need any physical substrate to exist; it is already present as an irreducible ontological primitive as an inherent aspect of the Unmanifest projected into the manifesting layers of Nature. The material aggregation appearing first as organic macromolecules, then as unicellular and multicellular organisms throughout an eonic biological process is the effect of the existence of a universal cosmic Consciousness. The emergence of the first proto-plasm was not the cause of the emergence of consciousness but, rather, the effect of the workings of the supraconscious Para-Prakriti in the realm of the inconscient Apara-Prakriti by the principle of unity in diversity.

In its very origin, there is the supramental creative Real-Idea which builds, in a spatio-temporal reality, its own names (powers, qualities, characters) and forms on the different planes of manifestation capable of containing an indwelling spirit. First, by its exclusive concentrating conscious Force, this creative Power works out new physical 'infinitesimals'—that is, particles, atoms, molecules, macromolecules—and which, by a grouping and association, builds a material basis to create and multiply an original protoplasm leading to the formation of the first living unit: the cell. Distinctive of a conscious Force playing with its own idea are the patterns with an identical rudimentary basis for all living organisms—a cosmic play of sameness and variation.

Let us begin by pointing out how, from the integral perspective, we can frame a spiritual cosmology that faces the combination and decombination problem affecting the modern idealistic philosophy of mind (recall Pt.II-II.3a&f) in a much more coherent manner.

On one hand, taking the bottom-up perspective, if we assume that even tiny subatomic particles have some primitive form of subjective experience, as the panpsychist contends, we must explain how these micro-level subjects can literally be summed up to constitute a new subject. That is, how can a bunch of particles which are supposed to have their own content of phenomenal experience build up atoms, molecules, unicellular, and multicellular organisms and, in the process, lose their original separate conscious identity coalescing to a new, single, individualized subjective conscious perception? On the other hand, taking the top-down approach, if we consider the universe to be the expression of a single and unique universal Consciousness, such as cosmopsychism, analytic idealism, non-duality, and, of course, the integral cosmology states, we are forced to explain the opposite problem: Why and how does the plurality of subjects come into being and why does our individual subjective experience feel separate and distinct from that of other subjects and from its original universal Consciousness? Kastrup advances the interesting hypothesis of the 'dissociated alters', but, as we have seen, it lacks an evolutionary dimension and thereby raises the teleological problem we outlined in Pt.II-II.3j –that is, it leaves an explanatory gap that fails to clarify how the first metabolic system of a cell could form from raw non-living matter if there was no individualized consciousness and only an instinctive Mind at Large.

Curiously, it is the same problem that plagues modern biology. Despite decades of research, the question of how such an amazingly complex structure like a biological cell, with all its functionalities and its ability to grow, divide, self-assemble, reproduce, and transfer genetic information by mechanisms of inheritance, emerged from prebiotic chemistry remains unanswered. How could such a complicated thing in Earth's natural history emerge so early (about 3.8 billion years ago) and, as testified to by fossil records, seemingly quite suddenly? It is now clear that this certainly did not happen just by the mixing of chemical compounds. We suppose that things may have happened in the infamous 'primordial soup' or in particular environments where organic compounds could build up. Some chemical pathways may have been catalyzed by energy sources of geothermal activity, like in deep-sea hydrothermal vents, near geysers, or because of volcanic eruptions. Others speculate that the prebiotic material might have been synthesized first in outer space and brought upon Earth by meteor, asteroid, or comet impacts—the famous *'panspermia'* theory. There are lots of ideas, speculations, and conjectures, but the processes that led from non-living matter to the first enclosed membrane, probably some primeval prokaryotic cell or bacteria, remain a deep unsolved mystery. As of the time of this writing, nothing indicates that it will be dispelled soon.

Of course, from the biochemical point of view, we won't dispel this mystery either. We leave the description of the strictly materialistic process that the emergence of life brings forth to biologists and biochemists.

From the perspective of an integral cosmology, as we elucidated by describing the planes and parts of being, making the distinction between evolution and involution and between the twofold vision of the action of a higher-Nature onto a lower-Nature, these material life-processes appear in a completely new light: They are not the cause but already the effect of a non-material but conscious Life-Force. It is by a progressive descent into the material manifestation of a conscious Life-Force from the cosmic Life-plane that the first material cellular aggregate forms. Life does not arise because of matter; rather, it is an immanent cosmic principle that emerges *through* matter. Life can eventually infuse a material structure to possess it but doesn't necessarily need matter to exist. Matter is a substrate through which a Life-principle expresses itself in the physical cosmic plane. Like electricity flows through an electric conductor but is not an epiphenomenon of the particles of the conductor, similarly, Life can organize and flow through matter but is not an epiphenomenon of matter. Of course, this is an exquisitely vitalist conception that, contrary to what is believed in the present orthodoxy, is perfectly in line with our current scientific understanding of life, as we have shown in Pt.III-II.5g. The same fits for mind and all the higher levels of consciousness, which progressively manifest in the material universe. It is in this light that the idealist's 'radio metaphor' or the prism analogy of W. James or A. Huxley's brain as a 'reducing valve' of a Mind at Large, and that we have discussed in Pt.I, acquire a deeper significance. Matter is a (still) opaque interface that allows the different cosmic planes of existence to 'shine through'. From the standpoint of integral cosmology, this gradual 'shining through' of a higher-Nature that transforms the lower-Nature from a Prakriti of ignorance into a Prakriti of Knowledge is what we commonly call 'evolution'.

However, according to Aurobindo's cosmology, the other planes of cosmic existence are non-evolutionary. Their primary function is not that of evolving anything. Their function, instead, is to furnish the in-between vertical layers of consciousness in the cosmic structure linking the supreme supramental consciousness with the physical plane whose destiny is to manifest it by a completely new Creation. The Supermind cannot manifest in the involved nescient matter without its necessary transformation, which it actualizes only indirectly, passing through these cosmic layers of consciousness. As we have seen in Pt.III-II.2, by dealing with the spiritual principle that leads from the One to the many, a multitude of infinitesimals, such as particles, atoms, and molecules, is a form of nescient involved consciousness arising due to the concentration of the overmind. However, everything was, is, and will remain under the control of the higher wisdom of Para-Prakriti, the higher-Nature, that determines by a Force of Knowledge the downpouring of the form of the Real-Idea in the progressively gross planes with its Consciousness-Force and Consciousness-Will directing the formation of the material aggregation from behind, by a concealed and implicit operation in the Apara-Prakriti, the lower-Nature manifesting it as a force of ignorance.

On the lower levels of existence, the infinitesimals associate themselves together by a slowly increasing organization of a seeking ignorance struggling for knowledge on the blank slate of the nescience. It is a process in which the subtle 'vibration' on the non-physical vital, or life-plane, determines the organization and functions of the physical. Matter does not determine life and mind; rather, mind and life determine matter's aggregation. Higher still, a divine Mind and Will determine mind and life. The material existence is the end, not the beginning, of Creation. It is by an indwelling presence of the Real-Idea that things emerge and take form, first involved and then evolving out of the unconscious in a dynamical and temporal material universe. Goethe's higher-mind way of seeing was not a coincidence. He already intuited, more or less subliminally, the possibility of a far-reaching Real-Idea beyond the appearances that he labeled as an archetype or Urphänomen and which is the common basis of all things and organisms.

If we view all existence from this standpoint, several inexplicable and seemingly mysterious, almost miraculous, events and facts of the cosmic play begin to take intelligible shape. From the perspective of an integral cosmology, it is not an aggregation, combination, decombination, or recombination of particles, cells, organs, or whatever material objects that leads to a new subject having experiences. It is consciousness, life, mind, and—at the highest levels of the hierarchy of consciousness and existence—the Real-Idea that organizes, combines, and decombines. Even if we could carefully position every cell, molecule, atom, and particle into a structure that perfectly copies the material brain and body of a living organism, that would not necessarily lead to a living and sentient being unless there were an indwelling life, indwelling mind, and indwelling spirit. It is the action of the spirit that panspiritism also intuitively captures.

d. From the Real-Idea to Biosemiotics

This also subverts how we think of meaning, purpose, aim, and goal-directedness. The materialist contends that these are concepts that tend to anthropomorphize Nature. According to this view, we should never think and talk in these terms because one posits a priori that the universe has no aim or goal or that there couldn't be any evolutionary goal-directedness. These are only human conceptions that should have no place in scientific considerations, as the quest for meaning is only a human impulse that we should not project outside of us.

The integral cosmologist, instead, contends the contrary. It is the scientific mind that, by clinging to a lower-level form of cognition, takes an anthropocentric human sense-mind point of view and makes itself incapable of recognizing that not only is there an aim and a meaning, but that these are the divine and original impulse of Creation and the most intimate truth of Nature itself. It is the human mind that lost the ability to seize the inherent télos in Nature's processes. Or, even if the mind remains open to this eventuality, reason captures only subliminally a more or less vague and pale redux and a fragment of a distant Real-Idea standing behind things. The Real-Idea has been 'percolated' into the physical, which, to the higher-mind of the intuitive scientists or philosophers, indeed expresses itself as an archetype that retains some of its original identity. However, the physicalist who sticks to the sense-mind cognition of mere name and form insists on eliminating and declaring this as an unnecessary illusion. While, from the perspective of the Real-Idea, every formation arising out of Nature is the expression of a preexisting Idea that our ordinary human mentality apprehends as names and forms. It is the primeval perfect, divine, and undifferentiated Knowledge, Consciousness, or 'Gnosis' that, however, throws itself into mutable names and forms–that is, into what the mind conceives in more or less accurate symbolic images of this Seed-Idea. Something that, at the human apprehending and comprehending level, the intuition captures as archetypes, symbols, signs, with their meaning and significance in or of things and phenomena.

We won't repeat here why science, no matter what, won't be able to expunge the questions for meaning and purpose. We will only point out how, paradoxically, it is taking small but interesting steps in the opposite direction.

An intriguing example of this is the modern finding in morphogenesis—that is, how a highly complex self-assembling pattern emerges from a single fertilized cell, developing the organism appropriate to its species. What are the mechanisms and underlying biophysical and chemical phenomena that preside over complex pattern formations such as the self-assembly of structures such as an eye, a limb, and the entire 'bodyplan' during embryogenesis? How do organs regenerate after an injury? How can large numbers of cells coordinate their individual activity to assemble themselves into organs and achieve geometric and functional goals defined at a macroscopic scale of the whole but can't be found anywhere at the cellular level, not even in the genetic code? Nowadays, we know that there are also correction mechanisms with adaptive decision-making capabilities of living tissues able to correct and adjust embryonic development despite forceful induced defects [335]. It is an inherent 'goal anatomy' that doesn't stop at the formation of an adult organism, as is well known from the salamander's amazing ability to regenerate an amputated limb (also recall the mindboggling self-regeneration of the freshwater flatworm in Pt.I-IV.2).

Contrary to common belief, biology is far from having any coherent theory capable of explaining how all that works out. What we know is that it looks like this self-monitoring and repair of complex multi-tissue organ systems and its pattern formation involves a bioelectric code that drives and changes the cell's transmembrane electric potentials. Large-scale anatomical pattern formations are regulated by very complicated networks of bioelectric signaling among cells, which determine differentiation and regulation of embryonic development, regeneration of injured tissue, and even the avoidance of tumor formation [336]. In developmental biology, old and despised concepts such as the *'morphogenetic field'* have been revived—that is, the idea that a group of cells leading to the development of specific morphological structures must be envisaged as an, in space and time, long-range dynamical information processing whole rather than mere units executing an internal genetic program.[106] Of course, conventional science hopes to reduce that to ordinary electro-chemistry by speaking of hugely complex 'long-range signals', 'planar polarity of proteins on cell surfaces', 'standing waves of gene expression', 'trans-membrane voltage potential and tensile forces', and *"chemical morphogen gradients carrying information about both the existing and the future pattern of the organism."* [337] Interestingly, however, they themselves now admit that the pervading 'teleophobia' that avoided intentional idioms which attributed goal-directedness to a system turned out to be all-or-nothing thinking that has prevented discovery and that *"has gone too far, putting biologists into a straitjacket that prevents them from exploring the most promising hypotheses."* [338] This does not mean that we will see them embrace a post-materialistic view anytime soon like the one we are advocating here. Nevertheless, the concession is interesting. The dream is no longer to reduce everything to genes but to electrochemical machinery that abolishes any metaphysical speculation. We guess that it is only a matter of time: Also, this will turn out to be yet another unfulfilled dream, and they will have to concede much more than that. Perhaps they will still not speak of a 'plan' but of a 'Will' or an 'Idea'?

Meanwhile, science is slowly but steadily developing into a discipline that studies and investigates the production of meaning in Nature. An interesting example of this could be *'biosemiotics'*, the study of signs and meaning in the biological realm (for an introductory essay, see [339]). Especially noteworthy is how, contrary to what a reductionist mind could have predicted, it finds its most interesting applications in molecular biology, particularly in cellular signaling processes. A cellular signal is, molecularly speaking, just a transmission of a molecule passed from one cell to another. This chemical exchange works as a message and an information transfer that triggers a cellular response and behavior, such as for survival and maintenance purposes. This allows cells to communicate with each other and behave in a coordinated

---

[106] It should be noted, however, that if and how the more popular term of 'morphogenetic field' introduced by the English best-selling author and biochemist Rupert Sheldrake, can be related to the conventional scientific term remains a highly controversial matter of debate. We won't take up this here.

way, seemingly driven by intentionality where these signals lend themselves as a means to an end.

These cellular signaling processes can be horrendously complex. By complicated signal pathways—that is, multistep signaling processes involving long chains of chemical reactions forming, for example, self-reinforcing cycles we might call 'habits'—life can be interpreted in a unified model as a process of transcription, transmission, and reception of signals, messages, and meanings between living beings. Applying system theory, cybernetics, and information theory, life could be conceived of as a self-organizing complex system in which molecular and intracellular signaling pathways acquire a biological meaning and represent significance ascribing a qualitative aspect to physical attributes which otherwise would not have revealed the role of signs and its meaning other than that of a fuzzy play of chemical exchanges which, in itself, would be without any apparent meaning or significance. In a certain sense, biosemiotics studies the language of life in its representation and meanings of biological codes—from genetic code sequences, of which DNA is the most known and paradigmatic example, to less known but perhaps even more interesting cases such as intercellular signaling processes to animal behavior and, last but not least, human language and abstract symbolic thought. It discovers that there is no real demarcation line between the human's generation of meaning, sense, and intentionality and the generation of signal and messages with an ascribed functionality—that is, 'purpose'—in the workings of a natural world which, in the eyes of materialistic naturalism, should have none. Words such as 'message', 'signal', 'code', and 'sign' are considered metaphoric and could someday be reduced to the mere chemical and physical interactions of complex cybernetic adaptive systems. However, these terms remained ineliminable, and the reductionist plan is becoming increasingly untenable, especially with the accelerating discoveries in cell signaling studies. Something inherent in Nature refuses to allow us to eliminate what we call 'meaning', 'purpose', and 'intention' as if they would represent and stand for an idea, a blurred image of an original idea, a Real-Idea .

Of course, the official biosemiotics would not adhere to our trans-physical orientations. The mechanistic and rationalistic conception of life is still too deeply entrenched in our scientific minds, and it will take time for the cultural background to change. However, it is to be expected that once it embraces a Goethean way of seeing, it might make further progress in exposing the fallacy of the reductionist aimless and meaningless universe.

In fact, the biosemiotics standpoint, which captures parallels between biochemistry and semantics, makes sense from an integral perspective. The integral cosmologist 'sees' life also as a natural and spontaneous expression of an indwelling meaning and purpose and can't even conceive of these as emerging only as by-products of mutual interactions among a myriad of elemental units alone. Rather, it 'sees' these interactions themselves as an expression and actualization of an intention that already carries meaning, namely, the Real-Idea that stands and works behind it. The reason that stands behind that strange and stubborn refusal of Nature to be reduced to a meaningless and undirected process must be found in the Real-Idea inherent to the higher-Nature, with its infinite creative possibilities reflected in the lower-Nature in its atomized and molecular and mechanical workings. The unsurmountable complexity of cell signaling is not a fuzzy by-product of a Darwinian selection process but, rather, is the expression of an idea in a world of polarities whose origin is that of a unitarian Consciousness. If signs, symbols, and meaningful signaling among living units, to which we would hardly ascribe a semantic ability, seem to appear so ineliminable and intractable in the real world; this is because they are already the reflection of a perfect, transcendent Sign, Symbol, and Meaning that mind can only vaguely seize or superficially intuit.

e. Integral abiogenesis

"You place matter before life and you decide that matter has existed for all eternity. How do you know that the incessant progress of science will not compel scientists to consider that life has existed for all eternity, and not matter? You pass from matter to life because your intelligence of today cannot conceive things otherwise. How do you know that in ten thousand years, one will not consider it more likely that matter has emerged from Eternal Life?"- Louis Pasteur [340]

Let us now focus on one particular step of this (apparently only) bottom-up creative process—that of the first cell. As already mentioned, the gap between the most complex macromolecule and the appearance of the first most primitive cell is so huge, almost like an abyss, that modern science is still struggling to come up with a reasonable explanation of how this could happen. Meanwhile, we might ask what such kind of mysterious process of the emergence of life means in the light of an integral teleology.

Aurobindo did not elaborate on these aspects at length, but from his writings, in particular from the 'Life Divine' and 'Savitri', one can sketch the kind of primordial evolutionary process as follows.

The first cell, maybe just a membrane enclosing some organic molecules, emerged in this difficult self-seeking blind process as a first 'point of life', a 'quanta of life-form' as the determinate of a determining conscious Life-Force from a non-physical cosmic Life-plane in the material earthly physical plane. By this mechanism, the Life-plane first determined this 'life-quanta' which, on the physical plane, we recognize as cells. However, a first intimation from the Mind-plane also manifests in the form of a primeval mental element. The living cell not only is infused by the vital energy from the vital plane but begins to contain already a first primordial 'mental-quanta'. It is no longer a completely nescient material aggregate, such as subatomic particles, atoms, or molecules, but begins to manifest the first primitive forms of subconscient mechanical cognitive awareness, the basal cognition we illustrated in Pt.I-IV.3&4. It goes beyond the mere mechanical blind reaction of the electron, atom, or molecule—not just a completely self-absorbed and utterly involuted mind and life that is completely unmanifest in the dead stone. Rather, it is the first localized expression of a Life-Force from the universal Life-plane which displays a primordial life instinct of a groping consciousness, a guideless sense that feels and clasps in a voiceless world, a seeking ignorance that tries to know—the first visible physical result of the lower-Nature's cosmic life entering matter by the pressure of a conscious Force.

Life evolved out of matter naturally because it is already latent and preexistent in matter from the beginning. Matter is itself a veiled form of life (while life is a veiled form of mind, and so on) because matter is the grosser manifestation of more subtle gradations of the same substance, such as the life-force, prana, which is already involved in all gross matter, even in a stone. Once the external conditions allow for it, matter begins to manifest this life-force (for the same reason that mind, being involved in the life-force, begins to evolve out of it as well). In this 'bio-cosmogony', one might even say that matter cannot exist without life and the first step in evolution is the liberation of the latent life.

But what do we mean by 'life' or a 'vital force' inherent in matter? We have seen that it is extremely difficult to find even a commonly accepted definition of life. We won't propose yet another definition here but it might be useful to highlight that in the 'integral abiogenesis', the properties that are commonly accepted as being characteristic of life are considered immanent will-forces and not derivative aspects, as biologists contend. For example, the impulse of existence, the will to self-preservation, the instinct of survival, emotions, feelings, and the pleasures in existence are not mere behavioral outcomes of some material metabolic machinery but are already intrinsic aspects originating from the vital cosmic plane that is precedent to the physical one and, ultimately, are the expression of a Spirit's Will to delight. The vital is the lower-Nature agent of Will in the physical evolution, while the capacity for knowledge and the will to know is an impulse coming from a cosmic mental plane. All these qualities and forces of being are prior to the physical. They emerge in and through the physical coming from other planes but are not originated by a bottom-up process as emergent material mechanistic properties.

According to the Vedic worldview, the way in which this vital force organizes its impact on the material plane is explicated by the three 'qualities' or 'forces' called '*gunas*'. The three gunas are called: '*sattva*' (equilibrium, balance, equipoise, adjustment, correction, adaptation, order, harmony, calmness), '*rajas*' (activity, dynamic movement, passion, energy, excitement, desire, egotism, restlessness), and *'tamas'* (inertia, inaction, inconscience, ignorance, passivity, stability, inactivity, lethargy). Each of these qualities is present everywhere in some combination and in form of mutual interaction and relationship, with one or the other dominating.

These are considered not just human psychological traits but much more universal principles inherent in all matter, life, mind, and every phenomenon in the universe. Aurobindo identifies the three gunas as the *'three modes of Nature'* or the *'three modes of cosmic force'*. They are the 'noumenal engines' of Nature's determinism. In the interplay between life and matter and in their physical appearance, sattva, rajas, and tamas manifest and effectuate physically their dynamical action in the evolutionary process as '**retention**', '**active reaction**', and '**passive reception**' to outward impacts, respectively. Sattva retains impressions as its inner self, rajas plays itself out as energy and force, while tamas has not necessarily a negative connotation, it can also absolve to a positive function, such as stability, and is a necessary ingredient in evolutionary processes as well.

Everything that exists in the lower Nature contains a combination of these three gunas. Every physical process or dynamic form is the result of these three qualitative powers interacting with each other. This has not to be interpreted as a metaphor or mere analogy but has a physical significance. A simple example is matter itself. In physics by 'material' one means everything having a mass, and mass is defined as a measure of inertia (the resistance to a change of motion) of a body–that is, tamas. Due to Einstein's famous matter-energy equivalence ($E=mc^2$), nowadays we know that matter is a form of condensed energy–that is rajas. And, ordinary matter has also the property of stability–that is, sattva.[107]

Receiving and retaining the impacts of the outer world shape the impressions of this first life-quanta. On the base of these impressions, it also reacts to the outer conditions and begins to maintain a two-way communication that forms, shapes and molds it as time passes.

*"This evolution is effected by the three gunas, the triple principle of reception, retention and reaction to outward impacts; as fresh forms of matter are evolved in which the **power of retaining** impacts **received** in the shape of impressions becomes more and more declared, consciousness slowly and laboriously develops; as the **power of reacting** on external objects becomes more pronounced and varied, organic life-growth begins its marvellous career; and the two, helping and enriching each other, evolve complete, well-organized and richly-endowed Life."* [270](Pt.II, pg.241)

Also, the physical senses of sound, touch, sight, taste and smell are not seen as being just a sensorial evolutionary material outgrowth of the organisms but, rather, as being themselves already inherent powers in and of more subtle

---

[107] Not all matter is stable. Nuclei and particles can be subjected to radioactive decay.

planes of being, each representing the more fundamental sense for vibration, contact, form, and 'substance-sensing'.

The Vedic Upanishad's description is reminiscent of what modern biology is becoming about. Reframing it in modern language, it sees the organization, sensing and development of the first life form as beginning on the basis of a bi-directional informational exchange with the world.

In fact, we don't need to refer to these ancient philosophical doctrines of which the modern skeptical mind might question how much was the result of genuine gnostic knowledge or whether they were no more than abstract philosophical ruminations. Modern biology speaks of information stored in the DNA being the result of a population adaption to the environment and the information about this environment being fixed-that is, retained–in the genome. It represents life as a collection of single-celled or multicellular agents that exchange and process information–that is, receive from and react to the world and among themselves. It even founded the areas of research of *'computational biology'* that applies information-theoretical concepts to describe gene expressions and transcriptions, metabolic networks, and protein sequence and structure as the result of an environmental interaction analysis. The fact that biology feels impelled to resort to concepts of information theory is no coincidence. It reflects a deeper truth that can be seized only on trans-material domains where the principles of retention, action and reaction, and reception of a conscious will, are fundamental, not derivative.

Moreover, this 'spiritual abiogenesis' based on the triple process of retention, reaction and reception is a well-known and characteristic function of living organisms: homeostasis. In modern terms, homeostasis is defined as a self-regulating process by which biological systems maintain stability and adapt (sattva) while adjusting to changing external conditions–for example, the regulation and balance of temperature, chemical concentrations, or other metabolic functions in response to the stimuli of the environment to maintain constant internal conditions. While homeostatic processes were already known in the middle of the 19th century, their centrality has been recognized only recently (for a modern review in a historical context, see [341]). Science is beginning to recognize that homeostasis is more than just one of the many functions of an organism; it increasingly appears to be a fundamental property of life. It is precisely homeostasis itself that distinguishes life from non-life. The concept of adaptation, the so-central principle of Darwinian evolution, is itself a homeostatic aspect of the organism. American physiologist J. Scott Turner even goes so far as to suggest that homeostasis is the naturalistic bedrock phenomenon of the emergence of cognitive systems [342]. Cognition, intentionality, purpose and desire are all a 'wanting' to attain a biological state by a homeostatic re-adaptation.

In this sense, cognition is seen as foundational to life. But the main difference between these new trends in evolutionary biology and the perspective of a spiritual emergentism is that the former takes life and mind for an effect, while the latter considers it a cause. According to this spiritual cosmology, information exchange and homeostasis are the outer expressions and a superficial manifestation of a deeper life-principle inherent in the material world, reflecting a relationship between spirit and matter. It is this way of seeing that is decisive and makes all the difference. Something reminiscent of what we saw in Pt.II-II.3b&c with biopsychism, the CBC theory and CBE model.

Fig. 136 shows diagrammatically the relation between the noumenal threefold guna qualities inherent in the vital life-quanta backed by the cosmic Life-Force and Life-Will, and its informational abstraction in biology–the information flow represented by the transcription of genetic information for protein synthesis in the nucleus–and a generic adaptive and stabilizing homeostasis scheme of cellular or organismic regulation.

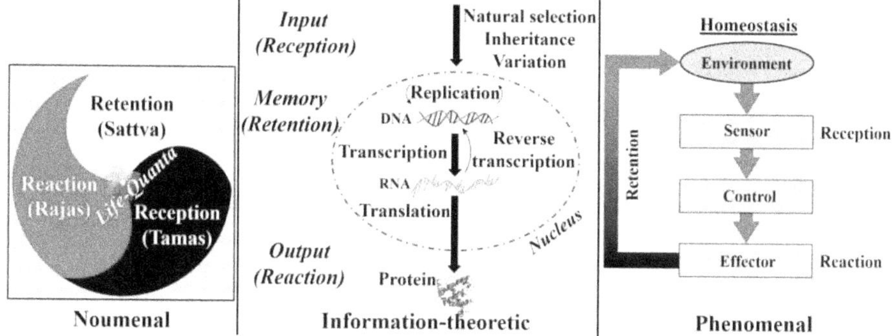

Fig. 136 The noumenal three qualities and their manifestation.

The information-theoretic and the metabolic aspect of life can be seen as the physical expression of the noumenal or archetypal qualities of reception, retention and reaction. The life-quanta becomes a physical cell, the Life-Force expresses itself as mitochondrial heat generation, and the Life-Will in the instinct of survival.

f. The Psychic Evolution and the Adventure of Consciousness

With the emergence of the first self-reproducing living cell, death also makes its first appearance on the stage of the cosmic drama. The cycle of physical birth, physical life, and physical dissolution of every living organism has an evolutionary spiritual aim and purpose.

A first aspect of the deeper meaning of reproduction and death appears as the trick of a Wisdom of an immanent Consciousness in Nature which resorts to the evolutionary process by maintaining alive a still veiled and secret Real-Idea in the multiplicity and unity unfolding in matter. This is what the species really manifests, though, in these evolutionary stages, only as a pale reflection that the physical allows for. In the night of the nescience of matter, the cycle of renewal performed by reproduction and death avoids the premature dissolution of the species, which, from the spiritual perspective, is the manifestation in matter of a trans-physical archetype inherent on other planes of existence. The secret aim of evolution is to express this original unmanifested Real-Idea with increasing perfection in the terrestrial physical.

This is not a refutation of modern biology. The integral vision does not abolish or contradict the scientific or any past spiritual knowledge. Rather, it extends, complements, and integrates it. In fact, it doesn't imply that all the biochemistry describing life's functions, from the first macromolecule such as DNA, RNA, and protein synthesis, through the unicellular organism with all its internal organelles and chemical-electric energetic processes or the multicellular interactions as described by modern biology, have no place in this description. It is not the processes as such that integral cosmology questions. It is the way of seeing the very same phenomenon that changes. In a certain sense, it is Goethe's way of seeing intensified at its utmost: No longer a higher-mind cognition but a supramental sight which seizes the phenomena with a Knowledge by Identity. All these workings of the lower-Nature, with all its huge complexity as described by modern biochemistry, become not the cause but, rather, the effect of a Life-Idea. The complex material combinations of biological entities are a material base, a physical substrate used by a Divinity to express its original creative Idea from planes above into other planes below, where it works by process of self-finding. The aggregation of neurons in a brain does not 'generate', 'create' or 'cause' consciousness; rather, it is the generation, creation, and causation of a consciousness that uses it as a means of self-expression by the appropriate combination of a multiplicity of material units. To use an analogy, the difference between the highest original supramental Real-Idea and its expression in a material subconscient domain could be metaphorically illustrated by the same difference between the image, the idea, and the mental picture an artist has compared to its creative realization on a material canvas. The material substrate, the painting on the canvas, reflects only a partial and dull image or representation of the real idea of the artist. The combination of the color pigments is the effect, not the cause, of the original artist's idea.[108]

However, in this complex cosmology, as there is a twofold cosmic impersonal action determined by the higher-Nature top-down and lower-Nature bottom-up processes, there is also an action determined by the unfoldment of a personal agency at work. The so-individuated organism, be it a cell, a plant, an animal, or a human, is not just a passive entity subjected to an external cosmic Will (what evolutionary biology labels as the

---

[108] Using this analogy one could describe physicalism as the doctrine which believes that the canvas' pigments in their combination and relation possess a causal power to produce the artist's thoughts, inspirations, ideas and imaginations.

'environment') but also contains an inner will, volition, and active power that comes from within. The individuation of the cosmic consciousness in a subjective lifeform and 'mindform' has its center in the central being—an individuation from the cosmic to the personal identity which has analogies with Schopenhauer's World as Will or Kastrup's dissociated alters but with much more explanatory power and emphasis on the evolutionary aspect. From this perspective, the integral cosmology sees the coming into existence of the individuality of a sentient subject. It is not a creation of an aggregate of material units but nevertheless something conditioned by a process in matter and also conditioning matter's aggregation.

As described in Pt.III-I.4e (refer to Fig. 122), one distinguishes between a non-evolutionary self-existent and timeless spiritual individuality beyond space and time, the Jiva, and a personal soul, the psychic entity, which, by evolving in the temporal manifestation, develops into the psychic being. Both the Jiva–the central being–and the psychic being constitute, the true individuality, but while the Jiva is our individuated spiritual core that always existed even before the beginning of any evolution, a psychic entity takes form in—and is formed by—the evolution in time, becoming the psychic being. The impersonal divine Absolute, the personal Jiva, and the psychic being are three terms of the Transcendent. The first two are immutable. The latter is a soul which is mutable and progresses by an evolutionary process. The Jiva 'portion' of the Absolute doesn't evolve or change. However, as we have tried to exemplify by the analogy of the hologram, it 'appears through' the created existence with increasing clarity, first in a single cell, then in a pluricellular combination and then in an organism with increasing organic complexity, and while the psychic develops, the Jiva becomes—or, more precisely, begins to 'appear through'—a finer grade of coarseness. And yet, it is the same Jiva, the same and immutable Whole. It always remains our inmost transcendent unchangeable non-evolutionary Soul.

However, despite its nature, the psychic being, is that of a transcendent personal identity. It is mutable and grows in and with the evolutionary process in matter. It changes by the accumulation of experiences resulting from a process of transmigration from one physical embodiment to another. According to its evolutionary state, it 'infuses through' and incarnates in a material aggregate which it eventually also determines. It is a determinant, not just a determination.

The complexity of the physical aggregate reflects not only a determination coming from other cosmic planes of existence but also the evolutionary state of the psychic. A 'newborn' transcendent soul that has just formed and is at the very beginning of its evolutionary journey through the eonic material processes of the physical Universe is still an almost complete unconscious entity. It will have to start from the bottom of the divided and differentiated nature of the subconscious material cosmos—that is, it may identify with the involved and almost nescient experiential phenomenality (say, a cell, a plant, or some primitive living form).

Because both our 'portions' of the Divine, the Jiva, as well as its projection as an individualized and mutable soul which progresses by a series of transmigrations—that is, our inmost evolutionary being, the psychic being—are both transcendent divine sparks: The individuation in a personality that cannot be the creation effected by a material aggregational process. Rather, it is the transcendent central being and psychic being that, by identification and fusion in and through matter, lead to the aggregation of a material structure that only reflects this individuality—something on the line of Taylor's panspiritism. However, it is not by combining or decombining material aggregates that a conscious subjective identity appears, as a personal psychic entity already exists in an unmanifest mode. The increasingly complex combinations of cells into multicellular life forms, such as algae, plants, trees, or vegetables, do not exist and live by a physical process alone but, rather, express an indwelling life-spirit that is penetrating from the vital to the physical plane an immanent idea of the higher-Nature's archetype which, at the highest cosmic plane, has originally been cast down into the manifestation by a Real-Idea.

This might raise some questions. For example, behavioural genetics claims that the DNA determines, or at least influences, our behavior, cognitive skills and mental illness. There is no soul or immaterial entity that makes you what you are. The research carried out typically on with twin-studies, adoption-studies, and other approaches, resorting to sophisticated statistical analyses, showed that there is consistent evidence of a correlation between one's genetic code and the individual personality. An idea that finds increasing acceptance among geneticists, is that your behavior, intelligence, sexual orientation, success in life, etc., are largely determined by your genetic pool.

This seems to suggest a quite gene-centric nature of our psychological identity, right?

First, some technical notes are necessary. All humans have about 99% of their genome in common. Here we are speaking only of the remaining 1% of the DNA code that varies between individuals. This 1% is subjected also to modifications coming from epigenetic and environmental factors making the data set 'noisy'. Moreover, there is no, or only scarce association between a single gene and a psychological trait, rather this genetic influence is determined by a very large number of genes. This makes genome-wide association studies rather complex, and where simplifying statistical assumptions and inaccurate interpretations of statistical analyses may easily creep in.

But let us assume that these findings and, especially, their interpretation are correct. If we allow ourselves to take a metaphysical perspective, as we have done with the mind-brain identity, we might equally question, if there isn't a correlation-causation fallacy, again? Taking the perspective of an 'integral metaphysics,' it is the incarnating psychic that determines the genetics at birth, according to its spiritual stage of development and the experiences it will have to go through in its imminent lifetime. The DNA code and all its genomic machinery may be seen as the expression of something on a physical plane—that is, a determination, not the determinant. If monozygotic twins have similar behavioural and psychological traits, we can equally postulate that this is because they are affine psychic beings that, consequently, express the same or similar genetic code. Or, if two heritably non-related individuals express a similar character and have a similar genetic code, we could equally well conjecture that this naturally reflects their similar psychic tendencies. Of course, from the point of view of our ordinary mentality, this is only speculation, but equally valid and not less speculative as that of a genetic reductionism that posits a priori a materialistic dogma without questioning it.

This multidimensional 'view from Infinity' suggests also other interpretations of Life. Another second aspect of the function of reproduction and death appears: It becomes a means of growth of the spirit. Despite their initial vital and embryonic mental element, cells are still subjected to the subconscient, inconscient, and hardest nescient planes of inertia and ignorance. Lacking sufficient plasticity and capacity to contain the full flow of the Spirit with all its conscious Force, cells are inevitably subjected to processes of degradation and, finally, dissolution. In its present state, matter isn't capable of embodying the full power of life, let alone the full Power of the Spirit, which, otherwise, would be able to cope with the process of the physical disaggregation caused by the cells' injuries and death. Death becomes an opportunity for renewal and further transformation of the psychic being. The soul's experience in matter determines its evolutionary leap, but matter in its present biological organization can't keep pace with the inner evolution of the spirit. The death of an organism is determined by its inability to transform quickly, flexibly, and with a plastic adaptation, not only to the physical environment but also to the indwelling spirit. Cells—and, therefore, our bodies—are not receptive enough to the top-down influx of the Spirit coming from the higher spiritual planes and must die out. However, the reproduction of the physical organism, besides its function to perpetuate the collective species in time, offers the opportunity to the indwelling individual psychic presence to continue to grow on the material plane, eventually in a physically more perfected next generation, and by participating in the cycle of life without having to begin from scratch.

After all, if we allow for a theory of transmigration, this is the most obvious logical conclusion. There can be no loss of consciousness by the physical dissolution. Otherwise, once a body of cells dissolves in its elementary constituents, the conscious subject would have to disintegrate, and no reincarnation or 're-ensoulment' into a future material aggregate would be possible. On the contrary, it is the physical dissolution we call 'death' that allows the spirit to enter new physical combinations and have new experiences on the material plane—experiences it otherwise couldn't have. Our individuality that grows, learns, changes, transmigrates, and thereby evolves is the transcendent psychic being that identifies with the material aggregation and participates in the life experience.

However, initially, at the very beginning of the adventure of consciousness, say, even before the creation of the first cell, the determinations are almost exclusively cosmic, as there is still no identity in the form of a personal soul. In some sense, this is where the integral cosmology meets with the panpsychist view. Particles, atoms, molecules, and their non-living aggregates are concentrations of cosmic consciousness, but the subjective experience is not inherent in these insentient expressions. These are only the physical substrate for the cosmic experience manifesting on that physical plane. It is not really the view of *'panexperientialism'*—that is, the view that individual particles, atoms, and molecules also have a degree of subjective interiority. The micro-subjective micro-consciousness arises and persists on a different plane of existence, independently from the electron, atom, or molecule in the cosmic mind and cosmic life planes. One might say that it is a Super-Subject—call it God—that experiences the clash of forces of every particle in the Universe as its own experience without a subjective individuation. In a rock, there is no Jivatman, let alone an evolving soul, as an aggregate of solid matter is not driven by any life force and is not subjected to any evolutionary process. However, it is an involved nescient form of consciousness—that is, an act of extreme concentration of the overmind projected into the physical plane—resulting in a collection of subatomic particles whose interaction is determined by a mechanical Will-Force. However, this Will-Force does not reveal any conscious action or self-awareness. If the molecule, atom, or electron had any primitive and elementary experience, as the panpsychist contends, it would be the localized experience of a cosmic Mind or universal consciousness in itself. Physical force is a blind force that senses, seeks, and feels, but it is not the force of an individualized subject identifying with a tiny particle. There is no helpless and passive sentient identity that has an experience in the electron, but there is nevertheless the cosmic consciousness that experiences its material 'universal body' *through* each electron. It is what the mystics report when they lose their sense of subjective separation and melt into the universal or cosmic consciousness, which is that of a cosmic spirit and cosmic Nature perceiving all the forces within itself—a cosmic consciousness that is as much conscious as a whole and with every event, all forces, impacts, movements, from clusters of galaxies down to the electrons, as an impersonal experience of and in itself.

It is this aspect that Spinoza perceived or the analytic idealist conjectured about—that of a Divinity which has yet to become conscious of itself by an apparently blind and instinctive self-finding process of evolution. In the integral cosmology, instead, this is only an aspect of the Divinity, not its entire nature. It is the projection in the domain of Para-Prakriti that renders the omniscient as seemingly ignorant. It is the Divine that puts up the undivine mask and plays its part, luring us into the illusion of Maya.

Therefore, initially, the influence and function of our soul-entity have yet to appear. An unevolved soul will be more prone to the clash of forces of the lower-Nature. Through this adventure of consciousness, the psychic accumulates experiences and evolves, beginning to rise from a nescient to a subconscient identity, and from the subconscient to the vital unit, from the vital to the mental personality, etc., which will reflect itself also on the physical plane, say, by an accumulation or, rather, a new material combination from the cell to the plant, from the plant to the animal, from the animal to the human, etc.

On the scale of multicellular organisms, such as plants, the spiritual identity of the being is still far from being formed. Nonetheless, plants express and have a sensibility to psychic properties. Their beauty in form and color and their tendency to grow towards light is not just a human interpretation and a purely mechanistic photosynthetic reaction but are a psychic expression on the physical plane. They express an aspiration towards the Light of the Spirit, whose Real-Idea on the physical plane is expressed as the electromagnetic force. Plants have a longing for harmony, beauty, and good. This is something Plotinus recognized [109]: Beauty is a power of the One flowing into things, a principle of unity emanating from the eternal Good and that we, as souls, recognize in forms because they reflect what is in us and, deep down, is like us. We might say that beauty in material forms mirrors God's delight. This is a perception and conception well in line with Goethe's way of seeing. In the frame of an integral understanding of life, plants are psychic but still don't possess a formed psychic. They are mainly 'channels' of vital energy, multicellular expressions of individuation on the physical plane of a vital cosmic plane with a psychic touch.

The more complex forms of life, such as animals and the human being, instead are the expression of a more formed and evolved psychic being. However, we might consider different gradations here also: The soul of a squirrel is less evolved than that of a chimpanzee, and that of a non-human primate is less than that of a human psychic being.

A common misunderstanding interprets this as a moral judgment or an anthropocentric sense of superiority. Let's not forget that in the integral perspective, humans are only a transitional species, as many others were before us, and we are far from being at the top of creation. We may appear to the gnostic being as the chimp does to us.

---

[109] The reader is invited to read one of the most beautiful philosophical inquiries on the origin and nature of beauty in Plotinus' Ennead I.6: "On Beauty".

Indeed, one could argue that, in a certain sense, the mental being has even regressed because, contrary to other animal species, its cruelties, wars, genocides, tortures, and so many dark chapters of history, only testify to its 'perversion.' On the other hand, humans have developed art, science, music, philosophy, religions, and a whole set of cultural activities that no other animal species developed. Moreover, there has been a change and evolution of the human culture throughout the centuries that was not imposed by the environment or some genetic mutations, but was purposefully triggered by the intention, will, and desire of humans themselves. Something that in the rest of the animal kingdom never occurs. This dichotomy has always sparked controversies between atheist materialists that, by pointing out the former aspect, conclude that humans are just a species as any other, and those with a more religious inclination, who emphasize the second aspect, claiming that humans are 'superior' or, at least, that humans are 'exceptional.'

While, from the integral perspective, there is no contradiction. Humans are not just a change in genotype and phenotype, but got a step further than other species, at least on this planet, by rising from a vital to a mental plane. Nevertheless, the homo sapiens has momentary disengaged from its psychic identity because of other evolutionary necessities, such as the exteriorization of consciousness that allowed us to focus on the control of the material world and, especially, to develop the mind. It is an apparent regression on one plane and a true progress on another plane. And yet, that doesn't make the human 'superior' (whatever that might mean), but rather confirms its transitional character. Whatever 'exceptionality' we possess, sooner or later, we will be surpassed by another species.

Thus, it is not about a presumed 'superiority' of a soul but about the degree of psychic progression, which is determined by the amount and quality of experiences a soul goes through and the lessons learned by pleasant or unpleasant life events during its bodily transmigrations. And, it can hardly go unnoticed that also among the same species, the soul maturation can be quite diverse. Young and old souls, both having human bodies, can be easily recognized independently from their physical age.

It is only later, at a much further developmental stage, that the human soul can come to the front by the psychic transformation; and the psychic will be able to guide the outer being. In its evolutionary journey, it will also acquire a progressive control of the mental and eventually contact the higher-than-mental sheets. Also, the material aggregate, the physical body, and its organic conditions will progressively get under the control of the higher Nature's domains of consciousness through the inmost psychic action together with the spiritual descent from above. But even before that important transformation, it determines or at least identifies itself with the structure of the physical organism as well. It is to be expected that a human psychic won't incarnate in a worm, plant, or animal. This is not because there is anything 'better' or 'superior' in the human but because it has already made the experience of these lifeforms. Every psychic individuality acquires a body according to its evolutionary necessities, which will be met by the new experiences. In this perspective, there are reasons to doubt the official narrative according to which not only our physical but also our psychological traits are determined by genetic factors. That a specific gene pool correlates with a particular psychological trait may have lured us, again, into the correlation-causation fallacy. In this integral life-cosmology, it is the psychic being which chooses our bodily existence, eventually also down to the microscopic level, by determining a specific genetic organization rather than the other way around.

Moreover, from the integral standpoint, natural selection is not just a sifting procedure. It is also determined by the subconscious energy in the type of species that answers to the need of the environment, while in another species, it remains unresponsive and, therefore, unable to survive. The reactions of a living creature to external stimuli do not depend only on its physical constitution and chemical status but also on how it feels, perceives, and experiences the environment from within. It is not only a nervous sense-based automatic reflex of the body to external contacts, but also a response that comes from an interior will, volition, and urge. In this sense, natural selection is seen as the clear sign of the workings of a life-energy and a psychology at work—a psychology that is influenced by the psychic element deep down. Natural selection is the repeated attempt of the lower-Nature to combine, mix and then dissolve several potentialities and types, like in an everlasting alchemic process that, however, subconsciously tries to realize an innate bliss, harmony, power, and delight that was lost in the involution in matter.

The conventional scientific neo-Darwinian theory of evolution is too full of unknown factors to be a definite theory because it completely ignores and even denies the existence of such 'occult' determinants. But from the perspective of an integral cosmology, life is not something formed only from the forces without; it is also a creative urge from within, an awakening and a call that descends and activates the myriad forms in which it casts itself in the material kingdom. It is a self-manifestation of a self-knowing Force in matter which has forgotten itself by a descent into ignorance. Because of this forgetfulness of its own origin, it is subjected to the blind play of physical forces. And yet there it secretly labors under its own impulse to manifest the form of itself. In its deepest significance, life is part of an adventure of consciousness that searches for the delight of being in the material realm of an all-powerful Omniscience.

7. Seeing Integrally with a Synthesis of Knowledge

While Aurobindo's integral cosmology can only be fully realized once the 'seer' has ascended the highest steps of the ladder of consciousness, we, as 'ordinary humans', can, nonetheless, recognize by a comprehensive seeing a la Goethe, the development of a human's knowledge represented by its philosophies, sciences, and history from a more integral perspective than before. The theories, ideas, realizations, or insights of the diverse natural philosopher, scientists, and mystics throughout the ages, cultures, and continents, which previously seemed irreconcilable and even mutually exclusive, now, in the light of an integral view, appear in a much more coherent picture that harmonizes all knowledge inside an integral paradigm.

It would be impossible to do this here for all knowledge, but that wouldn't be useful either. What it's all about, at this point, is not mere accumulation of content, but rather about developing another way of an expanded seeing—an integral seeing upon the human's path in the course of history and its intellectual, spiritual, and practical achievements as a whole in an evolutionary perspective.

To see what this means, let us work out this integral seeing by some examples that outline how we can look upon reality and the diverse philosophical and spiritual directions of the past to modernity from a more comprehensive perspective which can lead us toward a larger synthesis.

To a certain degree, we have already done this by integrating the concepts of evolution and of physics such as force, will, space and time inside an enlarged integral teleology in the previous chapters. This placed Aurobindo's cosmology in a modern perspective in line with the findings of contemporary science. Let us do something similar by standing on the shoulders of the past intellectual and spiritual giants of the Western tradition and see how these can be integrated with the Eastern tradition in a vaster synthesis of knowledge.

One aspect that always shows up throughout the history of human spiritual philosophy is that of an impersonal theistic monism. Heraclitus' maxim "from all things one and from one all things", Parmenides' uncreated and never-changing timeless and featureless "Being", Plotinus' uncaused but self-caused One, Meister Eckhart's God in us and in all living beings where "all things are God itself", and the vedic all-pervasive Brahman which is all there is, the same Vedantic "One without a second"—all these are examples of a recurring theme, a rather extreme form of impersonal monism (not to be confused with the personal God of monotheistic religions) much too ubiquitous throughout the continents and ages to be coincidental. Even the striking similarity between the Christian Trinity of Father, Son, and Holy Spirit, with the Upanishadic Satchitananda ('Existence-Consciousness-Beatitude') can hardly go unnoticed. We might suspect that there must have been a cultural exchange and cross-fertilization during prehistoric times, which could have determined these philosophical congruences, as indeed some historians claim. But it is unlikely that this could be the whole story. Because, if so, it is difficult to see why the Hellenistic, Christian, or Sufi mystics, who reported the same or similar states of transcendence that the Hindu or Buddhist were talking about, never felt it necessary to mention each other. These were not merely theoretical constructs or philosophical speculations, but the description of a lived experience, an authentic spiritual realization that some humans in several different cultures sooner or later discovered independently, and which don't necessarily need to be inspired by other external religious, spiritual or philosophical schools.

However, what these 'spiritual cosmologies' are lacking, or at least didn't emphasize, was the integration with the personal Godhead. The impersonality of an Absolute remains separated from the more common personal God. The mind thinks in terms of mutually exclusive polarities or considers one aspect superior (or inferior) to the other. In an integral metaphysics instead, the impersonal ultimate Reality of the mystic or the personal Divinity of the God-lover are two faces of the very same Oneness that is neither personal nor impersonal. It is Something that transcends these limitative human conceptions entirely. We can relate to the Divine as an almighty and all-loving personality or as a featureless ('nirguna') Brahman or an 'emptiness' that can be realized in the state of Nirvana. All these forms of self-transcendence lead to different realizations that reveal the different faces of the one and unique Divine—the Personal, Impersonal, and Beyond all are true; none excludes the others.

The same can be said for the transcendent and immanent dichotomy with which such different spiritual philosophies characterized all existence. Even though with different nuances and fortunes, what we label as 'panentheism' nowadays is a worldview that emerged throughout all times as well. Spinoza's Deity manifests its infinite modes of existence, of which matter (the 'res-extensa') and mind (the 'res-cogitans') are only two possible self-expressions of the one' substance'. The interpenetrating God of Meister Eckhart, the Anima Mundi that 'pulsates' through everything of Giordano Bruno, the universe of Berkeley where the universe is the Mind of God and all things a Thought of God, the World-Soul ('Weltgeist') of Schelling and Hegel or the 'World as Will and Representation' of Schopenhauer, Goethe's way of 'seeing' the wholeness of Nature that perceives in all living beings an archetype, are all different intuitions of a universal immanent or transcendent universal Presence. The Eastern tradition developed a similar panentheistic worldview that is most present in the philosophy of the Upanishads that, as we described earlier, conceives of a cosmology of universal and individual 'bodies' or 'sheets' ('koshas'), which, de facto, represent a hierarchical structure from the Transcendent to its immanence in the most subtle to the grossest forms of 'substance'. As we have seen, some modern philosophers of mind also delineate panentheistic models, such as cosmopsychism or Kastrup's analytic idealism.

But all these panentheistic cosmologies don't go much further than a generic conception of 'modes', 'substances', 'self-excitations' or more or less 'subtle bodies' whose function, properties, and raison d'être remain obscure. Furthermore, except for the Vedic philosophy, they remain models of low-dimensionality, which may make them appear incompatible and force us to choose one or the other. What does Spinoza's 'substance' have to do with Schopenhauer's Will? And how does this Will relate to mind, life and matter? Everything still looks a bit too fragmented if we conceive all these as separate powers and qualities, or too conflated if we conceive them as synonymous.

In the multidimensional integral cosmology, these questions dissolve ab initio because Consciousness, Will, Force, and Knowledge are all terms designating the very same Entity, which has self-determined itself in the universal manifestation as Supermind and by its Real-Ideas organizes the existence and dynamics of all things, from the overmind down to mind, life, matter and below, expressing all the qualities, forces and creative potentialities at each level of existence. The otherwise no-better-defined 'substances' or 'excitation modes' reveal themselves in an integral vertical and horizontal system of being, each with their function, power, and specific nature that altogether determine the cosmic evolution modern science explores.

Another recurring theme is that which perceives a 'double-Nature' working in and through us. Meister Eckhart's distinction between a "Nature that creates but is not created" complemented by Giordano Bruno's Anima Mundi, the World-Soul, contrasted to God's Soul, and, of course, Spinoza's "Natura naturans" vs. the "Natura naturata", sound familiar and look to be precursors to Aurobindo's higher Supernature ('Para-Prakriti') doubled by the lower-Nature ('Apara-Prakriti'). The integral vision of Nature, however, goes further. This perspective not only reveals these two polar manifestations in a multidimensional aspect, with each plane of being having its functions, powers, and qualities but also reconnects each other: The lowest nescient plane finds the highest divine Light. It is at the bottom of matter where the full Power of the Spirit resides and waits its turn to manifest fully once life and matter are ready. Meanwhile, in the previous traditions, the double aspect of Nature was just acknowledged but not explained; in the integral cosmology, it acquires a deeper meaning and purpose. It is the Spirit which, through the higher superconscient Nature, calls on the lower sub- and inconscient Nature and lifts mind, life, and matter—a 'call' whose effects we name 'evolution'.

Perhaps the most distinctive and largest extension of the integral cosmology pertains to the 'ladder of consciousness'. While most philosophers seem to not have ascended cognitive states beyond the mental or the higher-mental, and mystics were usually scarcely interested in natural philosophical questions and only secondarily reported higher forms of knowledge other than that which allowed for an escape into the Transcendent, the integral yoga of Aurobindo incorporates them all and goes much further than that. Plato realized a realm of Universals intuitively, Spinoza distinguished between knowledge of the first, second, and third kind, while Hegel's 'Idea-in-Itself' and Goethe's archetypal seeing seemed to have captured a glimpse of the Real-Idea in Nature as the workings of a World-Soul. Husserl intuited of 'essences', T. de Chardin envisioned a future emergence of a 'Christ-Consciousness', Gebser conjectured an integral structure which will integrate all the previous stages of consciousness that have determined human's history, Heidegger longed for the 'free unconcealment', Deleuze for an ontology to 'palpate the unknowable', Barfield spoke of the 'unrepresented' and Kastrup speculated that we perceive the world only as a 'desktop representation'. All these testify to the emergence of a trans-rational consciousness lurking behind thought and senses. But in the integral structure of Aurobindo, these are only the beginnings, an intimation of a first step out of mentality. There are cognitive functions that go far beyond the higher-mind or some intuitive flashes. The destiny of the human race is to realize the Supramental consciousness and infuse it in matter. Our future will go beyond a spiritualized philosophy or just an intuitive Goethian way of seeing the world. These have only scratched the surface of a much vaster realm of ascending states of consciousness. The integral yoga describes all the planes beyond mind in detail—that is, the higher-mind, illumined mind, intuition, overmind, and, at the highest summit, the Supermind.

But, as all the wisdom of all nations and peoples realized, there is also the knowledge of the heart. For Aristotle and Plato, there is a 'first energeia' — that is, a soul in every living being—while Plotinus felt the expression of the Good above all the ideas as the source and principle of beauty, not just a human figment without meaning. Spinoza recognized how there is in us something that can awaken to a constant love for God and to a state of blessedness. All this makes sense in an integral vision as well. It is our inmost soul, the psychic being, the center of love, beauty, and harmony behind the heart. But for the integral cosmology, the soul acquires a new scope in an evolutionary dimension–not just a coarse-grained metaphysics of soul and body, but a soul-centered being that grows by the accumulation of experiences in the cosmic adventure of consciousness. It is an emergentist and evolutionary spiritual cosmology extending R. Steiner's, T. de Chardin's, or J. Gebser's vision where the so-called 'evolution of consciousness' is twofold: gnostic and psychic.

From this integral perspective, the impulse, the will, the force, or the desire which emerges from within us reveals its deeper nature. Aristotle 's self-organizing principle, the entelechy, Bergson's vital and creative force that drives all living beings, Goethe's primal force in life ('Urkraft'), Deluze's 'machine desire' and Schopenhauer's Will which acts not only cosmically but also at the level of the individuated organism—all are manifestations, in one form or another, of a primal soul-force. The integral vision, most importantly, however, further distinguishes between a real selfless 'soul-desire' from the vital instinctive selfish desire of self-assertion and strife. Those who tried to reduce all life to the 'selfish gene' were not so off-track, after all. But, as it is so typical for human reason, once it discovers a partial, limited and fragmentary truth, it can't resist the temptation to elevate it to a final principle supposedly ruling all there is. It is here where one sees the power of the integrative cognitive act. It does not finalize, rather widens and integrates perspectives: in this case, the most material reductionist and deterministic reality within a much wider and comprehensive holistic intuitive vision.

Also, the distinction between the higher-Nature intimation behind the veil of appearances and the lower-Nature action must be kept in mind. That's where ordinary intuition so often gets confused while, here, clarity is reestablished. The psychic being is a transcendent entity that acts from behind only as long as the being is not sufficiently evolved to bring it to the front. Meanwhile, the 'desire-soul' that determines our character and personality is not a soul at all but, to a large degree, still a blind and mechanical instinct of the lower-Nature of Ignorance. This is not hairsplitting but, to the contrary, introduces clarity and an entirely new cosmology where there previously reigned a coarse-grained misrepresentation of our spiritual nature and the World, which has led much too often to inevitable paradoxes.

Also, the true nature and evolutionary role of matter are revealed in this integral vision. Intuition was not absent throughout ancient wisdom. Aristotle pointed out how matter looks like something that resists and opposes inertia to a forming force. Plotinus wrote of matter as the ultimate separation from the One, G. Bruno envisaged a spiritualization of our bodies and Bergson's 'creative impulse' meets the vital resistance and habits of an inert matter as well. These and other authors throughout history felt some undefinable role in the mechanical behavior of matter as if it is withholding a deeper truth and secret we only superficially perceive.

This secret is disclosed in the integral cosmology. While confirming this old wisdom, it reveals matter's mechanical resistance as the expression of an involved Spirit forgetful and obnubilated of and in itself. It is a Spirit that has plunged itself into the Night of the inconscient universe—the plane of being we call the physical universe—and is now re-emerging by process of self-discovery. The integrality of the metaphysics of Aurobindo not only embraces evolution which previous forms of spiritual practices or philosophies ignored or deemed of secondary importance but also complemented it by a preceding involution, which the Darwinian sciences ignore and even deny. The integral seer who practices cell yoga discovers that deep in the inconscience of our body is a mind of the cells—not a theory or philosophical abstraction but a lived experience that reveals the existence of a consciousness working deep down into the cells. This is something that modern science has now discovered, as we have amply documented in Pt.I on the basal cognition of the cells and plants. However, in this integral framework, one goes much further than a scientific discovery. A wholly new dimension which reveals an integral truth that was previously only a speculation or, at best, a distant and not well-formulated intuition now becomes the most solid and concrete fact: the existence of a divine Force that can be awakened in matter and lead us to a new species, the gnostic being.

Bergson had the right intuition about the principle of reproduction and death: reproduction is an invention of Nature to conquer death. The integral synthesis comes to the same conclusion and even goes further: Nature's aim is not just that to preserve a biological organism but to evolve a soul and manifest the immortality of the Spirit in an individuated material form. The organismic immortality is not an aim, but a means to express (the still unexpressed) Satchitananda's delight of Being on a material plane by the materialization of the psychic being.

One could continue with this 'integrative exercise', noting how panpsychism was yet another distinctive intuition that captured a partial reality which, however, gains a broader significance in the synthesis of knowledge. In integral cosmology, panpsychism is not irreconcilable with, say, Leibniz's monads, or cosmopsychism and analytic idealism. All are looking upon the same reality through different cognitive lenses. Indeed, here the 'monads', the 'psychic atoms', are not seen inside a natural framework in opposition to that of a holistic cosmic consciousness or Mind at Large, rather are comprehended both as two complementary aspects: They both are true because resulting from the 'forgetful exclusive self-concentration' of the cosmic Mind in itself. And Leibniz's idea of the monad's desire-driven appetition or the Aristotelian goal-oriented entelechy, is explained by the existence of the vital forces—that is, the Conscious-Force expressing itself on a vital cosmic plane of the Nature of ignorance, Apara-Prakriti. Kastrup's (very original but, perhaps, a bit too westernized psychoanalytic) idea of the dissociated alters with their dissociative boundaries, in the integral metaphysics are the reflection of the 'divine spark' (the central being–or Jiva–the psychic entity, and the psychic being) manifesting through an embodiment in and through the cosmic Life-plane and cosmic Matter-plane, respectively. The main difference between panpsychism, cosmopsychism, analytic idealism, and a more integral metaphysics is that the former lack a clear evolutionary dimension which is one of the central motives of the latter.

These were only some examples of how we can see comprehensively by adopting the integral cosmology of Aurobindo towards science, spirituality, psychology, and natural philosophy of all humankind throughout all times in a grand vision that makes a synthesis of knowledge.

This is not to claim any superiority of one teaching over the other. Also, the integral yoga, the integral science, and an integral vision of today may well be superseded by something that might go further. But, at the present evolutionary stage, this appears to be the most integrative theory that makes sense with modern findings. It is not necessary to know everything, and it is not even necessary to learn in a multidisciplinary manner–something we nowadays all cherish but rarely realize in practice. What is necessary is that we learn to see integrally, that we practice another way of looking at the world, Nature, and ourselves. At the end of the story, we always discover the same principle:

Knowledge isn't a matter of summing the parts; it is a matter of seeing the Whole.

III. Concluding Remarks: Modern Scientific Delusions and the Coming of the Subjective Age

1. The Pragmatic Side of Post-Materialism

"When you perceive nature only through the mind, through thinking, you cannot sense its aliveness, its beingness. You see the form only and are unaware of the life within the form - the sacred mystery. Thought reduces Nature to a commodity to be used in the pursuit of profit or knowledge or some other utilitarian purpose. The ancient forest becomes timber, the bird a research project, the mountain something to be mined or conquered. When you perceive Nature, let there be spaces of no thought, no mind. When you approach Nature in this way, it will respond to you and participate in the evolution of human and planetary consciousness."

Eckhart Tolle [343]

At this point, the question arises: What is all this prevalently theoretical and 'post-material philosophy' good for? For example, whatever answer we give to the question of whether the mind, the brain, and consciousness constitute an identity, we might regard these musings as having no practical consequence for our everyday lives. Replacing a one-dimensional physicalist ontology with a multidimensional emergentist spiritual worldview could look like an exclusively academic debate from which one might not expect anything having a practical fallout for ourselves or society.

For an analytic mind, the theoretical framework of an integral cosmology may look like a weird mystic phantasy. In particular, the scientific mind adhering to principles of parsimony will see in it a plethora of unnecessary pluralities and entities of a contrived cosmological, metaphysical construct to which it can't meaningfully relate. There seems to be no reason to resort to presumed 'higher-minds' or 'superminds', contrasting evolution with an 'involution' or to a soul another 'evolutionary soul', subdividing things in a hierarchy of vertical planes of consciousness and horizontal 'concentric beings' or exercise us in what seems to be only hairsplitting distinguishing between the subconscious and subliminal, etc. Why should we build up such a theoretical castle that seems to make things only more complicated? In a certain sense, one may perceive this almost as violence, a forced overinterpretation of reality—too much, too farfetched, too many ingredients making up a cocktail of a philosophical overdose, rather than a sound theory.

But this is only because we are accustomed to living in, and seeing through, the mind. There is still too large of a chasm between what we have been told and what Nature tells us, between what we perceive and what is, between our intuition and the order of things, between us and reality.

Imagine for a moment Nicolaus Copernicus, who speculated about a heliocentric cosmology envisaging the Sun, rather than the Earth, at the center of the Universe, being told that he should add other layers of complexity to his theory—that is, to consider the Sun also not at the center of the Universe because it is orbiting around the center of our galaxy. Furthermore, our galaxy is moving inside a local cluster of many other galaxies and these, in turn, are again only a cosmic substructure of a large cluster of local clusters, and so on. He would have legitimately rejected such a model as an unnecessary and too contrived hypothesis for which he could see no reason to resort—and rightly so because at those times, no evidence would have suggested such a reality.

But that's not the historical circumstance in which we find ourselves. This whole treatise went through several real-life facts, experiments, and scientific evidence that invites us with increasing pressure to make a choice. We are at a crossroads where we must choose between a linear mono-dimensional physicalism, which is now resulting in stagnation of a never-ending circle of a life where everything changes permanently and yet nothing seems to nurture our inner needs, or open ourselves to vaster vistas and expand our view beyond the linear thinking of a little mind.

The point is that reality is as it is, Nature's structure is as it is, and the Universe is as it is. It is futile to search desperately for shortcuts that reduce it to something simpler than it is to please our anthropomorphic concepts.

However, admittedly, this also doesn't answer the question: What could all this spiritual worldview be useful for in real practice, here and now?

After all, who cares if it will take another thousand years until a gnostic being appears upon Earth? If this is the case, isn't it a bit too early to think about gaining any profit from it? Are the philosophical, post-material, and spiritual reflections we outlined so far good only for some mystically inclined minds who secluded themselves in a solitary cave on top of a mountain, or could this deeper knowledge also have some more concrete applications?

We contend that the transition from a mind-matter-centric to soul-centric worldview will not only determine a new conceptual paradigm shift but also lead to very concrete, practical, material, and even economic transformations of humanity—eventually, even more than the scientific and industrial revolution did. Because what we do is largely determined by what we believe, feel and know and how we think. The quality of our social life, our interactions with others, the projects and dreams we nurture, the policies and social and economic choices we make, and the dynamics of the nations, depend first and foremost on the underlying metaphysical or spiritual assumptions with which we more or less subliminally work. Our social, economic, industrial, and educational systems are deeply entrenched in our way of seeing the world and are strongly affected by our more or less unaware assumptions based on philosophical and metaphysical concepts, and not least our ideological mindset.

If we believe that we are only 'meat-robots' without a soul and that what defines us is not much more than a complicated neural network in our skulls, we will develop a conception and a practical approach to reality, especially in the fields of medicine, psychology, and education, that will more or less implicitly embrace a particular type of practice to the exclusion of other options deemed as superfluous if not altogether nonsensical. If, however, we realize that there is also a spiritual dimension that does not exclude the materiality of our existence and extends it to a multidimensional reality, placing the soul at the center of human nature rather than a gray and wet blob of neurons, this will inevitably lead to a completely different type of practical approach to reality.

If we believe that a strictly material neo-Darwinist process of evolution explains where we come from within an apparently soulless universe that has no meaning and purpose, branding also the question for purpose itself as meaningless, we will develop an ideological background leading to an attitude that will also favor a certain conception of human nature and its social and economic dynamics. Meanwhile, if we go inward and discover that we are not just a physical organism but first and foremost an evolving soul in progress inside a universe that has a meaning and a life with a purpose, our conceptions of how a society develops, progresses, and must be organized in all its practical and dynamic aspects changes completely.

If we believe that we are only the outward expression of selfish genes such that our feelings and emotions of empathy, compassion, love, and commonality are only a biochemical reaction in a network of mechanistic cells whose main function, according to this narrative, is to guarantee an evolutionary advantage functional to group survival, we will inevitably conceive of models of society, economy and human interrelations that see the collaboration among humans as a necessary condition for individual survival, but not as a base for a true inner soul-identity. This inevitably will also cascade into myriad very concrete and material consequences at the practical level. Meanwhile, if we feel inwardly a pursuit of an ideal of human unity that reflects a deeper spiritual identity, our relationships, and very practical and outward behaviors and approaches towards other humans and animals, other nations, other cultures, and religions would be of a very different type and quality.

Of course, if this non-materialistic worldview isn't based on an inward spiritual realization or, at least, on an intellectual acceptance and openness that can go beyond preconceived and dogmatic thinking patterns, this won't lead us to a change for the better either. The detrimental effects of religious and anthropocentric authoritarian belief systems have been amply illustrated throughout human history and didn't rule better our destiny than materialistic industrialism did.

The worst option, unfortunately, has become a reality to a large degree: that of communion between a pragmatic secular materialism and a religious mentality that still sticks to blind faith. In fact, if we believe that Nature is only a giant automaton ruled exclusively by physical and biological mechanisms, eventually augmented by some religious dogmas that tell us how it has been given to us by a humanized God for our complete and unlimited disposal, we will treat it with a destructive and ultimately self-destructive approach that has led us to the present environmental crisis. However, if we learn to feel and realize an inner connection and intimate relationship with Nature as the unfoldment of one Being that is also our being, then a completely different understanding of how we could and should behave towards Mother Nature and how to treat its life forms would grow in us naturally. The preservation of the environment would expand from a still too technocratic conception that sees only in science, technology, and politics the solution to the problem, to a post-materialist and post-religious spiritual understanding that also recognizes the urgent need to re-find our inner connection, love, and oneness with Nature as part of our deepest self. Embracing worldwide renewable energies while leaving behind fossil fuels and protecting the environment is a necessary but insufficient strategy if it is not carried out with an inner perception and understanding of our place in this multidimensional evolutionary cosmos with a sense of connectedness with it and all its creatures.

Hard sciences like physics and biology, but especially medicine and psychology, economic models, and our present environmental mindset will change dramatically if we opt for a wider and richer view of reality. The human body will not be seen just as a machine that must be fixed only through chemical and surgical means but will also be recognized as having a body consciousness that responds to mental, subconscious, subliminal, and subtle physical stimuli. The human psyche will no longer be considered a mind-body-centered epiphenomenon but, rather, a soul-centered multidimensional being. The economy, finances, the exchange of goods will be recognized as a flow of power supposed to serve the multiplicity in unity, rather than a clash of selfish beings competing for resources, all aiming only for the survival of the fittest. The distribution of wealth and goods will be transformed from a system ruled by individual egoistic forces to that which envisages an economy of the common good, and yet allows for free individual entrepreneurship.

To put it more succinctly: Our theoretical, philosophical, and metaphysical frameworks determine our actions, which regulate our very individual and also collective social and material forms of practices and expressions. On one side, the choice humanity will have to make is between a perennially externalized physicalism in which only a strictly analytic and rationalistic reasoning is allowed that, at times, might come to terms with an equally narrowminded religiousism with no true light of knowledge or, on the other side, a non-religious spiritual worldview that, from within, expands its vision beyond. Neither sciences and humanities nor our social and economic structure will remain unaffected by this choice; in the long run, this will determine human destiny.

2. Nature's Unfathomable Complexity

What is escaping our awareness is our lack of awareness itself. For example, we still largely underestimate the complexity of life and the cosmos. We know things are complicated but nevertheless do not realize this fully. We have not really recognized the unfathomable complexity of Nature's workings.

Physicists, like Descartes and Laplace, once thought that the day would come when everything would be described by elementary particles—in the sense of minute material pieces of matter—bouncing around as in a huge billiard. This may not have allowed us to predict everything, but at least it would have been a description of the world reducible to simple elements of matter, sort of like marbles zigzagging all over the place and uncomplicated few elementary processes. The truth of the matter (no pun intended) is that this naïve worldview had to give way to the modern standard model of particle physics, which is one of the most complicated intellectual and mathematical abstract theories the human mind ever conceived. In particle physics, one works with the so-called '*Lagrangian*' functional—an equation that describes the dynamics of a system of particles

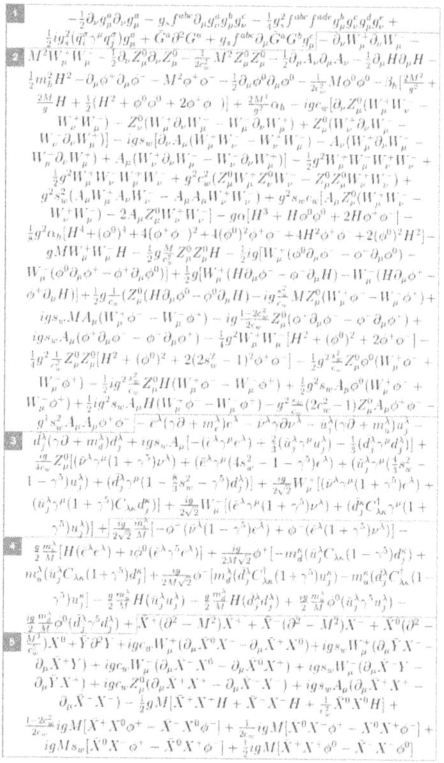

Fig. 137 The full Lagrangian equation of the standard model of particle physics. (For more details, see [377])

and its energy states. Fig. 137 is a snapshot of the full Lagrangian that contains the interactions of the nuclear forces and the electromagnetic forces, the Higgs boson—that is, the quantum field that gives particles a mass—and several other mathematical corrections. Of course, we won't bother the reader by trying to explain it. In the context of what we are discussing here, we believe it to be self-explanatory.

Biology met a not-too-dissimilar destiny. During the past hundred years, the extent of organization biologists discovered in living things has increased exponentially. We now know that organisms are zillions of times more organized and functionally complex and information-rich than Darwin ever dreamed. There is no sign of this trend slowing, and we have every reason to suspect that each decade will continue to reveal ever-more-astonishing levels of organization and dynamic functional wizardry in living things.

The model of a living cell being a simple cytoplasmatic bubble containing a nucleus with a DNA molecule supposedly containing all the information about what we are with mitochondria and a few other organelles performing some energetic and functional task also had to give way to a much more complicated picture. Nowadays, the sciences of biochemistry, molecular biology, genetics, and cell biology have become such a complicated discipline that probably few, if any, can master it in all its intricacies. Moreover, the more we examine the single constituents of life, the more exhaustingly difficult it becomes to overview and assess its structure, functions, and processes.

To begin with, the concept of the DNA molecule made up of serially aligned genes–that is, a discrete region of DNA producing a single protein in a cell– turned out to be a much too naïve and simplistic human idea. The concept designating a 'gene' has become uncertain and fuzzy. Now it is known that a single length of DNA–that is, what was formerly imagined as a unique sequence of genes–can be transcribed in multiple ways to produce many different RNAs. The same DNA section may code for proteins or transcribe regulatory RNAs, depending on where the transcription machinery starts and stops. This means that the genes for protein-coding overlap with the genes of regulatory RNA. Different strands of genetic information can overlap with each other. There is no such thing as a single definite gene. Moreover, the code for another biological function may run in the opposite direction. Curiously, the genes producing the regulatory RNAs are essential for several other genes: The DNA contains instructions that modify its own set of instructions. To make things even messier, the genetic code for a protein can be scattered far and wide around the genome, sometimes even on other chromosomes.

If this sounds confusing, be assured that it is. Imagine a book that can be read from the left to the right and from the right to the left at the same time, where words with different meanings overlap with each other, that can tell different stories according to where one starts in the text, and where the information is scattered non-serially throughout its pages. Such an apparently unreasonable messy way to encode and read information is taken by some as evidence contrary to an intelligent design in Nature. However, we may equally state that precisely this overwhelming complex language may have some advantage compared to our linear thinking and is the sign of a super-intelligence at work that may have no issues in encoding and reading information in such a manner.

Another classic example that illustrates the complexity of our biological nature, could be that of the protein folding problem. After over half a century of research, scientists are still struggling to find out how strings of amino acids wrap together to form intricate macromolecular three-dimensional shapes into functional proteins. There is a huge variety of proteins, each with its complicated 3D structure, responsible for some functions, such as repairing and building tissues, building cell

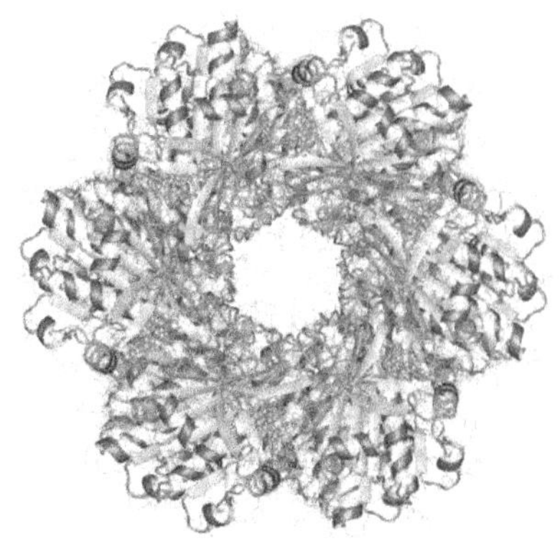

Fig. 138 *Molecular structure of the bacterial clock protein.*

membranes, controlling metabolic reactions, modulating cell signaling, transporting molecules or other proteins, etc. For example, Fig. 138 shows the structure of a protein called '*clock protein*', regulating the circadian rhythm of cyanobacteria. There are more than 20,000 types of protein molecules in the human body, each having a different functional specificity.

We regularly read of seemingly spectacular breakthroughs in this field. For example, in July 2022 Google announced that its AI program AlphaFold can 'predict' all the 200 million protein structures from 1 million species. [344] This is, indeed, an impressive achievement. In a sense, we consider it an example of what AI really could be useful for: Not so much to create humanoids or sci-fi conscious computers, but rather to fathom the complexity that the human mind can't. The announcement made headlines everywhere and provided for much excitement. But, in most cases, the headlines didn't tell the whole story. 'Predicting' means what it means: We have predictions, not real protein structures. In the context of IT, 'predicting' means that a computer program 'forecasts', and 'estimates' by making gazillions of (more or less approximate) calculations and simulations of something on the base of what we know. Sort of a weather forecast that might be pretty accurate on short-time periods, but we always must keep in mind that there could be considerable deviations between calculations and reality. It is about speculative calculations based on the information that has been fed into a machine. In fact, only about 35% of AlphaFold's predictions are deemed to be highly accurate, while 45% are accurate enough for applications, while 20% might turn out to be wrong. That is, it is a tool that gets it wrong 1 over 5 times. This can, nevertheless, be a quite useful (re-)search engine, giving indications, and suggesting directions of research, but it can't replace the good old empirical approach: Every single protein structure will have to be assessed via sophisticated X-Ray or electron-microscopy observation. Moreover, AI does not tell us how proteins fold. We are far from understanding why proteins fold like they do. Because we have no idea how the folding mechanism works and why a specific architecture leads to specific functionality. In fact, even once we will know all the protein structures, that will not automatically tell us how to design from it new drugs. Protein structure by itself will not be more informative to design new drugs as the mapping of the genome was for designing drugs against genetic diseases. As usual, again and again, it turns out that the map is not the territory.

Another quite mind-boggling protein is the mitochondrial *'ATP synthase'* enzyme. It is a much too complicated process to be described here. To make a long story short, it may only be said that one of the functions of this macromolecule is that, on the active sites (the 'cap' in Fig. 139), it generates ATP (adenosine triphosphate) by the diffusion of the adenosine diphosphate (ADP) or adenosine monophosphate (AMP) molecules. ATP is the energy-carrying molecule that releases energy, which is converted in our metabolic processes by combining

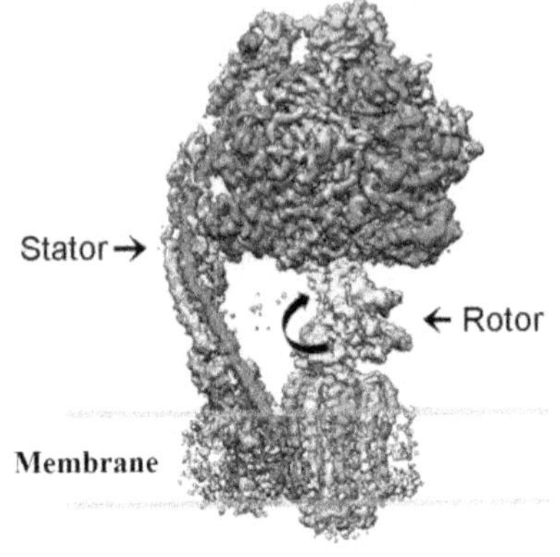

Fig. 139 Molecular structure of the ATP synthase. Credit: [390]

with the ADP or AMP, say, for instance, the energy necessary for muscle contraction. The most fascinating element of the ATP synthase is its 'rotor' structure, which spins like a water wheel while the rest of the structure is held in place by the 'stator'. This is literally a natural 'turbine' or 'rotary nano-motor'. It interfaces the mitochondria membrane with the cell's cytoplasm. Loosely speaking, the food we eat is stripped of its electrons and gets into the proteins forming the mitochondria's membrane. This causes an accumulation of positive charges—namely, protons—inside the mitochondria that must be pumped out to avoid an electrical overcharge. The ATP synthase's rotor function is to pump out the accumulating protons from the inside of the mitochondrion membrane, precisely as a rotating motor, and we have quadrillions of these in our body. The reader is invited to look for video animations that clarify the complicated dynamics (for example, see [345]).

As another example of Nature's inscrutable complexity, one could also mention the structure and function of the flagellum. The flagellum is a whip-like appendage that protrudes from the outer membrane body of (bacterial, archaea, eukaryotic) cells called *'flagellates'*, and has a locomotion and sensory function. The full structural and functional description is way beyond the scope of this treatise (for an engineering perspective see [346]). Putting it bluntly: The hook, the helical screw, propels the bacterium with the filament helix. A hollow structure runs between the hook through the rod and the basal body, passing through protein rings in the cell's membrane that act as supports (L-ring and P-ring of different compositions and the MS-ring and C-ring embedded in the plasma membrane). It is, de facto, a natural a nano-rotary engine powered by the ionic flow of the protons force across the cell membrane. It can operate at 200 to 1000 rpm. This flagellar motor can change direction of locomotion and be switched on and off almost instantaneously by a molecular protein switch in the rotor.

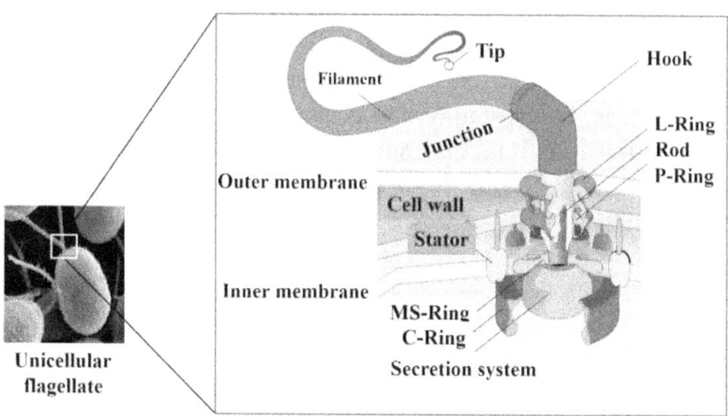

Fig. 140 Structure of a bacterial flagellum.

To get a deeper understanding of why a cell isn't a machine see also [191] or, for a more popular introduction, watch these two YouTube videos [347] and [348].

To explain the origin and formation of such hugely complicated structures that certainly can't be manufactured by randomly mixing atoms and molecules and that seem to be engineered by some unknown but quite smart architect is an almost impossible task. How such amazing objects resembling a human-made machine, only at much smaller scales, could self-assemble by natural processes is far beyond our present scientific knowledge.

Of course, as usual, we can always resort to the magic wands of the omnipotent natural selection, evolutionary advantages, fortuitous random events, plus a bonus of a billion years of time to get the job done and that, all put together, supposedly conspire to make the impossible possible. But this hypothesis has not a scratch of evidence supporting it. Of course, the evolutionary advantage of, say, the flagellum for cellular motility is self-explanatory and does not require proof, but this recognition only restates a fact without explaining it. The claim that such sophisticated and highly functional structures could come into existence only due to these processes alone remains purely hypothetical. The truth is, these are only wild speculations, ad hoc hypotheses, and far-stretched extrapolations to avoid contradictions with the dominant neo-Darwinian narrative. With natural selection and no better-defined 'evolutionary advantages', one can, in principle, explain everything and also the contrary of it. If pigs could fly, one could imagine some selective process and evolutionary advantage that could justify their existence as well.

Anyway, the question here is: How far and how well will the human mind be able to cope with Nature's complexity? We are far away not only from mastering the terrifying complexity of life at the microscopic scale but, to a large degree, also from being aware of it. The modern delusion in biology and medicine was the idea that the more we dig into the microscopic structure of living things, the more we will understand and potentially gain control over it.[110] But in too many instances, the contrary turned out to be true. Our discoveries reveal a complexity of Nature that increases much faster than our ability to fathom it. The structure and organization of Nature are much more complicated than we can imagine and simulate. Consequently, our sciences are also becoming exponentially complicated, leading to a compartmentalization that passionate appeals to multidisciplinarity won't overcome. We are quickly losing control over our own models of Nature's immense organization and processes. But the predominant culture continues to believe that it is only a matter of time, more mappings, more microscopic magnification, and more computing power before we will get to the bottom of things and finally understand and grasp the functions and processes of Life.

I believe this is only wishful thinking that is constantly dismissed by further evidence and discoveries. Every scientist knows that life is complicated but we still have no idea how complicated. We contend that it is the investigation tool—namely, mind—that is unable to cope with the full structure and the network of the process of Nature. It is not just about a lack of knowledge or progress; it is the very essence of our way of seeing that can't and will never be able to fathom the full complexity of life's origin, from the molecular to the biochemical, structural dynamics of a cell, let alone the workings of a brain. It is beyond the natural cognitive abilities of the human mind itself.

Ascending higher states of consciousness can lead us a step further instead. Once we see the world from a higher perspective where reductionism and holism, multiplicity and unity, the One and the many are not contrary, but the two aspects of that which transcends the One and the many, we might have more of a chance at coming to grips with Nature's complexity. The little scientific physical mind is a much too primitive tool, as it can't deal with such a huge atomized but interconnected and entangled reality.

Meanwhile, what could bring us closer to such a cognitive leap, is the acquisition of complex systemic thinking skills. A too simplistic linear thought

---

[110] This was also the delusion of physics in the past centuries. The above mentioned complexity of the standard model of particle physics and the spectacular failure of hugely complicated theories that were supposed to extend it to a theory of quantum gravity, eloquently showed how reductionism and simplicity do not at all go hand in hand.

still dominates our culture. From schools to research centers, or in politics, finance, and sciences, a nonlinear complex systemic view of reality still struggles to become the default paradigm.

For example, there is an intrinsic and paradigmatic fundamental difference between whatever complex machinery of a car engine and the dynamics of a living organism. The former is a structure whose functions and workings can still be described by a linear relationship of causes and effects; the latter is beyond any possible predictability because even a small change in only one of its parts triggers a sequence of effects that cascade throughout the whole system with many internal feedback processes complicating further the forecast of its behavior, let alone allow for its control. We still are dominated by a linear reductionist thought pattern that believes that modifying a system in point A will lead to result B. But Nature doesn't work like that. If we modify point A, this will cascade a plethora of effects B, C, D, E, F..., which in turn will trigger other modifications themselves throughout the whole system and eventually loop back to point A, and so on. That's why our idea to 'control Nature' or to establish an 'ecological balance' or 'help the environment to recover' by intervening in its processes only betrays our childish naivety and ignorance. What we should do is not to intervene or try to master or help Nature, but to learn to become aware of our systemic ignorance first, and then enter into the mentality of a nonlinear self-organizing interconnected complex system dynamics, eventually augmented by a holistic Goethean seeing. Once we are able to enter into this cognitive dimension, the solutions to our current worldwide and global issues will present themselves much more naturally.

Thus, theories of self-organizing complex systems and supercomputer simulation might help us, but that won't lead us to the real understanding, as neither these technological means nor these intellectual, theoretical constructs, as complicated and detailed as they might be, are not even in principle able to seize the Real-Idea standing behind all this fantastically complex natural machinery.

This is another aspect of the state of denial in which the modern scientific mind finds itself, and that can only postpone—and not avoid—an inevitable evolutionary impetus. We will have to recognize the narrowminded paradigm we are working with. The mechanistic view of life, biology, and especially medical science that seems unable to conceive of our health and medical treatments beyond complicated machinery is stagnating for this reason. The change of paradigm will come from within and then manifest externally. This change from a unidimensional, strictly reductionist, and material view of life to a multidimensional post-materialistic approach that recognizes the existence of other forces acting in a body consciousness that determines our biological wellbeing could change everything. It is not a mere question of theoretical metaphysics but about down-to-earth, very concrete, and pragmatic issues that can decide between health or illness and even life or death.

3. The Physicalist Health Delusion

To make such a step, we must get rid of some die-hard and still deeply engrained delusions—for example, that of a purely mechanistic and materialistic conception of life, our bodies, and health, which inevitably determines how medical sciences are practiced, not to mention how its related R&D is funded.

If we accept that there is a Life-Force and several other cosmic planes of existence with a concentric system of being that determines our personality, it should become pretty clear how limited and superficial a purely physicalist and mechanistic worldview of ourselves and our bodies must be. Our bodies are more than just complex machines. They are more than a compound of organs, bones, and a brain. The reductionist physicalist understanding of what determines our physical health and wellbeing, and which reduces the body's functions to genes, cells, and chemistry, carefully avoiding any wider perspective, has led to stagnation and a lack of progress in many fields. The popular belief deludes itself with supposed giant leaps of the medical sciences which, at closer scrutiny and critical analysis, largely evaporates or reveals itself to be much more limited than expected. But these exaggerations must not be blamed only, and eventually not even so much, on the general public media. Rather, they are already contained in the press releases of the academic institutions and are further inflated by the scientists involved in the research. The strong association between exaggeration and distortions in health-related science news and academic press releases has even been measured empirically [349].

The Covid-19 pandemic made it amply clear how, despite our theoretical microscopic knowledge of the structure and function of cells and viruses, when it comes to dealing with a worldwide pandemic at a practical level, we are still firmly in the hands of the forces of Nature. The coronavirus laid bare all the weaknesses and shortcomings of a science we believed to be much more 'exact' than it really is. Even by following strict empiric rules based on established models of cellular and molecular biology, science could not converge on a unique and unequivocal interpretation of the flood of data, let alone on the assessment and evaluation of the action that had to be taken. For at least a century, more or less veiled forms of scientism hammered into the collective consciousness the dogma that reason and a purely materialistic and reductionist science are the ultimate tools for knowledge on which everyone must agree. But this pandemic has laid bare the limitations of such an approach and skepticism is increasing toward the powers of the cherished analytic mind, and it now becomes almost impossible to explain to the public why this confusion, which also exists among scientists themselves, doesn't hide any conspiracy but is, and has always been, a normal state of affairs. The so-called 'Age of Reason' yearns more than ever for an 'Age of Wisdom'. It might well turn out that, in the long run, the collective consciousness has grown faster with this pandemic rather than without it. The question is: When will we have grown up sufficiently to no longer need the harsh lessons of Nature?

Similarly, the 'war on cancer' has been going on for about half a century but despite some substantial and notable advances, overall, the results did not meet the expectations. Cancer remains the most feared cause of death. After the mapping of the human genome and decades of intensive R&D in genetics, indeed, groundbreaking discoveries have been made, but these were mostly theoretical rather than practical, and the expected *personalized medicine* or therapeutic applications all hoped for did not materialize. Stem cell research was and still is a promise for organ transplants or regeneration, but so far, it hasn't delivered as expected, either.

This is not to say that these sciences are of no value. Allopathic medicine based on a pharmacological and surgical paradigm is useful and will continue to play a role. It led to a longer life expectancy, and several potentially deadly illnesses are nowadays curable. But an honest assessment of what was promised by past research projects and what was achieved to date reveals that, in most cases, their successes and progress have been largely overemphasized. This slow and disappointing progress has yet to come into the professional as collective surface awareness and is still denied. The sooner we recognize that there are several aspects of modern medical research that are a no-progress quest rather than a slow progressive-quest, the better for us all. We still delude ourselves by invoking more funds for R&D, more technology, even more microscopic and functional descriptions, asking for yet another two decades of time to build even more precise maps that will supposedly lead us to a bright future and, as usual, afterward realizing that these maps don't tell us much about the territory. But sooner or later, we will have to recognize how we are only scratching a thin surface underneath which a much vaster reality exists.

We have seen how our cells are more than a complex genetic automaton; they have a mind of their own. And our body is more than a machine; it has a consciousness of its own. Beyond our physical appearance, there is a subtle physical bodily existence. What rules between physical wellbeing and sickness is largely determined by the subconscious, psychic and supra-conscious forces, much more than what we believe the material layers do. A healing process depends on mental and submental individual self-suggestions as well as external suggestions that also come from a collective and universal domain through what Aurobindo called the environmental consciousness. Life is not only a biochemical process but also a cosmic life force with a vital dimension.

As long we dismiss these realities, ignoring and even denying them, further real progress will be excruciatingly slow or eventually also permanently delayed. Once we have fixed an illness, an infection, a genetic disease on one side, we will have to cope later with another new physical or mental disorder that was formerly unknown or unexpected. This is how the evolution of consciousness works: No real progress is possible without inner progress as well.

4. The AI and Simulation Delusion

"All the impressive achievements of deep learning amount to just curve fitting".
Judea Pearl [350]

For the same reason, we will have to recognize the AI delusion.

After undoubtable success stories, especially with the application of deep learning neural networks and *'large language models'* (LLM), such as the famous *"Generative Pre-trained Transformer"* (GPT) technology, and which, to date, seem to be the breakthrough and great promise for building machines mimicking the human mind, I maintain that, sooner or later, we will have to become aware how the original expectations to create a machine that really has a semantic awareness, let alone a conscious experience, were greatly exaggerated, thanks also to an ongoing media overhype.

The idea of a coming AI revolution is not new. It dates back to the Dartmouth Conference of 1956 in New Hampshire, which was organized to discuss the possibility of constructing intelligent machines. Claude Shannon, the father of information theory, prophetically predicted in 1961: *"I confidently expect that within a matter of 10 or 15 years, something will emerge from the laboratory which is not too far from the robot of science fiction fame."* During this workshop, the term 'artificial intelligence' was coined. In the 1980s, an AI programming language like *'Lisp'* and its related *'Lisp machines'* and the supercomputer AI company *'Thinking machines'* were supposed to open the gates to an AI Eldorado, but were soon absorbed by history; only the older generation or the historians of IT remember it. The Japanese *'Fifth-Generation-Computer-Systems'*, aimed at providing a platform for AI by creating massively parallel supercomputers, didn't go far and were soon shut down as well. Efforts to create *'expert systems'* faded because it turned out that they weren't experts at all. *'Connectionism'*, the science behind the first neural network models, was further developed but didn't meet expectations either. For a while, in a brief *'AI-winter'*, the spirits cooled down, but when the IBM chess-playing computer Deep Blue won against world champion Garry Kasparov in 1997, the irrational hopes of a coming AI revolution were revived. However, by 2001, the intelligent computer HAL 9000, which Arthur C. Clarke and Stanley Kubrick had imagined in their story and film "2001: A Space Odyssey", was not even nearly in sight. In 2004, DARPA sponsored a driverless car grand challenge and the funding for building automated translation and voice recognition systems skyrocketed. Again, a winter phase followed until 2016, when the program AlphaGo, running on the Google supercomputer DeepMind, won the ancient Go play against world champion

Lee Sedol and history repeated itself. New great expectations rose, and the AI-summer—that is, the AI delusion—was again in full swing, with corporations and governments throwing billions into deep learning neural networks, self-driving cars, and other AI research fields.

We were promised that in a few years, we would have cars driving completely autonomously from the US coast to coast, Skype sessions would have real-time natural speech automatic translation, or IBM Watson–a computer system capable of answering questions posed in natural language–would lead to *"cognitive systems that can know and reason with purpose"*. These were announced as imminent big breakthroughs and were abundantly hyped in the media with sensational-sounding article titles. Nowadays, nobody is talking about them anymore. These projects not only revealed themselves to be wishful thinking among the popular audience and in the news but they were also endorsed and defended by the majority of professionals in the field. We could read articles full of adjectives and proclamations such as 'acceleration', 'disruptive impacts', 'coming revolutions and apocalypses', 'breakthroughs', 'change of paradigm', etc., with lots of sci-fi applications, but with no concrete reasons to make one believe that we are anywhere near that. In other words, the border that separates weak AI from strong AI (or AGI) has been shown to be an impenetrable curtainwall of a seemingly unconquerable fortress—a no-progress quest.

But after more than seven decades of research since the Dartmouth Conference, AI algorithms, or brain-inspired neural network simulations, or LLMs and GPT, there has been almost no progress in building computers able to understand meaning other than by likelihood and curve fitting guessing procedures. As we have seen in Pt.I-II.3-5, semantics was and remains something utterly precluded to machine learning.

People are now impressed by the almost human dialog with chatbots–that is, AI programs able to entertain an on-line conversation via text or speech. Particularly impressive is the deep learning GPT technology, so-called ChatGPTs, of OpenAI officially launched in November 2022. It is trained according to LLMs on huge data sets and depend on billions of parameters to predict the next token.

ChatGPT describes itself: *"I am an AI language model developed by OpenAI. I have been trained on a large corpus of text data to generate human-like responses to text-based inputs. When a user provides an input, I analyze the input to understand its meaning and context, and then generate a response based on my training and the information I have learned. The training process involved training a deep neural network on a large dataset of text data, where the goal was to predict the next word in a sentence given the previous words. This process allowed the model to learn patterns and relationships in the text data, and to develop a representation of the text that can be used to generate responses to new inputs. Once trained, I can be used to generate responses to a wide range of text-based inputs, including questions, statements, and requests. The quality and specificity of my responses will depend on the complexity of the input and the amount of relevant information available in my training data. In the case of programming, I understand the syntax and semantics of the programming languages and can write programs accordingly."*

The result is impressive. In a sense, modern technology has almost realized the sort of conversations, in the form of a friendly natural language with robots, that we are so familiar with in sci-fi films. ChatGPT can even generate novel poems and songs and write sophisticated programs. This sparked a debate about whether chatbots could possess some form of conscious awareness. The short answer is no, for the following reasons.

First of all, we should ask if it really understands meaning and context? What these applications do is not based on logical inferences, intelligent reasoning on a text, or a rational evaluation of propositions. Rather, they try to make the best guess of what output could correspond to an input text. To do so, they are trained on billions of pages of textual data, fine-tuning their guess on a process dependent on hundreds of billions of parameters, modeling how the words appear in a certain sequence by identifying textual patterns that then are turned into rules and, therefrom, making the best guess as to what the answer could be based on conversations that humans extensively conducted on the World Wide Web, in articles and books, etc. Finally, it arranges the words according to the rules of syntax. The program looks for linguistic sequences that have the highest probability of being meaningful based on built-in grammatical assumptions and scanning all the available meaningful sentences contained in all documents. Deep down they don't understand a thing.

This might sound like a too harsh verdict, considering how LLMs seem to understand in many respects human language. Yet, one can see that the machine does not go beyond a statistical evaluation of what the most likely correct answer possibly could be. While most of the time it does answer correctly and simulates quite well human understanding, often with amazing results that seem to suggest a real AGI, when it fails, it does so sometimes with ridiculous answers. I use regularly ChatGPT, among other things to create sophisticated programs. It is amazing to see how it can replace much of the work of a programmer. However, when I asked it to create a very simple algorithm to draw a 5x5 grid, it only produced a non-sensical set of lines. Or when I asked it to count how many zeros and ones are contained in the string "0100011100101010001001001", it counted 12 zeros and 11 ones, while, as you can check by yourself, 15 zeros and 10 ones is the correct answer. That the answer is incorrect could be seen easily considering that the sum of ones and zeros should be a total of 25 characters, not 23. And, sometimes, I ask the same question in two different ways and obtain contradicting answers.

I don't get any impression that there is anybody 'in there' who understands what it is doing and who can evaluate its own output, realize how it is wrong, and try to amend and improve its answer.

Others made more sophisticated tests. For example, it could be shown that, when a creative task is given, such as designing new rockets, ChatGPT can't figure out facts and fails to understand basic concepts. This is clearly visible when it designs rocket motors without an engine nozzle or with pipes leading nowhere. Drowning the application with just more data, may fine-tune its skills but will not deliver a semantic agent.

These applications can, indeed, be good information retrieval tools, sort of more sophisticated search engines with a human-friendly interface, but, once again, they are machines that don't go much further than Searle's Chinese room or what Bender et al. called 'stochastic parrots' [351]. They hallucinate creating combinations of words according to rules they have learned mimicking the human language but remain unable to sort fact from fiction. Despite its very appealing human-likeness, it is doubtful that a system that works like an 'autofill on steroids' has any semantic understanding of what it is copying, disassembling, reassembling, and pasting. They frequently don't get it right simply because they are not designed to get things right, but rather are built to generate the most plausible answers by trying to predict the next words in the interaction with humans. Chatbots didn't convincingly pass the Turing test either and, most importantly, remain completely passive tools only waiting for input. There is no agency, no will, and no desire to act or create something new. These AI applications are more powerful and elegant symbol-crunching machines, but there is no sign that they will develop any conscious and active self-awareness. There is no reason to believe that anybody is 'in there,' let alone a sentient being that understands what it is doing. To reason over facts something more fundamental is needed. And that is a mind. A mind that might not be physical.

No matter how sophisticated and advanced an AI is, it shows no signs of understanding a sentence and the words in it. That's why, despite the intensive research dating back to the 1960s, we are still far from having an automatic language translation system that could replace human interpreters. Moreover, everything indicates that the more recent efforts to build self-driving cars are hitting the wall as well (again, no pun intended). An AI has no idea what a sentence in a book or the images of a dashcam means, as it does not know the world and not even what the patches of pixels representing objects signify. Changing a few pixels of an image can lure the best pattern recognition software to classify the image of a stop sign as a commercial show card or vice versa.

Thus, despite all the success of neural networks, LLM, GPT, and other AI related breakthroughs, I believe that we will sooner or later head again towards another AI winter.[111]

The question is whether a subjective experience, consciousness and sentience are necessary for understanding the world? I think that we will ultimately find out that without consciousness there could be no true semantic awareness.

This shouldn't be surprising. In humans, things aren't different. A congenitally blind person (recall also Mary's room and the knowledge argument) doesn't know what a color is or could possibly mean because ultimately the subjective experience is necessary.

Interestingly, however, LLMs show skills in understanding the significance related to lived experiences. But this is only because sensory relationships are already embedded in language. For example, AI can handle concepts like colors or 'transparency' as well as congenitally blind people can do. In fact, blind people can talk coherently about colors or can infer an abstraction of a sensory experience like 'transparency' only because they know that others report seeing colors and how they can see through windows. But in what sense is that a 'knowledge' or 'understanding' of what colors and transparency are?

And what about abstract but for humans very concrete and real concepts, like 'freedom', 'justice', 'beauty', 'love', 'emotions', or 'feelings'? These are concepts that cannot be understood only by rational, logical, and linguistic analysis, no matter how large and sophisticated the language model is. One can't understand these things without a subjective experience and without 'knowing' what it is like to be in these psychological states.

Modern AI machines still exhibit no sign of any ability for critical thinking skills and common sense, let alone wisdom. An AI can be useful for finding patterns in huge amounts of data, but, deep down, it doesn't associate to it any meaning or semantic content other than a stream of symbols to which it associates probabilities of occurrence. While advanced and complex AI can mine big data for patterns and structures, nothing indicates any ability in discriminating a meaningful occurrence from a meaningless coincidental spatial or temporal non-causal correlation of patterns that, in reality, are only random data. This last discriminative cognitive step was and remains a necessary human task.

Scientists learned that artificial neural networks are unable to process things as our mind does. Deep learning vision algorithms often fail to classify images because they take cues mostly from textures, not shapes (for example, they see the fur of a cat but not its shape—the binding problem, again) [352]. AI

---

[111] For a nice account of the 'summers' and 'winters' of AI, see the Wikipedia entry 'History of artificial intelligence'. For a more technical paper on the historical failures of AI, see also [376].

systems fail a vision test that a child could accomplish with ease, such as the *'double take'* situation in which one must look twice at an image that otherwise might appear confusing. [353] Machines don't agree at all in judging similarity. The images that humans consider to be visually similar don't match those selected by AI. Machines do not compare things as humans do [354]. And what a chatbot does, is that of transforming the world in tokens. For example, a picture is subdivided into squares and a certain number of pixels with a color and transformed into lexical tokens, such as "square x has y pixel in color z". The neural network then looks to the world, not even capturing surface patterns, but only through these gazillions of tokens trying to capture properties, regularities, and relationships that represent the world. In a sense, it lives in its own Plato cave, but without the emergence of meaning and a subject that perceives it.

Thus, it is misleading to speak of a computer that 'understands', has 'knowledge', or 'sees' something, especially in the sense that humans understand, know, and see the world.

Moreover, an aspect that is frequently overlooked is the psychological property of all living cognitive and conscious organisms: Will. There is no visible sign suggesting that a Chatbot might possess any internal will–that is a drive to act, an agency, an impulse to autonomously create, or a volition and intention to do something, unless explicitly instructed to do so. Otherwise, it is just a passive black box waiting for input.

What differentiates a human being from the best AI models is also the outrageous amount of data that the latter needs to reach comparable skills. A neural network and LLMs need to absorb millions of textual pages, scan all Wikipedia, and must be trained on extremely huge data sets, before becoming able to do what we can do, and have acquired with much less information and training. This alone should make it clear that something very fundamental is missing. ChatGPT is not much more than a colossal encyclopedia that crawls the net regurgitating what it finds on its way. It paraphrases information with a bulleted list of facts that can be easily found on the web or elsewhere, not always answering directly the question and often with a great deal of repetition. I'm impressed by the potential applications of this technology, and its possibility to change our way of retrieving and using information. But I have not the feeling that it really grasps what I'm saying, and that it simply parrots lots of tangential information absorbed elsewhere.

These were only a few examples of the divergences between machine learning and human cognition. I maintain my skepticism about AGI. It will turn out that deep neural network learning and LLM aren't deep enough and will be seen as yet another form of computation. Deep learning neural networks are not the magic wand initially thought and, despite the high hopes, self-driving cars that will allow us to take a nap while driving us safely to our destination are nowhere near in the future (for some easy-to-read but illuminating articles, see: [355], [312], [356]).

I believe that sooner or later, the bottom line will be that AI will tell us much more about us, about what our consciousness, our true inner nature, and our psychological and spiritual dimensions are *not*, rather than what consciousness is and how it supposedly emerges by a bottom-up process. We might get to a point where we will see that, no matter how much advanced and sophisticated AI will become, it will always lack what makes us human. If so, we will be forced to admit that mind and consciousness are not just a mere physical epiphenomenon, and might open our minds toward a metaphysical understanding of the world. Sometimes the path toward truth is found when we have exhausted all the other alternatives.

To exemplify where the problem lies in our thinking pattern, let us take a short interlude into another weird idea of modern naïve physicalism.

The idea that a computer simulation of the brain might one day lead to intelligent and eventually even conscious machines rests on unaware assumptions one finds rarely addressed by a commonsensical critical analysis. The narrative says that once we have sufficiently powerful supercomputers that simulate, in the tiniest detail, the process occurring in and between all the billions of neurons with the functions and signals they exchange with each other, this machine will become conscious and have mental states as our brains are supposed to have. For the unreflective physicalist, this seems an almost obvious and inevitable conclusion. However, it is not necessary to wait until the day when Skynet will become self-aware and invade us with terminators to see if that reasoning holds up with reality. The fallacy of this apparently logical and inescapable inference should be clear from the outset once we ponder more carefully on what a 'simulation' really is and what it really does.

Simulating a physical process, with whatever degree of detail and complexity, does not recreate the process itself. A simulation alone is not much more than a software program that performs numerical and symbolic computations internal to CPUs and memory chips. A computer simulation is physically nothing more than a temporal change of the internal electric states of microelectronic devices. It doesn't matter how precise, fast and accurate the simulation is; it will never instantiate the physical phenomenon it represents internally. It can't do that, not even in principle.

For example, as is well known, everyday powerful state-of-the-art supercomputers simulate the meteorological conditions that allow us to make weather forecasts. But no serious meteorologist would contend that inside the microchips of these computers, it is actually raining or that the sun is shining or that somewhere in the CPUs, it is snowing and the wind blowing. One thing is a symbolic (static or temporally evolving) representation of a physical process; a completely different thing is that of effectively reproducing the process itself with all its physical phenomenality. And it is pointless to insist on saying that it is only a matter of time before the technological progress will have gone far enough to simulate the whole process in all its details, that then we will finally produce, like magic, the physical process as well. Even if a supercomputer simulation would exactly represent the dynamics of every air molecule and atom of all the Earth's atmosphere internally, you won't see a drop of water or a sunray emerging from its CPUs.

As obvious as this might sound, the misconception that confuses a simulation with the re-creation of the physical process is a very common fallacy. For example, one has only to see how popular the so-called *'simulation hypothesis'* has become. The question of whether we are living in a simulation comes from a 2003 paper by Oxford philosopher Nick Bostrom [357]. This is a conjecture stating that, perhaps, all the universe and its galaxies, stars, planets, as well as the living beings in it, us included, is a gigantic computer simulation of a super civilization. According to this scenario, all our lives and our experiences, like all of the reality we perceive, are not real but are, instead, mere virtual reality. It is something like in the film "The Matrix", but with the decisive difference that there is no red pill to choose in order to awaken from the induced virtual digital hypnosis. We are literally only bits and bytes simulated in a hypothetical cosmic supercomputer simulation.

First of all, this should naturally raise the question of whether, also, the alien civilization simulating our universe is a simulation itself, or if it is 'real' (whatever 'real' might mean, at this point). If it is a simulation as well, then one is confronted with the same question: Is the simulating civilization that simulates us also another simulation? This leads to a logical infinite regression. On the other hand, if the computer simulating our universe is real and not a simulation itself, it can't be a physical machine in a material universe governed by the physical laws we observe, as in the simulation hypothesis, these are also part of the virtual reality it is simulating. Thus, the universe of the simulating civilization can't be a material realm ruled by physical laws—a curiously self-negating form of physicalism.

And how does one recreate a subjective experience? The alien super-civilization, no matter how advanced it is, must come to terms with the hard problem of consciousness. How? Certainly not by a numerical and symbolic simulation, however complicated and detailed it might be, because, again, it reproduces, phenomenally speaking, nothing.

Either way, the simulation hypothesis is itself an intellectual exercise that only obfuscates things and doesn't explain much. It raises more questions without answering any. It doesn't explain why and how anything comes into existence, be it us and our universe or the simulating civilizations, and offers no hint whatsoever as to the origin of consciousness. Addressing the hard problem of consciousness resorting to theoretical frameworks based on simulation hypotheses or functional arguments leads to a dead-end, as we have already elucidated in Pt.I-II.4&5.

If the simulation hypothesis, or popular movies like "The Matrix", have any meaning, it is that they subliminally intuit and superficially reflect a deeper truth. In fact, they inwardly feel and intuitively recognize how there is another logical possibility, namely, that the simulation and the Simulator are one and the same Being. In the integral cosmology, we have seen how an unborn and transcendent Self has plunged itself into the material manifestation of a Nature of ignorance and self-forgetfulness, and, thereby, subjected itself to the delusion of a universe of dualities–that is, the illusion of Maya. Brahman is the Simulator, while its projection out of itself is the simulation in which it dwells, and with us being its multifold emanations that mistake the simulation for real. But since this is a metaphysical ontology, the physicalists' mind can't accept it and must resort to contrived and logically flawed and unconvincing hypotheses.

Now, turning back to more earthly issues, we might be in a better position to understand what went wrong. It is now becoming clear that something is fundamentally missing or flawed. The underlying metaphysical assumption— that is, the physicalist's equation between mind and brain, thinking and computation, simulation, and phenomenal duplications—is not questioned and has always been taken for granted, even by the vast majority of high-ranking professionals in the field.

If we maintain for a moment the opposite view, things appear in a different light, and the paradigmatic glitch becomes almost obvious. If the brain is not the mind or, at least, does not exhaust all the mind's functions and skills, then, quite obviously, any simulation of the human brain, however complex and realistic it might be, will never instantiate a mental human-like activity. If semantic recognition, the understanding of meaning, is a mental function that occurs outside the brain, it should not be surprising that recreating with whatever detail and accuracy the brain's functions can't result in a device that understands the world. If consciousness is not an epiphenomenal secretion of neural activity, we would never be able to build a conscious AI, not even in principle. If life and mind are cosmic inherent conscious forces beyond physicality, it is futile to believe it could emerge by physical means only. Nature did not create mind, the higher-mind, and the supraconscious forms of cognition with brains, neural networks, or other forms of material organization.

AI is another example of how the practical results—or, as in this case, the lack thereof—betray our metaphysical assumptions. If we are not aware that we are working with a flawed abstract metaphysics that does not match reality, we will forever keep trying to achieve the impossible. We will ask for even more funds for R&D that afterward will not keep its promises. We will need permanent marketing hype to convince governments and foundations to keep these funds flowing. And, once the expectations have again not been met after a decade, we will need to ask for another decade of fund extension, keeping things running virtually forever. Sabine Hossenfelder[112], a German theoretical physicist, once summed up this state of affairs marvelously: *"Today, the task of scientists is no longer to understand nature. Instead, their task is to uphold an illusion of progress by wrapping incremental advances in false promise. Merchants they still are, all right. But now their job is not to bring enlightenment; it is to bring excitement. Nowhere is this more obvious than with big science initiatives. Quantum computing, personalized medicine, artificial intelligence, simulated brains, mega-scale particle colliders, and everything nano and neuro."* [358]

This is yet another example of how important it is to become aware of the philosophical and metaphysical background and ideological mindset with which we are working. Questioning the physicalist dogma is not just a theoretical musing for philosophers or mystics but also a down-to-earth, practical, material, and financially urgent necessity.

5. The Material Post-Scarcity Civilization and Transhumanist Delusions

A post-materialist, as post-religious, perspective on life, the universe, and especially ourselves would lead us also to a completely new understanding of how to manage our resources.

For example, people speak of a *'post-scarcity civilization'*—a time when humankind will be freed of the anxiety for material needs because, so goes the narrative, machines will take over in doing our jobs. Considering the huge technological, scientific, and industrial advances of the past centuries, at first, this might sound like a reasonable prediction of a future to come. However, a more cautious analysis reveals how this idea and hope simply ignores history and facts.

If we look at it from a historical perspective, we see that increasing mechanization and application of an ever-improving technology has led to the opposite effect. While it is true that machines have replaced many human activities, they have also created myriad new ones that need human intervention and maintenance. So far, technological progress has forced us to frequently change our jobs, but there is no evidence supporting the belief that it diminished our anxieties and increased our spare and leisure time. On the contrary, stress and burnout, especially in the most technologically oriented working places, are today more widespread than they have ever been before.

In the 1930s, the British economist John Keynes predicted that, due to the technological unemployment caused by industrialization and technological

---

[112] Hosselfelder is a die-hard physicalist and, obviously, would not share the spiritual worldview presented here, but that does not prevent us from appreciating her great insights and her valuable and serious contributions to popular science.

progress, in a couple of decades, we would be working only 15 hours a week. Almost a century later, most of us would be happy if we could keep our weekly working load limited to 40 hours.

If these simplistic beliefs survive and recur throughout generations, it is only because we work more or less subconsciously always with the same flawed premises. The hope that some complex machinery will reduce our stress levels and working hours rests on the assumption that once we have fixed our external and material conditions, the rest will follow. It is a materialistic belief system that cannot look inside and see the spiritual dimension of its own existence. With this materialistic mindset, we must inevitably search almost desperately outside of ourselves and ignore the facts in order to keep going. However, over and over again, this theory has been disproven. It is Nature that is forcing us to look inside. It won't allow us to find that rest with a greater accumulation of machinery and wealth. The idea that AI will take over our jobs and daily activities, making us free from material preoccupations, won't work. Facts have shown that while technology can free us from slavery, it can enslave us itself. A hyper-technological materially scarcity-free society won't free us from anxieties, fears, stress, and financial worries if we don't also work towards a spiritual scarcity-free society.

Post-scarcity expectations reflect the same reasoning that stands behind the famous *'Maslow's hierarchy of needs'*, a psychological theory proposed by the American psychologist Abraham Maslow in 1943. In this view, there is a pyramidal hierarchy of human needs: physiological needs, safety needs, social belonging, self-esteem, self-actualization, and transcendence. The latter means, among other things, also a spiritual transcendence of human consciousness. It must be noted, however, that Maslow added this last piece of the pyramid only a year before his departure. For two and a half decades, his conception of the human being was material and psychological, with no spiritual dimension—a detail that speaks volumes about how our materialistic conceptions are deeply entrenched in our collective consciousness.

For example, one of the most extreme forms of philosophical materialistic rationalism is *'transhumanism'*, a movement that believes sciences such as AI, nanotechnology, and other sophisticated emerging technologies will pave the way towards the transformation of the human condition, enhancing its physical and mental abilities. Brain chip implants may boost our brain activity, a partial replacement of human biological organs with electronic and mechanical advanced devices could enhance our senses and motoric abilities, nano-robots may repair parts of our bodies or hunt down cancer cells, and other futuristic sci-fi applications will supposedly lead to the transformation of our present body and psyche. This idea goes so far as to speculate that one day we might be able to download our mind and consciousness on a computer and eventually upload it back to a new bio-mechanical body. This would lead us to the creation of a new species, which would be a sort of hybrid between flesh and machine, and which may eventually become immortal.

This transhuman view has paradoxical similarities with the posthuman evolutionary ideal we presented here: We share the same vision of a humanity that will transcend itself. However, transhumanism and all the hyper-technological utopias that envisage a coming evolutionary leap restrict themselves to a purely one-dimensional physicalist and rationalistic conception of reality. Indeed, brain chip technologies like the '*Neuralink*' implants allowing paralyzed people to operate robotic limbs or interact with a PC with their thoughts might be a technology with interesting applications. Nonetheless, we contend that, other than potentially useful medical applications, transhumanist technologies have no future as a tool of evolutionary self-transcendence. The ability to crunch numbers like a supercomputer, develop super-intellectual skills, lift heavy weights like an elephant, run as fast as gazelle with mechanically enhanced limbs, see in the infrared or ultraviolet electromagnetic spectrum, or hear ultrasounds will neither change our nature nor make us more evolved beings. If our egoistic appetites and the lower vital instincts, and our primitive sense-mind intellect are not transformed, no posthuman evolution is possible. The idea that even these instincts and primitive reactive factors that still characterize homo sapiens' animality might be suppressed by meddling with the brain's chemistry and that we might ascend to higher states of consciousness with psychoactive drugs is also a naïve and simplistic belief ignorant of the fact that without a spiritual transformation and evolution of our inmost soul, a mere external and material change won't lead us much further than a magnification and multiplication of what we already are. And the idea that we shall be able to download and upload ourselves on a computer ignores that mind and consciousness are not in the brain and, therefore, there is nothing that can be downloaded or uploaded in the first place.

If we widen our horizons, embracing a multidimensional post-material view of Nature, life, and the cosmos, it becomes clear that transhumanism's vision is narrowminded, at least in its present formulation, and, therefore, isn't a particularly exciting idea. It is unimpressive because it conceives only of a cumulative improvement of the already known, focused on a purely technological, intellectual, and physiological transformation that our limited mind can think of. It limits itself to narrow material and mental boundaries and, by being neither trans-material nor trans-mental, leaves us with very little to transcend. Transhumanism is no real ascension to more evolved states of consciousness, leaving completely out of the equation the spiritual aspect, which its existence even denies. It may lead us to a sort of sci-fi Borg society a la "Star Trek", but with no or very little additional wisdom—something we badly need to avoid self-destruction.

But we don't worry that a future form of super AGI will take over and subjugate humanity. For the reasons we elucidated in the previous chapter, there won't be any *technological singularity*—that is, an AI capable of reproducing and self-improving itself by an exponential cognitive explosion of machines that will then menace us to kill us all and replace humanity. One can easily see how all these visions, utopias, or dystopias rest on unquestioned premises in the first place. If there will be a singularity, it will be spiritual in the sense of T. de Chardin's Omega-Point or the gnostic race that Sri Aurobindo envisaged.

At any rate, all these efforts to create or simulate in machines a human type of intelligence and all these speculations about a conscious AI and its threat to take over humanity may have a very positive side effect: Once its delusional character is recognized, it may force us to detach from an obsessive materialism. It may show us that the root problem is that we don't know ourselves and force us to turn our gaze towards an inward realm inside of us.

6. The Delusion of Intellectual Education

Perhaps the most urgent and pressing issue nowadays is that of education. None of the changes we envisage will be possible unless we are willing to replace our encrusted and century-old mind-centered educational paradigm. A concept of teaching and educating children that has its roots and raison d'être in the Enlightenment, with the glorification of reason and its industrial utilitarianism which, for about a couple of centuries, has served well the material and mental development of humankind but is now becoming the primary bottleneck that strangles the opportunities for further progress in all the other aspects of humans life.

As a former teacher, I couldn't fail to notice how more and more children are affected by anxiety, depression, so-called Attention Deficit Hyperactivity Disorder(ADHD), dyslexia, and other mental disorders. This should not be a surprise in a system that is still based on a mechanistic and rationalistic education paradigm. Once our inmost being, the psychic being, begins to ask to be allowed to manifest and develop itself on the surface, an inner longing of the child's need for a self-unfoldment, self-expression, and self-development is claiming to be acknowledged but is, most of the time, ignored if not even vehemently repressed. At the bottom, this is why teachers are having increasingly hard times explaining the aim and purpose of education itself to children and, even more, to high school teenagers. The answer that it supposedly helps us make a living and prepare ourselves for the future sounds increasingly unconvincing—and rightly so.

We must go beyond the fear- and certificate-based examination industry towards a soul-centered education in the light of an evolutionary concept in which curiosity, interests, passions, enthusiasm, and motivation from within will become the emotional drive in the learning process. The desire for inquiry, discovery, and exploration should be nurtured, and not that of learning stuff that does not relate to our inner calling, and that is only supposed to secure financial security or prepare us for a job that might soon not exist in a quickly and ever-changing world. Soul-based progress from within is the key to a modern concept of education that can cope with the challenges of the future. Unless chosen as a means of self-expression of the child, more IT and STEM or other, still more bookish knowledge won't fix the void of our educational system.

If we want to contrast the flood of pseudo-scientific and irrational conspiracy movements, we must first and foremost nurture the soul rather than the mind. The return to new forms of infra-rational thinking that can't discriminate, distinguish, and think clearly is also due to an educational concept and system that conceives of learning as an accumulation of abstract notions rather than an inquiry, research, and exploration of the world in the frame of a Platonic or Hegelian dialectic. It is not through cramming more math, science, and technology into the child's mind by a top-down teaching method that we will defeat these involutionary tendencies. This subculture is, among other things, also the result of a human collective that has not been allowed to know itself and to come into contact with one's own inner realms and that has been told, though only indirectly and subconsciously, that ultimately we are nothing more than sophisticated pieces of meat without meaning and purpose.

If schools, high schools, colleges, and universities don't change direction from a commercial mind-fear-certificate-based utilitarian idea of education, one can hardly expect future generations to conceive of, let alone set into practice, novel approaches in sciences and humanities that widen their view beyond an outdated, materialistic perception and conception of reality. Our education system increasingly feels like a straitjacket that hampers, discourages, and frustrates the spiritual evolution inherent in every human being.

What must come to the front now is the inmost soul that, having gone through a preparation stage through obedience and preconceived structures that made it grow, is now asking to be allowed to surface and determine itself. The intellect must not be abolished and not even replaced; rather, it is the soul-factor that must take the lead, honoring the mind's role but also maintaining its action inside its proper natural boundaries. The first and most urgent action that society must take is to create an educational context and structures in which each soul can unfold and progress freely according to its inherent and individual nature.

How and under what internal and external circumstances can our intuitive knowledge develop harmonically, and how can it be put into practice in a living research and learning environment? How can the individual potentialities be optimized and set free, evolving in a common activity without damaging the freedom and rights of other individuals? What learning and teaching techniques are the most effective in uniting the necessity of intellectual growth with a desire for individual spiritual, intuitive, and practical self-expression? Might Goethe's approach be useful? How can the practice of meditation be used to produce a scientific work?

The author has written extensively on this 'free progress education' paradigm and invites the interested reader to check it out [359].

7. The Mind is Dead, Long Live the Mind!

The human race is actually entering a transitional phase between the rational and spiritual ages. The rational, analytical mind is losing its grip on human consciousness, but a higher form of cognition has not yet established itself. The turmoil and chaos of our times are largely due to this void, which is characteristic of every evolutionary transition: The old no longer works properly while the new has yet to be established. If we resist and do not assent to this transformation, in the best-case scenario, this can take ridiculous forms; in the worst-case scenario, it can turn into a catastrophe. It depends on how and if we are willing to collaborate.

How this collective change of consciousness can manifest throughout society in times and countries all over the world can be seen, for example, in an apparently generalized weakening of critical thinking skills. We might perceive an almost viral expansion of pseudo-scientific theories, post-truth thinking, various forms of denialism, and a pervasive lack of logical reasoning. However, this is not really a new aspect. It is something that has always been there in a subconscious collective, but it was less evident in an age without social media. The flood of information and the confusion it triggers only lays bare what was already inherent in the race's still not fully developed mentality. The Covid-19 pandemic has shown how clear discriminative thinking is still missing in large portions of the population. Facts lose any value and don't count anymore. Denying facts that do not conform to one's worldview has become a common practice in the post-truth era. The ability to deal with facts and data, placing them into context, understanding what they imply or might imply vs. what they do not imply or might not imply has been shown to be something not within the reach of everyone's cognitive skills. Jumping to simplistic conclusions, making fallacious inferences, and being unaware that one's thought processes are skillfully manipulated by someone's else arguments turned out to be a quite common state of affairs.

It is the action of a force that is bringing to the surface these aspects of our still undeveloped mental. It is something that is mounting the pressure to manifest itself. A consciousness is preparing the ground to establish an intuitive mentality that will, if not replace, certainly augment rational thought.

It is no coincidence that people speak of 'intuitions', the 'reason of the heart', the 'emotional mind', the 'spiritual mind', being an 'empath', etc. Many have an undefinable feeling that what the sense-mind and analytic mind suggest is somehow incomplete and superficial. We intuitively feel that the truth of superficial facts hides a truth of causes. There is a growing sense that more subtle and invisible realities exist beyond immediate appearances and that something or someone else is the cause of the outer events.[113]

But, while this could be an opportunity to move beyond our ordinary thought processes and simplistic materialism, the spiritual pressure from above in the not yet fully formed mind can lead in the opposite direction. A mind that still struggles with discriminating, distinguishing, and discerning and that is new to spontaneous critical thought can easily be deceived into mistaking an artificial light for a spark of intuition, a lower vital impulse for the action of the spirit or size upon an intuitional flash and twist it into a generalized law. A mind that lacks a discerning power is particularly prone to deranging the descent of the higher consciousness and eventually even perverting it to its opposite form. If a part of the being isn't ready to receive the force from above, it can appropriate it for the sake of its own purposes. The paradox is that in an unregenerated mind, the Light from above can unleash the forces of darkness from below. That is, after all, the reason why a higher Wisdom only very rarely presents itself in its full Power and Splendor. It knows that we would make a mess out of it.

The number of people with psychic abilities will continue to grow, but this does not mean that they will form a more spiritual community. There is a plethora of so-called 'channelers' offering their services on the internet, but who or what are they channeling? In most cases, it is only a display of self-deceiving fantasies. Psychics and mediums pretend to have contact with spirits, angels, ghosts, or deceased people, but, if true, are those entities really what they claim to be? Are these (true or false) psychics not deceiving themselves?

And, guess what? A clear understanding of these extra-physical realms necessities the discriminative and analytic skills of the scientific mind! If mental and analytic discrimination isn't a fully established faculty in the psychic's mind, a premature opening to the subtle worlds may lead to a spiritual disaster. Confusion, misunderstandings, deception, and manipulation by forces one can't understand and is not even aware of eventually, is the probable outcome.

---

[113] Among other things, this is the origin of several conspiracy theories. The causal relation standing behind outer events is often so obscure that it lures us into imagining a plot arranged by invisible forces. This is true only insofar that these, rather than being some sort of secret society or cabal engineering intrigues and provocations, are non-physical hostile forces acting from other planes.

It would also be a lack of critical thinking from our side if, after all the 'mind-bashing' that we exercised so profusely, we would conclude that the mind is dead. Unfortunately, this is precisely what the spiritually inclined mind tends to believe: When a supra-rational power begins to manifest, even only slightly, many jump on it, throwing out the baby with the bathwater. The result is a fall into the infra-rational instead of ascension towards a supra-rational states of consciousness. Obviously, the rational materialist does not fail to notice the irrational nature of the spiritualist and, with the same lack of discrimination, jumps to conclusions by taking this as proof of the superiority of the materialist's paradigm. In reality, both are the sign of a not fully developed mind. The former is blind to the logic of matter, the other to that of the spirit.

The mind and its rationality dealing with facts and data and analytic and scientific thought will not become obsolete once one rises above a spiritualized mind. Quite the contrary, the discriminative mind will be necessary, twice as much, because it will have to deal with the material and spiritual as well.

This, by the way, may offer us insights into some of the worst human catastrophes throughout history. If a spiritual light descends into the human collective, which is unable to capture its real significance and lacks discriminative thought, this may eventually twist and pervert that light into its opposite. For example, the thought of many German philosophers, intellectuals, poets, and artists of the 19th century and the first half of the 20th century had a clear intuitional and spiritual hallmark. In particular, German idealists such as Hegel, Schopenhauer, and Shelling perceiving the World-Soul or World-Will, or the higher-mind approach to science and natural philosophy of Goethe, or the extreme existentialism a la Heidegger, who tried his best to push the intellect as far as possible without ever being able to cross the borders of the spiritual mind, were all signs of a supra-rational power that was getting into the reach of Western thinkers.

But the collective was not ready to embrace these heights and was unable to deal with the light coming from above perverting it into its opposite, ending up in calling the darkness from below. Two world wars and the rise of Nazi ideology followed. [114] The well-known example of how a philosophical ideal got perverted to favor Hitler's rise to power was Nietzsche's concept of the superman. As we already pointed out, Nietzsche's intuition of the superman came from a spiritual domain whose true meaning has been turned upside down by forces of destruction.

This is why Nature doesn't disclose all its powers at once. Otherwise, we would be overwhelmed (eventually, our physical existence might be

---

[114] It might be worth recalling that Heidegger was a convinced Nazi until the last of his days. This is yet another example of how a bright rational and analytic intellect is no guarantee against the worst falsehoods if it doesn't cross the frontier towards the spiritualized mind.

endangered as well), and we would be unable to deal with it without self-destructive consequences.

An interpretation of history that not everyone must agree with but that, whatever might be the case, it is clear that avoiding these recoils of mind to an infra-rational state or, worse, to a state of barbarism, by insisting on the virtues of analytic thinking and the scientific method, won't be enough. Celebrating the glories of the Age of Enlightenment and imposing more science, technology, and math in school curricula will not save us from falling back to an ignorant, anti-rational, and anti-scientific state of collective consciousness. These involutionary tendencies are not forgetfulness or lost abilities; they are the result of the refusal to integrate and assimilate a new consciousness. The unreflective impulse that denies the mind is the reaction to the stubborn and equally unreflecting impulse that denies the spirit. If we want to avoid a generalized irrational fallback of society, what we really need to do is (re-)open the analytic mind beyond its self-imposed physicalist pit. Only then can the analytic mind be transcended and kept as an elegant garb with its function and reason, which we will dress only under the command of a higher Wisdom.

8. The Coming of the Subjective Age and the Ideal of Human Unity

We will have to open the doors to this change. Resetting the mind means not abolishing it. Nevertheless, central remains the inner subjective and spiritual change: the next visible social and spiritual leap of humankind, the coming of a *'subjective age'*, as Aurobindo used to call it, describing the past, present, and future psychology of social development [360]. It is a subjective age that goes beyond utilitarian individualism but nevertheless will allow each individual to progress towards a larger fulfillment according to his or her true inner nature and eternal law.

Aurobindo's integral vision is 'integral' insofar as it doesn't stop at the psychological transformation of the classical forms of spirituality. The self-realization that aims at a personal ascension towards higher states of consciousness, such as the Buddhist Nirvana or the spiritual 'enlightenment', the realization of the Self of the Indian Vedanta tradition, is an important stage of one's psychological development, but it doesn't exhaust the process of evolutionary self-perfection. The goal is not personal salvation, let alone seclusion into an interior world that doesn't transform the rest of our being and the external conditions of our surroundings. Integral yoga instead goes beyond the mystical experience and demands the purification, transformation, and spiritualization of our whole being—that is, a cleansing and transmutation of the mental, vital and physical parts. It is an endeavor of self-perfection that does not ignore the dark subconscient side of our existence and takes up even this part towards the spiritual light. It is not an escape into a mystic and blissful heaven that leaves behind our emotional, mental, and bodily existence with all its untransformed imperfections. It is an intense discipline that takes up the whole being in all its aspects, even the most recalcitrant and ignorant corners of the subconscient and inconscient, opening it to a higher conscious force of transmutation.

Another distinctive character and aim of this integral practice is that it aims beyond personal salvation. It is not about the transformation of our individuality alone, but about something that can also change the collective as a center of radiation. The ultimate consummation would be to generalize yoga to the whole of humanity, elevating it to a gnostic community and a divine life.

We don't believe this to be unrealistic or wishful thinking detached from reality but, to the contrary, the inevitable and inescapable consequence of an evolutionary process that Nature is actually working on. It may appear to us to be a too-distant goal or even a fantasy, as we don't usually see the collective dimension from the point of view of the evolution of the Spirit. From the materialistic and intellectual perspective, the history of mankind may appear to be dominated almost exclusively by material necessities and egocentric motives. However, if we allow a more intuitive, post-material, and spiritual vision that goes beyond the obvious but too superficial aspects that rule society, we can see how human groups, nations, and the human collective as a whole are driven and determined by deeper motives, by a 'soul-factor'. Leaving out this soul-factor is the mistake of the intellectual conception that tries to reduce a complex, non-linear and multilayered individuality and collective existence into a linear, much-too-simple, mono-dimensional formula. But human society isn't driven only by physical and egoistic emotional animal appetites and instincts. Behind the veil of the crude, sometimes barbaric, historical events, deeper psychological and spiritual forces play a role. It is to be expected that these will become increasingly manifest with the emergence of a new consciousness. Not only is each individual the expression of a physical and also spiritual evolution, but also the whole of society, with all its historic happenings, is an unfolding of a collective spirit.

The psychological, spiritual, economic, political, and social development of humanity as an emergent 'soul-process' is rarely taken into consideration. The understanding of history, its social and political events, and the development of its arts, sciences, and culture in general, as an outward manifestation of a Spirit in progress, an evolution of consciousness or the growth of a 'collective psyche' beyond the mere material and economic interests and social or cultural factors, remains a difficult exercise for the positivistic secular-minded historians, anthropologists or sociologists. Though modern historians recognize the cultural and sociological factors that determine the political and economic history of nations, the tendency to reduce its happenings to a decision-making process of the strong powers represented by statesmen or personalities at the apex of a social and economic pyramid and that abstracts from the state of consciousness and the life and experiences of the collective basis, remains the prevailing interpretational paradigm with which we understand history. However, if we accept the fact that humans are more than biological machines driven by mere physical and emotional impulses, or genetic and biological determinants such as natural selection, adaptation, and survival, we will have to widen our mental horizon towards a deeper understanding of the psychology of the collective social development that integrates, besides the physical and instinctive natural drive, incarnated in modern society by the impulse to accumulate wealth and increase economic growth, also spiritual aspects with its soul-factors of the ensouled individual interacting with other souls in an overall play of subtle mental and vital forces that determine and mold society and its history throughout the ages.

Any analytical, rigorous, and formal analysis can't understand this play of forces that determine the collective destiny of humankind. It inevitably keeps out of the equation the deeper spiritual dimension in interpreting the historical events and those we get to know from the news. These are only the tip of an iceberg, the superficial appearance resulting from a long chain of causes and effects that take place at a much deeper level and that the ordinary analytic mind, no matter how developed and trained, can't penetrate. Previously inexplicable facts become almost obvious and self-evident once seen from the higher perspective that the spiritualized mind can furnish.

The spiritual transformation of society is the other side of the coin of personal enlightenment and individual spiritual, mental, vital, and physical perfection. From this perspective, Aurobindo describes humankind's history as grounded in an individual and collective purposeful and goal-driven progressive movement of spiritual evolution. In his description of the psychology of social development, entitled *"The Human Cycle"* [360], he took up Karl Lamprecht's idea to categorize human history into five developmental stages, maintaining its nomenclature but giving it an entirely different meaning.[115] The analogies with Jean Gebser's teleological concept of history, with its structures of consciousness, are also apparent. The five stages of human evolutionary history, according to Aurobindo, are as follows:

The prehistoric *'symbolic age'* was a predominantly religious and spiritual stage of evolution in which a widespread imaginative or intuitive religious feeling has a kinship with symbolism—an ancient spirit in which everything is mystically symbolic, such as the Divine, the Gods, the hidden living and mysterious nature of things, etc. It was a time when humanity had a close

---

[115] Aurobindo explicitly credited Lamprecht's intuition, which envisages history as a process conditioned also by psychological factors as a novel interpretation of history. However, Lamprecht was still centered on a too-economical perspective and didn't include the deeper spiritual dimension of the human being.

kinship with Nature. However, contrary to common belief, the prehistoric man was not a primitive, almost animal being, but had access to profound wisdom and spiritual knowledge and a harmonious relationship with the natural environment. We anticipated this interpretation of human prehistory in Pt.II-III.2e. These humans lived on an even higher state of consciousness than present humans have attained. In a certain sense, they were a more evolved species than the mentalized homo sapiens. It was a society reminiscent of the ancient legends of Atlantis or Lemuria, which, however, underwent an involutionary recoil because, while their spiritual stage of development was advanced, the mind was not fully formed. The integration and bridge between the physical-vital plane and the spiritual planes were missing. Therefore, a grounding was necessary, and a plunge back into a more physical and instinctive life was required to build up this link, we call 'mind'.

The following stage of human development was the *'typal age'* in which social forms, habits, and rituals began to take form. This was followed by a *'conventional age'* that further fixed these into strict laws, rules, and rigid and dogmatic forms of religion. These ages could be loosely identified beginning from known history to the Renaissance or the Age of Enlightenment, with the latter marking the *'individualistic age'* or *'rational age'*. This is, of course, the age we are still living in. It is the age of thought, in which the mind becomes the center of cognition, trying to impose itself on a collective scale by inquiring and questioning the conventions and dogmas that were previously so dominant. It is a developmental stage of human consciousness that not only apprehends the world with a mental view but also begins to stress the principles of equality and justice. However, this is precisely its strength and limitation. A strictly material and intellectual understanding of reality that tries to fix its problems by external means—that is, by scientific and technological methods and more or less democratic forms of government—but ignores and eventually even denies the deeper spiritual and trans-physical dimensions can build social structures that only partially satisfy its demands by unstable balances.

Aurobindo contends that only the coming of a *'subjective age'* or *'spiritual age'*, in which human beings go inward and realize their deeper and intimate spiritual nature and develop a comprehensive spiritual view of the world, can lead us to a more stable and higher order. This can lead to real change, realizing the principles of liberty, equality, and fraternity, which were so cherished by the French Revolution but actually remain far from becoming a reality.

On the horizon, we can see the coming of this spiritual age.[116] We are finally becoming aware of how the materialism, industrialism, and intellectualism of the last centuries allowed for an improvement of our material wealth but did

---

[116] The author assumes that, at this point, the reader understands that the term 'spiritual' has nothing to do with what is normally associated with a religion. The subjective/spiritual age will be neither a religious nor a materialistic age.

not seem to improve our inner psychological wealth. The spiritual age will conceive of all the social, political, and educational structures from a completely different perspective that integrates the spiritual aspects without excluding the material, vital, and physical dimension of the human being. Central to this will be the soul, the psychic being, the inmost voice telling us by an inner intimation what the right thing to do, think and feel is, thereby avoiding a trial and error path of struggle and strife. It will base its ideal not only on material progress but also on the inner and spiritual progress of the personality as well as that of the collective. It will be a society based not on the principles of division, competition, dominance, and conflict but on collaboration in unity in diversity.

The path towards a unity on the physical plane has already begun with our increasingly globalized and interconnected societies and nations. The internet, the efficiency of mass transportation systems, the interdependence of cities, nations and continents, the internationalized mass media, etc. are not just the outcome of economic and technological development; rather, they are the material manifestation of an inner psychic sense of commonality between the peoples, the nations and civilizations, which is the inevitable evolutionary destiny of the Spirit—a Spirit that is inherently One, the outward expression of the underlying Consciousness in all, and whose main trait is and has always been that of unity in diversity. Slowly but steadily, humanity is beginning to perceive this sense of connectedness, which Aurobindo described as *the ideal of human unity* back in 1919.

However, this must not be confused with modern conceptions of *globalism*, which usually don't care about the spiritual dimension of the collective and are prevalently centered on the commercial and material aspects of societies and nations, foreseeing the solution of conflicts only in regulatory and technocratic expedients, some of which have authoritarian tendencies and, most importantly, still don't perceive the importance of the multiplicity and variety in unity. This sort of 'unification' under the umbrella of a 'new world order' is the most feared in the collective and has recently gained a quite negative connotation. And rightly so. The future is not to be found in a new centralizing 'world order' that cuts away every diversity and imposes a global uniformity by a central power, eventually resorting to hierarchical top-down authoritarianism. This is the false unity of the mind. It is the mind which tends to look for measures that tend to uniform in a monotone sameness everyone in a more or less repressive system that compresses our diversity into a collective carbon-copied, lowest-order, common denominator personality. This would be contrary to the principle of spiritual unity. It is only a limited mental understanding that tends to imagine a collective unity as an aggregate made as uniformly as possible.

Whereas, a spiritualized mind immediately seizes upon the true significance of the real meaning of unity as diversity and multiplicity. It is already visible in the biological domain. It is the diversity of species that strengthens it, not a life in which everything adapts to sameness. It is a genetic variation that renders an organism stronger, not a genetically constant progeny. It is biodiversity that makes the forest more resilient to adverse conditions, such as droughts and fires, not monocultures imposed by a linear thinking animal with the so-called 'sustainable' industrial management. And yet, there is a commonality in each species, a genetic code that identifies it, a root system connecting all the trees in one organism. The Goethean way of seeing reveals to us that this is the external manifestation of a deeper truth—namely that, at the bottom, everything is One, which expresses itself as a multiplicity in unity by a Real-Idea. The beauty of a flower arises because of its color intensity and variety, its expression of symmetry in a form that crosses a unity. This beauty appears because of the multiplicity of its petals, which are still petals despite differing slightly from each other as a varying personality. A uniformly colored and featureless object has never been considered either beautiful or particularly functional. Beauty appears because of a unity that manifests itself in a variety and multiplicity without losing its underlying oneness and sweet psychic qualities.

The universe is made of hundreds of billions of galaxies that are all diverse– sometimes more similar and sometimes less. Each galaxy is made, again, of billions of stars that are all distinct from each other with regard to mass, luminosity, color, temperature, etc. And yet, one can classify all these galaxies into fewer than a dozen categories (the '*Hubble catalog*' of galaxy classification) and plot all the types of stars in a temperature and luminosity diagram (the '*Hertzsprung-Russel diagram*') that finely explicates their strong underlying commonality.

This unity in diversity is not a freak of Nature. On the contrary, Nature tends to favor and amplify multiplicity and variety everywhere to avoid decay, stagnation, or death. This all-pervading multiplicity in oneness is perceived as a contradiction only by a reductionist mind that must resort to a complicated material process to justify it[117] but becomes a completely natural fact for the spiritualized consciousness that needs no proof or argument to understand it. It is an inherent principle in Nature which in turn is the reflection of the transcendent unity of the Absolute that existed before the coming into existence of the cosmos, space, and time itself. It is something that exists beyond the laws of physics and biology and before the time of the Big Bang.

Sooner or later, we will have to accept this immanent universal law of unity in diversity as something that must also be integrated into our social order and anthropological thinking. Insisting on keeping alive an artificial ethnical,

---

[117] Of course, from the scientific point of view, one sees genetic variation as something allowing for an advantageous adaptation to a changing environment and thereby enabling survival. However, it sees this only as a given process, without asking where it comes from and failing to realize that it is a means for a goal.

cultural, and linguistic uniform identity and a spiritual mono-dimensional community detached from the rest of the world is, in the long run, a recipe for a society destined to become weak, crippled, and dysfunctional. Pretending to restore a lost identity with a monoethnic and monocultural nation and society is an idea doomed to failure, not because of an ideological political theory or because it is imposed by a globalized commercial state of affairs, but because it is contrary to the essence of Satchitananda, the unmanifest Absolute that is also the essence of our own being. Meanwhile, what is envisaged is a manifestation of diversity that retains its oneness. It is about a harmonious living together of all peoples and, eventually, some sort of confederate aggregation of all nations, as opposed to a low-level union in sameness and uniformity imposed by coercion, as also to a fictitious ideal clinging to a self-isolating, and independent cut-off entity retired on a cultural, mental and social fortress-island.

We can already see this in the world's events. There is a growing sense of unity that is no longer perceived as an abstract ideal of some lonely philosopher or mystic but is becoming evident as the years pass. It expresses itself in the environmental movements, the growing compassion for others, the fight against various forms of discrimination, or in the fact that some people care about other peoples and cultures that they don't even know about or take care of animals, plants, etc. There is a deepening of contact with the Earth as a whole, exemplified by the awe of astronauts who look down at the Earth and don't see borders, walls, or divisions but only one unique human family. The question is no longer whether we are heading towards unity in diversity but, rather, how we are supposed to build it.

Though the chaos and turmoil we see nowadays in the world might suggest otherwise, the divisions and polarizations, and all the conflicts, strife, and struggle are the pains and labor of the birth of a new social order that will have to progressively reflect what Nature asks to manifest. It is about the perfection of a social collective where its common inner Self is discovered and where each of our different individualities is a diverse expression. Also, the human social structure, as well as its activities and institutions, and the relationship between peoples and nations, will have to follow the inner spiritual growth as the reflection of a common Self that is one and many at the same time.

Therefore, a true human oneness can't be a superficial and external uniform and centralized Moloch of commercial or political power but will have to be structured according to the spiritual principles of a decentralized multiplicity in unity.

The hallmark of this subjective age will be the freedom of the individual, which, however, will nevertheless find a harmonious balance with the freedom and rights of the collective. The degree of freedom will be proportional to the assumption of responsibility (contrary to the false conception of individual freedom that does not care about the common good and must be controlled by repressing countermeasures, as it is still, to a large degree, necessary in our present social order). It will be characterized by the collective discovery of the soul and reflect itself as the urge and necessity of each individual for self-expression, self-realization, and self-perfection. The individual inner self, the spirit in us, will ask with increasing intensity for self-expression. However, this inner self will be recognized for the soul of others as well and will express itself in a movement of solidarity with all beings. The Oneness in us will be felt as the same Oneness in which all beings move and live and through which all can meet. Only then will the tension between personal freedoms and rights not clash with the general interests and not be seen as disrupting the common social order. On the contrary, it will enrich unity precisely because of its diversity.

But how can we reconcile this futuristic ideal with the facts on the ground? The present human condition seems to contradict this trend, as is amply documented by the manifestation of a resurgence of isolationistic and authoritarian tendencies, as well as divisive social and political struggles. From the higher perspective, precisely this recrudescence of the old and conservative forces from below is the distinctive sign of its reaction against a spiritual pressure for change that is coming from above. The spread of the ideals of unity in the human consciousness triggers the backlash of a collective subconscious that resists transformation. Because the change of authority from a mind-centered to a soul-centered society is such an extensive and far-reaching transformation and transmutation, obviously, the old forces that still retain control over the collective vital and mental consciousness will resist and fight it with whatever means necessary. Therefore, it is not surprising that, because of an increase of spiritual light in parts of human society, other parts resist the crumbling of old structures and values that were always taken for guaranteed but aren't anymore. It is the reaction of a collective subconscious cleansing. This is not only a global phenomenon but something we can observe in ourselves individually; at each step forward in our spiritual journey, we must deal with the 'dark side of the force' that we all have in us.[118] Looking at the broader picture, one sees the forces at work at a subconscious and subliminal level. Conflicts, crises, and struggles are always the sign of a transition from an old to a new social order, which sometimes succeeds and sometimes doesn't. It is the resistance of a consciousness that still clings to the past and refuses a new alignment with a spiritual truth that not only tries to realize something new but also triggers, by its light from above, a reaction from below. The forces that resist change pull back and profit from the weaknesses of our untransformed lower vital impulses and the non-spiritualized minds, such as

---

[118] That the film industry built its fortunes upon the 'dark forces' is no coincidence. It intuits a deeper truth and reality. We can only hope that the day will come when it will make similar fortunes by narrating the epics of the Forces of Light.

the desire for a nostalgic return to a romanticized past or the instinct to resort to authoritarian methods. Sometimes this leads to the recoil of the evolutionary progress. Evolution has never been a linear process and has always been constellated by ups and downs. Nevertheless, in the long run, the progress towards a new change of consciousness is inevitable.

The leitmotif is that the real change won't come from some new (more or less hyped) technological breakthroughs but by a collective ideal of human unity and a move towards a life within. It is a spiritual discovery met by an even vaster material development that a mind-based materialism cannot realize. A revolutionary reconstruction of all human activities, from the individual to the universal, is the last inevitable outcome that will follow the discovery of the truth and law of our deepest spirit. Once we better know ourselves, we will also discover unexpected potentials that will mold and transform our present condition. Once we become aware that the knowledge of the physical world is only a part of knowledge, our material, economic and social institutions will be transformed, replaced, and eventually, partially abolished. Knowing our own being will decide the individual and social destiny. An inner Light and self-consciousness going beyond the analytical superficialities is the key. If we give way to this ideal of intuitional and psychic knowledge and a deeper self-awareness, our inner life, as our outer life, will be dramatically affected for good. Living and acting out of our soul will lead us to the more or less secretly held dream of a grandiose change.

The journey towards a divine Life—that of a gnostic collectivity—will not be a sudden and abrupt leap. It will go through a series of transitions, one of which we are in the midst of. It is only a matter of time before the past will be surpassed, and the old methods that might have been successful and meaningful for centuries will no longer work and have to be replaced. The subjective age will embrace the ideal of human unity in diversity, leading to a collective spiritual awakening, and its transformation will be the inevitable destiny of the race. It will be an age of true self-expression, an ever-deepening subjectivism of a new collective self-consciousness with a true development towards a higher goal that will discover what the rational and objective cannot see. It will be a world that will progress towards a post-material synthesis of knowledge between matter and consciousness, Spirit and Nature, East and West, which, however, will not be just the sum of a matter-centered and a soul-centered ideal, but a third position: a divine materialism going to realize the delight of being in life and matter.

Acknowledgements

My special thanks to Tonya Blust and Philip Patrick; they are both exceptional proofreaders whom I surely recommend. Without their help, this book would have sounded ridiculous! (**www.fiverr.com/bluston** - **www.fiverr.com/inbox/philpen24**)

Further suggested Readings

"The End of Science: Facing the Limits of Knowledge in the Twilight of the Scientific Age," John Horgan, Abacus, 1996.

"Conscious Mind in the Physical World", by Euan Squires, Adam Hilger.

"The Outer Limits of Reason", by Noson S. Yanofsky, MIT Press.

"The Master and His Emissary: The Divided Brain and the Making of the Western World", by Iain McGilchrist, Yale University Press.

"The Spiritual Brain: A Neuroscientist's Case for the Existence of the Soul", by Mario Beauregard, HarperOne.

"Evolution 2.0: Breaking the Deadlock Between Darwin and Design", by Perry Marshall, PenBella Books.

"The Works of His Hands: A Scientist's Journey from Atheism to Faith", by Sy Garte Ph.D., Kregel Publications.

"The First Minds: Caterpillars, Karyotes, and Consciousness", by Arthur S. Reber, Oxford University Press.

"Purpose and Desire: What Makes Something 'Alive' and Why Modern Darwinism Has Failed to Explain It", by J. Scott Turner, HarperOne

"Evolution: A View from the 21st Century", by James A. Shapiro, Financial Times Press Science

"The Idea of the World: A multi-disciplinary argument for the mental nature of reality", by B. Kastrup, John Hunt Publishing.

"Choosing Reality: A Buddhist View of Physics and the Mind", by B. Allan Wallace, Snow Lion.

"Yoga Psychology and the Transformation of Consciousness: Seeing Through the Eyes of Infinity", by D. Salmon and J. Maslow, Paragon House.

"Beyond Physicalism: Toward Reconciliation of Science and Spirituality", by E. Kelly, C. Crabtree, P. Marshall, Rowman & Littlefield Publishers.

"Consciousness Unbound: Liberating Mind from the Tyranny of Materialism", by E. F. Kelly, P. Marshall, Rowman & Littlefield Publishers

"Understanding Consciousness", by Max Velmans, Routledge.

Bibliography

[1]   I. Newton, The Mathematical Principles of Natural Philosophy, 1729, p. 384.

[2]   J. Franklin, Ptolemy attributed by James Franklin in The Science of Conjecture: Evidence and Probability before Pascal., The Johns Hopkins University Press, 2001, p. 241.

[3]   A. Baker, "Occam's Razor in science: a case study from biogeography," *Biology and Philosophy,* vol. 22, p. 193–215, 2007.

[4]   R. Ziegler, "The discovery of non-Euclidean geometries and its consequences: observations on the history of consciousness in the nineteenth century," *Magazine of morphology,* vol. 1, pp. 3-32, 1998.

[5]   A. Scheinin and al., "Foundations of human consciousness: Imaging the twilight zone," *Journal of Neuroscience,* 2020.

[6]   E. Preston, "A 'Self-Aware' Fish Raises Doubts About a Cognitive Test," December 2018. [Online]. Available: https://www.quantamagazine.org/a-self-aware-fish-raises-doubts-about-a-cognitive-test-20181212.

[7]   J. Locke, An essay concerning human understanding - Bk II, ch. I, 1689.

[8]   R. Descartes, Search after Truth (Inquisitio Veritatis), 1701.

[9]   T. Nagel, "What is it like to be a bat?," *The Philosophical Review,* vol. 83, no. 4, p. 435–450, 1974.

[10] U. Winter and al., "Content-Free Awareness: EEG-fcMRI Correlates of Consciousness as Such in an Expert Meditator," *Front. Psychol.,* 2020.

[11] P. S. Laplace, A Philosophical Essay on Probabilities - 1814 - Translated into English from the original French 6th ed. by Truscott, F.W. and Emory, F.L. ed., Dover Publications, 1951.

[12] E. C. Hayden, "Life is complicated," *Nature,* vol. 464, p. 664, 1 April 2010.

[13] H. R. Max Roser, "Cancer," 2020. [Online]. Available: https://ourworldindata.org/cancer.

[14] D. Chalmers, "Facing up to the problem of consciousness," *Journal of Consciousness Studies,* vol. 2, pp. 200-19, 1995.

[15] M. Lenharo, "Decades-long bet on consciousness ends — and it's philosopher 1, neuroscientist 0," *Nature,* Vols. 14-15, p. 619, 2023.

[16] F. Jackson, "What Mary Didn't Know," *The Journal of Philosophy,* vol. 83, no. 5, p. 291–295, 1986.

[17] Y. Zhang and al., "The causal role of alpha oscillations in feature binding," *Proceedings of the National Academy of Sciences of the United States of America,* 2019.

[18] W. Seager, Theories of Consciousness. An introduction and assessment, Routledge, 2002.

[19] G. Smith, The AI Delusion, Oxford University Press.

[20] H. Bortoft, The Woleness of Nature, Lindisfarne Press.

[21] D. Buonomano, Brain Bugs, W. W. Norton & Company Inc, New York., 2001.

[22] P. A. Kolers and M. v. Grünau, "Shape and color in apparent motion," *Vision Research,* vol. 16, p. 329–35, 1976.

[23] F. A. Geldard and C. E. Sherrick, "Space, Time and Touch," *Scientific American,* July 1, 1986.

[24] F. A. Geldard and Sherrick, "The Cutaneous "Rabbit": A Perceptual Illusion," *Science,* vol. 178, no. 4057, p. 178–179, 1972.

[25] T. Bachmann, E. Poder and I. Luiga, "Illusory reversal of temporal order: the bias to report a dimmer stimulus," *Vision Research,* vol. 44, no. 3, p. 241–6, 2004.

[26] D. M. Eagleman, "Motion Integration and Postdiction in Visual Awareness," *Science,* vol. 287 (5460), p. 2036–2038, 2000.

[27] M. A. Khoei, G. S. Masson and L. U. Perrinet, "The Flash-Lag Effect as a Motion-Based Predictive Shift," *PLoS Comput Biol.,* vol. 13, no. 1, 2017.

[28] M. H. Herzog and al., "All in Good Time: Long-Lasting Postdictive Effects Reveal Discrete Perception," *Trends in Cognitive Sciences,* September 03, 2020.

[29] B. Tessel and al., "Predictions drive neural representations of visual events ahead of incoming sensory information," *Proceedings of the National Academy of Sciences,* 2020.

[30] G. Humphreys and M. J. Riddoch, "Interactions between objects and space systems revealed through neuropsychology.," *Attention and Performance - D.E. Meyer & Kornblum,* vol. XIV, pp. 183-218, 1993.

[31] T. Landis and al., "Visual recognition through kinaesthetic mediation," vol. 12, no. 3, pp. 515-31, 1982.

[32] M. v. Senden, Space and sight, Methuen, London, 1960.

[33] "Project Prakash," [Online]. Available: https://www.projectprakash.org/.

[34] R. Held, Y. Ostrovsky and al., "The newly sighted fail to match seen with felt," *Nature Neuroscience,* vol. 14, no. 5, p. 551–553, 2011.

[35] R. Velik, "From simple receptors to complex multimodal percepts: a first global picture on the mechanisms involved in perceptual binding," *Frontiers in Psychology - Sec. Cognitive Science,* vol. 3, 2012.

[36] M. Masi, Quantum Physics: An Overview of a Weird World - A Guide to the 21st Century Revolution, vol. II, 2020.

[37] J. Searle, "Minds, Brains and Programs," *Behavioral and Brain Sciences,* vol. 3, p. 417–57, 1980.

[38] S. Harnad, "The symbol grounding problem," *Physica D: Nonlinear Phenomena,* vol. 42, no. 1-3, pp. 335-346, 1990.

[39] R. Brette, "Is coding a relevant metaphor for the brain?," *Behavioral and Brain Sciences,* vol. 42, no. E215, 2019.

[40] J. Searle, "Is the Brain a Digital Computer?," *American Philosophical AssociationStable,* vol. 64, no. 3, pp. 21-37, 1990.

[41] D. Heaven, "Why deep-learning AIs are so easy to fool," Nature News Feature, 9 October 2019. [Online].

[42] T. Forese and S. Taguchi, "The Problem of Meaning in AI and Robotics: Still with Us after All These Years," *Philosophies - MDPI,* vol. 4, no. 2, p. 14, 2019.

[43] S. Ajina and B. Holly, "Blindsight and Unconscious Vision: What They Teach Us about the Human Visual System," *The Neuroscientist : a review journal bringing neurobiology, neurology and psychiatry,* vol. 23, no. 5, 2016.

[44] D. Robson, "Blindsight: the strangest form of consciousness," *BBC Future,* 28 September 2015.

[45] A. Vandenbroucke, J. Fahrenfort and I. Sligte, "Seeing without knowing: neural signatures of perceptual inference in the absence of report," *Jour. Cogn. Neuroscience,* vol. 26, no. 5, pp. 955-69, 2014.

[46] N. Tsuchiya, M. Wilke and al., "No-Report Paradigms: Extracting the True Neural Correlates of Consciousness," *Trends in Cognitive Science,* vol. 19, no. 12, pp. 757-770, 2015.

[47] D. Oudiette and al., "Dreamlike mentations during sleepwalking and sleep terrors in adults," *Sleep,* vol. 32, no. 12, pp. 1621-7, 2009.

[48] K. Konkoly and al., "Real-time dialogue between experimenters and dreamers during REM sleep," *Current Biology,* february 2021.

[49] J. M. Windt and al., "Does Consciousness Disappear in Dreamless Sleep?," *Trends in Cognitive Sciences,* vol. 20, no. 2, pp. 871-882, 2016.

[50] F. Siclari, G. Tononi and al., "Dreaming in NREM Sleep: A High-Density EEG Study of Slow Waves and Spindles," *Journal of Neuroscience,* vol. 38, no. 43, pp. 9175-9185, 2018.

[51] A. M. Owen and M. R. Coleman, "Detecting Awareness in the Vegetative State," *Annals of the New York Academy of Sciences,* vol. 1129, pp. 130-8, 2006.

[52] D. Cruse and al., "Bedside detection of awareness in the vegetative state: a cohort study," *The Lancet,* vol. 378, no. 9809, pp. 2088-2094, 2011.

[53] G. Mackenzie, "Can they Feel? The Capacity for Pain and Pleasure in Patients with Cognitive Motor Dissociation," *Neuroethics,* vol. 12, p. 153–169, 2019.

[54] Y. G. B. and e. al., "Cognitive Motor Dissociation in Disorders of Consciousness," *New England Journal of Medicine,* vol. 391, no. 7, pp. 598-608, 2024.

[55] "Consciousness is partly preserved during general anesthesia," ScienceDaily, 2018. [Online]. Available: https://www.sciencedaily.com/releases/2018/07/180703105631.htm.

[56] L. Radek, "Dreaming and awareness during dexmedetomidine- and propofol-induced unresponsiveness," *BJA British Journal of Anaesthesia,* vol. 121, no. 1, pp. 260-269, 2018.

[57] J. Claassen, K. Doyle, A. Matory and al., "Detection of Brain Activation in Unresponsive Patients with Acute Brain Injury," *N. Engl. J. Med.,* vol. 380, pp. 2497-2505, 2019.

[58] B. Libet and al., "Subjective Referral of the Timing for a Conscious Sensory Experience – A Functional Role for the Somatosensory Specific Projection System in Man," *Brain,* vol. 102, no. 1, p. 193–224, 1979.

[59] B. Libet, "Readiness-potentials preceding unrestricted 'spontaneous' vs. pre-planned voluntary acts," *Electroencephalography and Clinical Neurophysiology,* vol. 54, no. 3, pp. 322-335, 1982.

[60] B. Libet and al., "Time of Conscious Intention to Act in Relation to Onset of Cerebral Activity," *Brain,* vol. 106, no. 3, p. 623–642, 1983.

[61] B. Libet, "Unconscious cerebral initiative and the role of conscious will in voluntary action," *The Behavioral and Brain Sciences,* vol. 8, p. 529–566, 1985.

[62] P. Haggard and M. Eimer, "On the Relation between Brain Potentials and the Awareness of Voluntary Movements," *Experimental Brain Research,* vol. 126, p. 128–133, 1999.

[63] S. Kühn and M. Brass, "Retrospective construction of the judgement of free choice," *Consciousness and Cognition,* vol. 18, no. 1, p. 12–21, 2009.

[64] M. Schultze-Kraft and al., "The point of no return in vetoing self-initiated movements," *Proceedings of the National Academy of Sciences,* vol. 113, no. 4, pp. 1080-1085, 2016.

[65] S. Uithol and A. Schurger, "Reckoning the moment of reckoning," *Proceedings of the National Academy of Sciences,* vol. 113, no. 4, pp. 817-819, 2016.

[66] M. Brass, A. Furstenberg and A. R. Mele, "Why neuroscience does not disprove free will," *Neuroscience and Biobehavioral Reviews,* vol. 102, pp. 251-263, 2019.

[67] A. Schurger and al., "What Is the Readiness Potential?," *Trends in Cognitive Sciences,* vol. 25, no. 7, pp. 558-570, 2021.

[68] D. Bredikhin and al., "(Non)-experiencing the intention to move: On the comparisons between the Readiness Potential onset and Libet's W-time," *Neuropsychologia,* vol. 185, p. 108570, 2023.

[69] U. Maoz, G. Yaffe and C. M. L. Koch, "Neural precursors of decisions that matter—an ERP study of deliberate and arbitrary choice," *eLife,* Oct. 23, 2019.

[70] A. N. Whitehead, The function of reason, Beacon Pr; Highlighting/Underlining edition, 1971.

[71] I. Yaron and al., "The ConTraSt database for analysing and comparing empirical studies of consciousness theories," *Nature Human Behaviour,* vol. 6, p. 593–604, 2022.

[72] K. Kohitij and B. Krekelberg, "Transcranial electrical stimulation over visual cortex evokes phosphenes with a retinal origin," *Journal of Neurophysiology,* vol. 108, no. 8, p. 2173–2178, 2012.

[73] D. Yoshor and al., "Dynamic Stimulation of Human Visual Cortex Produces Useful Percepts of Visual Forms in Sighted and Blind Subjects," *Neurosurgery,* vol. 65, no. 1, p. 117, 2018.

[74] W. James, Human immortality, Houghton, Mifflin and Company, The Riverside Press, Cambridge, 1898.

[75] A. Huxley, The Doors of Perception, Chatto and Windus, 1954.

[76] Y. Feng and al., "A new case of complete primary cerebellar agenesis: clinical and imaging findings in a living patient," *Brain,* vol. 138, no. 6, p. 353, 2105.

[77] J. Y. J. H. N. Nakuci and e. al., "Multiple brain activation patterns for the same perceptual decision-making task.," *Nature Comm.,* vol. 16, p. 1785, 2025.

[78] L. v. Melchner, S. L. Pallas and M. Sur, "Visual behaviour mediated by retinal projections directed to the auditory pathway," *Nature,* vol. 404, p. 871–876 , 2000.

[79] M. Redinbaugh and al., "Thalamus Modulates Consciousness via Layer-Specific Control of Cortex," *Neuron,* vol. 106, no. 1, pp. 66-75, 2020.

[80] Z. Huang and al., "Anterior insula regulates brain network transitions that gate conscious access," *Cell Reports,* vol. 35, no. 5, 2021.

[81] M. Liang and al., "Primary sensory cortices contain distinguishable spatial patterns of activity for each sense," *Nature Communications,* vol. 4, no. Article number: 1979, 2013.

[82] L. B. Merabet and al., "Rapid and Reversible Recruitment of Early Visual Cortex for Touch," *Plos One,* 2008.

[83] J. Gonzalez-Castill and al., "Whole-brain, time-locked activation with simple tasks revealed using massive averaging and model-free analysis," *PNAS,* vol. 109, no. 14, pp. 5487-5492, 2012.

[84] M. Rule and al., "Causes and consequences of representational drift," *Computational Neuroscinece,* vol. 58, pp. 141-147, 2019.

[85] C. E. Schoonover and al., "Representational drift in primary olfactory cortex," *Nature,* vol. 594, p. 541–546, 2021.

[86] Y. Pinto and al., "Split brain: divided perception but undivided consciousness," *Brain,* vol. 140, no. 5, p. 1231–1237, 2017.

[87] Y. Pinto and al., "Unified Visual Working Memory without the Anterior Corpus Callosum," *Symmetry,* vol. 12, no. 12, p. 2106, 2020.

[88] B. J. Baars, A Cognitive Theory of Consciousness, New York: Cambridge University Press, 1988.

[89] M. Oizumi, L. Albantakis and G. Tononi, "From the Phenomenology to the Mechanisms of Consciousness: Integrated Information Theory 3.0," vol. 10, no. 5, 2014.

[90] G. Tononi, "Integrated information theory," Scholarpedia, [Online]. Available: http://www.scholarpedia.org/article/Integrated_information_theory.

[91] R. L. Kuhn, "A Landscape of Consciousness: Toward a Taxonomy of Explanations and Implication," *Progress in Biophysics and Molecular Biology,* 2024.

[92] A. K. Seth and T. Bayne, "Theories of consciousness," *Nat Rev Neurosci,* vol. 23, no. 7, pp. 439-452, 2022.

[93] H. H. Mørch, "Non-physicalist Theories of Consciousness," Cambridge, Cambridge University Press, 2024.

[94] M. Gazzaniga, "MRI assessment of human callosal surgery with neuropsychological correlates," *Neurology,* vol. 35, no. 12, pp. 1763-6, 1985.

[95] D. Kliemann and al., "Intrinsic Functional Connectivity of the Brain in Adults with a Single Cerebral Hemisphere," *Cell Reports,* vol. 29, no. 8, pp. 2398-2407, 2019.

[96] L. Muckli and al., "Bilateral visual field maps in a patient with only one hemisphere," *PNAS,* vol. 106, no. 31, pp. 13034-13039, 2009.

[97] R. A. McGovern and al., "Hemispherectomy in adults and adolescents: Seizure and functional outcomes in 47 patients," *Epilepsia,* vol. 60, p. 2416– 2427, 2019.

[98] M. C. Granovetter, S. Leah Ettensohn and M. Behrmann, "With Childhood Hemispherectomy, One Hemisphere Can Support--But is Suboptimal for--Word and Face Recognition," *PNAS,* vol. 119, no. 44, 2022.

[99] F. Vargha-Khadem and al., "Onset of speech after left hemispherectomy in a nine-year-old boy," *Brain,* vol. 120, no. 1, pp. 159-82, 1997.

[100] G. Tuckute and al., "Frontal language areas do not emerge in the absence of temporal language areas: A case study of an individual born without a left temporal lobe," *Neuropsychologia,* vol. 169, p. 108184, 2022.

[101] R. Lewin, "Is your brain really necessary?," *Science,* vol. 210, no. 4475, pp. 1232-1234, 1980.

[102] L. Feuillet and al., "Brain of a white-collar worker," *The Lancet,* vol. 370, no. 9583, p. 262, 2007.

[103] S. A. Heldstab and al., "When ontogeny recapitulates phylogeny: Fixed neurodevelopmental sequence of manipulative skills among primates," *Science Advances,* vol. 6, no. 30, 2020.

[104] C. Fichtel, K. Dinter and P. M. Kappeler, "The lemur baseline: how lemurs compare to monkeys and apes in the Primate Cognition Test Battery," *Zoological Science,* 2020.

[105] Y. Assaf and al., "Conservation of brain connectivity and wiring across the mammalian class," *Nature Neuroscience,* vol. 23, pp. 805-808, 2020.

[106] A. Nieder, L. Wagener and P. Rinnert, "A neural correlate of sensory consciousness in a corvid bird," *Science,* vol. 369, no. 6511, pp. 1626-1629, 2020.

[107] M. Stacho and al., "A cortex-like canonical circuit in the avian forebrain," *Science,* vol. 369, no. 6511, 2020.

[108] O. Güntürkün, R. Pusch and J. Rose, "Why birds are smart," *Trends in cognitive Science,* vol. 28, no. 3, pp. 197-209, 2024.

[109] V. Cox and al., "Crustacean & Cephalopod Sentience Briefing," 2021.

[110] American Academy of Nuerology, "Practice Guideline Update Systematic Review Summary: Disorders of Consciousness," 2018. [Online]. Available: https://www.aan.com/Guidelines/home/GuidelineDetail/927.

[111] D. Shewmon, G. Holmes and P. Byrne, "Consciousness in congenitally decorticate children: developmental vegetative state as self-fulfilling prophecy," *Dev Med Child Neurol.,* vol. 41, no. 6, pp. 364-74, 1999.

[112] B. Merker, "Consciousness without a cerebral cortex: A challenge for neuroscience and medicine," *Behavioral and Brain Sciences ,* vol. 30, no. 1, pp. 63 - 81, 2007.

[113] M. Solms and J. Panksepp, "The "Id" Knows More than the "Ego" Admits: Neuropsychoanalytic and Primal Consciousness Perspectives on the Interface Between Affective and Cognitive Neuroscience," *Brain Sciences,* vol. 2, no. 2, pp. 147-175, 2012.

[114] W. A. Zeiger and al., "Barrel cortex plasticity after photothrombotic stroke involves potentiating responses of pre-existing circuits but not functional remapping to new circuits," *Nature Communications,* vol. 12, p. 3972, 2021.

[115] T. Makin and J. Krakauer, "Against cortical reorganisation," *eLife,* vol. 84716, p. 12, 2023.

[116] M. Nahm, D. Rousseau and B. Greyson, "Discrepancy between cerebral structure and cognitive functioning," *J. Nerv. Ment. Dis.,* vol. 205, no. 12, pp. 967-972, 2017.

[117] W. Penfield, Mystery of the Mind: A Critical Study of Consciousness and the Human Brain, Princeton University Press, 1975.

[118] W. Men and al., "The corpus callosum of Albert Einstein's brain: another clue to his high intelligence?," *Brain,* vol. 137, no. 4, 2014.

[119] A. Jefferson, "What does it take to be a brain disorder?," *Synthese,* vol. 197, p. 249–262, 2020.

[120] J. A., "On Mental Illness and Broken Brains," *Think,* vol. 20, no. 58, pp. 103-112, 2021.

[121] "Are "mental disorders" "brain disorders"? - Neuoscience & Philosophy Salon," 2024. [Online]. Available: https://youtu.be/mFiJIaE77aw.

[122] L. Robert and al., "Neural correlates of the psychedelic state as determined by fMRI studies with psilocybin," *Proceedings of the National Academy of Sciences,* vol. 109, no. 6, pp. 2138-2143, 2012.

[123] F. Palhano-Fontes and al., "The Psychedelic State Induced by Ayahuasca Modulates the Activity and Connectivity of the Default Mode Network," *PLoS One,* 2015.

[124] L. Robin, "Neural correlates of the LSD experience revealed by multimodal neuroimaging," *PNAS,* vol. 113, no. 17, pp. 4853-4858, 2016.

[125] C. R. Lewis, "Two dose investigation of the 5-HT-agonist psilocybin on relative and global cerebral blood flow," *Neuroimage,* vol. 159, pp. 70-78, 2017.

[126] K. Hooks, J. P. Konsman and M. A. O'Malley, "Microbiota-gut-brain research: A critical analysis," *Behavioral and Brain Sciences,* vol. 12, pp. 1-40, 2018.

[127] G. Colombetti and E. Zavala, "Are emotional states based in the brain? A critique of affective brainocentrism from a physiological perspective," *Biol Philos,* vol. 34, no. 45, 2019.

[128] T. Bastiaanssen and al., "Making Sense of ... the Microbiome in Psychiatry.," *Int. J. Neuropsychopharmacol.,* vol. 22, no. 1, pp. 37-52, 2019.

[129] A. Ciaunica, E. Shmeleva and M. Levin, "The brain is not mental! coupling neuronal and immune cellular processing in human organisms," *Front. Integr. Neurosci.,* vol. 17, 2023.

[130] A. Batthyány and B. Greyson, "Spontaneous remission of dementia before death: Results from a study on paradoxical lucidity.," *Psychology of Consciousness,* vol. 8, no. 1, p. 1–8, 2021.

[131] T. Shomrat and M. Levin, "An automated training paradigm reveals long-term memory in planarians and its persistence through head regeneration," *Jour. of Experimental Biology,* vol. 216, pp. 3799-3810, 2013.

[132] A. Bédécarrats, S. Chen, K. Pearce, D. Cai and D. Glanzman, "RNA from trained Aplysia can induce an epigenetic engram for long-term sensitization in untrained Aplysia," *eNeuro,* vol. 5, no. 3, 2018.

[133] R. Moore and al., "The role of the Cer1 transposon in horizontal transfer of transgenerational memory," *Cell,* vol. 184, no. 18, pp. 4697-4712, 2021.

[134] S. Chen, D. L. Glanzman and al., "Reinstatement of long-term memory following erasure of its behavioral and synaptic expression in Aplysia," *eLife,* no. 3:e03896, 2014.

[135] J. Smeets and al., "Brain tissue plasticity: protein synthesis rates of the human brain," *Brain,* vol. 141, no. 4, p. 1122–1129, 2018.

[136] E. Millesi and al., "Hibernation effects on memory in European ground squirrels (Spermophilus citellus)," *J Biol Rhythms,* vol. 16, no. 3, pp. 264-71, 2001.

[137] L. Clemens, G. Heldmaier and C. Exner, "Keep cool: Memory is retained during hibernation in Alpine marmots," *Physiology & Behavior,* vol. 98, no. 1-2, pp. 78-84, 2009.

[138] I. Ruczynski and B. Siemers, "Hibernation does not affect memory retention in bats," *Biol. Lett.,* vol. 7, p. 153–155, 2011.

[139] J. Lázaro and al., "Profound reversible seasonal changes of individual skull size in a mammal," *Curr. Biol.,* vol. 27, no. 20, 2017.

[140] P. Hepper and B. Waldman, "Embryonic Olfactory Learning in Frogs," *Quarterly Journal of Experimental Psychology,* vol. 44, no. 3-4b, 1992.

[141] D. Blackiston, E. Casey and M. Weiss, "Retention of Memory through Metamorphosis: Can a Moth Remember What It Learned As a Caterpillar?," *PLOS ONE,* 2008.

[142] D. J. Blackiston, T. Shomrat and M. Levin, "The stability of memories during brain remodeling: A perspective," *Communicative & Integrative Biology,* vol. 8, no. 5, 2015.

[143] C. Carneiro and al., "Effect size and statistical power in the rodent fear conditioning literature – A systematic review," *PLOS ONE,* 2018.

[144] H. Bergson, Matter and Memory, New York: McMillan, 1896/1912.

[145] D. Forsdyke, "Wittgenstein's Certainty is Uncertain: Brain Scans of Cured Hydrocephalics Challenge Cherished Assumptions," *Biol Theory,* vol. 10, p. 336–342, 2015.

[146] S. Mancuso, "Federico Delpino and the foundation of plant biology," *Plant Signal Behav.,* vol. 5, no. 9, p. 1067–1071, 2010.

[147] C. Darwin and F. Darwin, The Power of Movements in Plants, New York: D. Appleton and Company, 1898.

[148] J. C. Bose, "The Nervous Mechanism of Plants," *Nature,* vol. 118, p. 654–655, 1926.

[149] P. Tomkins and C. Bird's, The Secret Life of Plants, Harper & Row, 1973.

[150] E. D. Brenner and al., "Plant neurobiology: an integrated view of plant signaling," *Trends in Plant Science,* vol. 11, no. 8, pp. 413-419, 2006.

[151] A. Alpi and al., "Plant neurobiology: no brain, no gain?," *Trends in Plant Science,* vol. 12, no. 4, pp. 135-136, 2007.

[152] P. Calvo, "The philosophy of plant neurobiology: a manifesto," *Synthese,* vol. 193, p. 1323–1343, 2016.

[153] M. Gagliano and al., "Learning by Association in Plants," *Sci. Rep.,* vol. 6:38427, 2016.

[154] S. Guerra, U. Castiello and al., "Flexible control of movement in plants," *Nature Sci. Rep.,* vol. 9, no. 16570, 2019.

[155] J. White and F. Yamashita, "Boquila trifoliolata mimics leaves of an artificial plastic host plant," *Plant Signaling & Behavior,* p. 1977530, 2021.

[156] S. Simard, "Mycorrhizal Networks Facilitate Tree Communication, Learning, and Memory," in *Memory and Learning in Plants*, Springer, 2018, pp. 191-213.

[157] I. Khait and al., "Sounds emitted by plants under stress are airborne and informative," *Cell,* vol. 186, no. 7, pp. 1328-1336, 2023.

[158] M. C. P. Segundo-Ortin, "Consciousness and cognition in plants," *WIREs Cognitive Science,* vol. 13, no. 2, 2021.

[159] P. Lyon and al., "Basal cognition: conceptual tools and the view from the single cell," *Philosophical Transactions of the Royal Society B,* vol. 376, no. 1820, 2021.

[160] L. Taiz and al., "Plants Neither Possess nor Require Consciousness," *Trends in Plant Science,* vol. 24, no. 8, pp. 677-687, 2019.

[161] J. Bielecki and al., "Associative learning in the box jellyfish Tripedalia cystophora," *Current Biology,* vol. 33, no. 19, pp. 4150-4159, 2023.

[162] I. De la Fuente, C. Bringas, I. Malaina and al., "Evidence of conditioned behavior in amoebae," *Nature Communications,* vol. 10, no. 3690, 2019.

[163] H. S. Jennings, Behavior of the Lower Organisms, New York: Columbia University Press, 1915.

[164] "New study hints at complex decision making in a single-cell organism," Science Daily - Harvard Medical School, 5 December 2019. [Online]. Available: www.sciencedaily.com/releases/2019/12/191205113129.htm.

[165] S. Gershman, P. Balbi, C. Gallistel and J. Gunawardena, "Reconsidering the evidence for learning in single cells," *eLife,* vol. 10:e61907., 2021.

[166] J. P. Dexter, S. Prabakaran and J. Gunawardena, "A Complex Hierarchy of Avoidance Behaviors in a Single-Cell Eukaryote," *Current Biology,* vol. 29, no. 24, pp. 4323-4329, 2019.

[167] T. Nakagaki, H. Yamada and Á. Tóth, "Maze-solving by an amoeboid organism," *Nature,* vol. 407, no. 470, 2000.

[168] T. Nakagaki, "Obtaining multiple separate food sources: behavioural intelligence in the Physarum plasmodium," *Proc. R. Soc. Lond. B,* vol. 271, p. 2305–2310, 2004.

[169] T. Saigusa and al., "Amoebae anticipate periodic events," *Phys Rev Lett.,* vol. 100(1), no. 018101, 2008.

[170] R. P. Boisseau, D. Vogel and A. Dussutour, "Habituation in non-neural organisms: evidence from slime moulds," *Proc. Roy. Soc. B,* vol. 283, no. 1829, 2016.

[171] M. Kramar and K. Alim, "Encoding memory in tube diameter hierarchy of living flow network," *PNAS,* vol. 118, no. 10, March 9, 2021.

[172] C. Reid, "Thoughts from the forest floor: a review of cognition in the slime mould Physarum polycephalum," *Animal Cognition,* 2023.

[173] Y. Fukasawa, M. Savoury and L. Boddy, "Ecological memory and relocation decisions in fungal mycelial networks: responses to quantity and location of new resources," *The ISME Journal,* vol. 14, p. 380–388, 2020.

[174] K. Aleklett and L. Boddy, "Fungal behaviour: a new frontier in behavioural ecology," *Trends in Ecology and Evolution,* vol. 36, no. 9, pp. 787-796, 2021.

[175] N. P. Money, "Hyphal and mycelial consciousness: the concept of the fungal mind," *Fungal Biology,* vol. 125, no. 4, pp. 257-259, 2021.

[176] A. Adamatzky, "Language of fungi derived from their electrical spiking activity," *The Royal Society Open Science,* vol. 9, no. 4, 2022.

[177] G. Gavelis and al., "Eye-like ocelloids are built from different endosymbiotically acquired components," *Nature,* vol. 523, p. 204–207, 2015.

[178] T. Phan and al., "Bacterial Route Finding and Collective Escape in Mazes and Fractals," *Phys. Rev.,* vol. X, no. 10, p. 031017, 2020.

[179] L. Tweedy and al., "Seeing around corners: Cells solve mazes and respond at a distance using attractant breakdown," *Science,* vol. 369, no. 6507, 2020.

[180] P. Lyon, "The cognitive cell: bacterial behavior reconsidered.," *Front Microbiol.,* vol. 6, p. 264, 2015.

[181] J. Carrasco-Pujante and al., "Associative Conditioning Is a Robust Systemic Behavior in Unicellular Organisms: An Interspecies Comparison," *Forntiers in Microbiology,* vol. 1968, p. 12, 2021.

[182] S. Biswas, W. Clawson and M. Levin, "Learning in Transcriptional Network Models: Computational Discovery of Pathway-Level Memory and Effective Interventions.," *International Journal of Molecular Sciences,* vol. 24, no. 1, 2023.

[183] M. Picard and C. Sandi, "The social nature of mitochondria: Implications for human health," *Neuroscience & Biobehavioral Reviews,* vol. 120, pp. 595-610, 2021.

[184] S. J. Gershman and al, "Reconsidering the evidence for learning in single cells," *eLife,* 2021.

[185] B. J. Ford, "Cellular intelligence: Microphenomenology and the realities of being," *Progress in Biophysics and Molecular Biology,* vol. 131, pp. 273-287, 2017.

[186] A. Boussard and al., "Adaptive behaviour and learning in slime moulds: the role of oscillations," *Philosophical transactions of the Royal Society B,* vol. 376, no. 1820, 2021.

[187] J. A. Shapiro, "All living cells are cognitive," *Biochemical and Biophysical Research Communications,* vol. 564, pp. 134-149, 2021.

[188] W. T. Fitch, "Information and the single cell," *Current Opinion in Neurobiology,* vol. 71, pp. 150-157, 2021.

[189] P. Marshall, "Biology transcends the limits of computation," *Progress in Biophysics and Molecular Biology,* 2021.

[190] M. Masi, "Vitalism and cognition in a conscious universe," *Communicative & Integrative Biology,* vol. 15, no. 1, 2022.

[191] D. J. Nicholson, "Is the cell really a machine?," *Journal of Theoretical Biology,* vol. 477, pp. 108-126, 2019.

[192] B. J. King, "Sapiens - Animal Grief Shows We Aren't Meant to Die Alone," 22 April 2020. [Online]. Available: https://www.sapiens.org/biology/animal-grief/.

[193] NIH news release, "NIH embraces bold, 12-year scientific vision for BRAIN Initiative," NIH, 2014. [Online]. Available: https://www.nih.gov/news-events/news-releases/nih-embraces-bold-12-year-scientific-vision-brain-initiative.

[194] M. Humphries, "Your Cortex Contains 17 Billion Computers," February 12 2018. [Online]. Available: https://medium.com/the-spike/your-cortex-contains-17-billion-computers-9034e42d34f2.

[195] D. Beniaguev, I. Segev and M. London, "Single cortical neurons as deep artificial neural networks," *Neuron,* vol. 109, no. 17, pp. 2727-2739, 2021.

[196] A. Gidon and al., "Dendritic action potentials and computation in human layer 2/3 cortical neurons," *Science,* vol. 367, no. 6473, pp. 83-87, 2020.

[197] A. Frances, "The lure of 'cool' brain research is stifling psychotherapy," *Aeon,* 4 March 2020.

[198] M. Nour, Y. Liu and R. Dolan, "Functional neuroimaging in psychiatry and the case for failing better," *Neuron,* vol. 110, no. 16, pp. 2524-2544., 2022.

[199] K. S. Button and al., "Power failure: why small sample size undermines the reliability of neuroscience," *Nature Reviews Neuroscience,* vol. 14, p. 365–376, 2013.

[200] C. Lazarus and al., "Classification and prevalence of spin in abstracts of non-randomized studies evaluating an intervention," *BMC Medical Research Methodology,* Vols. 15, 85, 2015.

[201] N. Haber and al., "Causal language and strength of inference in academic and media articles shared in social media (CLAIMS): A systematic review," *Plos One,* 2018.

[202] E. Vul and al., "Puzzlingly High Correlations in fMRI Studies of Emotion, Personality, and Social Cognition," *Perspectives on Psychological Science,* vol. 4, no. 3, p. 274, 2009.

[203] C. Bennett and al., "Neural correlates of interspecies perspective taking in the post-mortem Atlantic Salmon: an argument for multiple comparisons correction," *Neuroimage,* vol. 47, no. 1, 2009.

[204] R. Botvinik-Nezer and al., "Variability in the analysis of a single neuroimaging dataset by many teams," *Nature,* vol. 582, p. 84–88, 2020.

[205] L. E. Maxwell and al., "What Is the Test-Retest Reliability of Common Task-Functional MRI Measures? New Empirical Evidence and a Meta-Analysis," *Psychological Science,* vol. 31, no. 7, pp. 792-806, 2020.

[206] S. Marek and al., "Reproducible brain-wide association studies require thousands of individuals," *Nature,* vol. 603, p. 654–660, 2022.

[207] N. France, "How Not to Use Brain Scans in Neuroscience," Neuoscience News, 11 August 2022. [Online]. Available: https://neurosciencenews.com/neuroimaging-sample-21226/.

[208] E. Jonas and K. P. Kording, "Could a Neuroscientist Understand a Microprocessor?," *PLOS Computational Biology,* 2017.

[209] J. Gerafi, "A Unified Perspective of Unilateral Spatial Neglect," *Independent thesis,* 2012.

[210] A. Pais, "Einstein and the quantum theory," *Rev. Mod. Phys.,* vol. 51, p. 863 –1979, 1979.

[211] H. W. Lissmann and K. E. Machin, "The Mechanism of Object Location in Gymnarchus Niloticus and Similar Fish," *Journal of Experimental Biology,* vol. 35, pp. 451-486, 1958.

[212] H. W. Lissman, "Electrical location by fishes," *Scientific American,* no. 3, pp. 50-59, 1963.

[213] Y. Guo and M. Kawasaki, "The representation of accurate temporal information in the electrosensory system of the African electrical fish, Gymnarchus Niloticus," *Journal of Neuroscience,* vol. 17, no. 5, p. 1761–1768, 1997.

[214] G. v. d. Emde, "Distance and shape: perception of the 3-dimensional world by weakly electric fish," *Journal of Physiology-Paris,* vol. 98, no. 1–3, pp. 67-80, 2004.

[215] W. Seager, Theories of Consciousness, Routledge, 1999.

[216] B. Kastrup, The Idea of the World, iff Books, 2019, p. 45.

[217] O. Barfield, Saving the Appearances, Barfield Press UK, 1957.

[218] D. Salmon, *Personal communication.*

[219] D. Salmon and J. Maslow, Yoga Psychology and the Transformation of Consciousness: Seeing Through the Eyes of Infinity, Paragon House, 2007.

[220] A. Wallace, Choosing Reality - A Buddhist view of physics and the mind, 1995.

[221] M. Masi, Quantum Physics: An Overview of a Weird World - A primer on the Conceptual Foundations, vol. I, 2020.

[222] W. M. Itano and al., "The quantum Zeno Effect," *Phys. Rev.,* vol. A 41, no. 5, p. 2295, 1990.

[223] A. Einstein, B. Podolosky and N. Rosen, "Can Quantum-Mechanical Description of Physical Reality be Considered Complete?," *Physical Review,* vol. 47, no. 10, p. 777–780, 1935.

[224] B. Russel, An Outline of Philosophy, Routledge, 1927, pp. 154, 163.

[225] M. Planck, *The Observer,* 25 January 1931.

[226] M. Planck, "Das Wesen der Materie [The Nature of Matter]," *Archiv zur Geschichte der Max-Planck-Gesellschaft,* 1944.

[227] E. Schrödinger, My View of the World, Cambridge University Press, 1951.

[228] W. Heisenberg, Physics and Philosophy, 1958.

[229] W. Heisenberg, Das Naturgesetz und die Struktur der Materie, 1967.

[230] D. Bohm, Wholeness and the Implicate order, Roudledge, 1980.

[231] W. Hanegraaff, Hermetic Spirituality and the Historical Imagination: Altered States of Knowledge in Late Antiquity, Cambridge: Cambridge University Press., 2022.

[232] G. Brüntrup, B. P. Göcke and L. (. Jaskolla, Panentheism and Panpsychism, Leiden, The Netherlands: Brill | mentis., 2020.

[233] G. W. Leibniz, "Die Philosophischen Schriften von Gottfried Wilhelm Leibniz," vol. 6, no. C. I. Gerhardt. Weidmann, p. 609, 1875-90, reprint 1960.

[234] D. Hume, An Enquiry Concerning Human Understanding, 1748.

[235] A. N. Whitehead, Process and Reality: An Essay in Cosmology, D. R. Griffin, 1978.

[236] J. Jeans, The Mysterious Universe, Cambridge, 1931, p. 187.

[237] W. James, The Principles of Psychology, 1895.

[238] A. S. Reber and F. Baluška, "Cognition in some surprising places," *Biochem. Biophys.l Res. Commun.*, vol. 564, pp. 150-157, 2021.

[239] F. Baluška, W. B. Miller and A. S. Reber, "Biomolecular Basis of Cellular Consciousness via Subcellular Nanobrains," *Int. J. Mol. Sci.*, vol. 22, no. 5, p. 2545, 2021.

[240] W. B. Miller Jr, F. Baluška and J. S. Torday, "Cellular senomic measurements in cognition-based evolution," *Progr. Biophys. Mol. Biol.*, vol. 156, pp. 20-33, 2020.

[241] S. Arthur R., F. Baluska and W. Miller, The Sentient Cell: The Cellular Foundations of Consciousness, Oxford Academic, 2023.

[242] A. Reber, The First Minds: Caterpillars, Karyotes, and Consciousness, New York: Oxford University Press, 2019.

[243] W. Miller, J. Torday and F. Baluska, "The N-Space Episenome Unifies Cellular Information Space-Time within Cognition-Based Evolution," *Progress in Biophysics and Molecular Biology,* vol. 150, no. 2, 2019.

[244] F. Baluškaa and W. Miller, "Senomic view of the cell: Senome versus Genome," *Commun Integr Biol.,* vol. 11, no. 3, 2018.

[245] E. Thompson, "Could All Life Be Sentient?," *Journal of Consciousness Studies,* 2022.

[246] M. Velmans, Understanding Consciousness, Routledge.

[247] M. Velmans, "Reflexive Monism: psychophysical relations among mind, matter and consciousness," *Journal of Consciousness Studies,* vol. 19, no. 9-10, pp. 143-165, 2012.

[248] I. Shani, "Cosmopsychism: A Holistic Approach to the Metaphyscis of Experience," *Philosophical Papers,* vol. 44, no. 3, pp. 389-437, 2015.

[249] B. Kastrup, "The Universe in Consciousness," *Journal of Consciousness Studies,* vol. 25, no. 5–6, p. 125–55, 2018.

[250] B. Kastrup, Decoding Schopenhauer's Metaphysics, Iff Books, 2020.

[251] S. Taylor, "An introduction to panspiritism: an alternative to materialism and panpsychism," *Zygon - Journal of Religion & Science,* vol. 55, no. 4, 2020.

[252] R. Nuzzo, "How scientists fool themselves – and how they can stop," *Nature,* 07 October 2015.

[253] H. O. Proskauer, The Rediscovery of Color: Goethe Versus Newton, SteinerBooks, Inc, 1986.

[254] J. Bockemühl, Lebenszusammenhänge erkennen, erleben, gestalten, Dornach, 1980.

[255] J. W. Goethe, Naturwissenschaftliche Schriften, S. V. -. 2016, Hrsg., 1790.

[256] A. Dance, "What is a cell type, really? The quest to categorize life's myriad forms," *Nature,* no. 633, pp. 754-756, 2024.

[257] P. Mitchell, The Origin of Life on the Earth, Pergamon Press (eds Oparin, A. I. et al.), 1957 pg. 437–443.

[258] N. Trott, "Podcast: From marine biology to plant science: Plant cognition and bioacoustics with Monica Gagliano," 2020. [Online]. Available: https://goingconscious.libsyn.com/16-from-marine-biology-to-plant-science-plant-cognition-and-bioacoustics-with-monica-gagliano.

[259] R. W. Kimmerer, Braiding Sweetgrass: Indigenous Wisdom, Scientific Knowledge, and the Teachings of Plants, Milkweed Editions, 2013.

[260] J. W. Goethe, Zur Farbenlehre, Vols. 1- §718, Tübingen, 1810.

[261] M. Beauregard, N. L. Trent and G. E. Schwartz, "Toward a postmaterial psychology: Theroy, resaerch, and applications," *New Ideas in Psychology,* vol. 50, pp. 21-33, 2018.

[262] F. Crick, The Astonishing Hypothesis: The Scientific Search for the Soul, Scribner; Touchstone Edition, 1995.

[263] S. Weinberg, The First Three Minutes, Basic Books, 1996.

[264] R. Dawkins, River Out of Eden: A Darwinian View of Life, Basic Books, 1993.

[265] S. Aurobindo, The Synthesis of Yoga, Pondicherry: Sri Aurobindo Ashram Press.

[266] S. Aurobindo, The Life Divine, Pondicherry: Sri Aurobindo Ashram Press.

[267] S. Aurobindo, Savitri, Pondicherry: Sri Aurobindo Ashram Press.

[268] S. Aurobindo, Record of Yoga - Vol.I+II, Sri Aurobindo Ashram Press, Pondicherry, 2002.

[269] D. Banerji, Seven Quartets of Becoming: A Transformative Yoga Psychology Based on the Diaries of Sri Aurobindo, Nalanda International, 2012.

[270] S. Aurobindo, Isha Upanishad, Sri Aurobindo Ashram Pondicherry, 1914.

[271] S. Aurobindo, Letters on Yoga, Sri Aurobindo Ashram Press.

[272] N. Baldwin, As quoted in: Edison: Inventing the Century, NY: Hyperion, 1995, p. 376.

[273] S. Aurobindo, Thoughts and Aphorisms, Sri Aurobindo Ashram Press.

[274] A. S. Freiberg, "Why We Sleep: A Hypothesis for an Ultimate or Evolutionary Origin for Sleep and Other Physiological Rhythms," *J Circadian Rhythms,* vol. 18, no. 12, 2020.

[275] M. Frank and H. Heller, "The Function(s) of Sleep," in *Sleep-Wake Neurobiology and Pharmacology. Handbook of Experimental Pharmacology - Landolt HP., Dijk DJ. (eds)* , vol. 253, Springer, Cham., 2018.

[276] D. C. Rößler and al., "Regularly occurring bouts of retinal movements suggest an REM sleep–like state in jumping spiders," *PNAS,* vol. 119, no. 33, 2022.

[277] H. J. Kanaya and al., "A sleep-like state in Hydra unravels conserved sleep mechanisms during the evolutionary development of the central nervous system," *Science Advances,* vol. 6, no. 41, 2020.

[278] R. D. Nath and al., "The Jellyfish Cassiopea Exhibits a Sleep-like State," *Current Biology,* vol. 27, no. 19, pp. 2984-2990, 2017.

[279] R. C. Anafi and al., "Exploring phylogeny to find the function of sleep," *Nature Reviews Neuroscience,* vol. 20, pp. 109-116, 2019.

[280] A. L. Valencia and T. Froese, "What binds us? Inter-brain neural synchronization and its implications for theories of human consciousness," *Neuroscience of consciousness,* vol. 2020, no. 1, 2020.

[281] T. Merton, Conjectures of a Guilty Bystander, Penguin Random House, 1968.

[282] E. Teklinski, "A Matter of Heart and Soul: The Value of Positing a Personal Ontological Center for Developmental Psychology," *International Journal of Transpersonal Studies,* vol. 37, no. 1, pp. 90-119, 2019.

[283] E. Teklinski, *A matter of heart and soul: towards an integral psychology framework for postconventional development - Dissertation thesis,* California Institute of Integral Studies, 2016.

[284] M. Cornelissen, "The Self and the Structure of the Personality: An Overview of Sri Aurobindo's Topography of Consciousness," *International Journal of Transpersonal Studies,* vol. 37, no. 1, 2018.

[285] M. Alfassa and Satprem, Mother's Agenda - The Supramental Action upon Earth, Institut de Recherches Evolutives, 1951-1973.

[286] S. Aurobindo, *A God's Labour - Short Poems,* Sri Aurobindo Ashram Press, 1930-1950.

[287] A. Einstein, "Physics and reality," *Journal of the Franklin Institute,* vol. 221, no. 3, pp. 349-382, 1936.

[288] E. P. Wigner, "The unreasonable effectiveness of mathematics in the natural sciences," *Communications on Pure and Applied Mathematics,* vol. 13, pp. 1-14, 1959.

[289] J. J. C. Smart, "Free-Will, Praise and Blame," *Mind,* vol. 70, no. 279, pp. 291-306, 1961.

[290] B. Spinoza, Ethics: Demonstrated in Geometrical Order, 1677.

[291] T. Sider, "Free Will and Determinism," in *Riddles of Existence,* Oxford University Press Inc., New York, 2007.

[292] F. Faggin, "The primordial quantum language," Essentia Foundation, 2021. [Online]. Available: https://www.essentiafoundation.org/reading/the-primordial-quantum-language/.

[293] M. Masi, "Quantum Indeterminsim, Free Will, and Self-Causation," *Journal of Consciousness Studies,* vol. 30, no. 5-6, pp. 32-56(25), 2023.

[294] M. Masi, "Quantum Indeterminacy and Libertarian Panpsychism," *Mind and Matter,* vol. 2, no. 1, pp. 31-50(20), 2024.

[295] F. Dyson, Progress in Religion, 2000.

[296] P. Jordan, Die Physik und das Geheimnis des organischen Lebens, Braunschweig, Germany: Friedrich Vieweg & Sohn., 1941.

[297] F. Dyson, Disturbing the Universe, Basic Books, 1979.

[298] R. Penrose, Shadows of the Mind, Oxford University Press, 1994.

[299] S. Hameroff and R. Penrose, "Consciousness in the universe: A review of the 'Orch OR' theory," *Physics of Life Reviews,* vol. 11, no. 1, pp. 39-78, 2014.

[300] K. Pribram, "Quantum holography: Is it relevant to brain?," *Information Sciences,* vol. 115, no. 1-4, pp. 97-102, 1999.

[301] G. Vitiello and E. Alfinito, "The dissipative quantum model of brain: how does memory localize in correlated neuronal domains," *Information Sciences,* vol. 128, no. 3-4, pp. 217-229, 2000.

[302] A. Capolupo, W. Freeman and G. Vitiello, The Dissipative Many-Body Model and Phase Transitions in Brain Nonlinear Dynamics, Springer, 2014.

[303] H. P. Stapp, "Quantum theory and the role of mind in nature," *Foundations of Physics,* vol. 31, pp. 1465-1499, 2001.

[304] T. Palmer, "Human Creativity and Consciousness: Unintended Consequences of the Brain's Extraordinary Energy Efficiency?," *Entropy,* vol. 22, no. 3, p. 281, 2020.

[305] J. Keppler and I. Shani, "Cosmopsychism and Consciousness Research: A Fresh View on the Causal Mechanisms Underlying Phenomenal States," *Frontiers in Psychology,* vol. 11, p. 371, 2020.

[306] G. Mauro and F. Faggin, "Hard Problem and Free Will: An Information-Theoretical Approach," in *Artificial Intelligence Versus Natural Intelligence*, Scardigli, F. (eds) Springer, 2022, pp. 145-192.

[307] J. Cepelewicz, "Nature Versus Nurture? Add 'Noise' to the Debate.," March 23 2020. [Online]. Available: https://www.quantamagazine.org/nature-versus-nurture-add-noise-to-the-debate-20200323/.

[308] A. Eldar and M. B. Elowitz, "Functional roles for noise in genetic circuits," *Nature,* vol. 467 (7312), 2010.

[309] G. Vogt, "Stochastic developmental variation, an epigenetic source of phenotypic diversity with far-reaching biological consequences.," *J Biosci,* vol. 40, pp. 159-204, 2015.

[310] G. Balázsi, A. Oudenaarden and J. Collins, "Cellular Decision Making and Biological Noise: From Microbes to Mammals," *Cell,* vol. 144, no. 6, pp. 910 - 925, 2011.

[311] R. Noble and D. Npble, "Harnessing stochasticity: How do organisms make choices?," *Chaos,* vol. 28, no. 10, p. 106309, 2018.

[312] T. Simonite, "A Sobering Message About the Future at AI's Biggest Party," Wired, 13 December 2020. [Online]. Available: https://www.wired.com/story/sobering-message-future-ai-party/.

[313] S. Garte, "Evidence for phase transitions in replication fidelity and survival probability at the origin of life," *BioCosmos,* vol. 1, no. 1, pp. 2-10, 2021.

[314] R. V. Yampolskiy, "Why We Do Not Evolve Software? Analysis of Evolutionary Algorithms," *Evolutionary bioinformatics online,* vol. 14, 2018.

[315] S. A. Benner, "Defining Life," *Astrobiology,* vol. 10, no. 10, pp. 1021-1030, 2010.

[316] A. Christine, "Harnessing non-modernity: a case study in artificial life - Dissertation Thesis," 2010.

[317] A. Wagner and J. Wright, "Alternative routes and mutational robustness in complex regulatory networks," *Biosystems,* vol. 88, no. 1-2, pp. 163-172, 2007.

[318] J. A. Shapiro, "How life changes itself: The Read–Write (RW) genome," *Physics of Life Reviews,* vol. 10, no. 13, pp. 287-323, 2013.

[319] J. Wells, "Membrane Patterns Carry Ontogenetic Information that is Specified Independently of DNA," *Biophysical Journal,* vol. 106, no. 2 - Supp. 1 596A, 2014.

[320] R. Sapolsky, "YouTube video: "What Separates Us from Chimps? As It Turns Out, Not Much"," [Online]. Available: https://www.youtube.com/watch?v=AzDLkPFjev4.

[321] C. Horton and D. Adams, "The cortical column: a structure without a function," *Philos Trans R Soc Lond B Biol Sci.,* vol. 360, no. 1456, p. 837–862, 2005.

[322] D. Witvliet and al., "Connectomes across development reveal principles of brain maturation," *Nature,* vol. 596, p. 257–261, 2021.

[323] J. G. Monroe and al., "Mutation bias reflects natural selection in Arabidopsis thaliana," *Nature,* vol. 602, pp. 101-105, 2022.

[324] D. Melamed and al., "De novo mutation rates at the single-mutation resolution in a human HBB gene-region associated with adaptation and genetic disease," *Genome Research,* 2022.

[325] K. Laland and al., "The extended evolutionary synthesis: its structure, assumptions and predictions," *Proc. R. Soc. B,* vol. 282, p. 1813, 2015.

[326] Website, "The Third Way," 2021. [Online]. Available: https://www.thethirdwayofevolution.com/.

[327] J. Shapiro and D. Noble, "What prevents mainstream evolutionists teaching the whole truth about how genomes evolve?," *Progress in Biophysics and Molecular Biology,* 2021.

[328] K. Cutlip, "DNA may not be life's instruction book—just a jumbled list of ingredients," 22 April 2020. [Online]. Available: https://phys.org/news/2020-04-dna-life-bookjust-jumbled-ingredients.html.

[329] K. Richardson, "It's the End of the Gene As We Know It," October 2019. [Online]. Available: https://nautil.us/its-the-end-of-the-gene-as-we-know-it-237288/.

[330] D. Noble, "It's time to admit that genes are not the blueprint for life," *Nature,* 2024.

[331] R. Noble and D. Noble, "Physiology restores purpose to evolutionary biology," *Biological Journal of the Linnean Society,* vol. 139, no. 4, p. 357–369, 2022.

[332] A. R. M. Barwich, "Rage against the what? The machine metaphor in biology.," *Biol Philos,* vol. 39, no. 14, 2024.

[333] S. Aurobindo, Glossary of Terms in Sri Aurobindo's Writing, Sri Auribindo Ashram, Pondicherry, India.

[334] F. C. Adama, "The degree of fine-tuning in our universe — and others," *Physics Reports,* vol. 807, pp. 1-111, 2019.

[335] L. N. Vandenberg, D. S. Adams and M. Levin, "Normalized shape and location of perturbed craniofacial structures in the Xenopus tadpole reveal an innate ability to achieve correct morphology," *Developmental Dynamics,* vol. 241, no. 5, pp. 863-878, 2012.

[336] A. Tseng and M. Levin, "Cracking the bioelectric code," *Commun Integr Biol.,* vol. 6, no. 1, 2013.

[337] M. Levin, "Morphogenetic fields in embryogenesis, regeneration, and cancer: Non-local control of complex patterning," *Biosystems,* vol. 109, no. 3, pp. 243-261, 2012.

[338] M. Levin and D. D.C., "Cognition all the way down - Aeon," 13 October 2020. [Online]. Available: https://aeon.co/essays/how-to-understand-cells-tissues-and-organisms-as-agents-with-agendas.

[339] L. Else, "A meadowful of meaning," *New Scientist,* vol. 207, no. 2774, 2010.

[340] P. Pinet, Pasteur et la philosophie, Editions L'Harmattan, 2004, p. 63.

[341] G. E. Billman, "Homeostasis: The Underappreciated and Far Too Often Ignored Central Organizing Principle of Physiology," *Front. Physiol.,* 2020.

[342] J. S. Turner, Purpose & Desire: What Makes Something "Alive" and Why Modern Darwinism Has Failed to Explain It., Harper Collins Publishers., 2017.

[343] E. Tolle, Stillness Speaks - Ch. 7, New World Library, 2013.

[344] E. Callaway, "'The entire protein universe': AI predicts shape of nearly every known protein," *Nature,* vol. 608, 2022.

[345] "ATP synthase: Structure and Function," [Online]. Available: https://youtu.be/b_cp8MsnZFA.

[346] W. A. Schulz, "An Engineering Perspective on the Bacterial Flagellum: Part 1+2—Constructive and Analytic View," *Bio-Complexity,* vol. 2021, no. 1, p. 1, 2021.

[347] Veritasium, "Your Body's Molecular Machines," 2017. [Online]. Available: https://youtu.be/X_tYrnv_o6A.

[348] SubAnima, "How NOT To Think About Cells," 2022. [Online]. Available: https://youtu.be/jPhvic-eqbc.

[349] P. Summer and al., "The association between exaggeration in health related science news and academic press releases: retrospective observational study," *BMJ,* p. 349:g7015, 2014.

[350] "To Build Truly Intelligent Machines, Teach Them Cause and Effect," QuantaMagazine, 15 May 2018. [Online]. Available: https://www.quantamagazine.org/to-build-truly-intelligent-machines-teach-them-cause-and-effect-20180515/.

[351] E. M. Bender and al., "On the Dangers of Stochastic Parrots: Can Language Models Be Too Big?," *FAccT '21: Proceedings of the 2021 ACM Conference on Fairness, Accountability, and Transparency,* p. 610–623, 2021.

[352] J. Cepelewicz, "Where We See Shapes, AI Sees Textures," QuantaMagazine, 1 July 2019. [Online]. Available: https://www.quantamagazine.org/where-we-see-shapes-ai-sees-textures-20190701/.

[353] K. Hartnett, "Machine Learning Confronts the Elephant in the Room," 20 September 2018. [Online]. Available: https://www.quantamagazine.org/machine-learning-confronts-the-elephant-in-the-room-20180920/.

[354] A. Rosenfeld, M. D. Solbach and J. K. Tsotso, "Totally Looks Like - How Humans Compare, Compared to Machines," *Computer Vision – Lecture Notes in Computer Science-Springer, Cham.,* vol. 11361, 2019.

[355] S. Shead, "Researchers: Are we on the cusp of an 'AI winter'?," BBC News, 12 January 2020. [Online]. Available: https://www.bbc.com/news/technology-51064369.

[356] N. E. Boudette, "Despite High Hopes, Self-Driving Cars Are 'Way in the Future'," The New York Times, 17 July 2020. [Online].

Available: https://www.nytimes.com/2019/07/17/business/self-driving-autonomous-cars.html.

[357] N. Bostrom, "Are you living in a computer simulation?," *Philosophical Quarterly,* vol. 57, no. 211, pp. 243-255, 2003.

[358] S. Hosselfelder, "Backreaction - Merchants of Hype," 05 March 2019. [Online]. Available: http://backreaction.blogspot.com/2019/03/merchants-of-hype.html.

[359] M. Masi, Free Progress Education, Independent Publishing, 2019.

[360] S. Aurobindo, The Human Cycle - The Ideal of Human Unity - War and Self-Determination, Sir Aurobindo Ashram Press.

[361] M. Masi, The End of Materialism: Essays on Science, Philosophy, and Spirituality, for a Post-Material Future, Indy eds., 2023.

[362] S. Aurobindo, Essays on the Gita, Sri Aurobindo Ashram Trust, India.

[363] D. Hoffman, "Conscious Realism and the Mind-Body Problem," *Mind & Matter,* vol. 6, no. 1, p. 87–121, 2008.

[364] O. Sacks, The Man Who Mistook His Wife for a Hat, Pan Books, London, 1986.

[365] "Journey to the Microcosmos," [Online]. Available: https://www.youtube.com/channel/UCBbnbBWJtwsf0jLGUwX5Q3g.

[366] E. Hanson, "Torture and Truth in Renaissance England," *Representations,* vol. 34, p. 53–84, 1991.

[367] J. H. Langbein, Torture and the Law of Proof, The University of Chicago Press, 1976.

[368] Satprem, Sri Aurobindo or The Adventure of Consciousness, Lotus Press.

[369] K. Frankish, "The consciousness illusion," 2019. [Online]. Available: https://aeon.co/essays/what-if-your-consciousness-is-an-illusion-created-by-your-brain.

[370] M. Albahari, "Perennial Idealism: A Mystical Solution to the Mind-Body Problem," *Philosopher's Imprint,* vol. 19, no. 40, 2019.

[371] B. Kastrup, "Dr. Bernardo Kastrup, Idealism vs Materialism?," 2020. [Online]. Available: https://youtu.be/KVWP8h-KMEI?t=558.

[372] D. F. Skrbina, Panpsychism in the West, MIT Press, 2017.

[373] J. McFadden, "Integrating information in the brain's EM field: the cemi field theory of consciousness," *Neuroscience of Consciousness,* vol. 2020, no. 1, 2020.

[374] M. Levin, "The Computational Boundary of a "Self": Developmental Bioelectricity Drives Multicellularity and Scale-Free Cognition," *Front. Psychol.,* 13 December 2019.

[375] S. Manicka and M. Levin, "The Cognitive Lens: a primer on conceptual tools for analysing information processing in developmental and regenerative morphogenesis," *Phil. Trans. R. Soc.,* vol. B374:20180369, 2019.

[376] P. J. Scott and R. V. Yampolskiy, "Classification Schemas for Artificial Intelligence Failures," *Delphi - Interdisciplinary Review of Emerging Technologies,* vol. 2, no. 4, pp. 186-199, 2019.

[377] R. Shivni, "The deconstructed Standard Model equation," Symmetry, 28 July 2016. [Online]. Available: https://www.symmetrymagazine.org/article/the-deconstructed-standard-model-equation.

[378] R. Feigenbaum, "Toward a Nonanthropocentric Vision of Nature: Goethe's Discovery of the Intermaxillary Bone," *Goethe Yearbook,* vol. 22, pp. 73-93, 2015.

[379] D. Broderick, "Consciousness and Science Fiction," Springer, 2018, p. 38.

[380] "Microbes Don't Actually Look Like Anything," Journey to the Microcosmos, [Online]. Available: https://youtu.be/VBmzwM76V0o.

[381] S. Kauffman, "What Is Life, and Can We Create It?," *BioScience,* vol. 63, no. 8, p. 609–610, 2013.

[382] S. Aurobindo, Integral Yoga: Sri Aurobindo's Teaching & Method of Practice, Lotus Press, 1993.

[383] S. Kauffman, "No entailing laws, but enablement in the evolution of the biosphere," 2019. [Online]. Available: http://pirsa.org/19100055/.

[384] M. Kasey, "Lack of evidence for associative learning in pea plants," *eLife,* 2020.

[385] M. S. A. Graziano, "A conceptual framework for consciousness," *PNAS,* vol. 119, no. 18, 2022.

[386] B. R. Frieden and B. H. Soffer, "Lagrangians of physics and the game of Fisher-information transfer," *Phys. Rev.,* no. E 52, p. 2274, 1995.

[387] G. Ellis, "Emergence in Solid State Physics and Biology," *Foundations of Physics,* 2020.

[388] G. Ellis, "From chaos to free will," Aeon, 2020. [Online]. Available: https://aeon.co/essays/heres-why-so-many-physicists-are-wrong-about-free-will.

[389] G. Ellis, "The Causal Closure of Physics in Real World Contexts," *Foundations of Physics,* 2020.

[390] A. P. Srivastava and al., "High-resolution cryo-EM analysis of the yeast ATP synthase in a lipid membrane," *Science,* vol. 360, no. 6389, 2018.

[391] D. E. M. J. Nilsson, "Lens eyes in protists," *Cell Press - Current Biology Magazine,* p. 451–520, 2020.

[392] M. Kohda, R. Bshary and N. Kubo, "Cleaner fish recognize self in a mirror via self-face recognition like humans," *PNAS,* vol. 120, no. 7, 2023.

[393] M. Masi, "The Integral Cosmology of Sri Aurobindo: An Introduction from the Perspective of Consciousness Studies," *Integral Review,* vol. 18, no. 1, 2023.

[394] M. Masi, "An Evidence-Based Critical Review of the Mind-Brain Identity Theory," *Frontiers in Psychology - Sec ,* vol. 14, 2023.

About the author

Marco Masi (born 1965) attended the German School of Milan, Italy. He graduated in physics at the university of Padua, and later obtained a Ph.D. in physics at the university of Trento. He worked as a postdoc in universities in Italy, France, and Germany, and as a school teacher for three years. After he had authored some scientific papers (http://ow.ly/snz6u) his interests veered towards new forms of individual learning and a new concept of free progress education originated from his  activity both as a tutor in several universities and as a high school teacher, but especially from his direct, lived experience of what education should *not* be. This led him to author a book on *"Free progress Education"*. [359] He also wrote *a* two-volume series on quantum physics entitled *"Quantum Physics: An Overview of a Weird World", and* which tries to close a gap between the too high-level university textbooks and a too low level popular science approach [221] [36]. A good introduction to the present volume is his book *"The End of Materialism: Essays on Science, Philosophy, and Spirituality, for a Post-Material Future"* [361]

His interests in metaphysical and philosophical ruminations led him to the vision of Sri Aurobindo and the Mother. He loves walking in the woods, loves animals (and would never kill a cat to carry out Schrödinger's experiment).

And if you enjoyed this book, please support the post-material ideal by sharing it with friends and leaving a review wherever you bought it.

Index

9 783948 295134